Quintessence

" A Theoretical Fifth Force of Nature "

Edited by Paul F. Kisak

Contents

Chapter 1

Quintessence (physics)

For other forms of quintessence, see Quintessence (disambiguation). For theories and defunct or classical concepts named after the synonym "Aether", see Aether (disambiguation).

In physics, **quintessence** is a hypothetical form of dark energy, more precisely a scalar field, postulated as an explanation of the observation of an accelerating rate of expansion of the universe, rather than due to a true cosmological constant. The first example of this scenario was proposed by Ratra and Peebles (1988).[1] The concept was expanded to more general types of time-varying dark energy and the term "quintessence" was first introduced in a paper by R.R.Caldwell, Rahul Dave and Paul Steinhardt.[2] It has been proposed by some physicists to be a fifth fundamental force. Quintessence differs from the cosmological constant explanation of dark energy in that it is dynamic, that is, it changes over time, unlike the cosmological constant which always stays constant. It is suggested that quintessence can be either attractive or repulsive depending on the ratio of its kinetic and potential energy. Specifically, it is thought that quintessence became repulsive about ten billion years ago (the universe is approximately 13.8 billion years old).[3]

1.1 Scalar field

Quintessence is a scalar field with an equation of state where w_q, the ratio of pressure p_q and density ρ_q, is given by the potential energy $V(Q)$ and a kinetic term:

$$w_q = p_q/\rho_q = \frac{\frac{1}{2}\dot{Q}^2 - V(Q)}{\frac{1}{2}\dot{Q}^2 + V(Q)}$$

Hence, quintessence is dynamic, and generally has a density and w_q parameter that varies with time. By contrast, a cosmological constant is static, with a fixed energy density and $w_q = -1$.

1.2 Tracker behavior

Many models of quintessence have a *tracker* behavior, which according to Ratra and Peebles (1988) and Paul Steinhardt *et al.* (1999) partly solves the cosmological constant problem.[4] In these models, the quintessence field has a density which closely tracks (but is less than) the radiation density until matter-radiation equality, which triggers quintessence to start having characteristics similar to dark energy, eventually dominating the universe. This naturally sets the low scale of the dark energy.[5] When comparing the predicted expansion rate of the universe as given by the tracker solutions with cosmological data, a main feature of tracker solutions is that one needs four parameters to properly describe the behavior of their equation of state,[6][7] whereas it has been shown that at most a two-parameter model can optimally be constrained by mid-term future data (horizon 2015-2020).[8]

1.3 Specific models

Some special cases of quintessence are phantom energy, in which $w_q < -1$,[9] and k-essence (short for kinetic quintessence), which has a non-standard form of kinetic energy. If this type of energy were to exist, it would cause a big rip in the universe due to the growing energy density of dark energy which would cause the expansion of the universe to increase at a faster-than-exponential rate.

1.3.1 Holographic Dark Energy

Holographic Dark Energy models compared to Cosmological Constant models, imply a high degeneracy.[10] It has been suggested that dark energy might originate from quantum fluctuations of spacetime, and are limited by the event horizon of the universe.[11]

Studies with quintessence dark energy found that it dominates gravitational collapse in a spacetime simulation, based on the holographic thermalization. These results show that the smaller the state parameter of quintessence is, the harder it is for the plasma to thermalize.[12]

1.4 Quintom scenario

In 2004, when scientists fitted the evolution of dark energy with the cosmological data, they found that the equation of state had possibly crossed the cosmological constant boundary (w = −1) from above to below. A proven no-go theorem indicates this situation, called the Quintom scenario, requires at least two degrees of freedom for dark energy models.[13]

1.5 Terminology

The name comes from the classical elements in ancient Greece. The aether, a pure "fifth element" (*quinta essentia* in Latin), was thought to fill the universe beyond Earth. Similarly, modern quintessence would be the fifth known contribution to the overall mass-energy content of the universe. (The other four in the modern interpretation, different from the ancient ideas, are: baryonic matter; radiation – photons and the highly relativistic neutrinos, which may be considered hot dark matter; cold dark matter; and the term due to spatial curvature – loosely, gravitational self-energy.)

1.6 See also

- Dark-energy-dominated era

1.7 References

[1] Ratra, P.; Peebles, L. (1988). "Cosmological consequences of a rolling homogeneous scalar field". *Physical Review D* **37** (12): 3406. Bibcode:1988PhRvD..37.3406R. doi:10.1103/PhysRevD.37.3406.

[2] Caldwell, R.R.; Dave, R.; Steinhardt, P.J. (1998). "Cosmological Imprint of an Energy Component with General Equation-of-State". *Phys.Rev.Lett.* **80** (8): 1582–1585. arXiv:astro-ph/9708069. Bibcode:1998PhRvL..80.1582C. doi:10.1103/PhysRevLett.80.1582.

[3] Christopher Wanjek; "Quintessence, accelerating the Universe?"; http://www.astronomytoday.com/cosmology/quintessence.html

[4] Zlatev, I.; Wang, L.; Steinhardt, P. (1999). "Quintessence, Cosmic Coincidence, and the Cosmological Constant". *Physical Review Letters* **82** (5): 896–899.

arXiv:astro-ph/9807002. Bibcode:1999PhRvL..82..896Z. doi:10.1103/PhysRevLett.82.896.

[5] Steinhardt, P.; Wang, L.; Zlatev, I. (1999). "Cosmological tracking solutions". *Physical Review D* **59** (12): 123504. arXiv:astro-ph/9812313. Bibcode:1999PhRvD..59l3504S. doi:10.1103/PhysRevD.59.123504.

[6] Linden, Sebastian; Virey, Jean-Marc (2008). "Test of the Chevallier-Polarski-Linder parametrization for rapid dark energy equation of state transitions". *Physical Review D* **78** (2): 023526. arXiv:0804.0389. Bibcode:2008PhRvD..78b3526L. doi:10.1103/PhysRevD.78.023526.

[7] Ferramacho, L.; Blanchard, A.; Zolnierowsky, Y.; Riazuelo, A. (2010). "Constraints on dark energy evolution". *A&A* **514**: A20. arXiv:0909.1703. Bibcode:2010A&A...514A..20F. doi:10.1051/0004-6361/200913271.

[8] Linder, Eric V.; Huterer, Dragan (2005). "How many cosmological parameters". *Physical Review D* **72** (4): 043509. arXiv:astro-ph/0505330. Bibcode:2005PhRvD..72d3509L. doi:10.1103/PhysRevD.72.043509.

[9] Caldwell, R. R. (2002). "A phantom menace? Cosmological consequences of a dark energy component with super-negative equation of state". *Physics Letters B* **545** (1-2): 23–29. arXiv:astro-ph/9908168. Bibcode:2002PhLB..545...23C. doi:10.1016/S0370-2693(02)02589-3.

[10] Yazhou Hu, Miao Li, Nan Li, Zhenhui Zhang (2015). "Holographic Dark Energy with Cosmological Constant".

[11] Shan Gao (2013). "Explaining Holographic Dark Energy". Bibcode:2013Galax...1..180G. doi:10.3390/galaxies1030180.

[12] Xiao-Xiong Zeng, De-You Chen, Li-Fang Li (2014). "Holographic thermalization and gravitational collapse in the spacetime dominated by quintessence dark energy" (PDF). arXiv:1408.6632. Bibcode:2015PhRvD..91d6005Z. doi:10.1103/PhysRevD.91.046005.

[13] Hu, Wayne (2005). "Crossing the phantom divide: Dark energy internal degrees of freedom". *Physical Review D* **71** (4): 047301. arXiv:astro-ph/0410680. Bibcode:2005PhRvD..71d7301H. doi:10.1103/PhysRevD.71.047301.

1.7.1 Further reading

- Ostriker JP; Steinhardt P (January 2001). "The Quintessential Universe". *Scientific American* **284** (1): 46–53. doi:10.1038/scientificamerican0101-46.

- Lawrence M. Krauss (2000). *Quintessence: The Search for Missing Mass in the Universe*. Basic Books. ISBN 978-0465037414.

1.8 External links

Chapter 2

Fifth force

Modern physics describes physical reality in terms of four known fundamental forces. However, since physics has no accepted universal framework, occasionally physicists have postulated the existence of an additional fundamental **fifth force**. Most postulate a force of roughly the strength of gravity (i.e. it is much weaker than electromagnetism or the nuclear forces) and to have a range of anywhere from less than a millimeter to cosmological scales.

2.1 Experimental approaches

The idea of an additional fundamental force is difficult to test, because gravity is such a weak force: the gravitational interaction between two objects is only significant when one has a great mass. Therefore, it takes very precise equipment to measure gravitational interactions between objects that are small compared to the Earth. Nonetheless, in the late 1980s a fifth force, operating on municipal scales (i.e. with a range of about 100 meters), was reported by researchers (Fischbach *et al.*)[1] who were reanalyzing results of Loránd Eötvös from earlier in the century. The force was believed to be linked with hypercharge. Over a number of years, other experiments have failed to duplicate this result.[2]

There are at least three kinds of searches that can be undertaken, which depend on the kind of force being considered, and its range.

2.1.1 Equivalence principle

One way is to search for a fifth force with tests of the strong equivalence principle: this is one of the most powerful tests of Einstein's theory of gravity; general relativity. Alternative theories of gravity, such as Brans–Dicke theory, have a fifth force—possibly with infinite range. This is because gravitational interactions, in theories other than general relativity, have degrees of freedom other than the "metric", which dictates the curvature of space, and different kinds of degrees of freedom produce different effects. For example, a scalar field cannot produce the bending of light rays. The fifth force would manifest itself in an effect on solar system orbits, called the Nordtvedt effect. This is tested with Lunar Laser Ranging Experiment[3] and very long baseline interferometry.

2.1.2 Extra dimensions

Another kind of fifth force, which arises in Kaluza–Klein theory, where the universe has extra dimensions, or in supergravity or string theory is the Yukawa force, which is transmitted by a light scalar field (i.e. a scalar field with a long Compton wavelength, which determines the range). This has prompted a lot of recent interest, as a theory of supersymmetric large extra dimensions—dimensions with size slightly less than a millimeter—has prompted an experimental effort to test gravity on these very small scales. This requires extremely sensitive experiments which search for a deviation from the inverse square law of gravity over a range of distances.[4] Essentially, they are looking for signs that the Yukawa interaction is kicking in at a certain length.

Australian researchers, attempting to measure the gravitational constant deep in a mine shaft, found a discrepancy between the predicted and measured value, with the measured value being two percent too small. They concluded that the results may be explained by a repulsive fifth force with a range from a few centimetres to a kilometre. Similar experiments have been carried out on board a submarine, USS *Dolphin* (AGSS-555), while deeply submerged. A further experiment measuring the gravitational constant in a deep borehole in the Greenland ice sheet found discrepancies of a few percent, but it was not possible to eliminate a geological source for the observed signal.[5][6]

2.1.3 Earth's mantle

Another experiment uses the earth's mantle as a giant particle detector, focusing on geoelectrons.[7]

4

2.1.4 Cepheid variables

In 2012 Bhuvnesh Jain and others examined existing data on the rate of pulsation of cepheid variable stars in 25 galaxies comprising over a thousand stars in all. Theory suggests that the rate of pulsation would follow a different pattern in galaxies screened from a hypothetical 5th force by neighbourhood clusters from those that are not screened. They were unable to find any variation from Einstein's theory of gravity.[8][9]

2.1.5 Other approaches

Some experiments used a lake and a tower that is 320 m high.[10] A comprehensive review suggested there is no compelling evidence for the fifth force,[11] though scientists still search for it. Fishbach's article was written in 1992 and since then other evidence has come to light that may indicate a 5th force.[12]

The above experiments search for a fifth force that is, like gravity, independent of the composition of an object, so all objects experience the force in proportion to their masses. Forces that depend on the composition of an object can be very sensitively tested by torsion balance experiments of a type invented by Loránd Eötvös. Such forces may depend, for example, on the ratio of protons to neutrons in an atomic nucleus, nuclear spin,[13] or the relative amount of different kinds of binding energy in a nucleus (see the semi-empirical mass formula). Searches have been done from very short ranges, to municipal scales, to the scale of the Earth, the sun, and dark matter at the center of the galaxy.

2.2 Modified gravity

Also known as non-local gravity. A few physicists[14][15][16] think that Einstein's theory of gravity will have to be modified, not at small scales, but at large distances, or, equivalently, small accelerations. This would change the gravity force to a non-local force. They point out that dark matter and dark energy are unexplained by the Standard Model of particle physics and suggest that some modification of gravity, possibly arising from Modified Newtonian Dynamics or the holographic principle. This is fundamentally different from conventional ideas of a fifth force, as it grows stronger relative to gravity at longer distances. Most physicists, however, think that dark matter and dark energy are not *ad hoc*, but are supported by a large number of complementary observations and described by a very simple model.

2.3 See also

- Beyond the Standard Model
- Modified Gravity
- Modified Newtonian Dynamics
- Complex system
- Dark Energy
- Metric expansion of space
- Self-organization
- Quintessence (physics)
- Graviphoton
- Affine gauge theory

2.4 References

[1] Ephraim Fischbach, Daniel Sudarsky, Aaron Szafer, Carrick Talmadge, and S. H. Aronson, "Reanalysis of the Eötvös experiment", *Physical Review Letters* 56 3 (1986).

[2] University of Washington Eöt-Wash group, the leading group searching for a fifth force.

[3] Lunar Laser Ranging

[4] Satellite Energy Exchange (SEE) , which is set to test for a fifth force in space, where it is possible to achieve greater sensitivity.

[5] Ander, M. E., M. A. Zumberge, et al. (1989). "Test of Newton's inverse-square law in the Greenland ice cap." Physical Review Letters 62(9): 985–988

[6] Zumberge, M. A., M. E. Ander, et al. (1990). The Greenland gravitational constant experiment. Journal of Geophysical Research. 95: 15483–15501

[7] Aron, Jacob. (2013) Earth's mantle helps hunt for fifth force of nature

[8] Is There a 'Fifth Force' that Alters Gravity at Cosmos Scales?, Daily Galaxy, May 11, 2012

[9] Astrophysical Tests of Modified Gravity: Constraints from Distance Indicators in the Nearby Universe Bhuvnesh Jain, Vinu Vikram, Jeremy Sakstein, Cornell University Library, 7 May 2012

[10] Liu Y.C., Yang X.-S., Zhu H., Zhou W., Wang Q.-S., Zhao Z., Jiang W., Wu C.-Z.,"Testing non-Newtonian gravitation on a 320 m tower", Physics Letters A., vol. 169, 131–133 (1992).

[11] Fishbach E. and Talmadge C., "Six years of the fifth force", Nature, vol. 356, 207–215 (1992).

[12] Evidence for Correlations Between Nuclear Decay Rates and Earth–Sun Distance Jere H. Jenkins, Ephraim Fischbach, John B. Buncher, John T. Gruenwald, Dennis E. Krause, Joshua J. Mattes Astropart.Phys.32:42–46,2009

[13] A. M. Hall, H. Armbruster, E. Fischbach, C. Talmadge. Is the Eötvös experiment sensitive to spin? Progress in High Energy Phys. 325-339, W.-Y. Pauchy Hwang, et al., Eds., Elsevier Sci. Pub. Co., New York (1991)

[14] S. Dodelson, S. Park. "Nonlocal Gravity and Structure in the Universe". *arXiv.org*. arXiv.org. Retrieved 22 October 2013.

[15] Jaccard,Maggiore,mitsou. "A non-local theory of massive gravity". *arXiv.org*. arXiv.org. Retrieved 22 October 2013.

[16] Mashhoon, Bahram. "Nonlocal Gravity". *arXiv.org*. arXiv.org. Retrieved 22 October 2013.

Chapter 3

Extra dimensions

In physics, **extra dimensions** are proposed additional space or time dimensions beyond the (3 + 1) typical of our observed space-time, such as the first attempts based on the Kaluza–Klein theory. Among theories proposing extra dimension are:[1]

1. Large extra dimension, mostly motivated by the ADD model, by Nima Arkani-Hamed, Savas Dimopoulos, and Gia Dvali in 1998, in an attempt to solve the hierarchy problem. This theory requires that the fields of the Standard Model are confined to a four-dimensional membrane, while gravity propagates in several additional spatial dimensions that are large compared to the Planck scale.[2]

2. Warped extra dimensions, such as those proposed by the Randall–Sundrum model (RS), based on warped geometry where our universe is a five-dimensional anti-de Sitter space and the elementary particles except for the graviton are localized on a (3 + 1)-dimensional brane or branes.[3]

3. Universal extra dimension, proposed and first studied in 2000, assume, at variance with the ADD and RS approaches, that all fields propagate universally in the extra dimensions.

4. Multiple time dimensions, i.e. the possibility that there might be more than one dimension of time, has occasionally been discussed in physics and philosophy, although those models have to deal with the problem of causality.

3.1   References

[1] Rizzo, Thomas G. (2004). "Pedagogical Introduction to Extra Dimensions". *SLAC Summer Institute*. Retrieved 2016.

[2] For a pedagogical introduction, see M. Shifman (2009). *Large Extra Dimensions: Becoming acquainted with an alternative paradigm*. Crossing the boundaries: Gauge dynamics at strong coupling. Singapore: World Scientific. arXiv:0907.3074.

[3] Randall, Lisa; Sundrum, Raman (1999). "Large Mass Hierarchy from a Small Extra Dimension". *Physical Review Letters* **83** (17): 3370–3373. arXiv:hep-ph/9905221. Bibcode:1999PhRvL..83.3370R. doi:10.1103/PhysRevLett.83.3370.

Chapter 4

Quintom scenario

The **Quintom scenario** (derived from the words *quintessence* and *phantom*, as in phantom energy) is a hypothetical scenario regarding dark energy, with the equation of state crossing the cosmological constant boundary ($w = -1$) from above to below, or oppositely. Theoretically, a no-go theorem[1] has been proven, showing that to generate this scenario at least two degrees of freedom are required for dark energy models.

If this scenario exists, it may indicate how the universe avoids the dooms of time-like singularities, such as Big Bangs, Big Rips, and so on. For example, the Quintom Scenario was applied in 2008 to a model of inflationary cosmology without a Big Bang singularity, and in 2007 to a Big Bounce model of the universe.

4.1 External links

- Dark Energy Constraints from the Cosmic Age and Supernova by Bo Feng, Xiulian Wang and Xinmin Zhang

- Crossing the Phantom Divide by Martin Kunz and Domenico Sapone

- A Model Of Inflationary Cosmology Without Singularity by Yi-Fu Cai, Taotao Qiu, Jun-Qing Xia and Xinmin Zhang

- Quintom Cosmology: theoretical implications and observations by Yi-Fu Cai, Emmanuel N. Saridakis, Mohammad R. Setare and Jun-Qing Xia

- arXiv:0909.2776v2 [hep-th] 22 Apr 2010 Quintom Cosmology: theoretical implications and observations

4.2 References

[1] arXiv:0909.2776v2Quintom Cosmology - theoretical implications and observations

Chapter 5

Scale factor (cosmology)

The **scale factor, cosmic scale factor** or sometimes the **Robertson-Walker scale factor**[1] parameter of the Friedmann equations is a function of time which represents the relative expansion of the universe. It relates the proper distance (which can change over time, unlike the comoving distance which is constant) between a pair of objects, e.g. two galaxy clusters, moving with the Hubble flow in an expanding or contracting FLRW universe at any arbitrary time t to their distance at some reference time t_0. The formula for this is:

$$d(t) = a(t)d_0,$$

where $d(t)$ is the proper distance at epoch t, d_0 is the distance at the reference time t_0 and $a(t)$ is the scale factor.[2] Thus, by definition, $a(t_0) = 1$.

The scale factor is dimensionless, with t counted from the birth of the universe and t_0 set to the present age of the universe: 13.799 ± 0.021 Gyr [3] giving the current value of a as $a(t_0)$ or 1.

The evolution of the scale factor is a dynamical question, determined by the equations of general relativity, which are presented in the case of a locally isotropic, locally homogeneous universe by the Friedmann equations.

The Hubble parameter is defined:

$$H \equiv \frac{\dot{a}(t)}{a(t)}$$

where the dot represents a time derivative. From the previous equation $d(t) = d_0 a(t)$ one can see that $\dot{d}(t) = d_0\dot{a}(t)$, and also that $d_0 = \frac{d(t)}{a(t)}$, so combining these gives $\dot{d}(t) = \frac{d(t)\dot{a}(t)}{a(t)}$, and substituting the above definition of the Hubble parameter gives $\dot{d}(t) = Hd(t)$ which is just Hubble's law.

Current evidence suggests that the expansion rate of the universe is accelerating, which means that the second derivative of the scale factor $\ddot{a}(t)$ is positive, or equivalently that the first derivative $\dot{a}(t)$ is increasing over time.[4] This also implies that any given galaxy recedes from us with increasing speed over time, i.e. for that galaxy $\dot{d}(t)$ is increasing with time. In contrast, the Hubble parameter seems to be *decreasing* with time, meaning that if we were to look at some *fixed* distance d and watch a series of different galaxies pass that distance, later galaxies would pass that distance at a smaller velocity than earlier ones.[5]

According to the Friedmann–Lemaître–Robertson–Walker metric which is used to model the expanding universe, if at the present time we receive light from a distant object with a redshift of z, then the scale factor at the time the object originally emitted that light is $a(t) = \frac{1}{1+z}$.[6][7]

5.1 Chronology

Further information: Chronology of the universe

5.1.1 Radiation-dominated era

After Inflation), and until about 47,000 years after the Big Bang, the dynamics of the early universe were set by radiation (referring generally to the constituents of the universe which moved relativistically, principally photons and neutrinos).[8]

For a radiation-dominated universe the evolution of the scale factor in the Friedmann–Lemaître–Robertson–Walker metric is obtained solving the Friedmann equations:

$$a(t) \propto t^{1/2}.\ [9]$$

5.1.2 Matter-dominated era

Between about 47,000 years and 9.8 billion years after the Big Bang,[10] the energy density of matter exceeded both the energy density of radiation and the vacuum energy density.[11]

When the early universe was about 47,000 years old (redshift 3600), mass–energy density surpassed the radiation energy, although the universe remained optically thick to radiation until the universe was about 378,000 years old (redshift 1100). This second moment in time (close to the time of recombination) at which point the photons which compose the cosmic microwave background radiation were last scattered, is often mistaken as marking the end of the radiation era.

For a matter dominated universe the evolution of the scale factor in the Friedmann-Lemaitre-Robertson-Walker metric is easily obtained solving the Friedmann equations:

$$a(t) \propto t^{2/3}$$

5.1.3 Dark energy-dominated era

In physical cosmology, the **dark-energy-dominated era** refers to the last of the three phases of the known universe, the other two being the matter-dominated era and the radiation-dominated era. The dark-energy-dominated era began after the matter-dominated era, i.e. when the Universe was about 9.8 billion years old.[12] As other forms of the matter – dust and radiation – dropped to very low concentrations, the cosmological constant term started to dominate the energy density of the Universe.

For a dark-energy-dominated universe, the evolution of the scale factor in the Friedmann–Lemaître–Robertson–Walker metric is easily obtained solving the Friedmann equations:

$$a(t) \propto \exp(Ht)$$

Here, the coefficient H in the exponential, the Hubble constant, is

$$H = \sqrt{8\pi G \rho_{\text{full}}/3} = \sqrt{\Lambda/3}.$$

This exponential dependence on time makes the spacetime geometry identical to the de Sitter Universe, and only holds for a positive sign of the cosmological constant, the sign that was observed to be realized in Nature anyway. The current density of the observable universe is of the order of 9.44 x 10^{-27} kg m^{-3} and the age of the universe is of the order of 13.8 billion years, or 4.358 x 10^{17} s. The Hubble parameter, H, is ~70.88 km s^{-1} Mpc^{-1}. (The Hubble time is 13.79 billion years.) The value of the cosmological constant, Λ, is ~2 x 10^{-35} s^{-2}.

5.2 See also

- Cosmological principle
- Lambda-CDM model
- Redshift

5.3 References

[1] Steven Weinberg (2008). *Cosmology*. Oxford University Press. p. 3. ISBN 978-0-19-852682-7.

[2] Schutz, Bernard (2003). *Gravity from the Ground Up: An Introductory Guide to Gravity and General Relativity*. Cambridge University Press. p. 363. ISBN 978-0-521-45506-0.

[3] Planck Collaboration (2015). "Planck 2015 results. XIII. Cosmological parameters (See Table 4 on page 31 of pfd).". arXiv:1502.01589. Bibcode:2015arXiv150201589P.

[4] Jones, Mark H.; Robert J. Lambourne (2004). *An Introduction to Galaxies and Cosmology*. Cambridge University Press. p. 244. ISBN 978-0-521-83738-5.

[5] Is the universe expanding faster than the speed of light? (see final paragraph) Archived November 28, 2010, at the Wayback Machine.

[6] Davies, Paul (1992), *The New Physics*, p. 187.

[7] Mukhanov, V. F. (2005), *Physical Foundations of Cosmology*, p. 58.

[8] Ryden, Barbara, "Introduction to Cosmology", 2006, eqn. 5.25, 6.41

[9] Padmanabhan (1993), p. 64.

[10] Ryden, Barbara, "Introduction to Cosmology", 2006, eqn. 6.33, 6.41

[11] Zelik, M and Gregory, S: "Introductory Astronomy & Astrophysics", page 497. Thompson Learning, Inc. 1998

[12] Ryden, Barbara, "Introduction to Cosmology", 2006, eqn. 6.33

- Padmanabhan, Thanu (1993). *Structure formation in the universe*. Cambridge: Cambridge University Press. ISBN 0-521-42486-0.

- Spergel, D. N.; et al. (2003). "First-Year Wilkinson Microwave Anisotropy Probe (WMAP) Observations: Determination of Cosmological Parameters". *Astrophysical Journal Supplement* **148** (1): 175–194. arXiv:astro-ph/0302209. Bibcode:2003ApJS..148..175S. doi:10.1086/377226.

5.4 External links

- Relation of the scale factor with the cosmological constant and the Hubble constant

Chapter 6

Dark energy

Not to be confused with dark fluid, dark flow, or dark matter.

In physical cosmology and astronomy, **dark energy** is an unknown frm of   energy which is hypothesized to permeate all of space, tending to accelerate the expansion of the universe.[1] Dark energy is the most accepted hypothesis to explain since the 1990s the observations indicating that the universe is expanding at an accelerating rate.

Assuming that the standard model of cosmology is correct, the best current measurements indicate that dark energy contributes 68.3% of the total energy in the present-day observable universe. The mass–energy of dark matter and ordinary (baryonic) matter contribute 26.8% and 4.9%, respectively, and other components such as neutrinos and photons contribute a very small amount.[2][3][4][5] Again on a mass–energy equivalence basis, the density of dark energy ($\sim 7 \times 10^{-30}$ g/cm^3) is very low, much less than the density of ordinary matter or dark matter within galaxies. However, it comes to dominate the mass–energy of the universe because it is uniform across space.[6][7][8]

Two proposed forms for dark energy are the cosmological constant,[9] a *constant* energy density filling space homogeneously,[10] and scalar fields such as quintessence or moduli, *dynamic* quantities whose energy density can vary in time and space. Contributions from scalar fields that are constant in space are usually also included in the cosmological constant. The cosmological constant can be formulated to be equivalent to vacuum energy. Scalar fields that do change in space can be difficult to distinguish from a cosmological constant because the change may be extremely slow.

High-precision measurements of the expansion of the universe are required to understand how the expansion rate changes over time and space. In general relativity, the evolution of the expansion rate is parameterized by the cosmological equation of state (the relationship between temperature, pressure, and combined matter, energy, and vacuum energy density for any region of space). Measuring the equation of state for dark energy is one of the biggest efforts in observational cosmology today. Adding the cosmological constant to cosmology's standard FLRW metric leads to the Lambda-CDM model, which has been referred to as the "*standard model of cosmology*" because of its precise agreement with observations. Dark energy has been used as a crucial ingredient in a recent attempt to formulate a cyclic model for the universe.[11]

6.1   Nature of dark energy

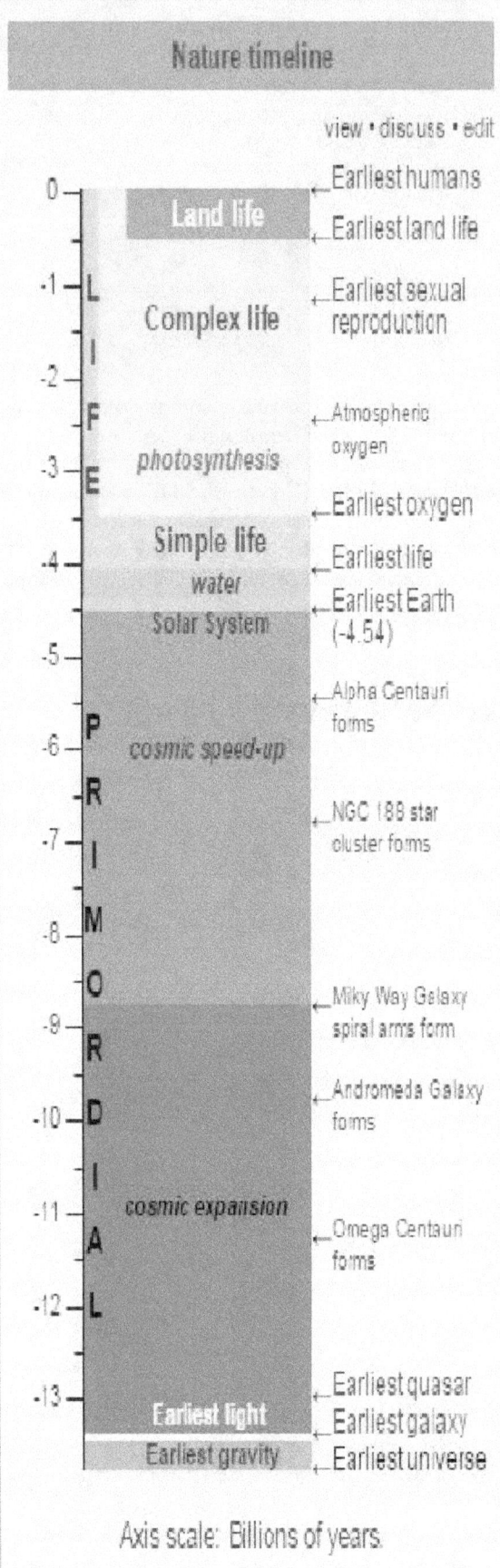

Axis scale: Billions of years.

Many things about the nature of dark energy remain matters of speculation.[12] The evidence for dark energy is indirect but comes from three independent sources:

- Distance measurements and their relation to redshift, which suggest the universe has expanded more in the last half of its life.[13]

- The theoretical need for a type of additional energy that is not matter or dark matter to form the observationally flat universe (absence of any detectable global curvature).

- It can be inferred from measures of large scale wave-patterns of mass density in the universe.

Dark energy is thought to be very homogeneous, not very dense and is not known to interact through any of the fundamental forces other than gravity. Since it is quite rarefied — roughly 10^{-27} kg/m^3 — it is unlikely to be detectable in laboratory experiments. Dark energy can have such a profound effect on the universe, making up 68% of universal density, only because it uniformly fills otherwise empty space. The two leading models are a cosmological constant and quintessence. Both models include the common characteristic that dark energy must have negative pressure.

6.1.1 Effect of dark energy: a small constant negative pressure of vacuum

Independently of its actual nature, dark energy would need to have a strong negative pressure (acting repulsively) like radiation pressure in a metamaterial[14] in order to explain the observed acceleration of the expansion of the universe. According to general relativity, the pressure within a substance contributes to its gravitational attraction for other things just as its mass density does. This happens because the physical quantity that causes matter to generate gravitational effects is the stress–energy tensor, which contains both the energy (or matter) density of a substance and its pressure and viscosity. In the Friedmann–Lemaître–Robertson–Walker metric, it can be shown that a strong constant negative pressure in all the universe causes an acceleration in universe expansion if the universe is already expanding, or a deceleration in universe contraction if the universe is already contracting. This accelerating expansion effect is sometimes labeled "gravitational repulsion", which is a colorful but possibly confusing expression. In fact a negative pressure does not influence the gravitational interaction between masses—which remains attractive—but rather alters the overall evolution of the universe at the cosmological scale, typically resulting in the accelerating expansion of the universe despite the attraction among the masses present in the universe. The acceleration is simply a function of dark energy density. Dark energy is persistent: its density remains constant (experimentally, within a factor of 1:10), i.e. it does not get diluted when space expands.

6.2   Evidence of existence

6.2.1   Supernovae

A Type Ia supernova (bright spot on the bottom-left) near a galaxy

In 1998, the High-Z Supernova Search Team[15] published observations of Type Ia ("one-A") supernovae. In 1999, the Supernova Cosmology Project[16] followed by suggesting that the expansion of the universe is accelerating.[17] The 2011 Nobel Prize in Physics was awarded to Saul Perlmutter, Brian P. Schmidt and Adam G. Riess for their leadership in the discovery.[18][19]

Since then, these observations have been corroborated by several independent sources. Measurements of the cosmic microwave background, gravitational lensing, and the large-scale structure of the cosmos as well as improved measurements of supernovae have been consistent with the Lambda-CDM model.[20] Some people argue that the only indication for the existence of dark energy is observations of distance measurements and associated redshifts. Cosmic microwave background anisotropies and baryon acoustic oscillations are only observations that distances to a given redshift are larger than expected from a "dusty" Friedmann–Lemaître universe and the local measured Hubble constant.[21]

Supernovae are useful for cosmology because they are excellent standard candles across cosmological distances. They allow the expansion history of the universe to be measured by looking at the relationship between the distance to an object and its redshift, which gives how fast it is receding from us. The relationship is roughly linear, according to Hubble's law. It is relatively easy to measure redshift, but finding the distance to an object is more difficult. Usually, astronomers use standard candles: objects for which the intrinsic brightness, the absolute magnitude, is known. This allows the object's distance to be measured from its actual observed brightness, or apparent magnitude. Type Ia supernovae are the best-known standard candles across cosmological distances because of their extreme and consistent luminosity.

Recent observations of supernovae are consistent with a universe made up 71.3% of dark energy and 27.4% of a combination of dark matter and baryonic matter.[22]

6.2.2   Cosmic microwave background

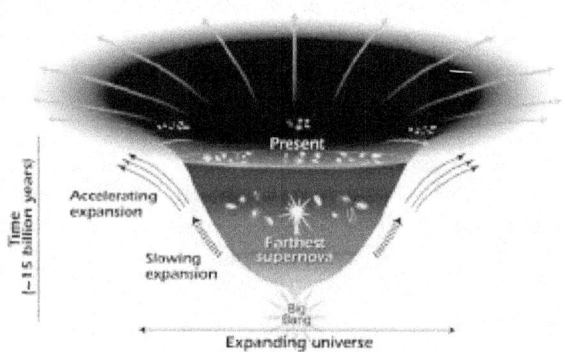

This diagram reveals changes in the rate of expansion since the universe's birth 15 billion years ago. The more shallow the curve, the faster the rate of expansion. The curve changes noticeably about 7.5 billion years ago, when objects in the universe began flying apart at a faster rate. Astronomers theorize that the faster expansion rate is due to a mysterious dark force that is pushing galaxies apart.

Diagram representing the accelerated expansion of the universe due to dark energy.

The existence of dark energy, in whatever form, is needed to reconcile the measured geometry of space with the total amount of matter in the universe. Measurements of cosmic microwave background (CMB) anisotropies indicate that the universe is close to flat. For the shape of the universe to be flat, the mass/energy density of the universe must be equal to the critical density. The total amount of matter in the universe (including baryons and dark matter), as measured from the CMB spectrum, accounts for only about 30% of the critical density. This implies the existence of an additional form of energy to account for the remaining 70%.[20] The Wilkinson Microwave Anisotropy Probe (WMAP) spacecraft seven-year analysis estimated a universe made up of 72.8% dark energy, 22.7% dark matter and 4.5% ordinary matter.[4] Work done in 2013 based on the Planck spacecraft observations of the CMB gave a more accurate estimate of 68.3% of dark energy, 26.8% of dark matter and 4.9% of ordinary matter.[23]

6.2.3 Large-scale structure

The theory of large-scale structure, which governs the formation of structures in the universe (stars, quasars, galaxies and galaxy groups and clusters), also suggests that the density of matter in the universe is only 30% of the critical density.

A 2011 survey, the WiggleZ galaxy survey of more than 200,000 galaxies, provided further evidence towards the existence of dark energy, although the exact physics behind it remains unknown.[24][25] The WiggleZ survey from the Australian Astronomical Observatory scanned the galaxies to determine their redshift. Then, by exploiting the fact that baryon acoustic oscillations have left voids regularly of ~150 Mpc diameter, surrounded by the galaxies, the voids were used as standard rulers to determine distances to galaxies as far as 2,000 Mpc (redshift 0.6), which allowed astronomers to determine more accurately the speeds of the galaxies from their redshift and distance. The data confirmed cosmic acceleration up to half of the age of the universe (7 billion years) and constrain its inhomogeneity to 1 part in 10.[25] This provides a confirmation to cosmic acceleration independent of supernovae.

6.2.4 Late-time integrated Sachs-Wolfe effect

Accelerated cosmic expansion causes gravitational potential wells and hills to flatten as photons pass through them, producing cold spots and hot spots on the CMB aligned with vast supervoids and superclusters. This so-called late-time Integrated Sachs–Wolfe effect (ISW) is a direct signal of dark energy in a flat universe.[26] It was reported at high significance in 2008 by Ho *et al.*[27] and Giannantonio *et al.*[28]

6.2.5 Observational Hubble constant data

A new approach to test evidence of dark energy through observational Hubble constant (H(z)) data (OHD) has gained significant attention in recent years.[29][30][31][32] The Hubble constant is measured as a function of cosmological redshift. OHD directly tracks the expansion history of the universe by taking passively evolving early-type galaxies as "cosmic chronometers".[33] From this point, this approach provides standard clocks in the universe. The core of this idea is the measurement of the differential age evolution as a function of redshift of these cosmic chronometers. Thus, it provides a direct estimate of the Hubble parameter $H(z) = -1/(1+z)dz/dt \approx -1/(1+z)\Delta z/\Delta t$. The merit of this approach is clear: the reliance on a differential quantity, $\Delta z/\Delta t$, can minimize many common issues and systematic effects; and as a direct measurement of the Hubble parameter instead of its integral, like supernovae and baryon acoustic oscillations (BAO), it brings more information and is appealing in computation. For these reasons, it has been widely used to examine the accelerated cosmic expansion and study properties of dark energy.

6.3 Theories of explanation

6.3.1 Cosmological constant

Main article: Cosmological constant
For more details on this topic, see Equation of state (cosmology).

The simplest explanation for dark energy is that it is simply

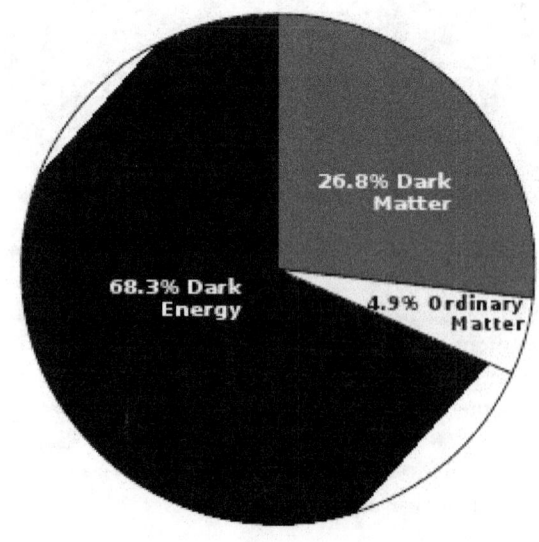

Estimated distribution of matter and energy in the universe[34]

the "cost of having space": that is, a volume of space has some intrinsic, fundamental energy. This is the cosmological constant, sometimes called Lambda (hence Lambda-CDM model) after the Greek letter Λ, the symbol used to represent this quantity mathematically. Since energy and mass are related by $E = mc^2$, Einstein's theory of general relativity predicts that this energy will have a gravitational effect. It is sometimes called a vacuum energy because it is the energy density of empty vacuum. In fact, most theories of particle physics predict vacuum fluctuations that would give the vacuum this sort of energy. This is related to the Casimir effect, in which there is a small suction into regions where virtual particles are geometrically inhibited from forming (e.g. between plates with tiny separation). The cosmological constant is estimated by cosmologists to be on the order of 10^{-29} g/cm^3, or about 10^{-120} in reduced Planck units. Particle physics predicts a natural value of 1 in reduced Planck units, leading to a large discrepancy.

Lambda, the letter that represents the cosmological constant

The cosmological constant has negative pressure equal to its energy density and so causes the expansion of the universe to accelerate. The reason why a cosmological constant has negative pressure can be seen from classical thermodynamics; Energy must be lost from inside a container to do work on the container. A change in volume dV requires work done equal to a change of energy $-P\,dV$, where P is the pressure. But the amount of energy in a container full of vacuum actually increases when the volume increases (dV is positive), because the energy is equal to ϱV, where ϱ (rho) is the energy density of the cosmological constant. Therefore, P is negative and, in fact, $P = -\varrho$.

A major outstanding problem is that most quantum field theories predict a huge cosmological constant from the en-

ergy of the quantum vacuum, more than 100 orders of magnitude too large.[10] This would need to be cancelled almost, but not exactly, by an equally large term of the opposite sign. Some supersymmetric theories require a cosmological constant that is exactly zero,[35] which does not help because supersymmetry must be broken. The present scientific consensus amounts to extrapolating the empirical evidence where it is relevant to predictions, and fine-tuning theories until a more elegant solution is found. Technically, this amounts to checking theories against macroscopic observations. Unfortunately, as the known error-margin in the constant predicts the fate of the universe more than its present state, many such "deeper" questions remain unknown.

In spite of its problems, the cosmological constant is in many respects the most economical solution to the problem of cosmic acceleration. One number successfully explains a multitude of observations. Thus, the current standard model of cosmology, the Lambda-CDM model, includes the cosmological constant as an essential feature.

6.3.2 Quintessence

Main article: Quintessence (physics)

In quintessence models of dark energy, the observed acceleration of the scale factor is caused by the potential energy of a dynamical field, referred to as quintessence field. Quintessence differs from the cosmological constant in that it can vary in space and time. In order for it not to clump and form structure like matter, the field must be very light so that it has a large Compton wavelength.

No evidence of quintessence is yet available, but it has not been ruled out either. It generally predicts a slightly slower acceleration of the expansion of the universe than the cosmological constant. Some scientists think that the best evidence for quintessence would come from violations of Einstein's equivalence principle and variation of the fundamental constants in space or time.[36] Scalar fields are predicted by the *Standard Model of particle physics* and string theory, but an analogous problem to the cosmological constant problem (or the problem of constructing models of cosmological inflation) occurs: renormalization theory predicts that scalar fields should acquire large masses.

The coincidence problem asks why the acceleration of the Universe began when it did. If acceleration began earlier in the universe, structures such as galaxies would never have had time to form and life, at least as we know it, would never have had a chance to exist. Proponents of the anthropic principle view this as support for their arguments. However, many models of quintessence have a so-called **tracker** behavior, which solves this problem. In these models, the quintessence field has a density which closely tracks (but is less than) the radiation density until matter-radiation equality, which triggers quintessence to start behaving as dark energy, eventually dominating the universe. This naturally sets the low energy scale of the dark energy.).[37][38]

In 2004, when scientists fit the evolution of dark energy with the cosmological data, they found that the equation of state had possibly crossed the cosmological constant boundary (w=−1) from above to below. A No-Go theorem has been proved that gives this scenario at least two degrees of freedom as required for dark energy models. This scenario is so-called Quintom scenario.

Some special cases of quintessence are phantom energy, in which the energy density of quintessence actually increases with time, and k-essence (short for kinetic quintessence) which has a non-standard form of kinetic energy such as a negative kinetic energy.[39] They can have unusual properties: phantom energy, for example, can cause a Big Rip.

6.4 Alternative ideas

Some alternatives to dark energy aim to explain the observational data by a more refined use of established theories, focusing, for example, on the gravitational effects of density inhomogeneities, or on consequences of electroweak symmetry breaking in the early universe. If we are located in an emptier-than-average region of space, the observed cosmic expansion rate could be mistaken for a variation in time, or acceleration.[40][41][42][43] A different approach uses a cosmological extension of the equivalence principle to show how space might appear to be expanding more rapidly in the voids surrounding our local cluster. While weak, such effects considered cumulatively over billions of years could become significant, creating the illusion of cosmic acceleration, and making it appear as if we live in a Hubble bubble.[44][45][46]

Another class of theories attempts to come up with an all-encompassing theory of both dark matter and dark energy as a single phenomenon that modifies the laws of gravity at various scales. An example of this type of theory is the theory of dark fluid. Another class of theories that unifies dark matter and dark energy are suggested to be covariant theories of modified gravities. These theories alter the dynamics of the space-time such that the modified dynamic stems what have been assigned to the presence of dark energy and dark matter.[48]

A 2011 paper in the journal *Physical Review D* by Christos Tsagas, a cosmologist at Aristotle University of Thessaloniki in Greece, argued that it is likely that the accelerated expansion of the universe is an illusion caused by the

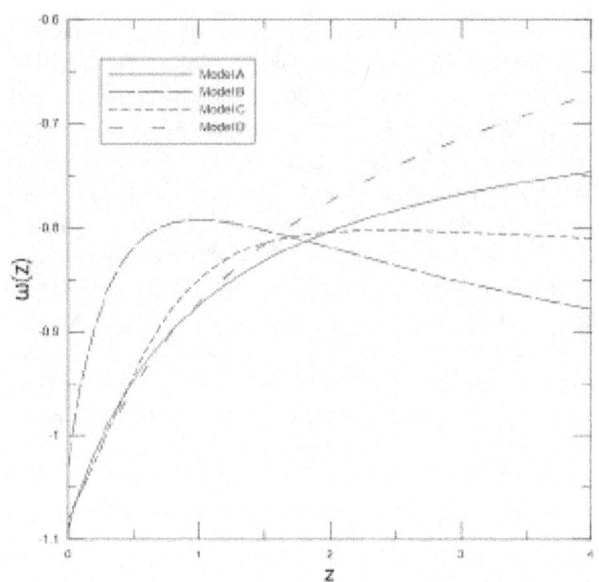

The equation of state of Dark Energy for 4 common models by Redshift.[47]

A: CPL Model,

B: Jassal Model,

C: Barboza & Alcaniz Model,

D: Wetterich Model

relative motion of us to the rest of the universe. The paper cites data showing that the 2.5 billion light-years-wide region of space we are inside of is moving very quickly relative to everything around it. If the theory is confirmed, then dark energy would not exist (but the "dark flow" still might).[49][50]

Some theorists think that dark energy and cosmic acceleration are a failure of general relativity on very large scales, larger than superclusters. However most attempts at modifying general relativity have turned out to be either equivalent to theories of quintessence, or inconsistent with observations. Other ideas for dark energy have come from string theory, brane cosmology and the holographic principle, but have not yet proved as compelling as quintessence and the cosmological constant. In other hand, M.R. Khoshbin-e-Khoshnazar believes that a model discretization of the universe could explain the origin of dark energy.[51]

On string theory, an article in the journal *Nature* described:[52]

> String theories, popular with many particle physicists, make it possible, even desirable, to think that the observable universe is just one of 10^{500} universes in a grander multiverse, says Leonard Susskind, a cosmologist at Stanford University in California. The vacuum energy will have different values in different universes, and in many or

most it might indeed be vast. But it must be small
in ours because it is only in such a universe that
observers such as ourselves can evolve.

Paul Steinhardt in the same article criticizes string theory's explanation of dark energy stating "...Anthropics and randomness don't explain anything... I am disappointed with what most theorists are willing to accept".[52]

Another set of proposals is based on the possibility of a double metric tensor for space-time.[53][54] It has been argued that time reversed solutions in general relativity require such double metric for consistency, and that both dark matter and dark energy can be understood in terms of time reversed solutions of general relativity.[55]

It has been shown that if inertia is assumed to be due to the effect of horizons on Unruh radiation then this predicts galaxy rotation and a cosmic acceleration similar to that observed.[56]

6.4.1 Variable Dark Energy models

In general, the dark energy can be variable. Modern observational data have determined the density of dark energy in the present. Using baryon acoustic oscillations, we can investigate the effect of dark energy in the history of the Universe and we can constrain parameters of the equation of state of dark energy. One of the proposed solutions to get closer to answering the question of dark energy, is to assume that it is variable. To that end, several models have been proposed. One of their most popular models is Chevallier–Polarski–Linder model (CPL).[57][58] Some other common models are, (Barboza & Alcaniz. 2008),[59] (Jassal et al. 2005),[60] (Wetterich. 2004).[61]

6.5 Implications for the fate of the universe

Cosmologists estimate that the acceleration began roughly 5 billion years ago. Before that, it is thought that the expansion was decelerating, due to the attractive influence of dark matter and baryons. The density of dark matter in an expanding universe decreases more quickly than dark energy, and eventually the dark energy dominates. Specifically, when the volume of the universe doubles, the density of dark matter is halved, but the density of dark energy is nearly unchanged (it is exactly constant in the case of a cosmological constant).

If the acceleration continues indefinitely, the ultimate result will be that galaxies outside the local supercluster will have a line-of-sight velocity that continually increases with time,

eventually far exceeding the speed of light.[62] This is not a violation of special relativity because the notion of "velocity" used here is different from that of velocity in a local inertial frame of reference, which is still constrained to be less than the speed of light for any massive object (see Uses of the proper distance for a discussion of the subtleties of defining any notion of relative velocity in cosmology). Because the Hubble parameter is decreasing with time, there can actually be cases where a galaxy that is receding from us faster than light does manage to emit a signal which reaches us eventually.[63][64] However, because of the accelerating expansion, it is projected that most galaxies will eventually cross a type of cosmological event horizon where any light they emit past that point will never be able to reach us at any time in the infinite future[65] because the light never reaches a point where its "peculiar velocity" toward us exceeds the expansion velocity away from us (these two notions of velocity are also discussed in Uses of the proper distance). Assuming the dark energy is constant (a cosmological constant), the current distance to this cosmological event horizon is about 16 billion light years, meaning that a signal from an event happening *at present* would eventually be able to reach us in the future if the event were less than 16 billion light years away, but the signal would never reach us if the event were more than 16 billion light years away.[64]

As galaxies approach the point of crossing this cosmological event horizon, the light from them will become more and more redshifted, to the point where the wavelength becomes too large to detect in practice and the galaxies appear to vanish completely[66][67] (*see* Future of an expanding universe). The Earth, the Milky Way, and the Virgo Supercluster, however, would remain virtually undisturbed while the rest of the universe recedes and disappears from view. In this scenario, the local supercluster would ultimately suffer heat death, just as was thought for the flat, matter-dominated universe before measurements of cosmic acceleration.

There are some very speculative ideas about the future of the universe. One suggests that phantom energy causes *divergent* expansion, which would imply that the effective force of dark energy continues growing until it dominates all other forces in the universe. Under this scenario, dark energy would ultimately tear apart all gravitationally bound structures, including galaxies and solar systems, and eventually overcome the electrical and nuclear forces to tear apart atoms themselves, ending the universe in a "Big Rip". On the other hand, dark energy might dissipate with time or even become attractive. Such uncertainties leave open the possibility that gravity might yet rule the day and lead to a universe that contracts in on itself in a "Big Crunch".[68] Some scenarios, such as the cyclic model, suggest this could be the case. It is also possible the universe may never have an end and continue in its present state forever (see The

Second Law as a law of disorder). While these ideas are not supported by observations, they are not ruled out.

6.6 History of discovery and previous speculation

The cosmological constant was first proposed by Einstein as a mechanism to obtain a solution of the gravitational field equation that would lead to a static universe, effectively using dark energy to balance gravity.[69] Not only was the mechanism an inelegant example of fine-tuning but it was also later realized that Einstein's static universe would actually be unstable because local inhomogeneities would ultimately lead to either the runaway expansion or contraction of the universe. The equilibrium is unstable: If the universe expands slightly, then the expansion releases vacuum energy, which causes yet more expansion. Likewise, a universe which contracts slightly will continue contracting. These sorts of disturbances are inevitable, due to the uneven distribution of matter throughout the universe. More importantly, observations made by Edwin Hubble in 1929 showed that the universe appears to be expanding and not static at all. Einstein reportedly referred to his failure to predict the idea of a dynamic universe, in contrast to a static universe, as his greatest blunder.[70]

Alan Guth and Alexei Starobinsky proposed in 1980 that a negative pressure field, similar in concept to dark energy, could drive cosmic inflation in the very early universe. Inflation postulates that some repulsive force, qualitatively similar to dark energy, resulted in an enormous and exponential expansion of the universe slightly after the Big Bang. Such expansion is an essential feature of most current models of the Big Bang. However, inflation must have occurred at a much higher energy density than the dark energy we observe today and is thought to have completely ended when the universe was just a fraction of a second old. It is unclear what relation, if any, exists between dark energy and inflation. Even after inflationary models became accepted, the cosmological constant was thought to be irrelevant to the current universe.

Nearly all inflation models predict that the total (matter+energy) density of the universe should be very close to the critical density. During the 1980s, most cosmological research focused on models with critical density in matter only, usually 95% cold dark matter and 5% ordinary matter (baryons). These models were found to be successful at forming realistic galaxies and clusters, but some problems appeared in the late 1980s: notably, the model required a value for the Hubble constant lower than preferred by observations, and the model under-predicted observations of large-scale galaxy clustering. These difficul-

ties became stronger after the discovery of anisotropy in the cosmic microwave background by the COBE spacecraft in 1992, and several modified CDM models came under active study through the mid-1990s: these included the Lambda-CDM model and a mixed cold/hot dark matter model. The first direct evidence for dark energy came from supernova observations in 1998 of accelerated expansion in Riess et al.[15] and in Perlmutter et al.,[16] and the Lambda-CDM model then became the leading model. Soon after, dark energy was supported by independent observations: in 2000, the BOOMERanG and Maxima cosmic microwave background experiments observed the first acoustic peak in the CMB, showing that the total (matter+energy) density is close to 100% of critical density. Then in 2001, the 2dF Galaxy Redshift Survey gave strong evidence that the matter density is around 30% of critical. The large difference between these two supports a smooth component of dark energy making up the difference. Much more precise measurements from WMAP in 2003–2010 have continued to support the standard model and give more accurate measurements of the key parameters.

The term "dark energy", echoing Fritz Zwicky's "dark matter" from the 1930s, was coined by Michael Turner in 1998.[71]

As of 2013, the Lambda-CDM model is consistent with a series of increasingly rigorous cosmological observations, including the Planck spacecraft and the Supernova Legacy Survey. First results from the SNLS reveal that the average behavior (i.e., equation of state) of dark energy behaves like Einstein's cosmological constant to a precision of 10%.[72] Recent results from the Hubble Space Telescope Higher-Z Team indicate that dark energy has been present for at least 9 billion years and during the period preceding cosmic acceleration.

6.7 See also

- Conformal gravity

- De Sitter relativity

- Illustris project

- *The Dark Energy Survey*

- *Quintessence: The Search for Missing Mass in the Universe*

- Vacuum state

6.8   References

[1] Peebles, P. J. E. and Ratra, Bharat (2003). "The cosmological constant and dark energy". *Reviews of Modern Physics* **75** (2): 559–606. arXiv:astro-ph/0207347. Bibcode:2003RvMP...75..559P. doi:10.1103/RevModPhys.75.559.

[2] Ade, P. A. R.; Aghanim, N.; Armitage-Caplan, C.; et al. (Planck Collaboration), C.; Arnaud, M.; Ashdown, M.; Atrio-Barandela, F.; Aumont, J.; Aussel, H.; Baccigalupi, C.; Banday, A. J.; Barreiro, R. B.; Barrena, R.; Bartelmann, M.; Bartlett, J. G.; Bartolo, N.; Basak, S.; Battaner, E.; Battye, R.; Benabed, K.; Benoît, A.; Benoit-Lévy, A.; Bernard, J.-P.; Bersanelli, M.; Bertincourt, B.; Bethermin, M.; Bielewicz, P.; Bikmaev, I.; Blanchard, A.; et al. (22 March 2013). "Planck 2013 results. I. Overview of products and scientific results – Table 9". *Astronomy and Astrophysics* **571**: A1. arXiv:1303.5062. Bibcode:2014A&A...571A...1P. doi:10.1051/0004-6361/201321529.

[3] Ade, P. A. R.; Aghanim, N.; Armitage-Caplan, C.; et al. (Planck Collaboration), C.; Arnaud, M.; Ashdown, M.; Atrio-Barandela, F.; Aumont, J.; Aussel, H.; Baccigalupi, C.; Banday, A. J.; Barreiro, R. B.; Barrena, R.; Bartelmann, M.; Bartlett, J. G.; Bartolo, N.; Basak, S.; Battaner, E.; Battye, R.; Benabed, K.; Benoît, A.; Benoit-Lévy, A.; Bernard, J.-P.; Bersanelli, M.; Bertincourt, B.; Bethermin, M.; Bielewicz, P.; Bikmaev, I.; Blanchard, A.; et al. (31 March 2013). "Planck 2013 Results Papers". *Astronomy and Astrophysics* **571**: A1. arXiv:1303.5062. Bibcode:2014A&A...571A...1P. doi:10.1051/0004-6361/201321529.

[4] "First Planck results: the Universe is still weird and interesting".

[5] Sean Carroll, Ph.D., Caltech, 2007, The Teaching Company, *Dark Matter, Dark Energy: The Dark Side of the Universe*, Guidebook Part 2 page 46. Retrieved Oct. 7, 2013, "...dark energy: A smooth, persistent component of invisible energy, thought to make up about 70 percent of the current energy density of the universe. Dark energy is known to be smooth because it doesn't accumulate preferentially in galaxies and clusters..."

[6] Paul J. Steinhardt, Neil Turok (2006). "Why the cosmological constant is small and positive". *Science* **312** (5777): 1180–1183. arXiv:astro-ph/0605173. Bibcode:2006Sci...312.1180S. doi:10.1126/science.1126231.

[7] "Dark Energy". *Hyperphysics*. Retrieved January 4, 2014.

[8] Ferris, Timothy. "Dark Matter(Dark Energy)". Retrieved 2015-06-10.

[9] http://www.ft.com/intl/cms/s/2/493de45a-8bef-11e0-854c-00144feab49a.html#axzz3m9WSVVkC

[10] Carroll, Sean (2001). "The cosmological constant". *Living Reviews in Relativity* **4**. arXiv:astro-ph/0004075. Bibcode:2001LRR.....4....1C. doi:10.12942/lrr-2001-1. Retrieved 2006-09-28.

[11] Baum, L. and Frampton, P.H. (2007). "Turnaround in Cyclic Cosmology". *Physical Review Letters* **98** (7): 071301. arXiv:hep-th/0610213. Bibcode:2007PhRvL..98g1301B. doi:10.1103/PhysRevLett.98.071301. PMID 17359014.

[12] Overbye, Dennis. "Astronomers Report Evidence of 'Dark Energy' Splitting the Universe". The New York Times. Retrieved August 5, 2015.

[13] Durrer, R. (2011). "What do we really know about Dark Energy?". *Philosophical Transactions of the Royal Society A: Mathematical, Physical and Engineering Sciences* **369** (1957): 5102. arXiv:1103.5331. Bibcode:2011RSPTA.369.5102D. doi:10.1098/rsta.2011.0285.

[14] Zhong-Yue Wang (2016). "Modern Theory for Electromagnetic Metamaterials". *Plasmonics* **11** (2): 503–508. doi:10.1007/s11468-015-0071-7.

[15] Riess, Adam G.; Filippenko; Challis; Clocchiatti; Diercks; Garnavich; Gilliland; Hogan; Jha; Kirshner; Leibundgut; Phillips; Reiss; Schmidt; Schommer; Smith; Spyromilio; Stubbs; Suntzeff; Tonry (1998). "Observational evidence from supernovae for an accelerating universe and a cosmological constant". *Astronomical Journal* **116** (3): 1009–38. arXiv:astro-ph/9805201. Bibcode:1998AJ....116.1009R. doi:10.1086/300499.

[16] Perlmutter, S.; Aldering; Goldhaber; Knop; Nugent; Castro; Deustua; Fabbro; Goobar; Groom; Hook; Kim; Kim; Lee; Nunes; Pain; Pennypacker; Quimby; Lidman; Ellis; Irwin; McMahon; Ruiz-Lapuente; Walton; Schaefer; Boyle; Filippenko; Matheson; Fruchter; et al. (1999). "Measurements of Omega and Lambda from 42 high redshift supernovae". *Astrophysical Journal* **517** (2): 565–86. arXiv:astro-ph/9812133. Bibcode:1999ApJ...517..565P. doi:10.1086/307221.

[17] The first paper, using observed data, which claimed a positive Lambda term was Paál, G.; et al. (1992). "Inflation and compactification from galaxy redshifts?". *Astrophysics and Space Science* **191**: 107–24. Bibcode:1992Ap&SS.191..107P. doi:10.1007/BF00644200.

[18] "The Nobel Prize in Physics 2011". Nobel Foundation. Retrieved 2011-10-04.

[19] The Nobel Prize in Physics 2011. Perlmutter got half the prize, and the other half was shared between Schmidt and Riess.

[20] Spergel, D. N. (WMAP collaboration); et al. (March 2006). "Wilkinson Microwave Anisotropy Probe (WMAP) three year results: implications for cosmology".

[21] Durrer, R. (2011). "What do we really know about dark energy?". *Philosophical Transactions of the Royal Society A* **369** (1957): 5102–5114. arXiv:1103.5331. Bibcode:2011RSPTA.369.5102D. doi:10.1098/rsta.2011.0285.

[22] Kowalski, Marek; Rubin, David; Aldering, G.; Agostinho, R. J.; Amadon, A.; Amanullah, R.; Balland, C.; Barbary, K.; Blanc, G.; Challis, P. J.; Conley, A.; Connolly, N. V.; Covarrubias, R.; Dawson, K. S.; Deustua, S. E.; Ellis, R.; Fabbro, S.; Fadeyev, V.; Fan, X.; Farris, B.; Folatelli, G.; Frye, B. L.; Garavini, G.; Gates, E. L.; Germany, L.; Goldhaber, G.; Goldman, B.; Goobar, A.; Groom, D. E.; et al. (October 27, 2008). "Improved Cosmological Constraints from New, Old and Combined Supernova Datasets". *The Astrophysical Journal* (Chicago: University of Chicago Press) **686** (2): 749–778. arXiv:0804.4142. Bibcode:2008ApJ...686..749K. doi:10.1086/589937.. They find a best fit value of the dark energy density, $\Omega\Lambda$ of 0.713+0.027–0.029(stat)+0.036–0.039(sys), of the total matter density, ΩM, of 0.274+0.016–0.016(stat)+0.013–0.012(sys) with an equation of state parameter w of –0.969+0.059–0.063(stat)+0.063–0.066(sys).

[23] "Big Bang's afterglow shows universe is 80 million years older than scientists first thought". *The Washington Post*. Retrieved 22 March 2013.

[24] "New method 'confirms dark energy'". BBC News. 2011-05-19.

[25] Dark energy is real, Swinburne University of Technology, 19 May 2011

[26] Crittenden; Neil Turok (1995). "Looking for Λ with the Rees-Sciama Effect". *Physical Review Letters* **76** (4): 575–578. arXiv:astro-ph/9510072. Bibcode:1996PhRvL..76..575C. doi:10.1103/PhysRevLett.76.575. PMID 10061494.

[27] Shirley Ho; Hirata; Nikhil Padmanabhan; Uros Seljak; Neta Bahcall (2008). "Correlation of CMB with large-scale structure: I. ISW Tomography and Cosmological Implications". *Physical Review D* **78** (4): 043519. arXiv:0801.0642. Bibcode:2008PhRvD..78d3519H. doi:10.1103/PhysRevD.78.043519.

[28] Tommaso Giannantonio; Ryan Scranton; Crittenden; Nichol; Boughn; Myers; Richards (2008). "Combined analysis of the integrated Sachs-Wolfe effect and cosmological implications". *Physical Review D* **77** (12): 123520. arXiv:0801.4380. Bibcode:2008PhRvD..77l3520G. doi:10.1103/PhysRevD.77.123520.

[29] Zelong Yi; Tongjie Zhang (2007). "Constraints on holographic dark energy models using the differential ages of passively evolving galaxies". *Modern Physics Letters A* **22** (1): 41. arXiv:astro-ph/0605596. Bibcode:2007MPLA...22...41Y. doi:10.1142/S0217732307020889.

[30] Haoyi Wan; Zelong Yi; Tongjie Zhang; Jie Zhou (2007). "Constraints on the DGP Universe Using Observational Hubble parameter". *Physics Letters B* **651** (5): 352. arXiv:0706.2723. Bibcode:2007PhLB..651..352W. doi:10.1016/j.physletb.2007.06.053.

[31] Cong Ma; Tongjie Zhang (2010). "Power of Observational Hubble Parameter Data: a Figure of Merit Exploration". *Astrophysical Journal* **730** (2): 74. arXiv:1007.3787. Bibcode:2011ApJ...730...74M. doi:10.1088/0004-637X/730/2/74.

[32] Tongjie Zhang; Cong Ma; Tian Lan (2010). "Constraints on the Dark Side of the Universe and Observational Hubble Parameter Data". *Advances in Astronomy* **2010** (1): 1. arXiv:1010.1307. Bibcode:2010AdAst2010E..81Z. doi:10.1155/2010/184284.

[33] Joan Simon; Licia Verde; Raul Jimenez (2005). "Constraints on the redshift dependence of the dark energy potential". *Physical Review D* **71** (12): 123001. arXiv:astro-ph/0412269. Bibcode:2005PhRvD..7113001S. doi:10.1103/PhysRevD.71.123001.

[34] "Planck reveals an almost perfect universe". *Planck*. ESA. 2013-03-21. Retrieved 2013-03-21.

[35] Wess, Julius; Bagger, Jonathan. *Supersymmetry and Supergravity*. ISBN 978-0691025308.

[36] Carroll, Sean M. (1998). "Quintessence and the Rest of the World: Suppressing Long-Range Interactions". *Physical Review Letters* **81** (15): 3067–3070. arXiv:astro-ph/9806099. Bibcode:1998PhRvL..81.3067C. doi:10.1103/PhysRevLett.81.3067. ISSN 0031-9007.

[37] Ratra, Bharat; Peebles, P.J.E. "Cosmological consequences of a rolling homogeneous scalar field". *Phys. Rev.* **D37**: 3406. Bibcode:1988PhRvD..37.3406R. doi:10.1103/PhysRevD.37.3406.

[38] Steinhardt, Paul J.; Wang, Li-Min; Zlatev, Ivaylo. "Cosmological tracking solutions". *Phys. Rev.* **D59**: 123504. arXiv:astro-ph/9812313. Bibcode:1999PhRvD..59l3504S. doi:10.1103/PhysRevD.59.123504.

[39] R.R.Caldwell (2002). "A phantom menace? Cosmological consequences of a dark energy component with super-negative equation of state". *Physics Letters B* **545** (1-2): 23–29. arXiv:astro-ph/9908168. Bibcode:2002PhLB..545...23C. doi:10.1016/S0370-2693(02)02589-3.

[40] Wiltshire, David L. (2007). "Exact Solution to the Averaging Problem in Cosmology". *Physical Review Letters* **99** (25): 251101. arXiv:0709.0732. Bibcode:2007PhRvL..99y1101W. doi:10.1103/PhysRevLett.99.251101. PMID 18233512.

[41] Ishak, Mustapha; Richardson, James; Garred, David; Whittington, Delilah; Nwankwo, Anthony; Sussman, Roberto (2007). "Dark Energy or Apparent Acceleration

Due to a Relativistic Cosmological Model More Complex than FLRW?". *Physical Review D* **78** (12): 123531. arXiv:0708.2943. Bibcode:2008PhRvD..7813531I. doi:10.1103/PhysRevD.78.123531.

[42] Mattsson, Teppo (2007). "Dark energy as a mirage". *Gen. Rel. Grav.* **42** (3): 567–599. arXiv:0711.4264. Bibcode:2010GReGr..42..567M. doi:10.1007/s10714-009-0873-z.

[43] Clifton, Timothy; Ferreira, Pedro (April 2009). "Does Dark Energy Really Exist?". *Scientific American* **300** (4): 48–55. doi:10.1038/scientificamerican0409-48. PMID 19363920. Retrieved April 30, 2009.

[44] Wiltshire, D. (2008). "Cosmological equivalence principle and the weak-field limit". *Physical Review D* **78** (8): 084032. arXiv:0809.1183. Bibcode:2008PhRvD..78h4032W. doi:10.1103/PhysRevD.78.084032.

[45] Gray, Stuart. "Dark questions remain over dark energy". ABC Science Australia. Retrieved 27 January 2013.

[46] Merali, Zeeya (March 2012). "Is Einstein's Greatest Work All Wrong—Because He Didn't Go Far Enough?". *Discover magazine*. Retrieved 27 January 2013.

[47] by Ehsan Sadri M.A Ap

[48] Exirifard, Q. (2010). "Phenomenological covariant approach to gravity". *General Relativity and Gravitation* **43**: 93–106. arXiv:0808.1962. Bibcode:2011GReGr..43...93E. doi:10.1007/s10714-010-1073-6.

[49] Wolchover, Natalie (27 September 2011) 'Accelerating universe' could be just an illusion, MSNBC

[50] Tsagas, Christos G. (2011). "Peculiar motions, accelerated expansion, and the cosmological axis". *Physical Review D* **84** (6): 063503. arXiv:1107.4045. Bibcode:2011PhRvD..84f3503T. doi:10.1103/PhysRevD.84.063503.

[51] Khoshbin-e-Khoshnazar, M.R. (2013). "Binding Energy of the Very Early Universe: Abandoning Einstein for a Discretized Three–Torus Poset.A Proposal on the Origin of Dark Energy". *Gravitation and Cosmology* **19** (2): 106–113. doi:10.1134/s0202289313020059.

[52] Hogan, Jenny (2007). "Unseen Universe: Welcome to the dark side". *Nature* **448** (7151): 240–245. Bibcode:2007Natur.448..240H. doi:10.1038/448240a. PMID 17637630.

[53] Hossenfelder, S. (2008). "A Bi-Metric Theory with Exchange Symmetry". *Physical Review D* **78** (4): 044015. arXiv:0807.2838. Bibcode:2008PhRvD..78d4015H. doi:10.1103/PhysRevD.78.044015.

[54] Henry-Couannier, F. (2005). "Discrete Symmetries and General Relativity, the Dark Side of Gravity". *International Journal of Modern Physics A* **20** (11): 2341. arXiv:gr-qc/0410055. Bibcode:2005IJMPA..20.2341H. doi:10.1142/S0217751X05024602.

[55] Ripalda, Jose M. (1999). "Time reversal and negative energies in general relativity". arXiv:gr-qc/9906012.

[56] McCulloch, M.E. (2010). "Minimum accelerations from quantised inertia". *EPL* **90** (2): 29001. arXiv:1004.3303. Bibcode:2010EL......9029001M. doi:10.1209/0295-5075/90/29001.

[57] Chevallier, M; Polarski, D (2001). "Accelerating Universes with Scaling Dark Matter". *International Journal of Modern Physics D* **10**: 213–224. arXiv:gr-qc/0009008. Bibcode:2001IJMPD..10..213C. doi:10.1142/S0218271801000822.

[58] Linder, Eric V. (3 March 2003). "Exploring the Expansion History of the Universe". *Physical Review Letters* **90** (9). arXiv:astro-ph/0208512v1. Bibcode:2003PhRvL..90i1301L. doi:10.1103/PhysRevLett.90.091301.

[59] Alcaniz, E.M.; Alcaniz, J.S. (2008). "A parametric model for dark energy". *Physics Letters B* **666**: 415–419. arXiv:0805.1713. Bibcode:2008PhLB..666..415B. doi:10.1016/j.physletb.2008.08.012.

[60] Jassal, H.K; Bagla, J.S (2010). "Understanding the origin of CMB constraints on Dark Energy". *Monthly Notices of the Royal Astronomical Society* **405**: 2639–2650. arXiv:astro-ph/0601389. Bibcode:2010MNRAS.405.2639J. doi:10.1111/j.1365-2966.2010.16647.x.

[61] Wetterich, C. (2004). "Phenomenological parameterization of quintessence". arXiv:astro-ph/0403289v1.

[62] Krauss, Lawrence M. and Scherrer, Robert J. (March 2008). "The End of Cosmology?". *Scientific American* **82**. Retrieved 2011-01-06.

[63] Is the universe expanding faster than the speed of light? (see the last two paragraphs)

[64] Lineweaver, Charles; Tamara M. Davis (2005). "Misconceptions about the Big Bang" (PDF). *Scientific American*. Retrieved 2008-11-06.

[65] Loeb, Abraham (2002). "The Long-Term Future of Extragalactic Astronomy". *Physical Review D* **65** (4): 047301. arXiv:astro-ph/0107568. Bibcode:2002PhRvD..65d7301L. doi:10.1103/PhysRevD.65.047301.

[66] Krauss, Lawrence M.; Robert J. Scherrer (2007). "The Return of a Static Universe and the End of Cosmology". *General Relativity and Gravitation* **39** (10): 1545–1550. arXiv:0704.0221. Bibcode:2007GReGr..39.1545K. doi:10.1007/s10714-007-0472-9.

[67] Using Tiny Particles To Answer Giant Questions. Science Friday, 3 Apr 2009. According to the transcript, Brian Greene makes the comment "And actually, in the far future,

everything we now see, except for our local galaxy and a region of galaxies will have disappeared. The entire universe will disappear before our very eyes, and it's one of my arguments for actually funding cosmology. We've got to do it while we have a chance."

[68] *How the Universe Works 3*. End of the Universe. Discovery Channel. 2014.

[69] Harvey, Alex (2012). "How Einstein Discovered Dark Energy". arXiv:1211.6338.

[70] Gamow, George (1970) *My World Line: An Informal Autobiography*. p. 44: "Much later, when I was discussing cosmological problems with Einstein, he remarked that the introduction of the cosmological term was the biggest blunder he ever made in his life." – Here the "cosmological term" refers to the cosmological constant in the equations of general relativity, whose value Einstein initially picked to ensure that his model of the universe would neither expand nor contract; if he hadn't done this he might have theoretically predicted the universal expansion that was first observed by Edwin Hubble.

[71] The first appearance of the term "dark energy" is in the article with another cosmologist and Turner's student at the time, Dragan Huterer, "Prospects for Probing the Dark Energy via Supernova Distance Measurements", which was posted to the ArXiv.org e-print archive in August 1998 and published in Huterer, D.; Turner, M. (1999). "Prospects for probing the dark energy via supernova distance measurements". *Physical Review D* **60** (8). arXiv:astro-ph/9808133. Bibcode:1999PhRvD..60h1301H. doi:10.1103/PhysRevD.60.081301., although the manner in which the term is treated there suggests it was already in general use. Cosmologist Saul Perlmutter has credited Turner with coining the term in an article they wrote together with Martin White, where it is introduced in quotation marks as if it were a neologism. Perlmutter, S.; Turner, M.; White, M. (1999). "Constraining Dark Energy with Type Ia Supernovae and Large-Scale Structure". *Physical Review Letters* **83** (4): 670. arXiv:astro-ph/9901052. Bibcode:1999PhRvL..83..670P. doi:10.1103/PhysRevLett.83.670.

[72] Astier, Pierre (Supernova Legacy Survey); Guy; Regnault; Pain; Aubourg; Balam; Basa; Carlberg; Fabbro; Fouchez; Hook; Howell; Lafoux; Neill; Palanque-Delabrouille; Perrett; Pritchet; Rich; Sullivan; Taillet; Aldering; Antilogus; Arsenijevic; Balland; Baumont; Bronder; Courtois; Ellis; Filiol; et al. (2006). "The Supernova legacy survey: Measurement of ΩM, $\Omega \Lambda$ and W from the first year data set". *Astronomy and Astrophysics* **447**: 31–48. arXiv:astro-ph/0510447. Bibcode:2006A&A...447...31A. doi:10.1051/0004-6361:20054185.

- Dark Energy on *In Our Time* at the BBC. (listen now)

- Dark energy Eric Linder Scholarpedia 3(2):4900. doi: 10.4249/scholarpedia.4900

- Dark energy: how the paradigm shifted Physicsworld.com

- Dennis Overbye (November 2006). "9 Billion-Year-Old 'Dark Energy' Reported". *The New York Times*.

- "Mysterious force's long presence" BBC News online (2006) More evidence for dark energy being the cosmological constant

- "Astronomy Picture of the Day" one of the images of the Cosmic Microwave Background which confirmed the presence of dark energy and dark matter

- SuperNova Legacy Survey home page The Canada-France-Hawaii Telescope Legacy Survey Supernova Program aims primarily at measuring the equation of state of Dark Energy. It is designed to precisely measure several hundred high-redshift supernovae.

- "Report of the Dark Energy Task Force"

- "HubbleSite.org – Dark Energy Website" Multimedia presentation explores the science of dark energy and Hubble's role in its discovery.

- "Surveying the dark side"

- "Dark energy and 3-manifold topology" Acta Physica Polonica 38 (2007), p. 3633–3639

- The Dark Energy Survey

- The Joint Dark Energy Mission

- Harvey: Dark Energy Found Stifling Growth in Universe, primary source

- April 2010 Smithsonian Magazine Article

- HETDEX Dark energy experiment

- Dark Energy FAQ

- "The Hunt for Dark Energy" George FR Ellis, Peter Cameron and David Tong discuss the presence of dark energy in the Universe

- Euclid ESA Satellite, a mission to map the geometry of the dark universe

6.9 External links

-

Chapter 7

Dark matter

Not to be confused with antimatter, dark energy, dark fluid, or dark flow. For other uses, see Dark Matter (disambiguation)

Dark matter is a hypothetical type of matter composing the approximately 27% of the mass and energy in the observable universe [1] that is not accounted for by dark energy, baryonic matter, and neutrinos.[2] The name refers to the fact that it does not emit or interact with electromagnetic radiation, such as light, and is thus invisible to the entire electromagnetic spectrum.[3] Although dark matter cannot be directly observed with conventional electromagnetic telescopes, its existence and properties are inferred from its various gravitational effects such as the motions of visible matter, via gravitational lensing, its influence on the universe's large-scale structure, and its effects in the cosmic microwave background. Dark matter is transparent to electromagnetic radiation and/or is so dense and small that it fails to absorb or emit enough radiation to be detectable with current imaging technology.

Estimates of masses for galaxies and larger structures via dynamical and general relativistic means are much greater than those based on the mass of the visible "luminous" matter.[4]

The standard model of cosmology indicates that the total mass–energy of the universe contains 4.9% ordinary matter, 26.8% dark matter and 68.3% dark energy.[5][6] Thus, dark matter constitutes 84.5%[note 1] of total mass, while dark energy plus dark matter constitute 95.1% of total mass–energy content.[7][8][9][10] The great majority of ordinary matter in the universe is also unseen, since visible stars and gas inside galaxies and clusters account for less than 10% of the ordinary matter contribution to the mass-energy density of the universe.[11]

The dark matter hypothesis plays a central role in current modeling of cosmic structure formation and galaxy formation and evolution and on explanations of the anisotropies observed in the cosmic microwave background (CMB). All these lines of evidence suggest that galaxies, clusters of galaxies, and the universe as a whole contain far more matter than that which is observable via electromagnetic signals.[12]

The most widely accepted hypothesis on the form for dark matter is that it is composed of weakly interacting massive particles (WIMPs) that interact only through gravity and the weak force.[13]

Although the existence of dark matter is generally accepted by most of the astronomical community, a minority of astronomers [14] argue for various modifications of the standard laws of general relativity, such as MOND, TeVeS, and Conformal gravity[15] that attempt to account for the observations without invoking additional matter.[16]

Many experiments to detect proposed dark matter particles through non-gravitational means are under way.[17]

7.1 History

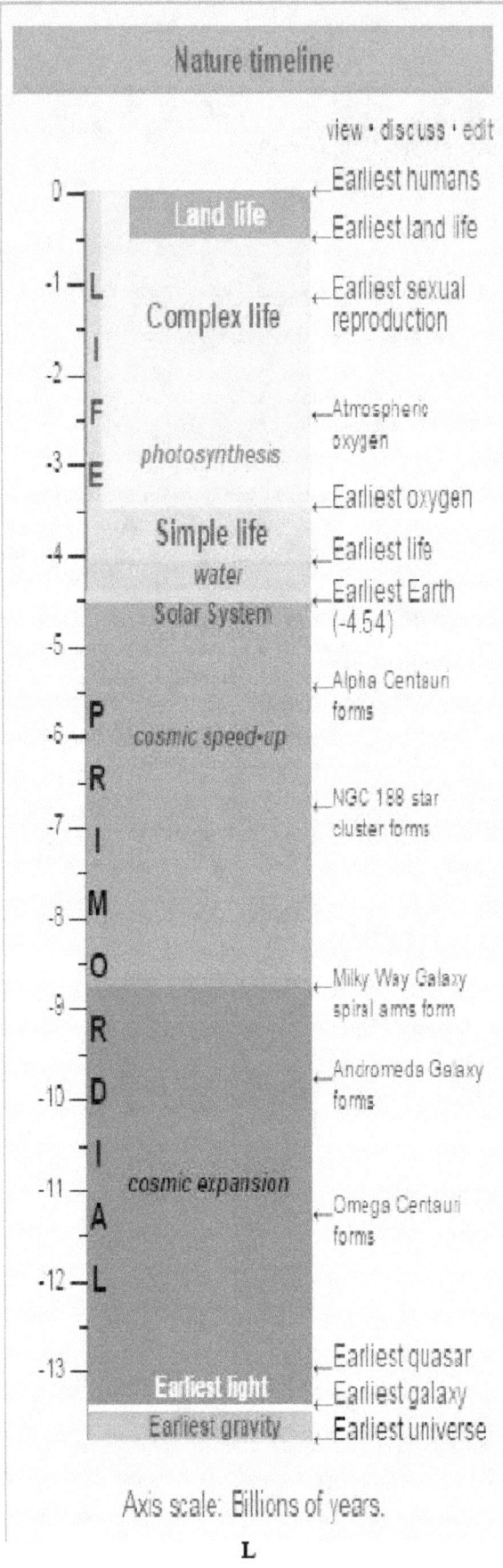

The first to suggest the existence of dark matter (using stellar velocities) was Dutch astronomer Jacobus Kapteyn in 1922.[18][19] Fellow Dutchman and radio astronomy pioneer Jan Oort also hypothesized the existence of dark matter in 1932.[19][20][21] Oort was studying stellar motions in the local galactic neighborhood and found that the mass in the galactic plane must be greater than what was observed, but this measurement was later determined to be erroneous.[22]

In 1933, Swiss astrophysicist Fritz Zwicky, who studied galactic clusters while working at the California Institute of Technology, made a similar inference.[23][24][25] Zwicky applied the virial theorem to the Coma galaxy cluster and obtained evidence of unseen mass that he called *dunkle Materie* 'dark matter'. Zwicky estimated its mass based on the motions of galaxies near its edge and compared that to an estimate based on its brightness and number of galaxies. He estimated that the cluster had about 400 times more mass than was visually observable. The gravity effect of the visible galaxies was far too small for such fast orbits, thus mass must be hidden from view. Based on these conclusions, Zwicky inferred that some unseen matter provided the mass and associated gravitation attraction to hold the cluster together. This was the first formal inference about the existence of dark matter.[26] Zwicky's estimates were off by more than an order of magnitude, mainly due to an obsolete value of the Hubble constant;[27] the same calculation today shows a smaller fraction, using greater values for luminous mass. However, Zwicky did correctly infer that the bulk of the matter was dark.[26]

The first robust indications that the mass to light ratio was anything other than unity came from measurements of galaxy rotation curves. In 1939, Horace W. Babcock reported the rotation curve for the Andromeda nebula, which suggested that the mass-to-luminosity ratio increases radially.[28] He attributed it to either light absorption within the galaxy or modified dynamics in the outer portions of the spiral and not to missing matter.

Vera Rubin and Kent Ford in the 1960s–1970s provided further strong evidence, also using galaxy rotation curves.[29][30] Rubin worked with a new spectrograph to measure the velocity curve of edge-on spiral galaxies with greater accuracy.[30] This result was independently confirmed in 1978.[31] An influential paper presented Rubin's results in 1980.[32] Rubin found that most galaxies must contain about six times as much dark as visible mass; thus, by around 1980 the apparent need for dark matter was widely recognized as a major unsolved problem in astronomy.

A stream of independent observations in the 1980s indicated its presence, including gravitational lensing of background objects by galaxy clusters, the temperature distribution of hot gas in galaxies and clusters, and the pattern of anisotropies in the cosmic microwave background. According to consensus among cosmologists, dark matter is composed primarily of a not yet characterized type of subatomic particle.[13][33] The search for this particle, by a variety of means, is one of the major efforts in particle physics.[17]

7.1.1   Cosmic microwave background radiation

In cosmology, the CMB is explained as relic radiation which has travelled freely since the era of recombination, around 375,000 years after the Big Bang. The CMB's anisotropies are explained as the result of small primordial density fluctuations, and subsequent acoustic oscillations in the photon-baryon plasma whose restoring force is gravity.[34]

The Cosmic Background Explorer (COBE) satellite found the CMB spectrum to be a very precise blackbody spectrum with a temperature of 2.726 K. In 1992, COBE detected CMB fluctuations (anisotropies) at a level of about one part in 10^5.[35]

In the following decade, CMB anisotropies were investigated by ground-based and balloon experiments. Their primary goal was to measure the angular scale of the first acoustic peak of the anisotropies' power spectrum, for which COBE had insufficient resolution. During the 1990s, the first peak was measured with increasing sensitivity, and in 2000 the BOOMERanG experiment[36] reported that the highest power fluctuations occur at scales of approximately one degree, showing that the Universe is close to flat. These measurements were able to rule out cosmic strings as the leading theory of cosmic structure formation, and suggested cosmic inflation was the correct theory.

Ground-based interferometers provided fluctuation measurements with higher accuracy, including the Very Small Array, the Degree Angular Scale Interferometer (DASI) and the Cosmic Background Imager (CBI). DASI first detected the CMB polarization,[37][38] and CBI provided the first E-mode polarization spectrum with compelling evidence that it is out of phase with the T-mode spectrum.[39] COBE's successor, the Wilkinson Microwave Anisotropy Probe (WMAP) provided the most detailed measurements of (large-scale) anisotropies in the CMB in 2003 - 2010.[40] ESA's Planck spacecraft returned more detailed results in 2013-2015.

WMAP's measurements played the key role in establishing the Standard Model of Cosmology, namely the Lambda-CDM model, which posits a dark energy-dominated flat universe, supplemented by dark matter and atoms with density fluctuations seeded by a Gaussian, adiabatic, nearly scale invariant process. Its basic properties are determined by six adjustable parameters: dark matter density, baryon (atom) density, the universe's age (or equivalently, the Hubble constant), the initial fluctuation amplitude and their scale dependence.

7.2   Observational evidence

This artist's impression shows the expected distribution of dark matter in the Milky Way galaxy as a blue halo of material surrounding the galaxy.[41]

Much of the evidence comes from the motions of galaxies.[42] Many of these appear to be fairly uniform, so by the virial theorem, the total kinetic energy should be half the galaxies' total gravitational binding energy. Observationally, the total kinetic energy is much greater. In particular, assuming the gravitational mass is due to only visible matter, stars far from the center of galaxies have much higher velocities than predicted by the virial theorem. Galactic rotation curves, which illustrate the velocity of rotation versus the distance from the galactic center, show the "excess" velocity. Dark matter is the most straightforward way of accounting for this discrepancy.[43]

The distribution of dark matter in galaxies required to explain the motion of the observed matter suggests the presence of a roughly spherically symmetric, centrally concentrated halo of dark matter with the visible matter concentrated in a central disc.

Low surface brightness dwarf galaxies are important sources of information for studying dark matter. They have an uncommonly low ratio of visible to dark matter, and have few bright stars at the center that would otherwise impair observations of the rotation curve of outlying stars.

Gravitational lensing observations of galaxy clusters allow direct estimates of the gravitational mass based on its effect on light coming from background galaxies, since large collections of matter (dark or otherwise) gravitationally deflect light. In clusters such as Abell 1689, lensing observations confirm the presence of considerably more mass than is indicated by the clusters' light. In the Bullet Cluster, lensing observations show that much of the lensing mass is separated from the X-ray-emitting baryonic mass. In July 2012, lensing observations were used to identify a "filament" of dark matter between two clusters of galaxies, as cosmological simulations predicted.[44]

7.2.1 Galaxy rotation curves

Main article: Galaxy rotation curve
 A galaxy rotation curve is a plot of the orbital velocities

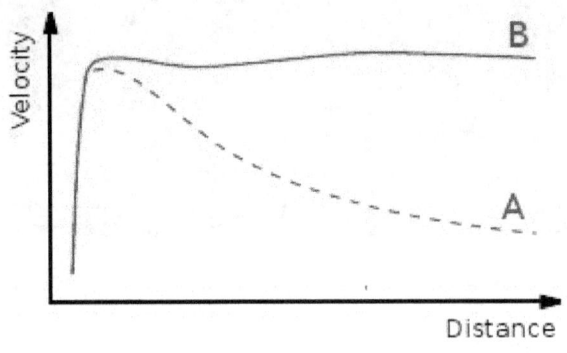

Rotation curve of a typical spiral galaxy: predicted (A) and observed (B). Dark matter can explain the 'flat' appearance of the velocity curve out to a large radius

(i.e., the speeds) of visible stars or gas in that galaxy versus their radial distance from that galaxy's center. The rotational/orbital speed of galaxies/stars does not decline with distance, unlike other orbital systems such as stars/planets and planets/moons that also have most of their mass at the centre. In the latter cases, this reflects the mass distributions within those systems. The mass observations for galaxies based on the light that they emit are far too low to explain the velocity observations. The dark matter hypothesis accounts for the missing mass, explaining the anomaly.[28]

A universal rotation curve can be expressed as the sum of an exponential distribution of visible matter that tapers to zero with distance from the center, and a spherical dark matter halo with a flat core of radius r_0 and density $\rho_0 = 4.5 \times 10^{-2}(r_0/kpc)^{-2/3} \, M\odot pc^{-3}$.[45]

Low-surface-brightness (LSB) galaxies have a much larger visible mass deficit than others. This property simplifies the disentanglement of the dark and visible matter contributions to the rotation curves.[17]

Rotation curves for some elliptical galaxies do display low velocities for outlying stars (tracked for example by the motion of embedded planetary nebulae). A dark-matter compliant hypothesis proposes that some stars may have been torn by tidal forces from disk-galaxy mergers from their original galaxies during the first close passage and put on outgoing trajectories, explaining the low velocities of the remaining stars even in the presence of a halo.[17][46]

7.2.2 Velocity dispersions of galaxies

Velocity dispersion estimates of elliptical galaxies,[47] with some exceptions, generally indicate a relatively high dark matter content.

Diffuse interstellar gas measurements of galactic edges indicate missing ordinary matter beyond the visible boundary, but that galaxies are virialized (i.e., gravitationally bound and orbiting each other with velocities that correspond to predicted orbital velocities of general relativity) up to ten times their visible radii.[48] This has the effect of pushing up the dark matter as a fraction of the total matter from 50% as measured by Rubin to the now accepted value of nearly 95%.

Dark matter seems to be a small component or absent in some places. Globular clusters show little evidence of dark matter.[49] Star velocity profiles seemed to indicate a concentration of dark matter in the disk of the Milky Way. It now appears, however, that the high concentration of baryonic matter in the disk (especially in the interstellar medium) can account for this motion. Galaxy mass and light profiles appear to not match. The typical model for dark matter galaxies is a smooth, spherical distribution in virialized halos. This avoids small-scale (stellar) dynamical effects. A 2006 study explained the warp in the Milky Way's disk by the interaction of the Large and Small Magellanic Clouds and the 20-fold increase in predicted mass from dark matter.[50]

In 2005, astronomers claimed to have discovered a galaxy made almost entirely of dark matter, 50 million light years away in the Virgo Cluster, which was named VIRGOHI21.[51] Unusually, VIRGOHI21 does not appear to contain visible stars: it was discovered with radio frequency observations of hydrogen. Based on rotation profiles, scientists estimate that this object contains approximately 1000 times more dark matter than hydrogen and has a mass of about 1/10 that of the Milky Way. The Milky Way is estimated to have roughly 10 times as much dark matter as ordinary matter. Models of the Big Bang and structure formation suggested that such dark galaxies should be very common, but VIRGOHI21 was the first to be detected.

The velocity profiles of some galaxies such as NGC 3379 indicate an absence of dark matter.[52]

7.2.3 Galaxy clusters and gravitational lensing

Strong gravitational lensing as observed by the Hubble Space Telescope in Abell 1689 indicates the presence of dark matter—enlarge the image to see the lensing arcs.

Clusters of galaxies are particularly important for dark matter studies since their masses can be estimated in three independent ways:

- From the scatter in radial velocities of the galaxies within clusters

- From X-rays emitted by hot gas in the clusters. From the X-ray energy spectrum and flux, the gas temperature and density can be estimated, hence giving the pressure: assuming pressure and gravity balance determines the cluster's mass profile. Many Chandra X-ray Observatory experiments use this technique to independently determine cluster masses. These observations generally indicate that baryonic mass is approximately 12–15 percent, in reasonable agreement with the Planck spacecraft cosmic average of 15.5–16 percent.[53]

Dark matter *is invisible. Based on the effect of gravitational lensing, a ring of* dark matter *has been inferred in this image of a galaxy cluster (CL0024+17) and has been represented in blue.*[54]

- Gravitational lensing (usually of more distant galaxies) can measure cluster masses without relying on observations of dynamics (e.g., velocity). There are two types of lensing: strong lensing produces multiple images or giant arcs near the cluster core, while weak lensing is observed as small shape distortions around the outer regions. Multiple Hubble projects have used this method to measure cluster masses.

Generally, these three methods are in reasonable agreement that dark matter outweighs visible matter by approximately 5 to 1.

Gravity acts as a lens to bend the light from a more distant source (such as a quasar) around a massive object (such as a cluster of galaxies) lying between the source and the observer in accordance with general relativity.

Strong lensing is the observed distortion of background galaxies into arcs when their light passes through such a gravitational lens. It has been observed around many distant clusters including Abell 1689.[55] By measuring the distortion geometry, the mass of the intervening cluster can be obtained. In the dozens of cases where this has been done, the mass-to-light ratios obtained correspond to the dynamical dark matter measurements of clusters.[56]

Weak gravitational lensing investigates minute distortions of galaxies, using statistical analyses from vast galaxy surveys. By examining the apparent shear deformation of the adjacent background galaxies, astrophysicists can characterize the mean distribution of dark matter. The mass-to-light ratios correspond to dark matter densities predicted by other large-scale structure measurements.[57]

Galaxy cluster Abell 2029 comprises thousands of galaxies enveloped in a cloud of hot gas and dark matter equivalent to more than 10^{14} $M\odot$. At the center of this cluster is an enormous elliptical galaxy likely formed from many smaller galaxies.[58]

The Bullet Cluster: HST image with overlays. The total projected mass distribution reconstructed from strong and weak gravitational lensing is shown in blue, while the X-ray emitting hot gas observed with Chandra is shown in red.

The most direct observational evidence comes from the Bullet Cluster. In most regions dark and visible matter are found together,[59] due to their gravitational attraction. In the Bullet Cluster however, the two matter types split apart, due to a past collision between two smaller clusters. Electromagnetic interactions between gas particles has caused the gas to slow and concentrate near the point of impact. The galaxies, stars and dark matter continued through with negligible collisions. Lensing observations show two dark matter peaks near the galaxy peaks, as expected in dark matter theory. Since the gas peaks contain more ordinary matter than the stars, modified-gravity theories should show the lensing peaks near the gas peaks, contrary to the observations.

X-ray observations show that much of the luminous matter (in the form of 10^7–10^8 Kelvin[60] gas or plasma) is concentrated in the cluster's center. Weak gravitational lensing observations show that much dark matter resides outside the central region. Unlike galactic rotation curves, this evidence is independent of the details of Newtonian gravity, directly supporting dark matter.[60] Dark matter's observed behavior constrains whether and how much it scatters off

other dark matter particles, quantified as its self-interaction cross section. If dark matter has no pressure, it can be described as a perfect fluid that has no damping.[61] The distribution of mass in galaxy clusters has been used to argue both for[62][63] and against[64] the significance of self-interaction.

An ongoing survey using the Subaru Telescope uses weak lensing to analyze background light, bent by dark matter, to determine the statistical distribution of dark matter in the foreground. The survey studies galaxies more than a billion light-years distant, across an area greater than a thousand square degrees (about one fortieth of the entire sky).[65][66]

7.2.4 Cosmic microwave background

Main article: Cosmic microwave background
Angular CMB fluctuations provide evidence for dark mat-

The cosmic microwave background by WMAP

ter. The typical angular scales of CMB oscillations, measured as the power spectrum of the CMB anisotropies, reveal the different effects of baryonic and dark matter. Ordinary matter interacts strongly via radiation whereas dark matter particles (WIMPs) do not; both affect the oscillations by way of their gravity, so the two forms of matter have different effects.

The spectrum shows a large first peak and smaller successive peaks.[40] The first peak mostly shows the density of baryonic matter, while the third peak relates mostly to the density of dark matter, measuring the density of matter and the density of atoms.

7.2.5 Sky surveys and baryon acoustic oscillations

Main article: Baryon acoustic oscillations

The early universe's acoustic oscillations in the photon-baryon fluid are observed as the prominent acoustic peaks in the CMB spectrum. This set up a preferred length scale for baryons in the early universe which is determined as 147 megaparsec (comoving) by the Planck spacecraft. As the dark matter and baryons clumped together after recombination, the effect is much weaker in the galaxy distribution in the nearby universe, but is detectable as a subtle (~ 1 percent) preference for pairs of galaxies to be separated by 147 Mpc, rather than 130 or 160 Mpc, called the BAO feature. This feature was predicted theoretically in the 1990s and then discovered in 2005, in two large galaxy redshift surveys, the Sloan Digital Sky Survey and the 2dF Galaxy Redshift Survey.[67] Combining the CMB observations with BAO measurements from galaxy redshift surveys provides a precise estimate of the Hubble constant and the average matter density in the Universe.[34]

7.2.6 Redshift-space distortions

Large galaxy redshift surveys may be used to make a three-dimensional map of the galaxy distribution. These maps are slightly distorted because distances are estimated from observed redshifts; the redshift contains a contribution from the galaxy's so-called peculiar velocity in addition to the dominant Hubble expansion term. On average, superclusters are expanding but more slowly than the cosmic mean due to their gravity, while voids are expanding faster than average. In a redshift map, galaxies in front of a supercluster have excess radial velocities towards it and have redshifts slightly higher than their distance would imply, while galaxies behind the supercluster have redshifts slightly low for their distance. This effect causes superclusters to appear "squashed" in the radial direction, and likewise voids are "stretched"; angular positions are unaffected. The effect is not detectable for any one structure since the true shape is not known, but can be measured by averaging over many structures assuming we are not at a special location in the Universe.

The effect was predicted quantitatively by Nick Kaiser in 1987, and first decisively measured in 2001 by the 2dF Galaxy Redshift Survey.[68] Results are in agreement with the Lambda-CDM model.

7.2.7 Type Ia supernova distance measurements

Main article: Type Ia supernova

Type Ia supernovae can be used as "standard candles" to measure extragalactic distances. Extensive data sets of these supernovae can be used to constrain cosmological models.[69] They constrain the dark energy density $\Omega\Lambda$ = ~0.713 for a flat, Lambda CDM universe and the parameter

w for a quintessence model. The results are roughly consistent with those derived from the WMAP observations and further constrain the Lambda CDM model and (indirectly) dark matter.[34]

7.2.8 Lyman-alpha forest

Main article: Lyman-alpha forest

In astronomical spectroscopy, the Lyman-alpha forest is the sum of the absorption lines arising from the Lyman-alpha transition of neutral hydrogen in the spectra of distant galaxies and quasars. Lyman-alpha forest observations can also constrain cosmological models.[70] These constraints agree with those obtained from WMAP data.

7.2.9 Structure formation

Main article: Structure formation
Structure formation refers to the serial transformations of

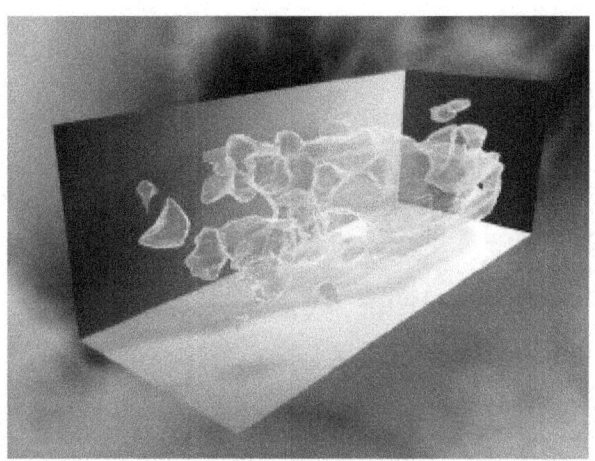

3D map of the large-scale distribution of dark matter, reconstructed from measurements of weak gravitational lensing with the Hubble Space Telescope.[71]

the universe following the Big Bang. Prior to structure formation, e.g., Friedmann cosmology solutions to general relativity describe a homogeneous universe. Later, small anisotropies gradually grew and condensed the homogeneous universe into stars, galaxies and larger structures.

Observations suggest that structure formation proceeds hierarchically, with the smallest structures collapsing first, followed by galaxies and then galaxy clusters. As the structures collapse in the evolving universe, they begin to "light up" as baryonic matter heats up through gravitational contraction and approaches hydrostatic pressure balance.

CMB anisotropy measurements fix models in which most matter is dark. Dark matter also closes gaps in models of large-scale structure. The dark matter hypothesis corresponds with statistical surveys of the visible structure and precisely to CMB predictions.

Initially, baryonic matter's post-Big Bang temperature and pressure were too high to collapse and form smaller structures, such as stars, via the Jeans instability. The gravity from dark matter increase the compaction force, allowing the formation of these structures.

Computer simulations of billions of dark matter particles[72] seem to confirm that the "cold" dark matter model of structure formation is consistent with the structures observed through galaxy surveys, such as the Sloan Digital Sky Survey and 2dF Galaxy Redshift Survey, as well as observations of the Lyman-alpha forest.

Tensions separate observations and simulations. Observations have turned up 90-99% fewer small galaxies than permitted by dark matter-based predictions.[73][74] In addition, simulations predict dark matter distributions with a dense cusp near galactic centers, but the observed halos are smoother than predicted.

7.3 Composition

The composition of dark matter remains uncertain. Possibilities include dense baryonic (interacts with electromagnetic force) matter and non-baryonic matter (interacts with its surroundings only through gravity).

7.3.1 Baryonic vs. nonbaryonic matter

Fermi-LAT observations of dwarf galaxies provide new insights on dark matter.

Baryonic matter

Baryonic matter is made of baryons (protons and neutrons) that make up stars and planets. It also encompasses less common black holes, neutron stars, faint old white dwarfs and brown dwarfs, collectively known as massive compact halo objects (MACHOs).

Multiple lines of evidence suggest the majority of dark matter is not made of baryons:

- Sufficient diffuse, baryonic gas or dust would be visible when backlit by stars.

- The theory of Big Bang nucleosynthesis predicts the observed abundance of the chemical elements;[75][76] agreement with observed abundances requires that baryonic matter makes up between 4–5% of the universe's critical density. In contrast, large-scale structure and other observations indicate that the total matter density is about 30% of the critical density (with dark energy providing the remaining 70%).

- Large astronomical searches for gravitational microlensing in the Milky Way found that at most a small fraction of the dark matter may be in dark, compact, conventional objects (MACHOs, etc.); the excluded range of object masses is from half the Earth's mass up to 30 solar masses, which covers nearly all the plausible candidates.[77][78][79][80][81][82]

- Detailed analysis of the small irregularities (anisotropies) in the cosmic microwave background observed by WMAP and Planck shows that around five-sixths of the total matter is in a form that interacts significantly with ordinary matter or photons only through gravitational effects.

Non-baryonic matter

Candidates for nonbaryonic dark matter are hypothetical particles such as axions or supersymmetric particles; neutrinos can only supply a small fraction of dark matter, due to limits derived from large-scale structure and high-redshift galaxies.[83]

Unlike baryonic matter, nonbaryonic matter did not contribute to the formation of the elements in the early universe ("Big Bang nucleosynthesis")[13] and so its presence is revealed only via its gravitational effects. In addition, if the particles of which it is composed are supersymmetric, they can undergo annihilation interactions with themselves, possibly resulting in observable by-products such as gamma rays and neutrinos ("indirect detection").[83]

7.3.2   Classification: cold/warm/hot

Dark matter can be divided into *cold*, *warm* and *hot* categories.[84] These categories refer to velocity rather than an actual temperature, indicating how far corresponding objects moved due to random motions in the early universe, before they slowed due to cosmic expansion – this is an important distance called the "free streaming length" (FSL). Primordial density fluctuations smaller than this length get washed out as particles spread from overdense to underdense regions, while larger fluctuations are unaffected; therefore this length sets a minimum scale for later structure formation. The categories are set with respect to the size of a protogalaxy (an object that later evolves into a dwarf galaxy): dark matter particles are classified as cold, warm, or hot according as their FSL; much smaller (cold), similar (warm), or much larger (hot) than a protogalaxy.[85][86]

Mixtures of the above are also possible: a theory of mixed dark matter was popular in the mid-1990s, but was rejected following the discovery of dark energy.

Cold dark matter leads to a "bottom-up" formation of structure while hot dark matter would result in a "top-down" formation scenario; the latter is excluded by high-redshift galaxy observations.[17]

Alternative definitions

These categories also correspond to fluctuation spectrum effects and the interval following the Big Bang at which each type became non-relativistic. Davis *et al.* wrote in 1985:

> Candidate particles can be grouped into three categories on the basis of their effect on the fluctuation spectrum (Bond *et al.* 1983). If the dark matter is composed of abundant light particles which remain relativistic until shortly before recombination, then it may be termed "hot". The best candidate for hot dark matter is a neutrino ... A second possibility is for the dark matter particles to interact more weakly than neutrinos, to be less abundant, and to have a mass of order 1 keV. Such particles are termed "warm dark matter", because they have lower thermal velocities than massive neutrinos ... there are at present few candidate particles which fit this description. Gravitinos and photinos have been suggested (Pagels and Primack 1982; Bond, Szalay and Turner 1982) ... Any particles which became nonrelativistic very early, and so were able to diffuse a negligible distance, are termed "cold" dark matter (CDM). There are many candidates for CDM including supersymmetric particles.[87]

Another approximate dividing line is that "warm" dark matter became non-relativistic when the universe was approximately 1 year old and 1 millionth of its present size and in the radiation-dominated era (photons and neutrinos), with a photon temperature 2.7 million K. Standard physical cosmology gives the particle horizon size as 2ct (speed of light multiplied by time) in the radiation-dominated era, thus 2 light-years. A region of this size would expand to 2 million light years today (absent structure formation). The actual FSL is roughly 5 times the above length, since it continues to grow slowly as particle velocities decrease inversely with the scale factor after they become non-relativistic. In this example the FSL would correspond to 10 million light-years or 3 Mpc today, around the size containing an average large galaxy.

The 2.7 million K photon temperature gives a typical photon energy of 250 electron-volts, thereby setting a typical mass scale for "warm" dark matter: particles much more massive than this, such as GeV – TeV mass WIMPs, would become non-relativistic much earlier than 1 year after the Big Bang and thus have FSLs much smaller than a protogalaxy, making them "cold". Conversely, much lighter particles, such as neutrinos with masses of only a few eV, have FSLs much larger than a protogalaxy, thus qualifying them as "hot".

7.3.3   Cold dark matter

Main article: Cold dark matter

"Cold" dark matter offers the simplest explanation for most cosmological observations. It is dark matter composed of constituents with an FSL much smaller than a protogalaxy. This is the focus for dark matter research, as hot dark matter does not seem to be capable of supporting galaxy or galaxy cluster formation, and most particle candidates slowed early.

The constituents of "cold" dark matter are unknown. Possibilities range from large objects like MACHOs (such as black holes[88]) or RAMBOs (such as clusters of brown dwarfs), to new particles such as WIMPs and axions.

Studies of Big Bang nucleosynthesis and gravitational lensing convinced most cosmologists[17][89][90][91][92][93] that MACHOs[89][91] cannot make up more than a small fraction of dark matter.[13][89] According to A. Peter: "... the only *really plausible* dark-matter candidates are new particles."[90]

The 1997 DAMA/NaI experiment and its successor DAMA/LIBRA in 2013, claimed to directly detect dark matter particles passing through the Earth, but many researchers remain skeptical, as negative results from similar experiments seem incompatible with the DAMA results.

Many supersymmetric models offer dark matter candidates in the form of the WIMPy Lightest Supersymmetric Particle (LSP).[94] Separately, heavy sterile neutrinos exist in non-supersymmetric extensions to the standard model that explain the small neutrino mass through the seesaw mechanism.

7.3.4 Warm dark matter

Main article: Warm dark matter

"Warm" dark matter refers to particles with an FSL comparable to the size of a protogalaxy. Predictions based on warm dark matter are similar to those for cold dark matter on large scales, but with less small-scale density perturbations. This reduces the predicted abundance of dwarf galaxies and may lead to lower density of dark matter in the central parts of large galaxies; some researchers consider this to be a better fit to observations. A challenge for this model is the lack of particle candidates with the required mass ~ 300 eV to 3000 eV.

No known particles can be categorized as "warm" dark matter. A postulated candidate is the sterile neutrino: a heavier, slower form of neutrino that does not interact through the weak force, unlike other neutrinos. Some modified gravity theories, such as scalar-tensor-vector gravity, require "warm" dark matter to make their equations work.

7.3.5 Hot dark matter

Main article: Hot dark matter

"Hot" dark matter consists of particles whose FSL is much larger than the size of a protogalaxy. The neutrino qualifies as such particle. They were discovered independently, long before the hunt for dark matter: they were postulated in 1930, and detected in 1956. Neutrinos' mass is less than 10^{-6} that of an electron. Neutrinos interact with normal matter only via gravity and the weak force, making them difficult to detect (the weak force only works over a small distance, thus a neutrino triggers a weak force event only if it hits a nucleus head-on). This makes them 'weakly interacting light particles' (WILPs), as opposed to WIMPs.

The three known flavours of neutrinos are the *electron*, *muon* and *tau*. Their masses are slightly different. Neutrinos oscillate among the flavours as they move. It is hard to determine an exact upper bound on the collective average mass of the three neutrinos (or for any of the three individually). For example, if the average neutrino mass were

over 50 eV/c^2 (less than 10^{-5} of the mass of an electron), the universe would collapse. CMB data and other methods indicate that their average mass probably does not exceed 0.3 eV/c^2. Thus, observed neutrinos cannot explain dark matter.[95]

Because galaxy-size density fluctuations get washed out by free-streaming, "hot" dark matter implies that the first objects that can form are huge supercluster-size pancakes, which then fragment into galaxies. Deep-field observations show instead that galaxies formed first, followed by clusters and superclusters as galaxies clump together.

7.4 Detection

If dark matter is made up of WIMPs, then millions, possibly billions, of WIMPs must pass through every square centimeter of the Earth each second.[96][97] Many experiments aim to test this hypothesis. Although WIMPs are popular search candidates,[17] the Axion Dark Matter eXperiment (ADMX) searches for axions. Another candidate is heavy hidden sector particles that only interact with ordinary matter via gravity.

These experiments can be divided into two classes: direct detection experiments, which search for the scattering of dark matter particles off atomic nuclei within a detector; and indirect detection, which look for the products of WIMP annihilations.[83]

7.4.1 Direct detection

Direct detection experiments operate deep underground to reduce the interference from cosmic rays. Detectors include the Stawell mine, the Soudan mine, the SNOLAB underground laboratory at Sudbury, the Gran Sasso National Laboratory, the Canfranc Underground Laboratory, the Boulby Underground Laboratory, the Deep Underground Science and Engineering Laboratory and the Particle and Astrophysical Xenon Detector.

These experiments mostly use either cryogenic or noble liquid detector technologies. Cryogenic detectors operating at temperatures below 100mK, detect the heat produced when a particle hits an atom in a crystal absorber such as germanium. Noble liquid detectors detect scintillation produced by a particle collision in liquid xenon or argon. Cryogenic detector experiments include: CDMS, CRESST, EDELWEISS, EURECA. Noble liquid experiments include ZEPLIN, XENON, DEAP, ArDM, WARP, DarkSide, PandaX, and LUX, the Large Underground Xenon experiment. Both of these techniques distinguish background particles (that scatter off electrons) from dark matter particles (that scatter off nuclei). Other experiments include SIMPLE and PICASSO.

The DAMA/NaI, DAMA/LIBRA experiments detected an annual modulation in the event rate[98] that they claim is due to dark matter. (As the Earth orbits the Sun, the velocity of the detector relative to the dark matter halo will vary by a small amount). This claim is so far unconfirmed and unreconciled with negative results of other experiments.[99]

Directional detection is a search strategy based on the motion of the Solar System around the Galactic Center.[100][101][102][103]

A low pressure time projection chamber makes it possible to access information on recoiling tracks and constrain WIMP-nucleus kinematics. WIMPs coming from the direction in which the Sun is travelling (roughly towards Cygnus) may then be separated from background, which should be isotropic. Directional dark matter experiments include DMTPC, DRIFT, Newage and MIMAC.

Results

In 2009, CDMS researchers reported two possible WIMP candidate events. They estimate that the probability that these events are due to background (neutrons or misidentified beta or gamma events) is 23%, and conclude "this analysis cannot be interpreted as significant evidence for WIMP interactions, but we cannot reject either event as signal."[104]

In 2011, researchers using the CRESST detectors presented evidence of 67 collisions occurring in detector crystals from subatomic particles.[105] They calculated the probability that all were caused by known sources of interference/contamination was 1 in 10^5.

7.4.2 Indirect detection

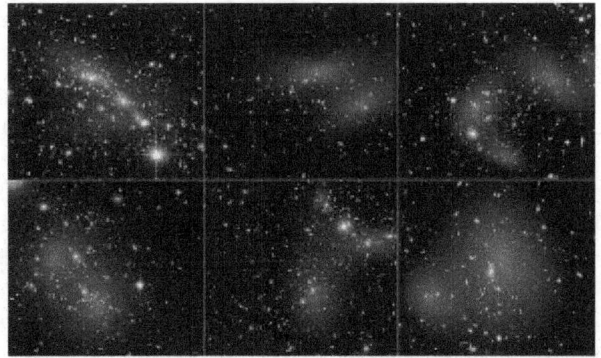

Collage of six cluster collisions with dark matter maps. The clusters were observed in a study of how dark matter in clusters of galaxies behaves when the clusters collide.[106]

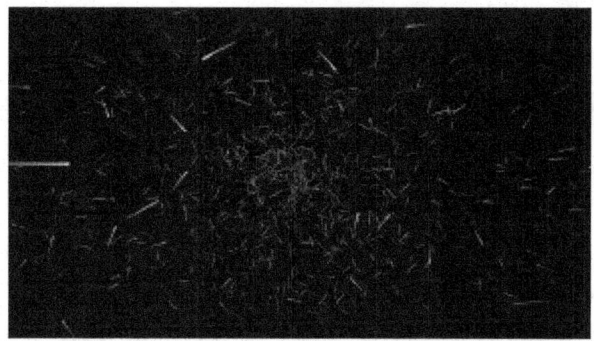

Video about the potential gamma-ray detection of dark matter annihilation around supermassive black holes. (Duration 3:13, also see file description.)

Indirect detection experiments search for the products of WIMP annihilation/decay. If WIMPs are Majorana particles (their own antiparticle) then two WIMPs could annihilate to produce gamma rays or Standard Model particle-antiparticle pairs. If the WIMP is unstable, WIMPs could decay into standard model (or other) particles. These processes could be detected indirectly through an excess of gamma rays, antiprotons or positrons emanating from high density regions. The detection of such a signal is not conclusive evidence, as the sources of gamma ray production are not fully understood.[17][83]

A few of the WIMPs passing through the Sun or Earth may scatter off atoms and lose energy. Thus WIMPs may accumulate at the center of these bodies, increasing the chance of collision/annihilation. This could produce a distinctive signal in the form of high-energy neutrinos.[107] Such a signal would be strong indirect proof of WIMP dark matter.[17] High-energy neutrino telescopes such as AMANDA, IceCube and ANTARES are searching for this signal.

WIMP annihilation from the Milky Way galaxy as a whole may also be detected in the form of various annihilation products.[108] The Galactic Center is a particularly good place to look because the density of dark matter may be higher there.[109]

The recent detection by LIGO in February 2016, of gravity waves, opens the possibility of observing dark matter in a new way. Dark matter seems to have no effects except gravitational, and so the actual observation of gravitational waves provides scientists with a new way of observing the phenomenon.[110]

Results

The EGRET gamma ray telescope observed more gamma rays in 2008 than expected from the Milky Way, but scientists concluded that this was most likely due to incorrect estimation of the telescope's sensitivity.[111]

The Fermi Gamma-ray Space Telescope is searching for similar gamma rays.[112] In April 2012, an analysis of previously available data from its Large Area Telescope instrument produced statistical evidence of a 130 GeV signal in the gamma radiation coming from the center of the Milky Way.[113] WIMP annihilation was seen as the most probable explanation.[114]

At higher energies, ground-based gamma-ray telescopes have set limits on the annihilation of dark matter in dwarf spheroidal galaxies[115] and in clusters of galaxies.[116]

The PAMELA experiment (launched 2006) detected excess positrons. They could be from dark matter annihilation or from pulsars. No excess antiprotons were observed.[117]

In 2013 results from the Alpha Magnetic Spectrometer on the International Space Station indicated excess high-energy cosmic rays that could be due to dark matter annihilation.[118][119][120][121][122][123]

7.5   Synthesis

An alternative approach to the detection of WIMPs in nature is to produce them in a laboratory. Experiments with the Large Hadron Collider (LHC) may be able to detect WIMPs produced in collisions of the LHC proton beams. Because a WIMP has negligible interaction with matter, it may be detected indirectly as (large amounts of) missing energy and momentum that escape the detectors, provided other (non-negligible) collision products are detected.[124] These experiments could show that WIMPs can be formed, but a direct detection experiment must still show that they exist in sufficient numbers to account for dark matter.

7.6   Alternative theories

7.6.1   Mass in extra dimensions

In some multidimensional theories, the force of gravity is the only force with effect across all dimensions.[125] This explains the relative weakness of gravity compared to the other forces of nature that cannot cross into extra dimensions. In that case, dark matter could exist in a "Hidden Valley" in other dimensions that only interact with the matter in our dimensions through gravity. That dark matter could potentially aggregate in the same way as ordinary matter, forming other-dimensional galaxies.[12][126]

7.6.2   Topological defects

Dark matter could consist of primordial defects ("birth defects") in the topology of quantum fields, which would contain energy and therefore gravitate. This hypothesis may be investigated by the use of an orbital network of atomic clocks that would register the passage of topological defects by changes to clock synchronization. The Global Positioning System may be able to operate as such a network.[127]

7.6.3   Modified gravity

Some theories modify the laws of gravity. The earliest was Mordehai Milgrom's Modified Newtonian Dynamics (MOND) in 1983, which adjusts Newton's laws to increase gravitational field strength where gravitational acceleration becomes tiny (such as near the rim of a galaxy). It had some success explaining rotational velocity curves of elliptical and dwarf elliptical galaxies, but not galaxy cluster gravitational lensing. MOND was not relativistic: it was an adjustment of the Newtonian account. Attempts were made to bring MOND into conformity with general relativity; this spawned competing MOND-based hypotheses—including TeVeS, MOG or STV gravity and the phenomenological covariant approach.[128]

In 2007, John Moffat proposed a modified gravity hypothesis based on nonsymmetric gravitational theory (NGT) that claims to account for the behavior of colliding galaxies.[129] This model requires the presence of non-relativistic neutrinos or other cold dark matter, to work.

Another proposal uses a gravitational backreaction from a theory that explains gravitational force between objects as an action, a reaction and then a back-reaction. Thus, an object A affects an object B, and the object B then re-affects object A, and so on, creating a feedback loop that strengthens gravity.[130]

In 2008, a group proposed "dark fluid", a modification of large-scale gravity. It hypothesized that attractive gravitational effects are instead a side-effect of dark energy. Dark fluid combines dark matter and dark energy in a single energy field that produces different effects at different scales. This treatment is a simplification of a previous fluid-like model called the generalized Chaplygin gas model in which the whole of spacetime is a compressible gas.[131] Dark fluid can be compared to an atmospheric system. Atmospheric pressure causes air to expand and air regions can collapse to form clouds. In the same way, the dark fluid might generally disperse, while collecting around galaxies.[131]

7.6.4 Spacetime fractality

Applying relativity to fractal, non-differentiable spacetime, Nottale suggests that potential energy may arise due to the fractality of spacetime, which would account for the missing mass-energy observed at cosmological scales.[132][133]

7.7 Popular culture

Main article: Dark matter in fiction

Mention of dark matter is made in some video games and other works of fiction. In such cases, it is usually attributed extraordinary physical or magical properties. Such descriptions are often inconsistent with the hypothesized properties of dark matter in physics and cosmology.

7.8 See also

- Chameleon particle
- Conformal gravity
- DEAP, a dark matter experiment
- DAMPE, a space mission
- General Antiparticle Spectrometer
- Illustris project, astrophysical simulations
- Light dark matter
- Mirror matter
- Multidark, a research program
- Scalar field dark matter
- Self-interacting dark matter
- SIMP, hypothetical particles of dark matter
- Unparticle physics

7.9 Notes

[1] Since dark energy, by convention, does not count as "matter", this is 26.8/(4.9 + 26.8)=0.845

7.10 References

[1] "Planck Mission Brings Universe Into Sharp Focus". *NASA Mission Pages*. 21 March 2013.

[2] "Dark Energy, Dark Matter". *NASA Science: Astrophysics*. 5 June 2015.

[3] "Dark Matter". *CERN Physics*. 20 January 2012.

[4] Trimble, V. (1987). "Existence and nature of dark matter in the universe". *Annual Review of Astronomy and Astrophysics* 25: 425–472. Bibcode:1987ARA&A..25..425T. doi:10.1146/annurev.aa.25.090187.002233.

[5] Ade, P. A. R.; Aghanim, N.; Armitage-Caplan, C.; (Planck Collaboration); et al. (22 March 2013). "Planck 2013 results. I. Overview of products and scientific results – Table 9". *Astronomy and Astrophysics* 1303: 5062. arXiv:1303.5062. Bibcode:2014A&A...571A...1P. doi:10.1051/0004-6361/201321529.

[6] Francis, Matthew (22 March 2013). "First Planck results: the Universe is still weird and interesting". *Arstechnica*.

[7] "Planck captures portrait of the young Universe, revealing earliest light". University of Cambridge. 21 March 2013. Retrieved 21 March 2013.

[8] Sean Carroll, Ph.D., Cal Tech, 2007, The Teaching Company, *Dark Matter, Dark Energy: The Dark Side of the Universe*, Guidebook Part 2 page 46, Accessed Oct. 7, 2013, "...dark matter: An invisible, essentially collisionless component of matter that makes up about 25 percent of the energy density of the universe... it's a different kind of particle... something not yet observed in the laboratory..."

[9] Ferris, Timothy. "Dark Matter". Retrieved 2015-06-10.

[10] Jarosik, N.; et al. (2011). "Seven-Year Wilson Microwave Anisotropy Probe (WMAP) Observations: Sky Maps, Systematic Errors, and Basic Results". *Astrophysical Journal Supplement* 192 (2): 14. arXiv:1001.4744. Bibcode:2011ApJS..192...14J. doi:10.1088/0067-0049/192/2/14.

[11] Persic, Massimo; Salucci, Paolo (1992-09-01). "The baryon content of the Universe". *Monthly Notices of the Royal Astronomical Society* 258 (1): 14P–18P. arXiv:astro-ph/0502178. Bibcode:1992MNRAS.258P..14P. doi:10.1093/mnras/258.1.14P. ISSN 0035-8711.

[12] Siegfried, T. (5 July 1999). "Hidden Space Dimensions May Permit Parallel Universes, Explain Cosmic Mysteries". *The Dallas Morning News*.

[13] Copi, C. J.; Schramm, D. N.; Turner, M. S. (1995). "Big-Bang Nucleosynthesis and the Baryon Density of the Universe". *Science* 267 (5195): 192–199. arXiv:astro-ph/9407006. Bibcode:1995Sci...267..192C. doi:10.1126/science.7809624. PMID 7809624.

[14] Kroupa, P.; et al. (2010). "Local-Group tests of dark-matter Concordance Cosmology: Towards a new paradigm for structure formation". *Astronomy and Astrophysics* **523**: 32–54. arXiv:1006.1647. Bibcode:2010A&A...523A..32K. doi:10.1051/0004-6361/201014892.

[15] Conformal theory: New light on dark matter, dark energy, and dark galactic halos." (PDF) Robert K. Nesbet. IBM Almaden Research Center, June 17, 2014.

[16] Angus, G. (2013). "Cosmological simulations in MOND: the cluster scale halo mass function with light sterile neutrinos". *Monthly Notices of the Royal Astronomical Society* **436**: 202–211. arXiv:1309.6094. Bibcode:2013MNRAS.436..202A. doi:10.1093/mnras/stt1564.

[17] Bertone, G.; Hooper, D.; Silk, J. (2005). "Particle dark matter: Evidence, candidates and constraints". *Physics Reports* **405** (5–6): 279–390. arXiv:hep-ph/0404175. Bibcode:2005PhR...405..279B. doi:10.1016/j.physrep.2004.08.031.

[18] Kapteyn, Jacobus Cornelius (1922). "First attempt at a theory of the arrangement and motion of the sidereal system". *Astrophysical Journal* **55**: 302–327. Bibcode:1922ApJ....55..302K. doi:10.1086/142670. It is incidentally suggested that when the theory is perfected it may be possible to determine *the amount of dark matter* from its gravitational effect. (emphasis in original)

[19] Rosenberg, Leslie J (30 June 2014). *Status of the Axion Dark-Matter Experiment (ADMX)* (PDF). 10th PATRAS Workshop on Axions, WIMPs and WISPs. p. 2.

[20] Oort, J.H. (1932) "The force exerted by the stellar system in the direction perpendicular to the galactic plane and some related problems," *Bulletin of the Astronomical Institutes of the Netherlands*, **6** : 249-287.

[21] "The Hidden Lives of Galaxies: Hidden Mass". *Imagine the Universe!*. NASA/GSFC.

[22] Kuijken, K.; Gilmore, G. (July 1989). "The Mass Distribution in the Galactic Disc - Part III - the Local Volume Mass Density" (PDF). *Monthly Notices of the Royal Astronomical Society* **239** (2): 651–664. Bibcode:1989MNRAS.239..651K. doi:10.1093/mnras/239.2.651.

[23] Zwicky, F. (1933). "Die Rotverschiebung von extragalaktischen Nebeln". *Helvetica Physica Acta* **6**: 110–127. Bibcode:1933AcHPh...6..110Z.

[24] Zwicky, F. (1937). "On the Masses of Nebulae and of Clusters of Nebulae". *The Astrophysical Journal* **86**: 217. Bibcode:1937ApJ....86..217Z. doi:10.1086/143864.

[25] Zwicky, F. (1933), "Die Rotverschiebung von extragalaktischen Nebeln", *Helvetica Physica Acta* **6**: 110–127, Bibcode:1933AcHPh...6..110Z See also Zwicky, F. (1937),

"On the Masses of Nebulae and of Clusters of Nebulae", *Astrophysical Journal* **86**: 217, Bibcode:1937ApJ....86..217Z, doi:10.1086/143864

[26] Some details of Zwicky's calculation and of more modern values are given in Richmond, M., *Using the virial theorem: the mass of a cluster of galaxies*, retrieved 2007-07-10

[27] Freese, Katherine (4 May 2014). *The Cosmic Cocktail: Three Parts Dark Matter*. Princeton University Press. ISBN 978-1-4008-5007-5.

[28] Babcock, H, 1939, "The rotation of the Andromeda Nebula", Lick Observatory bulletin ; no. 498

[29] First observational evidence of dark matter. Darkmatter-physics.com. Retrieved 6 August 2013.

[30] Rubin, Vera C.; Ford, W. Kent, Jr. (February 1970). "Rotation of the Andromeda Nebula from a Spectroscopic Survey of Emission Regions". *The Astrophysical Journal* **159**: 379–403. Bibcode:1970ApJ...159..379R. doi:10.1086/150317.

[31] Bosma, A. (1978). "The distribution and kinematics of neutral hydrogen in spiral galaxies of various morphological types" (Ph.D. Thesis). Rijksuniversiteit Groningen.

[32] Rubin, V.; Thonnard, W. K. Jr.; Ford, N. (1980). "Rotational Properties of 21 Sc Galaxies with a Large Range of Luminosities and Radii from NGC 4605 ($R = 4$kpc) to UGC 2885 ($R = 122$kpc)". *The Astrophysical Journal* **238**: 471. Bibcode:1980ApJ...238..471R. doi:10.1086/158003.

[33] Bergstrom, L. (2000). "Non-baryonic dark matter: Observational evidence and detection methods". *Reports on Progress in Physics* **63** (5): 793–841. arXiv:hep-ph/0002126. Bibcode:2000RPPh...63..793B. doi:10.1088/0034-4885/63/5/2r3.

[34] Komatsu, E.; et al. (2009). "Five-Year Wilkinson Microwave Anisotropy Probe Observations: Cosmological Interpretation". *The Astrophysical Journal Supplement* **180** (2): 330–376. arXiv:0803.0547. Bibcode:2009ApJS..180..330K. doi:10.1088/0067-0049/180/2/330.

[35] Boggess, N. W.; et al. (1992). "The COBE Mission: Its Design and Performance Two Years after the launch". *The Astrophysical Journal* **397**: 420. Bibcode:1992ApJ...397..420B. doi:10.1086/171797.

[36] Melchiorri, A.; et al. (2000). "A Measurement of Ω from the North American Test Flight of Boomerang". *The Astrophysical Journal Letters* **536** (2): L63–L66. arXiv:astro-ph/9911445. Bibcode:2000ApJ...536L..63M. doi:10.1086/312744.

[37] Leitch, E. M.; et al. (2002). "Measurement of polarization with the Degree Angular Scale Interferometer". *Nature* **420** (6917): 763–771. arXiv:astro-ph/0209476. Bibcode:2002Natur.420..763L. doi:10.1038/nature01271. PMID 12490940.

[38] Leitch, E. M.; et al. (2005). "Degree Angular Scale Inter- ferometer 3 Year Cosmic Microwave Background Polariza- tion Results". *The Astrophysical Journal* **624** (1): 10–20. arXiv:astro-ph/0409357.   Bibcode:2005ApJ...624...10L. doi:10.1086/428825.

[39] Readhead, A. C. S.; et al. (2004). "Polarization Observations   with   the   Cosmic   Background   Im- ager". *Science* **306** (5697): 836–844. arXiv:astro- ph/0409569.   Bibcode:2004Sci...306..836R. doi:10.1126/science.1105598. PMID 15472038.

[40] Hinshaw, G.; et al. (2009). "Five-Year Wilkinson Mi- crowave Anisotropy Probe Observations: Data Processing, Sky Maps, and Basic Results". *The Astrophysical Jour- nal Supplement* **180** (2): 225–245.   arXiv:0803.0732. Bibcode:2009ApJS..180..225H.   doi:10.1088/0067- 0049/180/2/225.

[41] "Serious Blow to Dark Matter Theories?" (Press release). European Southern Observatory. 18 April 2012.

[42] Freeman, K.; McNamara, G. (2006). *In Search of Dark Matter*. Birkhäuser. p. 37. ISBN 0-387-27616-5.

[43] Randall, Lisa (2015). *Dark matter and the dinosaurs: The astounding interconnectedness of the universe*. Harper Collins Publishers. ISBN 978-0-06-232847-2.

[44] Jörg, D.; et al. (2012). "A filament of dark matter be- tween two clusters of galaxies". *Nature* **487** (7406): 202– 204. arXiv:1207.0809. Bibcode:2012Natur.487..202D. doi:10.1038/nature11224.

[45] Salucci, P.; Borriello, A. (2003). "The Intriguing Dis- tribution of Dark Matter in Galaxies". *Lecture Notes in Physics*. Lecture Notes in Physics **616**: 66–77. arXiv:astro-ph/0203457.   Bibcode:2003LNP...616...66S. doi:10.1007/3-540-36539-7_5. ISBN 978-3-540-00711-1.

[46] Dekel, A.; et al. (2005). "Lost and found dark mat- ter in elliptical galaxies". *Nature* **437** (7059): 707–710. arXiv:astro-ph/0501622.   Bibcode:2005Natur.437..707D. doi:10.1038/nature03970. PMID 16193046.

[47] Faber, S. M.; Jackson, R. E. (1976). "Velocity dispersions   and   mass-to-light   ratios   for   elliptical galaxies". *The Astrophysical Journal* **204**: 668–683. Bibcode:1976ApJ...204..668F. doi:10.1086/154215.

[48] Collins, G. W. (1978). "The Virial Theorem in Stellar As- trophysics". Pachart Press.

[49] Rejkuba, M.; Dubath, P.; Minniti, D.; Meylan, G. (2008). "Masses and M/L Ratios of Bright Globular Clusters in NGC 5128". *Proceedings of the International Astronomi- cal Union* **246**: 418–422. Bibcode:2008IAUS..246..418R. doi:10.1017/S1743921308016074.

[50] Weinberg, M. D.; Blitz, L. (2006). "A Magellanic Ori- gin for the Warp of the Galaxy". *The Astrophysical Jour- nal Letters* **641** (1): L33–L36. arXiv:astro-ph/0601694. Bibcode:2006ApJ...641L..33W. doi:10.1086/503607.

[51] Minchin, R.; et al.   (2005).   "A Dark Hydrogen Cloud in the Virgo Cluster". *The Astrophysical Jour- nal Letters* **622**: L21–L24.   arXiv:astro-ph/0502312. Bibcode:2005ApJ...622L..21M. doi:10.1086/429538.

[52] Ciardullo, R.; Jacoby, G. H.; Dejonghe, H. B. (1993). "The radial velocities of planetary nebulae in NGC 3379". *The Astrophysical Journal* **414**: 454–462. Bibcode:1993ApJ...414..454C. doi:10.1086/173092.

[53] Vikhlinin, A.; et al.   (2006).   "Chandra Sample of Nearby Relaxed Galaxy Clusters: Mass, Gas Fraction, and Mass–Temperature Relation". *The Astrophysical Journal* **640** (2):   691–709.   arXiv:astro-ph/0507092. Bibcode:2006ApJ...640..691V. doi:10.1086/500288.

[54] "Hubble Finds Dark Matter Ring in Galaxy Cluster".

[55] Taylor, A. N.; et al. (1998). "Gravitational Lens Mag- nification and the Mass of Abell 1689". *The Astrophys- ical Journal* **501** (2): 539–553. arXiv:astro-ph/9801158. Bibcode:1998ApJ...501..539T. doi:10.1086/305827.

[56] Wu, X.; Chiueh, T.; Fang, L.; Xue, Y. (1998). "A com- parison of different cluster mass estimates: consistency or discrepancy?". *Monthly Notices of the Royal Astronomical Society* **301** (3):   861–871.   arXiv:astro-ph/9808179. Bibcode:1998MNRAS.301..861W.   doi:10.1046/j.1365- 8711.1998.02055.x.

[57] Refregier, A.   (2003).   "Weak gravitational lensing by large-scale structure". *Annual Review of Astron- omy and Astrophysics* **41** (1): 645–668. arXiv:astro- ph/0307212.   Bibcode:2003ARA&A..41..645R. doi:10.1146/annurev.astro.41.111302.102207.

[58] "Abell 2029: Hot News for Cold Dark Matter". Chandra X-ray Observatory. 11 June 2003.

[59] Massey, R.; et al.   (2007).   "Dark matter maps re- veal cosmic scaffolding". *Nature* **445** (7125): 286–290. arXiv:astro-ph/0701594.   Bibcode:2007Natur.445..286M. doi:10.1038/nature05497. PMID 17206154.

[60] Clowe, D.; et al. (2006). "A direct empirical proof of the existence of dark matter". *The Astrophysical Journal* **648** (2):   109–113.   arXiv:astro-ph/0608407. Bibcode:2006ApJ...648L.109C. doi:10.1086/508162.

[61] Tiberiu, H.; Lobo, F. S. N. (2011). "Two-fluid dark matter models". *Physical Review D* **83** (12): 124051. arXiv:1106.2642.   Bibcode:2011PhRvD..8314051H. doi:10.1103/PhysRevD.83.124051.

[62] Spergel, D. N.; Steinhardt, P. J. (2000). "Obser- vational evidence for self-interacting cold dark mat- ter". *Physical Review Letters* **84** (17): 3760–3763. arXiv:astro-ph/9909386. Bibcode:2000PhRvL..84.3760S. doi:10.1103/PhysRevLett.84.3760.

[63] Markevitch, M.; et al. (2004). "Direct Constraints on the Dark Matter Self-Interaction Cross Section from the

Merging Galaxy Cluster 1E 0657-56". *The Astrophysical Journal* **606** (2): 819–824. arXiv:astro-ph/0309303. Bibcode:2004ApJ...606..819M. doi:10.1086/383178.

[64] Allen, S. W.; Evrard, A. E.; Mantz, A. B. (2011). "Cosmological Parameters from Observations of Galaxy Clusters". *Annual Review of Astronomy & Astrophysics* **49**: 409–470. arXiv:1103.4829. Bibcode:2011ARA&A..49..409A. doi:10.1146/annurev-astro-081710-102514.

[65] "Press Release - Dark Matter Map Begins to Reveal the Universe's Early History - Subaru Telescope". *www.subarutelescope.org*. Retrieved 2015-07-03.

[66] Miyazaki, Satoshi; Oguri, Masamune; Hamana, Takashi; Tanaka, Masayuki; Miller, Lance; Utsumi, Yousuke; Komiyama, Yutaka; Furusawa, Hisanori; Sakurai, Junya (2015-07-01). "Properties of Weak Lensing Clusters Detected on Hyper Suprime-Cam's 2.3 deg2 field". *The Astrophysical Journal* **807** (1): 22. arXiv:1504.06974. Bibcode:2015ApJ...807...22M. doi:10.1088/0004-637X/807/1/22. ISSN 0004-637X.

[67] Percival, W. J.; et al. (2007). "Measuring the Baryon Acoustic Oscillation scale using the Sloan Digital Sky Survey and 2dF Galaxy Redshift Survey". *Monthly Notices of the Royal Astronomical Society* **381** (3): 1053–1066. arXiv:0705.3323. Bibcode:2007MNRAS.381.1053P. doi:10.1111/j.1365-2966.2007.12268.x.

[68] Peacock, J.; et al. (2001). "A measurement of the cosmological mass density from clustering in the 2dF Galaxy Redshift Survey". *Nature* **410**: 169. arXiv:astro-ph/0103143. Bibcode:2001Natur.410..169P.

[69] Kowalski, M.; et al. (2008). "Improved Cosmological Constraints from New, Old, and Combined Supernova Data Sets". *The Astrophysical Journal* **686** (2): 749–778. arXiv:0804.4142. Bibcode:2008ApJ...686..749K. doi:10.1086/589937.

[70] Viel, M.; Bolton, J. S.; Haehnelt, M. G. (2009). "Cosmological and astrophysical constraints from the Lyman α forest flux probability distribution function". *Monthly Notices of the Royal Astronomical Society* **399** (1): L39–L43. arXiv:0907.2927. Bibcode:2009MNRAS.399L..39V. doi:10.1111/j.1745-3933.2009.00720.x.

[71] "Hubble Maps the Cosmic Web of "Clumpy" Dark Matter in 3-D" (Press release). NASA. 7 January 2007.

[72] Springel, V.; et al. (2005). "Simulations of the formation, evolution and clustering of galaxies and quasars". *Nature* **435** (7042): 629–636. arXiv:astro-ph/0504097. Bibcode:2005Natur.435..629S. doi:10.1038/nature03597. PMID 15931216.

[73] Mateo, M. L. (1998). "Dwarf Galaxies of the Local Group". *Annual Review of Astronomy and Astrophysics* **36** (1): 435–506. arXiv:astro-ph/9810070. Bibcode:1998ARA&A..36..435M. doi:10.1146/annurev.astro.36.1.435.

[74] Moore, B.; et al. (1999). "Dark Matter Substructure within Galactic Halos". *The Astrophysical Journal Letters* **524** (1): L19–L22. arXiv:astro-ph/9907411. Bibcode:1999ApJ...524L..19M. doi:10.1086/312287.

[75] Achim Weiss, "Big Bang Nucleosynthesis: Cooking up the first light elements" in: Einstein Online Vol. 2 (2006), 1017

[76] Raine, D.; Thomas, T. (2001). *An Introduction to the Science of Cosmology*. IOP Publishing. p. 30. ISBN 0-7503-0405-7.

[77] Tisserand, P.; Le Guillou, L.; Afonso, C.; Albert, J. N.; Andersen, J.; Ansari, R.; Aubourg, É.; Bareyre, P.; Beaulieu, J. P.; Charlot, X.; Coutures, C.; Ferlet, R.; Fouqué, P.; Glicenstein, J. F.; Goldman, B.; Gould, A.; Graff, D.; Gros, M.; Haissinski, J.; Hamadache, C.; De Kat, J.; Lasserre, T.; Lesquoy, É.; Loup, C.; Magneville, C.; Marquette, J. B.; Maurice, É.; Maury, A.; Milsztajn, A.; Moniez, M. (2007). "Limits on the Macho content of the Galactic Halo from the EROS-2 Survey of the Magellanic Clouds". *Astronomy and Astrophysics* **469** (2): 387–404. arXiv:astro-ph/0607207. Bibcode:2007A&A...469..387T. doi:10.1051/0004-6361:20066017.

[78] Graff, D. S.; Freese, K. (1996). "Analysis of a *Hubble Space Telescope* Search for Red Dwarfs: Limits on Baryonic Matter in the Galactic Halo". *The Astrophysical Journal* **456**. arXiv:astro-ph/9507097. Bibcode:1996ApJ...456L..49G. doi:10.1086/309850.

[79] Najita, J. R.; Tiede, G. P.; Carr, J. S. (2000). "From Stars to Superplanets: The Low-Mass Initial Mass Function in the Young Cluster IC 348". *The Astrophysical Journal* **541** (2): 977–1003. arXiv:astro-ph/0005290. Bibcode:2000ApJ...541..977N. doi:10.1086/309477.

[80] Wyrzykowski, Lukasz et al. (2011) The OGLE view of microlensing towards the Magellanic Clouds – IV. OGLE-III SMC data and final conclusions on MACHOs, MNRAS, 416, 2949

[81] Freese, Katherine; Fields, Brian; Graff, David (2000). "Death of Stellar Baryonic Dark Matter Candidates". arXiv:astro-ph/0007444 [astro-ph].

[82] Freese, Katherine; Fields, Brian; Graff, David (2000). "Death of Stellar Baryonic Dark Matter". *The First Stars*. ESO Astrophysics Symposia. p. 18. arXiv:astro-ph/0002058. Bibcode:2000fist.conf...18F. doi:10.1007/10719504_3. ISBN 3-540-67222-2.

[83] Bertone, G.; Merritt, D. (2005). "Dark Matter Dynamics and Indirect Detection". *Modern Physics Letters A* **20** (14): 1021–1036. arXiv:astro-ph/0504422. Bibcode:2005MPLA...20.1021B. doi:10.1142/S0217732305017391.

[84] Silk, Joseph (6 December 2000). "IX". *The Big Bang: Third Edition*. Henry Holt and Company. ISBN 978-0-8050-7256-3.

[85] Vittorio, N.; J. Silk (1984). "Fine-scale anisotropy of the cosmic microwave background in a universe dominated by cold dark matter". *Astrophysical Journal Letters* 285: L39–L43. Bibcode:1984ApJ...285L..39V. doi:10.1086/184361.

[86] Umemura, Masayuki; Satoru Ikeuchi (1985). "Formation of Subgalactic Objects within Two-Component Dark Matter". *Astrophysical Journal* 299: 583–592. Bibcode:1985ApJ...299..583U. doi:10.1086/163726.

[87] Davis, M.; Efstathiou, G., Frenk, C. S., & White, S. D. M. (May 15, 1985). "The evolution of large-scale structure in a universe dominated by cold dark matter". *Astrophysical Journal* 292: 371–394. Bibcode:1985ApJ...292..371D. doi:10.1086/163168.

[88] Hawkins, M. R. S. (2011). "The case for primordial black holes as dark matter". *Monthly Notices of the Royal Astronomical Society* 415 (3): 2744–2757. arXiv:1106.3875. Bibcode:2011MNRAS.415.2744H. doi:10.1111/j.1365-2966.2011.18890.x.

[89] Carr, B. J.; et al. (May 2010). "New cosmological constraints on primordial black holes" (PDF). *Physical Review D* 81 (10): 104019. arXiv:0912.5297. Bibcode:2010PhRvD..81j4019C. doi:10.1103/PhysRevD.81.104019.

[90] Peter, A. H. G. (2012). "Dark Matter: A Brief Review". arXiv:1201.3942 [astro-ph.CO].

[91] Garrett, Katherine; Dūda, Gintaras (2011). "Dark Matter: A Primer". *Advances in Astronomy* 2011: 1–22. arXiv:1006.2483. Bibcode:2011AdAst2011E...8G. doi:10.1155/2011/968283. MACHOs can only account for a very small percentage of the nonluminous mass in our galaxy, revealing that most dark matter cannot be strongly concentrated or exist in the form of baryonic astrophysical objects. Although microlensing surveys rule out baryonic objects like brown dwarfs, black holes, and neutron stars in our galactic halo, can other forms of baryonic matter make up the bulk of dark matter? The answer, surprisingly, is no...

[92] Bertone, G. (2010). "The moment of truth for WIMP dark matter". *Nature* 468 (7322): 389–393. arXiv:1011.3532. Bibcode:2010Natur.468..389B. doi:10.1038/nature09509. PMID 21085174.

[93] Olive, Keith A. (2003). "TASI Lectures on Dark Matter". p. 21

[94] Jungman, Gerard; Kamionkowski, Marc; Griest, Kim (1996-03-01). "Supersymmetric dark matter". *Physics Reports* 267 (5–6): 195–373. arXiv:hep-ph/9506380. Bibcode:1996PhR...267..195J. doi:10.1016/0370-1573(95)00058-5.

[95] "Neutrinos as Dark Matter". Astro.ucla.edu. 21 September 1998. Retrieved 6 January 2011.

[96] Gaitskell, Richard J. (2004). "Direct Detection of Dark Matter". *Annual Review of Nuclear and Particle Science* 54: 315–359. Bibcode:2004ARNPS..54..315G. doi:10.1146/annurev.nucl.54.070103.181244.

[97] "NEUTRALINO DARK MATTER". Retrieved 26 December 2011. Griest, Kim. "WIMPs and MACHOs" (PDF). Retrieved 26 December 2011.

[98] Drukier, A.; Freese, K. and Spergel, D. (1986). "Detecting Cold Dark Matter Candidates". *Physical Review D 33* (12): 3495–3508. Bibcode:1986PhRvD..33.3495D. doi:10.1103/PhysRevD.33.3495.

[99] Bernabei, R.; Belli, P.; Cappella, F.; Cerulli, R.; Dai, C. J.; d'Angelo, A.; He, H. L.; Incicchitti, A.; Kuang, H. H.; Ma, J. M.; Montecchia, F.; Nozzoli, F.; Prosperi, D.; Sheng, X. D.; Ye, Z. P. (2008). "First results from DAMA/LIBRA and the combined results with DAMA/NaI". *Eur. Phys. J. C 56* (3): 333–355. arXiv:0804.2741. doi:10.1140/epjc/s10052-008-0662-y.

[100] Stonebraker, Alan (2014-01-03). "Synopsis: Dark-Matter Wind Sways through the Seasons". *Physics - Synopses* (American Physical Society). Retrieved 6 January 2014.

[101] Lee, Samuel K.; Mariangela Lisanti, Annika H. G. Peter, and Benjamin R. Safdi (2014-01-03). "Effect of Gravitational Focusing on Annual Modulation in Dark-Matter Direct-Detection Experiments". *Phys. Rev. Lett.* (American Physical Society) 112 (1): 011301 (2014) [5 pages]. arXiv:1308.1953. Bibcode:2014PhRvL.112a1301L. doi:10.1103/PhysRevLett.112.011301.

[102] The Dark Matter Group. "An Introduction to Dark Matter". *Dark Matter Research* (Sheffield, UK: University of Sheffield). Retrieved 7 January 2014.

[103] "Blowing in the Wind". *Kavli News* (Sheffield, UK: Kavli Foundation). Retrieved 7 January 2014. Scientists at Kavli MIT are working on...a tool to track the movement of dark matter.

[104] The CDMS II Collaboration; Ahmed, Z.; Akerib, D. S.; Arrenberg, S.; Bailey, C. N.; Balakishiyeva, D.; Baudis, L.; Bauer, D. A.; Brink, P. L.; Bruch, T.; Bunker, R.; Cabrera, B.; Caldwell, D. O.; Cooley, J.; Cushman, P.; Daal, M.; Dejongh, F.; Dragowsky, M. R.; Duong, L.; Fallows, S.; Figueroa-Feliciano, E.; Filippini, J.; Fritts, M.; Golwala, S. R.; Grant, D. R.; Hall, J.; Hennings-Yeomans, R.; Hertel, S. A.; Holmgren, D.; Hsu, L. (2010). "Dark Matter Search Results from the CDMS II Experiment". *Science* 327 (5973): 1619–1621. arXiv:0912.3592. Bibcode:2010Sci...327.1619C. doi:10.1126/science.1186112. PMID 20150446.

[105] Angloher, G.; Bauer; Bavykina; Bento; Bucci; Ciemniak; Deuter; von Feilitzsch; Hauff; Huff (2011). "Results from 730kg days of the CRESST-II Dark Matter Search". arXiv:1109.0702v1 [astro-ph.CO].

[106] "Dark matter even darker than once thought". Retrieved 16 June 2015.

[107] Freese, K. (1986). "Can Scalar Neutrinos or Massive Dirac Neutrinos be the Missing Mass?". *Physics Letters B* **167** (3): 295–300. Bibcode:1986PhLB..167..295F. doi:10.1016/0370-2693(86)90349-7.

[108] Ellis, J.; Flores, R. A.; Freese, K.; Ritz, S.; Seckel, D.; Silk, J. (1988). "Cosmic ray constraints on the annihilations of relic particles in the galactic halo". *Physics Letters B* **214** (3): 403–412. Bibcode:1988PhLB..214..403E. doi:10.1016/0370-2693(88)91385-8.

[109] Bertone, Gianfranco (2010). "Dark Matter at the Centers of Galaxies". *Particle Dark Matter: Observations, Models and Searches*. Cambridge University Press. pp. 83–104. arXiv:1001.3706. ISBN 978-0-521-76368-4.

[110] Sokol, Joshua et al (2016), "Surfing gravity's waves" (New Scientist) (New Scientist 20 February)

[111] Stecker, F.W.; Hunter, S; Kniffen, D (2008). "The likely cause of the EGRET GeV anomaly and its implications". *Astroparticle Physics* **29** (1): 25–29. arXiv:0705.4311. Bibcode:2008APh....29...25S. doi:10.1016/j.astropartphys.2007.11.002.

[112] Atwood, W.B.; Abdo, A. A.; Ackermann, M.; Althouse, W.; Anderson, B.; Axelsson, M.; Baldini, L.; Ballet, J.; et al. (2009). "The large area telescope on the Fermi Gamma-ray Space Telescope Mission". *Astrophysical Journal* **697** (2): 1071–1102. arXiv:0902.1089. Bibcode:2009ApJ...697.1071A. doi:10.1088/0004-637X/697/2/1071.

[113] Weniger, Christoph (2012). "A Tentative Gamma-Ray Line from Dark Matter Annihilation at the Fermi Large Area Telescope". *Journal of Cosmology and Astroparticle Physics* **2012** (8): 7. arXiv:1204.2797v2. Bibcode:2012JCAP...08..007W. doi:10.1088/1475-7516/2012/08/007.

[114] Cartlidge, Edwin (24 April 2012). "Gamma rays hint at dark matter". Institute Of Physics. Retrieved 23 April 2013.

[115] Albert, J.; Aliu, E.; Anderhub, H.; Antoranz, P.; Backes, M.; Baixeras, C.; Barrio, J. A.; Bartko, H.; Bastieri, D.; Becker, J. K.; Bednarek, W.; Berger, K.; Bigongiari, C.; Biland, A.; Bock, R. K.; Bordas, P.; Bosch-Ramon, V.; Bretz, T.; Britvitch, I.; Camara, M.; Carmona, E.; Chilingarian, A.; Commichau, S.; Contreras, J. L.; Cortina, J.; Costado, M. T.; Curtef, V.; Danielyan, V.; Dazzi, F.; De Angelis, A. (2008). "Upper Limit for γ-Ray Emission above 140 GeV from the Dwarf Spheroidal Galaxy Draco". *The Astrophysical Journal* **679**: 428–431. arXiv:0711.2574. Bibcode:2008ApJ...679..428A. doi:10.1086/529135.

[116] Aleksić, J.; Antonelli, L. A.; Antoranz, P.; Backes, M.; Baixeras, C.; Balestra, S.; Barrio, J. A.; Bastieri, D.; González, J. B.; Bednarek, W.; Berdyugin, A.; Berger, K.; Bernardini, E.; Biland, A.; Bock, R. K.; Bonnoli, G.; Bordas, P.; Tridon, D. B.; Bosch-Ramon, V.; Bose, D.; Braun, I.; Bretz, T.; Britzger, D.; Camara, M.; Carmona, E.; Carosi, A.; Colin, P.; Commichau, S.; Contreras, J. L.; Cortina, J. (2010). "Magic Gamma-Ray Telescope Observation of the Perseus Cluster of Galaxies: Implications for Cosmic Rays, Dark Matter, and Ngc 1275". *The Astrophysical Journal* **710**: 634–647. arXiv:0909.3267. Bibcode:2010ApJ...710..634A. doi:10.1088/0004-637X/710/1/634.

[117] Adriani, O.; Barbarino, G. C.; Bazilevskaya, G. A.; Bellotti, R.; Boezio, M.; Bogomolov, E. A.; Bonechi, L.; Bongi, M.; Bonvicini, V.; Bottai, S.; Bruno, A.; Cafagna, F.; Campana, D.; Carlson, P.; Casolino, M.; Castellini, G.; De Pascale, M. P.; De Rosa, G.; De Simone, N.; Di Felice, V.; Galper, A. M.; Grishantseva, L.; Hofverberg, P.; Koldashov, S. V.; Krutkov, S. Y.; Kvashnin, A. N.; Leonov, A.; Malvezzi, V.; Marcelli, L.; Menn, W. (2009). "An anomalous positron abundance in cosmic rays with energies 1.5–100 GeV". *Nature* **458** (7238): 607–609. arXiv:0810.4995. Bibcode:2009Natur.458..607A. doi:10.1038/nature07942. PMID 19340076.

[118] Aguilar, M. (AMS Collaboration); et al. (3 April 2013). "First Result from the Alpha Magnetic Spectrometer on the International Space Station: Precision Measurement of the Positron Fraction in Primary Cosmic Rays of 0.5–350 GeV". *Physical Review Letters* **110**. Bibcode:2013PhRvL.110n1102A. doi:10.1103/PhysRevLett.110.141102. Retrieved 3 April 2013.

[119] "First Result from the Alpha Magnetic Spectrometer Experiment". *AMS Collaboration*. 3 April 2013. Retrieved 3 April 2013.

[120] Heilprin, John; Borenstein, Seth (3 April 2013). "Scientists find hint of dark matter from cosmos". Associated Press. Retrieved 3 April 2013.

[121] Amos, Jonathan (3 April 2013). "Alpha Magnetic Spectrometer zeroes in on dark matter". *BBC*. Retrieved 3 April 2013.

[122] Perrotto, Trent J.; Byerly, Josh (2 April 2013). "NASA TV Briefing Discusses Alpha Magnetic Spectrometer Results". *NASA*. Retrieved 3 April 2013.

[123] Overbye, Dennis (3 April 2013). "New Clues to the Mystery of Dark Matter". *New York Times*. Retrieved 3 April 2013.

[124] Kane, G. and Watson, S. (2008). "Dark Matter and LHC:. what is the Connection?". *Modern Physics Letters A* **23** (26): 2103–2123. arXiv:0807.2244. Bibcode:2008MPLA...23.2103K. doi:10.1142/S0217732308028314.

[125] Extra dimensions, gravitons, and tiny black holes. CERN. Retrieved on 17 November 2014.

[126] Dark matter. CERN. Retrieved on 17 November 2014.

[127] Rzetelny, Xaq (19 November 2014). "Looking for a different sort of dark matter with GPS satellites". *Ars Technica*. Retrieved 24 November 2014.

[128] Exirifard, Q. (2010). "Phenomenological covariant approach to gravity". *General Relativity and Gravitation* **43** (1): 93–106. arXiv:0808.1962. Bibcode:2011GReGr..43...93E. doi:10.1007/s10714-010-1073-6.

[129] Brownstein, J.R.; Moffat, J. W. (2007). "The Bullet Cluster 1E0657-558 evidence shows modified gravity in the absence of dark matter". *Monthly Notices of the Royal Astronomical Society* **382** (1): 29–47. arXiv:astro-ph/0702146. Bibcode:2007MNRAS.382...29B. doi:10.1111/j.1365-2966.2007.12275.x.

[130] Anastopoulos, C. (2009). "Gravitational backreaction in cosmological space-times". *Physical Review D* **79** (8): 084029. arXiv:0902.0159. Bibcode:2009PhRvD..79h4029A. doi:10.1103/PhysRevD.79.084029.

[131] "New Cosmic Theory Unites Dark Forces". SPACE.com. 11 February 2008. Retrieved 6 January 2011.

[132] Nottale, Laurent (May 29, 2009). "Scale relativity and fractal space-time: theory and applications" (PDF).

[133] Nottale, Laurent (17 June 2011). *Scale Relativity and Fractal Space-Time: A New Approach to Unifying Relativity and Quantum Mechanics*. World Scientific. p. 516. ISBN 978-1-908977-87-8.

- Sample, Ian (17 December 2009). "Dark Matter Detected". London: Guardian. Retrieved 1 May 2010.

- Video lecture on dark matter by Scott Tremaine, IAS professor

- Science Daily story "Astronomers' Doubts About the Dark Side ..."

- Gray, Meghan; Merrifield, Mike; Copeland, Ed (2010). "Dark Matter". *Sixty Symbols*. Brady Haran for the University of Nottingham.

7.11 External links

- Dark matter at DMOZ

- Dark matter (Astronomy) at *Encyclopædia Britannica*

- What is dark matter? at cosmosmagazine.com

- The Dark Matter Crisis 18 August 2010 by Pavel Kroupa, posted in General

- The European astroparticle physics network

- Helmholtz Alliance for Astroparticle Physics

- "NASA Finds Direct Proof of Dark Matter" (Press release). NASA. 21 August 2006.

- Tuttle, Kelen (22 August 2006). "Dark Matter Observed". SLAC (Stanford Linear Accelerator Center) Today.

- "Astronomers claim first 'dark galaxy' find". New Scientist. 23 February 2005.

Chapter 8

Hot dark matter

Hot dark matter (HDM) is a form of dark matter which consists of particles that travel with ultrarelativistic velocities.

Dark matter is matter that does not interact with, and therefore cannot be detected by, electromagnetic radiation, hence *dark*. It is postulated to exist to explain how clusters and superclusters of galaxies formed after the Big Bang. Data from galaxy rotation curves indicate that around 90% of the mass of a galaxy cannot be seen. It can only be detected by its gravitational effect.

Hot dark matter cannot explain how individual galaxies formed from the Big Bang. The cosmic microwave background radiation as measured by the COBE satellite is very smooth and fast moving particles cannot form clumps as small as galaxies beginning from such a smooth initial state. Due to theory,[1] in order to explain small scale structure in the Universe, it is necessary to invoke cold dark matter (CDM) or warm dark matter (WDM). Hot dark matter as the sole explanation of dark matter is no longer viable,[1] therefore, it is nowadays considered only as part of a mixed dark matter (MDM) theory.

8.1 Neutrinos

The best example of a hot dark matter particle is the neutrino. Neutrinos have very small masses, and do not take part in two of the four fundamental forces, the electromagnetic interaction and the strong interaction. They *do* interact by the weak interaction, and gravity, but due to the feeble strength of these forces, they are difficult to detect. A number of projects, such as the Super-Kamiokande neutrino observatory, in Gifu, Japan are currently studying these neutrinos.

8.2 See also

- Lambda-CDM model

- Modified Newtonian Dynamics

8.3 References

[1] Frenk, Carlos S.; White, Simon D. M. (2012). "Dark matter and cosmic structure". arXiv:1210.0544 [astro-ph.CO].

8.4 Further reading

- Bertone, Gianfranco (2010). *Particle Dark Matter: Observations, Models and Searches*. Cambridge University Press. p. 762. ISBN 978-0-521-76368-4.

8.5 External links

- Hot dark matter by Berkeley

- Dark Matter

Chapter 9

Cold dark matter

In cosmology and physics, **cold dark matter (CDM)** is a hypothetical form of matter (a kind of dark matter) whose particles moved slowly compared to the speed of light (the *cold* in CDM) since the universe was approximately one year old (a time when the cosmic particle horizon contained the mass of one typical galaxy); and interact very weakly with ordinary matter and electromagnetic radiation (the *dark* in CDM). It is believed that approximately 84.54% of matter in the Universe is dark matter, with only a small fraction being the ordinary (baryonic) matter that composes stars, planets and living organisms. Since the late 1980s or 1990s, most cosmologists favor the cold dark matter theory (specifically the modern Lambda-CDM model) as a description of how the Universe went from a smooth initial state at early times (as shown by the cosmic microwave background radiation) to the lumpy distribution of galaxies and their clusters we see today — the large-scale structure of the Universe. The theory sees the role that dwarf galaxies played as crucial, as they are thought to be natural building blocks that form larger structures, created by small-scale density fluctuations in the early Universe.[1]

The theory was originally published in 1982 by three independent groups of cosmologists; James Peebles at Princeton,[2] J. Richard Bond, Alex Szalay and Michael Turner;[3] and George Blumenthal, H. Pagels and Joel Primack.[4] An influential review article in 1984 by Blumenthal, Sandra Moore Faber, Primack and British scientist Martin Rees developed the details of the theory.[5]

In the cold dark matter theory, structure grows hierarchically, with small objects collapsing under their self-gravity first and merging in a continuous hierarchy to form larger and more massive objects. In the hot dark matter paradigm, popular in the early 1980s, structure does not form hierarchically (*bottom-up*), but rather forms by fragmentation (*top-down*), with the largest superclusters forming first in flat pancake-like sheets and subsequently fragmenting into smaller pieces like our galaxy the Milky Way. The predictions of the hot dark matter theory disagree with observations of large-scale structures, whereas the cold dark matter paradigm is in general agreement with the observations.

9.1 Composition

Dark matter is detected through its gravitational interactions with ordinary matter and radiation. As such, it is very difficult to determine what the constituents of cold dark matter are. The candidates fall roughly into three categories:

- Axions are very light particles with a specific type of self-interaction that makes them a suitable CDM candidate.[6][7] Axions have the theoretical advantage that their existence solves the Strong CP problem in QCD, but have not been detected.

- MACHOs or *Massive Compact Halo Objects* are large, condensed objects such as black holes, neutron stars, white dwarfs, very faint stars, or non-luminous objects like planets. The search for these consists of using gravitational lensing to see the effect of these objects on background galaxies. Most experts believe that the constraints from those searches rule out MACHOs as a viable dark matter candidate.[8][9][10][11][12][13]

- WIMPs: Dark matter is composed of *Weakly Interacting Massive Particles*. There is no currently known particle with the required properties, but many extensions of the standard model of particle physics predict such particles. The search for WIMPs involves attempts at direct detection by highly sensitive detectors, as well as attempts at production by particle accelerators. WIMPs are generally regarded as the most promising dark matter candidates.[9][11][13] The DAMA/NaI experiment and its successor DAMA/LIBRA have claimed to directly detect dark matter particles passing through the Earth, but many scientists remain skeptical, as no results from similar experiments seem compatible with the DAMA results.

9.2 Challenges

Several discrepancies between the predictions of the particle cold dark matter paradigm and observations of galaxies and their clustering have arisen:

- The cuspy halo problem: the density distributions of dark matter halos in cold dark matter simulations are much more peaked than what is observed in galaxies by investigating their rotation curves.[14]

- The missing satellites problem: cold dark matter simulations predict much larger numbers of small dwarf galaxies than are observed around galaxies like the Milky Way.[15]

- The disk of satellites problem: dwarf galaxies around the Milky Way and Andromeda galaxies are observed to be orbiting in thin, planar structures whereas the simulations predict that they should be distributed randomly about their parent galaxies.[16]

Some of these problems have proposed solutions but it remains unclear whether they can be solved without abandoning the CDM paradigm.[17]

9.3 See also

- Fuzzy cold dark matter

- Meta-cold dark matter

- Dark matter

  - Hot dark matter (HDM)
  - Warm dark matter (WDM)
  - Self-interacting dark matter

- Lambda-CDM model

- Modified Newtonian dynamics

9.4 References

[1] Battinelli, P.; S. Demers (2005-10-06). "The C star population of DDO 190: 1. Introduction" (PDF). *Astronomy and Astrophysics* (Astronomy & Astrophysics) **447**: 1. Bibcode:2006A&A...447..473B. doi:10.1051/0004-6361:20052829. Archived from the original on 2005-10-06. Retrieved 2012-08-19. Dwarf galaxies play a crucial role in the CDM scenario for galaxy formation, having been suggested to be the natural building blocks from which larger structures are built up by merging processes. In this scenario dwarf galaxies are formed from small-scale density fluctuations in the primeval Universe.

[2] Peebles, P. J. E. (December 1982). "Large-scale background temperature and mass fluctuations due to scale-invariant primeval perturbations". *The Astrophysical Journal* **263**: L1. Bibcode:1982ApJ...263L...1P. doi:10.1086/183911.

[3] "Formation of galaxies in a gravitino-dominated universe". *Physical Review Letters* **48**: 1636–1639. Bibcode:1982PhRvL..48.1636B. doi:10.1103/PhysRevLett.48.1636.

[4] Blumenthal, George R.; Pagels, Heinz; Primack, Joel R. (2 September 1982). "Galaxy formation by dissipationless particles heavier than neutrinos". *Nature* **299** (5878): 37–38. Bibcode:1982Natur.299...37B. doi:10.1038/299037a0.

[5] Blumenthal, G. R.; Faber, S. M.; Primack, J. R.; Rees,, M. J. (1984). "Formation of galaxies and large-scale structure with cold dark matter". *Nature* **311** (517): 517–525. Bibcode:1984Natur.311..517B. doi:10.1038/311517a0.

[6] e.g. M. Turner (2010). "Axions 2010 Workshop". U. Florida, Gainesville, USA.

[7] e.g. Pierre Sikivie (2008). "Axion Cosmology". Lect. Notes Phys. 741, 19-50.

[8] Carr, B. J.; et al. (May 2010). "New cosmological constraints on primordial black holes". *Physical Review D* **81** (10): 104019. arXiv:0912.5297. Bibcode:2010PhRvD..81j4019C. doi:10.1103/PhysRevD.81.104019.

[9] Peter, A. H. G. (2012). "Dark Matter: A Brief Review". arXiv:1201.3942.

[10] Bertone, Gianfranco; Hooper, Dan; Silk, Joseph (January 2005). "Particle dark matter: evidence, candidates and constraints". *Physics Reports* **405** (5–6): 279–390. arXiv:hep-ph/0404175. Bibcode:2005PhR...405..279B. doi:10.1016/j.physrep.2004.08.031.

[11] Garrett, Katherine; Dūda, Gintaras. "Dark Matter: A Primer". *Advances in Astronomy* **2011**: 968283. arXiv:1006.2483. Bibcode:2011AdAst2011E...8G. doi:10.1155/2011/968283.. p. 3: "MACHOs can only account for a very small percentage of the nonluminous mass in our galaxy, revealing that most dark matter cannot be strongly concentrated or exist in the form of baryonic astrophysical objects. Although microlensing surveys rule out baryonic objects like brown dwarfs, black holes, and neutron stars in our galactic halo, can other forms of baryonic matter make up the bulk of dark matter? The answer, surprisingly, is no..."

[12] Gianfranco Bertone, "The moment of truth for WIMP dark matter," Nature 468, 389–393 (18 November 2010)

[13] Olive, Keith A (2003). "TASI Lectures on Dark Matter". *Physics* 54: 21.

[14] Gentile, G.; P., Salucci (2004). "The cored distribution of dark matter in spiral galaxies". *Monthly Notices of the Royal Astronomical Society* 351: 903–922. arXiv:astro-ph/0403154. Bibcode:2004MNRAS.351..903G. doi:10.1111/j.1365-2966.2004.07836.x.

[15] Klypin, Anatoly; Kravtsov, Andrey V.; Valenzuela, Octavio; Prada, Francisco (1999). "Where Are the Missing Galactic Satellites?". *ApJ* 522: 82–92. arXiv:astro-ph/9901240. Bibcode:1999ApJ...522...82K. doi:10.1086/307643.

[16] Marcel Pawlowski et al., "Co-orbiting satellite galaxy structures are still in conflict with the distribution of primordial dwarf galaxies" MNRAS (2014) http://arxiv.org/abs/1406.1799

[17] Kroupa, P.; Famaey, B.; de Boer, Klaas S.; Dabringhausen, Joerg; Pawlowski, Marcel; Boily, Christian; Jerjen, Helmut; Forbes, Duncan; Hensler, Gerhard (2010). "Local-Group tests of dark-matter Concordance Cosmology: Towards a new paradigm for structure formation". *Astronomy and Astrophysics* 523: 32–54. arXiv:1006.1647. Bibcode:2010A&A...523A..32K. doi:10.1051/0004-6361/201014892.

9.5 Further reading

- Bertone, Gianfranco (2010). *Particle Dark Matter: Observations, Models and Searches.* Cambridge University Press. p. 762. ISBN 978-0-521-76368-4.

Chapter 10

Accelerating expansion of the universe

The **accelerating expansion of the universe** is the observation that the universe appears to be expanding at an increasing rate. In formal terms, this means that the cosmic scale factor $a(t)$ has a positive second derivative,[1] so that the velocity at which a distant galaxy is receding from the observer is continuously increasing with time.[2]

The expansion of the universe has been accelerating since the universe entered its dark-energy-dominated era, at redshift z≈0.4 (roughly 5 billion years ago).[3] Within the framework of general relativity, an accelerating expansion can be accounted for by a positive value of the cosmological constant Λ, equivalent to the presence of a positive vacuum energy, dubbed "dark energy". While there are alternative possible explanations, the description assuming dark energy (positive Λ) is used in the current standard model of cosmology, known as ΛCDM ("Lambda cold dark matter").

The accelerated expansion was discovered in 1998, when two independent projects, the Supernova Cosmology Project and the High-Z Supernova Search Team simultaneously obtained results suggesting a totally unexpected acceleration in the expansion of the universe by using distant type Ia supernovae as standard candles.[4][5][6] The discovery was unexpected, cosmologists at the time expecting a deceleration in the expansion of the universe, and amounts to the realization that the universe is currently in a "dark-energy-dominated era". Three members of these two groups have subsequently been awarded Nobel Prizes for their discovery.[7] Confirmatory evidence has been found in baryon acoustic oscillations and other new results about the clustering of galaxies.

10.1 Background

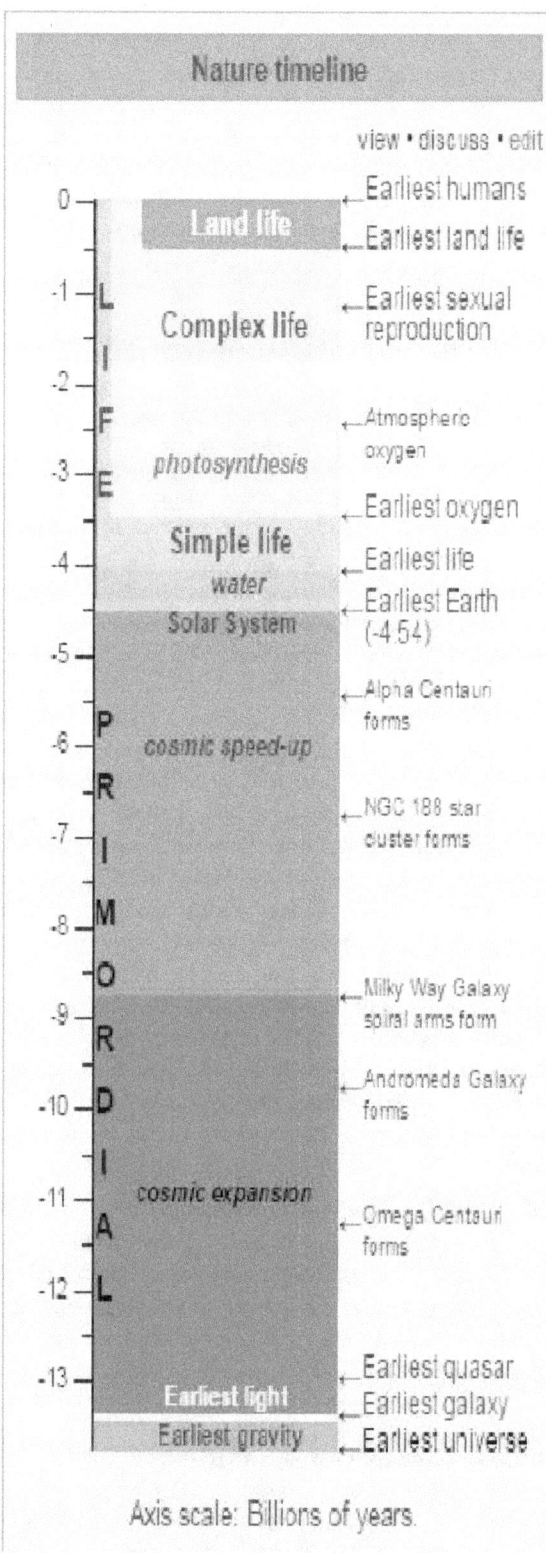

Axis scale: Billions of years.

equation defines how the energy in the universe drives its expansion.

$$H^2 = \left(\frac{\dot{a}}{a}\right)^2 = \frac{8\pi G}{3}\rho - \frac{Kc^2}{R^2 a^2}$$

where K represents the curvature of the universe, $a(t)$ is the scale factor, ρ is the total energy density of the universe, and H is the Hubble parameter.[9]

We define a critical density

$$\rho_c = \frac{3H^2}{8\pi G}$$

and the density parameter

$$\Omega = \frac{\rho}{\rho_c}$$

We can then rewrite the Hubble parameter as

$$H(a) = H_0 \sqrt{\Omega_k a^{-2} + \Omega_m a^{-3} + \Omega_r a^{-4} + \Omega_{DE} a^{-3(1+w)}}$$

where the four currently hypothesized contributors to the energy density of the universe are curvature, matter, radiation and dark energy.[10] Each of the components decreases with the expansion of the universe (increasing scale factor), except perhaps the dark energy term. It is the values of these cosmological parameters which physicists use to determine the acceleration of the universe.

The acceleration equation describes the evolution of the scale factor with time

$$\frac{\ddot{a}}{a} = -\frac{4\pi G}{3}\left(\rho + \frac{3P}{c^2}\right)$$

where the pressure P is defined by the cosmological model chosen. (see explanatory models below)

Physicists at one time were so assured of the deceleration of the universe's expansion that they introduced a so-called deceleration parameter q_0 .[11] Current observations point towards this deceleration parameter being negative.

10.2   Evidence for acceleration

Since Hubble's discovery of the expansion of the universe in 1929,[8] the Big Bang model has become the accepted explanation for the origin of our universe. The Friedmann

To learn about the rate of expansion of the universe we look at the magnitude-redshift relationship of astronomical objects using standard candles, or their distance-redshift relationship using standard rulers. We can also look at the growth of large-scale structure, and find that the observed values of the cosmological parameters are best described by models which include an accelerating expansion.

10.2.1 Supernova observation

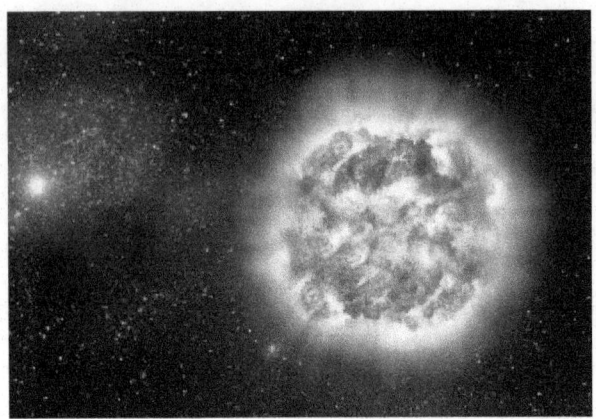

Artist's impression of a Type Ia supernova, as revealed by spectropolarimetry observations

The first evidence for acceleration came from the observation of Type Ia supernovae, which are exploding white dwarfs that have exceeded their stability limit. Because they all have similar masses, their intrinsic luminosity is standardizable. Repeated imaging of selected areas of sky is used to discover the supernovae, then follow-up observations give their peak brightness, which is converted into a quantity known as luminosity distance (see distance measures in cosmology for details).[12] Spectral lines of their light can be used to determine their redshift.

For supernovae at redshift less than around 0.1, or light travel time less than 10 percent of the age of the universe, this gives a nearly linear distance-redshift relation due to Hubble's law. At larger distances, since the expansion rate of the universe has changed over time, the distance-redshift relation deviates from linearity, and this deviation depends on how the expansion rate has changed over time. The full calculation requires integration of the Friedmann equation, but a simple derivation can be given as follows: the redshift z directly gives the cosmic scale factor at the time the supernova exploded.

$$a(t) = \frac{1}{1+z}$$

So a supernova with a measured redshift z = 0.5 implies the universe was 1/(1+0.5) = 2/3 of its present size when the supernova exploded. In an accelerating universe, the universe was expanding more slowly in the past than it is today, which means it took a longer time to expand from 2/3

to 1.0 times its present size compared to a non-accelerating universe. This results in a larger light-travel time, larger distance and fainter supernovae, which corresponds to the actual observations. Riess found that "the distances of the high-redshift SNe Ia were, on average, 10% to 15% farther than expected in a low mass density $\Omega_M = 0.2$ universe without a cosmological constant".[13] This means that the measured high-redshift distances were too large, compared to nearby ones, for a decelerating universe.[14]

10.2.2 Baryon acoustic oscillations

Main article: Baryon acoustic oscillations

In the early universe before recombination and decoupling took place, photons and matter existed in a primordial plasma. Points of higher density in the photon-baryon plasma would contract, being compressed by gravity until the pressure became too large and they expanded again.[11] This contraction and expansion created vibrations in the plasma analogous to sound waves. Since dark matter only interacts gravitationally it stayed at the centre of the sound wave, the origin of the original overdensity. When decoupling occurred, approximately 380,000 years after the Big Bang,[15] photons separated from matter and were able to stream freely through the universe, creating the cosmic microwave background as we know it. This left shells of baryonic matter at a fixed radius from the overdensities of dark matter, a distance known as the **sound horizon**. As time passed and the universe expanded, it was at these anisotropies of matter density where galaxies started to form. So by looking at the distances at which galaxies at different redshifts tend to cluster, it is possible to determine a standard angular diameter distance and use that to compare to the distances predicted by different cosmological models.

Peaks have been found in the correlation function (the probability that two galaxies will be a certain distance apart) at $100h^{-1}$ Mpc,[10] indicating that this is the size of the sound horizon today, and by comparing this to the sound horizon at the time of decoupling (using the CMB), we can confirm that the expansion of the universe is accelerating.[16]

10.2.3 Clusters of galaxies

Measuring the mass functions of galaxy clusters, which describe the number density of the clusters above a threshold mass, also provides evidence for dark energy.[17] By comparing these mass functions at high and low redshifts to those predicted by different cosmological models, values for w and Ω_m are obtained which confirm a low matter density and a non zero amount of dark energy.[14]

10.2.4   Age of the universe

See also: Age of the universe

Given a cosmological model with certain values of the cosmological density parameters, it is possible to integrate the Friedmann equations and derive the age of the universe.

$$t_0 = \int\limits_0^1 \frac{da}{\dot{a}}$$

By comparing this to actual measured values of the cosmological parameters, we can confirm the validity of a model which is accelerating now, and had a slower expansion in the past.[14]

10.3   Explanatory models

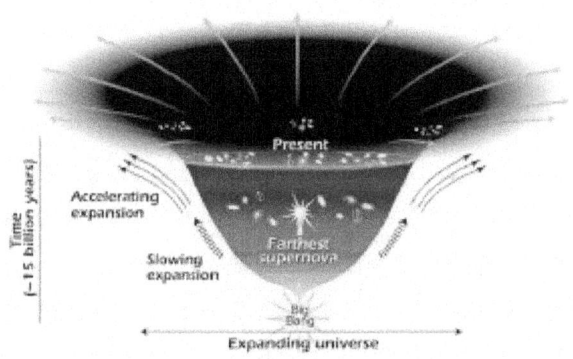

The expansion of the Universe accelerating. Time flows from bottom to top

10.3.1   Dark energy

Main article: Dark energy

The most important property of dark energy is that it has negative pressure which is distributed relatively homogeneously in space.

$$P = wc^2\rho$$

where c is the speed of light, ρ is the energy density. Different theories of dark energy suggest different values of w, with $w < -1/3$ for cosmic acceleration (this leads to a positive value of $\ddot{a}$ in the acceleration equation above).

The simplest explanation for dark energy is that it is a cosmological constant or vacuum energy; in this case $w = -1$. This leads to the Lambda-CDM model, which has generally been known as the Standard Model of Cosmology from 2003 through the present, since it is the simplest model in good agreement with a variety of recent observations. Riess found that their results from supernovae observations favoured expanding models with positive cosmological constant ($\Omega_\lambda > 0$) and a current acceleration of the expansion ($q_0 < 0$).[13]

10.3.2   Phantom energy

Main article: Phantom energy

Current observations allow the possibility of a cosmological model containing a dark energy component with equation of state:

$$w < -1$$

This phantom energy density would become infinite in finite time, causing such a huge gravitational repulsion that the universe would lose all structure and end in a Big Rip.[18] For example, for $w = -3/2$ and $H_0 = 70$ km·s^{-1}·Mpc^{-1}, the time remaining before the universe ends in this "Big Rip" is 22 billion years.[19]

10.3.3   Alternative theories

Other explanations for the accelerating universe include quintessence, a proposed form of dark energy with a non-constant state equation, whose density decreases with time. Dark fluid is an alternative explanation for accelerating expansion which attempts to unite dark matter and dark energy into a single framework.[20] Alternatively, some authors have argued that the universe expansion acceleration could be due to a repulsive gravitational interaction of anti-matter.[21][22][23]

Another type of model, the backreaction conjecture,[24][25] was proposed by cosmologist Syksy Räsänen:[26] the rate of expansion is not homogenous, but we are in a region where expansion is faster than the background. Inhomogeneities in the early universe cause the formation of walls and bubbles, where the inside of a bubble has less matter than on average. According to general relativity, space is less curved than on the walls, and thus appears to have more volume and a higher expansion rate. In the denser regions, the expansion is retarded by a higher gravitational attraction. Therefore, the inward collapse of the denser regions looks the same as an accelerating expansion of the bubbles, leading us to conclude that the universe is expanding at an accelerating rate.[27] The benefit is that it does not require any new physics such as dark energy. Räsänen does not consider the model likely, but without any falsification, it must remain a possibility. It would require rather large density fluctuations (20%) to work.[26]

10.4   Theories for the consequences to the universe

See also: Future of an expanding universe

As the universe expands, the density of radiation and ordinary and dark matter declines more quickly than the density of dark energy (see equation of state) and, eventually, dark energy dominates. Specifically, when the scale of the universe doubles, the density of matter is reduced by a factor of 8, but the density of dark energy is nearly unchanged (it is exactly constant if the dark energy is a cosmological constant).[11]

In models where dark energy is a cosmological constant, the universe will expand exponentially with time from now on, coming closer and closer to a de Sitter spacetime. This will eventually lead to all evidence for the Big Bang disappearing, as the cosmic microwave background is redshifted to lower intensities and longer wavelengths. Eventually its frequency will be low enough that it will be absorbed by the interstellar medium, and so be screened from any observer within the galaxy. This will occur when the universe is less than 50 times its current age, leading to the end of cosmology as we know it as the distant universe turns dark.[28]

Alternatives for the ultimate fate of the universe include the Big Rip mentioned above, a Big Bounce, Big Freeze, or Big Crunch.

10.5   See also

- Cosmological constant
- Friedmann–Lemaître–Robertson–Walker metric
- High-z Supernova Search Team
- Lambda-CDM model
- List of multiple discoveries

- Metric expansion of space
- Scale factor (cosmology)
- Supernova Cosmology Project

10.6   References

[1] Jones, Mark H.; Robert J. Lambourne (2004). *An Introduction to Galaxies and Cosmology*. Cambridge University Press. p. 244. ISBN 978-0-521-83738-5.

[2] Is the universe expanding faster than the speed of light? (see final paragraph)

[3] Frieman et al. (2008) p. 6: "The Universe has gone through three distinct eras: radiation-dominated, $z \gtrsim 3000$; matter-dominated, $3000 \gtrsim z \gtrsim 0.5$; and dark-energy dominated, $z \lesssim 0.5$. The evolution of the scale factor is controlled by the dominant energy form: $a(t) \propto t^{2/3(1+w)}$ (for constant w). During the radiation-dominated era, $a(t) \propto t^{1/2}$; during the matter-dominated era, $a(t) \propto t^{2/3}$; and for the dark energy-dominated era, assuming $w = -1$, asymptotically $a(t) \propto \exp(Ht)$." p. 44: "Taken together, all the current data provide strong evidence for the existence of dark energy; they constrain the fraction of critical density contributed by dark energy, 0.76 ± 0.02, and the equation-of-state parameter, $w \approx -1 \pm 0.1$ (stat) ± 0.1 (sys), assuming that w is constant. This implies that the Universe began accelerating at redshift $z \sim 0.4$ and age $t \sim 10$ Gyr. These results are robust – data from any one method can be removed without compromising the constraints – and they are not substantially weakened by dropping the assumption of spatial flatness."

[4] "Nobel physics prize honours accelerating universe find". BBC News. October 4, 2011.

[5] "The Nobel Prize in Physics 2011". Nobelprize.org. Retrieved 2011-10-06.

[6] Peebles, P. J. E. & Ratra, Bharat (2003). "The cosmological constant and dark energy". *Reviews of Modern Physics* 75 (2): 559–606. arXiv:astro-ph/0207347. Bibcode:2003RvMP...75..559P. doi:10.1103/RevModPhys.75.559.

[7] *Cosmology*, Steven Weinberg, Oxford University Press, 2008

[8] Hubble, Edwin (1929). "A relation between distance and radial velocity among extra-galactic nebulae". *PNAS* 15 (3): 168–173. Bibcode:1929PNAS...15..168H. doi:10.1073/pnas.15.3.168. PMC 522427. PMID 16577160.

[9] Nemiroff, Robert J.; Patla, Bijunath. "Adventures in Friedmann cosmology: A detailed expansion of the cosmological Friedmann equations". *American Journal of Physics* 76 (3): 265. arXiv:astro-ph/0703739. Bibcode:2008AmJPh..76..265N. doi:10.1119/1.2830536.

[10] Lapuente, P.. "Baryon Acoustic Oscillations." Dark energy: observational and theoretical approaches. Cambridge, UK: Cambridge University Press, 2010.

[11] Ryden, Barbara. "Introduction to Cosmology." Physics Today: 77. Print.

[12] Albrecht, A., Bernstein, G., Cahn, R., et al. Report of the Dark Energy TaskForce. ArXiv Astrophysics e-prints, September 2006.

[13] Riess, Adam G.; Filippenko, Alexei V.; Challis, Peter; Clocchiatti, Alejandro; Diercks, Alan; Garnavich, Peter M.; Gilliland, Ron L.; Hogan, Craig J.; Jha, Saurabh; Kirshner, Robert P.; Leibundgut, B.; Phillips, M. M.; Reiss, David; Schmidt, Brian P.; Schommer, Robert A.; Smith, R. Chris; Spyromilio, J.; Stubbs, Christopher; Suntzeff, Nicholas B.; Tonry, John. "Observational Evidence from Supernovae for an Accelerating Universe and a Cosmological Constant". *The Astronomical Journal* **116** (3): 1009–1038. arXiv:astro-ph/9805201.   Bibcode:1998AJ....116.1009R. doi:10.1086/300499.

[14] Pain, Reynald. "Observational evidence of the accelerated expansion of the Universe." Comptes Rendus Physique: 521-538.

[15] Hinshaw, G. (2014). "Five-Year Wilkinson Microwave Anisotropy Probe (WMAP) Observations: Data Processing, Sky Maps, and Basic Results". *Astrophysical Journal Supplement* **180**: 225–245. arXiv:0803.0732. Bibcode:2009ApJS..180..225H.   doi:10.1088/0067-0049/180/2/225.

[16] Eisenstein, Daniel J.; Zehavi, Idit; Hogg, David W.; Scoccimarro, Roman; Blanton, Michael R.; Nichol, Robert C.; Scranton, Ryan; Seo, Hee-Jong; Tegmark, Max; Zheng, Zheng; Anderson, Scott F.; Annis, Jim; Bahcall, Neta; Brinkmann, Jon; Burles, Scott; Castander, Francisco J.; Connolly, Andrew; Csabai, Istvan; Doi, Mamoru; Fukugita, Masataka; Frieman, Joshua A.; Glazebrook, Karl; Gunn, James E.; Hendry, John S.; Hennessy, Gregory; Ivezić, Željko; Kent, Stephen; Knapp, Gillian R.; Lin, Huan; Loh, Yeong-Shang; Lupton, Robert H.; Margon, Bruce; McKay, Timothy A.; Meiksin, Avery; Munn, Jeffery A.; Pope, Adrian; Richmond, Michael W.; Schlegel, David; Schneider, Donald P.; Shimasaku, Kazuhiro; Stoughton, Christopher; Strauss, Michael A.; SubbaRao, Mark; Szalay, Alexander S.; Szapudi, Istvan; Tucker, Douglas L.; Yanny, Brian; York, Donald G. (10 November 2005). "Detection of the Baryon Acoustic Peak in the Large-Scale Correlation Function of SDSS Luminous Red Galaxies". *The Astrophysical Journal* **633** (2): 560–574. arXiv:astro-ph/0501171. Bibcode:2005ApJ...633..560E. doi:10.1086/466512.

[17] Dekel, Avishai. Formation of structure in the Universe. New York: Cambridge University Press, 1999.

[18] Caldwell, Robert; Kamionkowski, Marc; Weinberg, Nevin (August 2003). "Phantom Energy: Dark Energy with w-1 Causes a Cosmic Doomsday". *Phys-*

ical Review Letters **91** (7): 071301. arXiv:astro-ph/0302506.   Bibcode:2003PhRvL..91g1301C. doi:10.1103/PhysRevLett.91.071301. PMID 12935004.

[19] Caldwell, R.R. "A phantom menace? Cosmological consequences of a dark energy component with super-negative equation of state". *Physics Letters B* **545** (1-2): 23–29. arXiv:astro-ph/9908168.   Bibcode:2002PhLB..545...23C. doi:10.1016/S0370-2693(02)02589-3.

[20] Anaelle Halle, HongSheng Zhao, Baojiu Li (2008) "Perturbations in a non-uniform dark energy fluid: equations reveal effects of modified gravity and dark matter "

[21] A. Benoit-Lévy and G. Chardin, Introducing the Dirac-Milne universe, Astronomy and Astrophysics 537, A78 (2012)

[22] D.S. Hajdukovic, Quantum vacuum and virtual gravitational dipoles: the solution to the dark energy problem?, Astrophysics and Space Science 339(1), 1-–5 (2012)

[23] M. Villata, On the nature of dark energy: the lattice Universe, 2013, Astrophysics and Space Science 345, 1. Also available here

[24] "Backreaction: directions of progress". *Classical and Quantum Gravity* **28**:   164008.   arXiv:1102.0408. Bibcode:2011CQGra..28p4008R.   doi:10.1088/0264-9381/28/16/164008.

[25] "Backreaction in Late-Time Cosmology". *Annual Review of Nuclear and Particle Science* **62**:   57–79. arXiv:1112.5335.   Bibcode:2012ARNPS..62...57B. doi:10.1146/annurev.nucl.012809.104435.

[26] https://www.newscientist.com/article/dn11498-is-dark-energy-an-illusion/

[27] "A Cosmic 'Tardis': What the Universe Has In Common with 'Doctor Who'". *Space.com*.

[28] Krauss, Lawrence M.; Scherrer, Robert J. (28 June 2007). "The return of a static universe and the end of cosmology". *General Relativity and Gravitation* **39** (10): 1545–1550. arXiv:0704.0221. Bibcode:2007GReGr..39.1545K. doi:10.1007/s10714-007-0472-9.

- Frieman, Joshau A.; Turner, Michael S.; Huterer, Dragan (2008). "Dark Energy and the Accelerating Universe" (PDF). *Annu. Rev. Astron. Astrophys.* arXiv:0803.0982v1. Retrieved April 1, 2016.

Chapter 11

Big Rip

The **Big Rip** is a cosmological hypothesis about the ultimate fate of the universe, in which the matter of the universe, from stars and galaxies to atoms and subatomic particles, and even spacetime itself, is progressively torn apart by the expansion of the universe at a certain time in the future. According to the hypothesis, first published in 2003, the scale factor of the universe and with it all distances in the universe will become infinite at a finite time in the future. The possibility of sudden singularities and crunch or rip singularities at late times occur only for hypothetical matter with implausible physical properties.[1]

11.1 Definition and overview

The hypothesis relies crucially on the type of dark energy in the universe. The key value is the equation of state parameter w, the ratio between the dark energy pressure and its energy density. If $w < -1$, the universe will eventually be pulled apart. Such energy is called phantom energy, an extreme form of quintessence.

A universe dominated by phantom energy expands at an ever-increasing rate. However, this implies that the size of the observable universe is continually shrinking; the distance to the edge of the observable universe which is moving away at the speed of light from any point moves ever closer. When the size of the observable universe becomes smaller than any particular structure, no interaction by any of the fundamental forces (gravitational, electromagnetic, weak, or strong) can occur between the most remote parts of the structure. When these interactions become impossible, the structure is "ripped apart". The model implies that after a finite time there will be a final singularity, called the "Big Rip", in which all distances diverge to infinite values.

The authors of this hypothesis, led by Robert Caldwell of Dartmouth College, calculate the time from the present to the end of the universe as we know it for this form of energy to be

$$t_{rip} - t_0 \approx \frac{2}{3|1+w|H_0\sqrt{1-\Omega_m}}$$

where w is defined above, H_0 is Hubble's constant and Ω_m is the present value of the density of all the matter in the universe.

In their paper, the authors consider an example with $w = -1.5$, $H_0 = 70$ km/s/Mpc and $\Omega_m = 0.3$, in which case the end of the universe is approximately 22 billion years from the present. This is not considered a prediction, but a hypothetical example. The authors note that evidence indicates w to be very close to -1 in our universe, which makes w the dominating term in the equation. The closer that w is to -1, the closer the denominator is to zero and the further the Big Rip is in the future. If w were exactly equal to -1, the Big Rip could not happen, regardless of the values of H_0 or Ω_m.

11.2 What will happen

In their scenario for $w = -1.5$, the galaxies would first be separated from each other. About 60 million years before the end, gravity would be too weak to hold the Milky Way and other individual galaxies together. Approximately three months before the end, the Solar System (or systems similar to our own at this time, as the fate of the Solar System 22 billion years in the future is questionable) would be gravitationally unbound. In the last minutes, stars and planets would be torn apart, and an instant before the end, atoms would be destroyed. At the end of the universe, spacetime itself will be ripped apart. [2]

11.3 Experimental data

According to the latest cosmological data available, the uncertainties are still too large to discriminate among the three cases $w < -1$, $w = -1$, and $w > -1$.[3][4]

11.4   See also

- Accelerating universe

- Big Bounce

- Big Crunch

- Big Freeze

- Cyclic Model

- Dark energy

- Entropy (arrow of time)

- Future of an expanding universe

- Heat death of the universe

- List of astronomical topics

- Phantom energy

- Timeline of the far future

- Ultimate fate of the universe

11.5   References

[1] Ellis, George F. R., R. Maartens, and M. A. H. MacCallum. Relativistic Cosmology. Cambridge: Cambridge UP, 2012. 146-47. Print.

[2] Caldwell, Robert R.; Kamionkowski, Marc; Weinberg, Nevin N. (2003).   "Phantom Energy and Cosmic Doomsday". *Physical Review Letters* **91** (7): 071301. arXiv:astro-ph/0302506. Bibcode:2003PhRvL..91g1301C. doi:10.1103/PhysRevLett.91.071301. PMID 12935004.

[3] http://wmap.gsfc.nasa.gov/news/

[4] Allen, S. W.; Rapetti, D. A.; Schmidt, R. W.; Ebeling, H.; Morris, R. G.; Fabian, A. C. (2008). "Improved constraints on dark energy from Chandra X-ray observations of the largest relaxed galaxy clusters". *Monthly Notices of the Royal Astronomical Society* **383** (3): 879. arXiv:0706.0033. Bibcode:2008MNRAS.383..879A. doi:10.1111/j.1365-2966.2007.12610.x.

11.6   External links

- New York Times article

-

-

Chapter 12

Standard Model

This article is about the Standard Model of particle physics. For other uses, see Standard model (disambiguation).

This article is a non-mathematical general overview of the Standard Model. For a mathematical description, see the article Standard Model (mathematical formulation).

For the Standard Model of Big Bang cosmology, Lambda-CDM model.

The **Standard Model** of particle physics is a theory con-

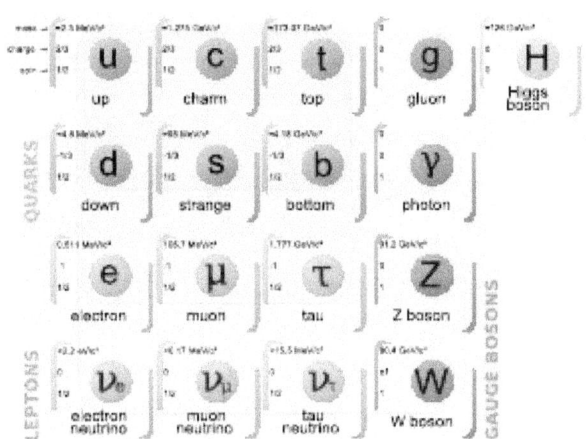

The Standard Model of elementary particles (more schematic depiction), with the three generations of matter, gauge bosons in the fourth column, and the Higgs boson in the fifth.

cerning the electromagnetic, weak, and strong nuclear interactions, as well as classifying all the subatomic particles known. It was developed throughout the latter half of the 20th century, as a collaborative effort of scientists around the world.[1] The current formulation was finalized in the mid-1970s upon experimental confirmation of the existence of quarks. Since then, discoveries of the top quark (1995), the tau neutrino (2000), and more recently the Higgs boson (2012), have given further credence to the Standard Model. Because of its success in explaining a wide variety of experimental results, the Standard Model is sometimes regarded as the "theory of almost everything".

Although the Standard Model is believed to be theoretically self-consistent[2] and has demonstrated huge and continued successes in providing experimental predictions, it does leave some phenomena unexplained and it falls short of being a complete theory of fundamental interactions. It does not incorporate the full theory of gravitation[3] as described by general relativity, or account for the accelerating expansion of the universe (as possibly described by dark energy). The model does not contain any viable dark matter particle that possesses all of the required properties deduced from observational cosmology. It also does not incorporate neutrino oscillations (and their non-zero masses).

The development of the Standard Model was driven by theoretical and experimental particle physicists alike. For theorists, the Standard Model is a paradigm of a quantum field theory, which exhibits a wide range of physics including spontaneous symmetry breaking, anomalies and non-perturbative behavior. It is used as a basis for building more exotic models that incorporate hypothetical particles, extra dimensions, and elaborate symmetries (such as supersymmetry) in an attempt to explain experimental results at variance with the Standard Model, such as the existence of dark matter and neutrino oscillations.

12.1 Historical background

The first step towards the Standard Model was Sheldon Glashow's discovery in 1961 of a way to combine the electromagnetic and weak interactions.[4] In 1967 Steven Weinberg[5] and Abdus Salam[6] incorporated the Higgs mechanism[7][8][9] into Glashow's electroweak interaction, giving it its modern form.

The Higgs mechanism is believed to give rise to the masses of all the elementary particles in the Standard Model. This includes the masses of the W and Z bosons, and the masses of the fermions, i.e. the quarks and leptons.

After the neutral weak currents caused by Z boson exchange were discovered at CERN in 1973,[10][11][12][13] the electroweak theory became widely accepted and Glashow, Salam, and Weinberg shared the 1979 Nobel Prize in Physics for discovering it. The W and Z bosons were discovered experimentally in 1981, and their masses were found to be as the Standard Model predicted.

The theory of the strong interaction, to which many contributed, acquired its modern form around 1973–74, when experiments confirmed that the hadrons were composed of fractionally charged quarks.

12.2   Overview

At present, matter and energy are best understood in terms of the kinematics and interactions of elementary particles. To date, physics has reduced the laws governing the behavior and interaction of all known forms of matter and energy to a small set of fundamental laws and theories. A major goal of physics is to find the "common ground" that would unite all of these theories into one integrated theory of everything, of which all the other known laws would be special cases, and from which the behavior of all matter and energy could be derived (at least in principle).[14]

12.3   Particle content

The Standard Model includes members of several classes of elementary particles (fermions, gauge bosons, and the Higgs boson), which in turn can be distinguished by other characteristics, such as color charge.

12.3.1   Fermions

The Standard Model includes 12 elementary particles of spin-½ known as fermions. According to the spin-statistics theorem, fermions respect the Pauli exclusion principle. Each fermion has a corresponding antiparticle.

The fermions of the Standard Model are classified according to how they interact (or equivalently, by what charges they carry). There are six quarks (up, down, charm, strange, top, bottom), and six leptons (electron, electron neutrino, muon, muon neutrino, tau, tau neutrino). Pairs from each classification are grouped together to form a generation, with corresponding particles exhibiting similar physical behavior (see table).

The defining property of the quarks is that they carry color charge, and hence, interact via the strong interaction. A phenomenon called color confinement results in quarks being very strongly bound to one another, forming color-neutral composite particles (hadrons) containing either a

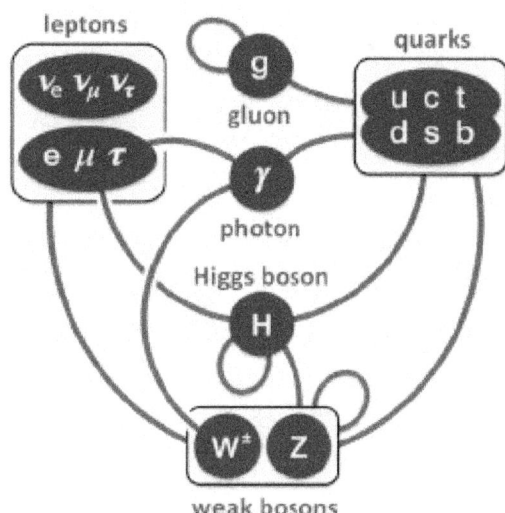

Summary of interactions between particles described by the Standard Model.

quark and an antiquark (mesons) or three quarks (baryons). The familiar proton and the neutron are the two baryons having the smallest mass. Quarks also carry electric charge and weak isospin. Hence they interact with other fermions both electromagnetically and via the weak interaction.

The remaining six fermions do not carry colour charge and are called leptons. The three neutrinos do not carry electric charge either, so their motion is directly influenced only by the weak nuclear force, which makes them notoriously difficult to detect. However, by virtue of carrying an electric charge, the electron, muon, and tau all interact electromagnetically.

Each member of a generation has greater mass than the corresponding particles of lower generations. The first generation charged particles do not decay; hence all ordinary (baryonic) matter is made of such particles. Specifically, all atoms consist of electrons orbiting around atomic nuclei, ultimately constituted of up and down quarks. Second and third generation charged particles, on the other hand, decay with very short half lives, and are observed only in very high-energy environments. Neutrinos of all generations also do not decay, and pervade the universe, but rarely interact with baryonic matter.

12.3.2   Gauge bosons

In the Standard Model, gauge bosons are defined as force carriers that mediate the strong, weak, and electromagnetic fundamental interactions.

Interactions in physics are the ways that particles influence

Standard Model Interactions
(Forces Mediated by Gauge Bosons)

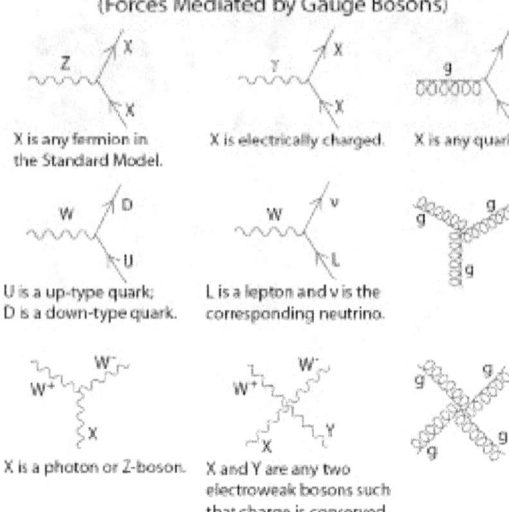

The above interactions form the basis of the standard model. Feynman diagrams in the standard model are built from these vertices. Modifications involving Higgs boson interactions and neutrino oscillations are omitted. The charge of the W bosons is dictated by the fermions they interact with; the conjugate of each listed vertex (i.e. reversing the direction of arrows) is also allowed.

- Photons mediate the electromagnetic force between electrically charged particles. The photon is massless and is well-described by the theory of quantum electrodynamics.

- The W+, W−, and Z gauge bosons mediate the weak interactions between particles of different flavors (all quarks and leptons). They are massive, with the Z being more massive than the W±. The weak interactions involving the W± exclusively act on *left-handed* particles and *right-handed* antiparticles. Furthermore, the W± carries an electric charge of +1 and −1 and couples to the electromagnetic interaction. The electrically neutral Z boson interacts with both left-handed particles and antiparticles. These three gauge bosons along with the photons are grouped together, as collectively mediating the electroweak interaction.

- The eight gluons mediate the strong interactions between color charged particles (the quarks). Gluons are massless. The eightfold multiplicity of gluons is labeled by a combination of color and anticolor charge (e.g. red–antigreen).[nb 1] Because the gluons have an effective color charge, they can also interact among themselves. The gluons and their interactions are described by the theory of quantum chromodynamics.

other particles. At a macroscopic level, electromagnetism allows particles to interact with one another via electric and magnetic fields, and gravitation allows particles with mass to attract one another in accordance with Einstein's theory of general relativity. The Standard Model explains such forces as resulting from matter particles exchanging other particles, generally referred to as *force mediating particles*. When a force-mediating particle is exchanged, at a macroscopic level the effect is equivalent to a force influencing both of them, and the particle is therefore said to have *mediated* (i.e., been the agent of) that force. The Feynman diagram calculations, which are a graphical representation of the perturbation theory approximation, invoke "force mediating particles", and when applied to analyze high-energy scattering experiments are in reasonable agreement with the data. However, perturbation theory (and with it the concept of a "force-mediating particle") fails in other situations. These include low-energy quantum chromodynamics, bound states, and solitons.

The gauge bosons of the Standard Model all have spin (as do matter particles). The value of the spin is 1, making them bosons. As a result, they do not follow the Pauli exclusion principle that constrains fermions: thus bosons (e.g. photons) do not have a theoretical limit on their spatial density (number per volume). The different types of gauge bosons are described below.

The interactions between all the particles described by the Standard Model are summarized by the diagrams on the right of this section.

12.3.3 Higgs boson

Main article: Higgs boson

The Higgs particle is a massive scalar elementary particle theorized by Robert Brout, François Englert, Peter Higgs, Gerald Guralnik, C. R. Hagen, and Tom Kibble in 1964 (see 1964 PRL symmetry breaking papers) and is a key building block in the Standard Model.[7][8][9][15] It has no intrinsic spin, and for that reason is classified as a boson (like the gauge bosons, which have integer spin).

The Higgs boson plays a unique role in the Standard Model, by explaining why the other elementary particles, except the photon and gluon, are massive. In particular, the Higgs boson explains why the photon has no mass, while the W and Z bosons are very heavy. Elementary particle masses, and the differences between electromagnetism (mediated by the photon) and the weak force (mediated by the W and Z bosons), are critical to many aspects of the structure of microscopic (and hence macroscopic) matter. In electroweak theory, the Higgs boson generates the masses of the leptons (electron, muon, and tau) and quarks. As the Higgs boson is massive, it must interact with itself.

Because the Higgs boson is a very massive particle and also decays almost immediately when created, only a very high-energy particle accelerator can observe and record it. Experiments to confirm and determine the nature of the Higgs boson using the Large Hadron Collider (LHC) at CERN began in early 2010, and were performed at Fermilab's Tevatron until its closure in late 2011. Mathematical consistency of the Standard Model requires that any mechanism capable of generating the masses of elementary particles becomes visible at energies above 1.4 TeV;[16] therefore, the LHC (designed to collide two 7 to 8 TeV proton beams) was built to answer the question of whether the Higgs boson actually exists.[17]

On 4 July 2012, the two main experiments at the LHC (ATLAS and CMS) both reported independently that they found a new particle with a mass of about 125 GeV/c^2 (about 133 proton masses, on the order of 10^{-25} kg), which is "consistent with the Higgs boson." Although it has several properties similar to the predicted "simplest" Higgs,[18] they acknowledged that further work would be needed to conclude that it is indeed the Higgs boson, and exactly which version of the Standard Model Higgs is best supported if confirmed.[19][20][21][22][23]

On 14 March 2013 the Higgs Boson was tentatively confirmed to exist.[24]

12.3.4   Total particle count

Counting particles by a rule that distinguishes between particles and their corresponding antiparticles, and among the many color states of quarks and gluons, gives a total of 61 elementary particles.[25] (If neutrinos are their own antiparticles,[26] then by the same counting conventions the total number of elementary particles would be 58.)

12.4   Theoretical aspects

Main article: Standard Model (mathematical formulation)

12.4.1   Construction of the Standard Model Lagrangian

Technically, quantum field theory provides the mathematical framework for the Standard Model, in which a Lagrangian controls the dynamics and kinematics of the theory. Each kind of particle is described in terms of a dynamical field that pervades space-time. The construction of

the Standard Model proceeds following the modern method of constructing most field theories: by first postulating a set of symmetries of the system, and then by writing down the most general renormalizable Lagrangian from its particle (field) content that observes these symmetries.

The global Poincaré symmetry is postulated for all relativistic quantum field theories. It consists of the familiar translational symmetry, rotational symmetry and the inertial reference frame invariance central to the theory of special relativity. The local SU(3)×SU(2)×U(1) gauge symmetry is an internal symmetry that essentially defines the Standard Model. Roughly, the three factors of the gauge symmetry give rise to the three fundamental interactions. The fields fall into different representations of the various symmetry groups of the Standard Model (see table). Upon writing the most general Lagrangian, one finds that the dynamics depend on 19 parameters, whose numerical values are established by experiment. The parameters are summarized in the table above (note: with the Higgs mass is at 125 GeV, the Higgs self-coupling strength $\lambda \sim 1/8$).

Quantum chromodynamics sector

Main article: Quantum chromodynamics

The quantum chromodynamics (QCD) sector defines the interactions between quarks and gluons, with SU(3) symmetry, generated by T^a. Since leptons do not interact with gluons, they are not affected by this sector. The Dirac Lagrangian of the quarks coupled to the gluon fields is given by

$$\mathcal{L}_{QCD} = i\overline{U}(\partial_\mu - ig_s G_\mu^a T^a)\gamma^\mu U + i\overline{D}(\partial_\mu - ig_s G_\mu^a T^a)\gamma^\mu D.$$

G_μ^a is the SU(3) gauge field containing the gluons, γ^μ are the Dirac matrices, D and U are the Dirac spinors associated with up- and down-type quarks, and g_s is the strong coupling constant.

Electroweak sector

Main article: Electroweak interaction

The electroweak sector is a Yang–Mills gauge theory with the simple symmetry group U(1)×SU(2)L,

$$\mathcal{L}_{EW} = \sum_\psi \bar{\psi}\gamma^\mu \left(i\partial_\mu - g'\frac{1}{2}Y_W B_\mu - g\frac{1}{2}\vec{\tau}_L \vec{W}_\mu\right)\psi$$

where $B\mu$ is the U(1) gauge field; YW is the weak hyper-charge—the generator of the U(1) group; $\vec{W}_\mu$ is the three-component SU(2) gauge field; $\vec{\tau}_L$ are the Pauli matrices—infinitesimal generators of the SU(2) group. The subscript L indicates that they only act on left fermions; g' and g are coupling constants.

Higgs sector

Main article: Higgs mechanism

In the Standard Model, the Higgs field is a complex scalar of the group SU(2)L:

$$\varphi = \frac{1}{\sqrt{2}} \begin{pmatrix} \varphi^+ \\ \varphi^0 \end{pmatrix} \,,$$

where the indices + and 0 indicate the electric charge (Q) of the components. The weak isospin (YW) of both components is 1.

Before symmetry breaking, the Higgs Lagrangian is:

$$\mathcal{L}_H = \varphi^\dagger \left(\partial^\mu - \frac{i}{2} \left(g' Y_W B^\mu + g \vec{\tau} \vec{W}^\mu \right) \right) \left(\partial_\mu + \frac{i}{2} \left(g' \right. \right.$$

$$\left. \left. Y_W B_\mu + g \vec{\tau} \vec{W}_\mu \right) \right) \varphi - \frac{\lambda^2}{4} \left(\varphi^\dagger \varphi - v^2 \right)^2 \,,$$

which can also be written as:

$$\mathcal{L}_H = \left| \left(\partial_\mu + \frac{i}{2} \left(g' Y_W B_\mu + g \vec{\tau} \vec{W}_\mu \right) \right) \varphi \right|^2 -$$

$$\frac{\lambda^2}{4} \left(\varphi^\dagger \varphi - v^2 \right)^2 \,.$$

12.5   Fundamental forces

Main article: Fundamental interaction

The Standard Model classified all four fundamental forces in nature. In the Standard Model, a force is described as an exchange of bosons between the objects affected, such as a photon for the electromagnetic force and a gluon for the strong interaction. Those particles are called force carriers.[27]

12.6   Tests and predictions

The Standard Model (SM) predicted the existence of the W and Z bosons, gluon, and the top and charm quarks before these particles were observed. Their predicted properties were experimentally confirmed with good precision.

To give an idea of the success of the SM, the following table compares the measured masses of the W and Z bosons with the masses predicted by the SM:

The SM also makes several predictions about the decay of Z bosons, which have been experimentally confirmed by the Large Electron-Positron Collider at CERN.

In May 2012 BaBar Collaboration reported that their recently analyzed data may suggest possible flaws in the Standard Model of particle physics.[29][30] These data show that a particular type of particle decay called "B to D-star-tau-nu" happens more often than the Standard Model says it should. In this type of decay, a particle called the B-bar meson decays into a D meson, an antineutrino and a tau-lepton. While the level of certainty of the excess (3.4 sigma) is not enough to claim a break from the Standard Model, the results are a potential sign of something amiss and are likely to impact existing theories, including those attempting to deduce the properties of Higgs bosons.[31]

On December 13, 2012, physicists reported the constancy, over space and time, of a basic physical constant of nature that supports the *standard model of physics*. The scientists, studying methanol molecules in a distant galaxy, found the change ($\Delta\mu/\mu$) in the proton-to-electron mass ratio μ to be equal to "(0.0 ± 1.0) × 10^{-7} at redshift z = 0.89" and consistent with "a null result".[32][33]

12.7   Challenges

See also: Physics beyond the Standard Model

Self-consistency of the Standard Model (currently formulated as a non-abelian gauge theory quantized through path-integrals) has not been mathematically proven. While regularized versions useful for approximate computations (for example lattice gauge theory) exist, it is not known whether they converge (in the sense of S-matrix elements) in the limit that the regulator is removed. A key question related to the consistency is the Yang–Mills existence and mass gap problem.

Experiments indicate that neutrinos have mass, which the classic Standard Model did not allow.[34] To accommodate this finding, the classic Standard Model can be modified to include neutrino mass.

If one insists on using only Standard Model particles, this can be achieved by adding a non-renormalizable interaction of leptons with the Higgs boson.[35] On a fundamental level, such an interaction emerges in the seesaw mechanism where heavy right-handed neutrinos are added to the theory. This is natural in the left-right symmetric extension of the Standard Model[36][37] and in certain grand unified theories.[38]

As long as new physics appears below or around 10^{14} GeV, the neutrino masses can be of the right order of magnitude.

Theoretical and experimental research has attempted to extend the Standard Model into a Unified field theory or a Theory of everything, a complete theory explaining all physical phenomena including constants. Inadequacies of the Standard Model that motivate such research include:

- The model does not explain gravitation, although physical confirmation of a theoretical particle known as a graviton would account for it to a degree. Though it addresses strong and electroweak interactions, the Standard Model does not consistently explain the canonical theory of gravitation, general relativity, in terms of quantum field theory. The reason for this is, among other things, that quantum field theories of gravity generally break down before reaching the Planck scale. As a consequence, we have no reliable theory for the very early universe.

- Some physicists consider it to be *ad hoc* and inelegant, requiring 19 numerical constants whose values are unrelated and arbitrary.[39] Although the Standard Model, as it now stands, can explain why neutrinos have masses, the specifics of neutrino mass are still unclear. It is believed that explaining neutrino mass will require an additional 7 or 8 constants, which are also arbitrary parameters.

- The Higgs mechanism gives rise to the hierarchy problem if some new physics (coupled to the Higgs) is present at high energy scales. In these cases, in order for the weak scale to be much smaller than the Planck scale, severe fine tuning of the parameters is required; there are, however, other scenarios that include quantum gravity in which such fine tuning can be avoided.[40] There are also issues of Quantum triviality, which suggests that it may not be possible to create a consistent quantum field theory involving elementary scalar particles.

- The model is inconsistent with the emerging "Standard Model of cosmology." More common contentions include the absence of an explanation in the Standard Model for the observed amount of cold dark matter (CDM) and its contributions to dark energy, which are many orders of magnitude too large. It is also difficult to accommodate the observed predominance of matter over antimatter (matter/antimatter asymmetry). The isotropy and homogeneity of the visible universe over large distances seems to require a mechanism like cosmic inflation, which would also constitute an extension of the Standard Model.

- The existence of ultra-high-energy cosmic rays are difficult to explain under the Standard Model.

Currently, no proposed Theory of Everything has been widely accepted or verified.

12.8 See also

- Fundamental interaction:
  - Quantum electrodynamics
  - Strong interaction: Color charge, Quantum chromodynamics, Quark model
  - Weak interaction: Electroweak theory, Fermi theory of beta decay, Weak hypercharge, Weak isospin
- Gauge theory: Nontechnical introduction to gauge theory
- Generation
- Higgs mechanism: Higgs boson, Higgsless model
- J. C. Ward
- J. J. Sakurai Prize for Theoretical Particle Physics
- Lagrangian
- Open questions: BTeV experiment, CP violation, Neutrino masses, Quark matter, Quantum triviality
- Penguin diagram
- Quantum field theory
- Standard Model: Mathematical formulation of, Physics beyond the Standard Model

12.9 Notes and references

[1] Technically, there are nine such color–anticolor combinations. However, there is one color-symmetric combination that can be constructed out of a linear superposition of the nine combinations, reducing the count to eight.

12.10 References

[1] R. Oerter (2006). *The Theory of Almost Everything: The Standard Model, the Unsung Triumph of Modern Physics* (Kindle ed.). Penguin Group. p. 2. ISBN 0-13-236678-9.

[2] In fact, there are mathematical issues regarding quantum field theories still under debate (see e.g. Landau pole), but the predictions extracted from the Standard Model by current methods applicable to current experiments are all self-consistent. For a further discussion see e.g. Chapter 25 of R. Mann (2010). *An Introduction to Particle Physics and the Standard Model*. CRC Press. ISBN 978-1-4200-8298-2.

[3] Sean Carroll, Ph.D., Cal Tech, 2007, The Teaching Company, *Dark Matter, Dark Energy: The Dark Side of the Universe*, Guidebook Part 2 page 59, Accessed Oct. 7, 2013, "...Standard Model of Particle Physics: The modern theory of elementary particles and their interactions ... It does not, strictly speaking, include gravity, although it's often convenient to include gravitons among the known particles of nature..."

[4] S.L. Glashow (1961). "Partial-symmetries of weak interactions". *Nuclear Physics* **22** (4): 579–588. Bibcode:1961NucPh..22..579G. doi:10.1016/0029-5582(61)90469-2.

[5] S. Weinberg (1967). "A Model of Leptons". *Physical Review Letters* **19** (21): 1264–1266. Bibcode:1967PhRvL..19.1264W. doi:10.1103/PhysRevLett.19.1264.

[6] A. Salam (1968). N. Svartholm, ed. *Elementary Particle Physics: Relativistic Groups and Analyticity*. Eighth Nobel Symposium. Stockholm: Almquist and Wiksell. p. 367.

[7] F. Englert, R. Brout (1964). "Broken Symmetry and the Mass of Gauge Vector Mesons". *Physical Review Letters* **13** (9): 321–323. Bibcode:1964PhRvL..13..321E. doi:10.1103/PhysRevLett.13.321.

[8] P.W. Higgs (1964). "Broken Symmetries and the Masses of Gauge Bosons". *Physical Review Letters* **13** (16): 508–509. Bibcode:1964PhRvL..13..508H. doi:10.1103/PhysRevLett.13.508.

[9] G.S. Guralnik, C.R. Hagen, T.W.B. Kibble (1964). "Global Conservation Laws and Massless Particles". *Physical Review Letters* **13** (20): 585–587. Bibcode:1964PhRvL..13..585G. doi:10.1103/PhysRevLett.13.585.

[10] F.J. Hasert; et al. (1973). "Search for elastic muon-neutrino electron scattering". *Physics Letters B* **46** (1): 121. Bibcode:1973PhLB...46..121H. doi:10.1016/0370-2693(73)90494-2.

[11] F.J. Hasert; et al. (1973). "Observation of neutrino-like interactions without muon or electron in the Gargamelle neutrino experiment". *Physics Letters B* **46** (1): 138. Bibcode:1973PhLB...46..138H. doi:10.1016/0370-2693(73)90499-1.

[12] F.J. Hasert; et al. (1974). "Observation of neutrino-like interactions without muon or electron in the Gargamelle neutrino experiment". *Nuclear Physics B* **73** (1): 1. Bibcode:1974NuPhB..73....1H. doi:10.1016/0550-3213(74)90038-8.

[13] D. Haidt (4 October 2004). "The discovery of the weak neutral currents". *CERN Courier*. Retrieved 8 May 2008.

[14] "Details can be worked out if the situation is simple enough for us to make an approximation, which is almost never, but often we can understand more or less what is happening." from *The Feynman Lectures on Physics*, Vol 1. pp. 2–7

[15] G.S. Guralnik (2009). "The History of the Guralnik, Hagen and Kibble development of the Theory of Spontaneous Symmetry Breaking and Gauge Particles". *International Journal of Modern Physics A* **24** (14): 2601–2627. arXiv:0907.3466. Bibcode:2009IJMPA..24.2601G. doi:10.1142/S0217751X09045431.

[16] B.W. Lee, C. Quigg, H.B. Thacker (1977). "Weak interactions at very high energies: The role of the Higgs-boson mass". *Physical Review D* **16** (5): 1519–1531. Bibcode:1977PhRvD..16.1519L. doi:10.1103/PhysRevD.16.1519.

[17] "Huge $10 billion collider resumes hunt for 'God particle'". CNN. 11 November 2009. Retrieved 2010-05-04.

[18] M. Strassler (10 July 2012). "Higgs Discovery: Is it a Higgs?". Retrieved 2013-08-06.

[19] "CERN experiments observe particle consistent with long-sought Higgs boson". CERN. 4 July 2012. Retrieved 2012-07-04.

[20] "Observation of a New Particle with a Mass of 125 GeV". CERN. 4 July 2012. Retrieved 2012-07-05.

[21] "ATLAS Experiment". ATLAS. 1 January 2006. Retrieved 2012-07-05.

[22] "Confirmed: CERN discovers new particle likely to be the Higgs boson". *YouTube*. Russia Today. 4 July 2012. Retrieved 2013-08-06.

[23] D. Overbye (4 July 2012). "A New Particle Could Be Physics' Holy Grail". *New York Times*. Retrieved 2012-07-04.

[24] "New results indicate that new particle is a Higgs boson". CERN. 14 March 2013. Retrieved 2013-08-06.

[25] S. Braibant, G. Giacomelli, M. Spurio (2009). *Particles and Fundamental Interactions: An Introduction to Particle Physics*. Springer. pp. 313–314. ISBN 978-94-007-2463-1.

[26] Kayser, Boris (2009). "Are neutrinos their own antiparticles?" (PDF). *Journal of Physics: Conference Series* (IOP Publishing) **173** (1): 012013. arXiv:0903.0899. Bibcode:2009JPhCS.173a2013K. doi:10.1088/1742-6596/173/1/012013.

[27] http://home.web.cern.ch/about/physics/standard-model Official CERN website

[28] http://www.pha.jhu.edu/~{}dfehling/particle.gif

[29] "BABAR Data in Tension with the Standard Model". SLAC. 31 May 2012. Retrieved 2013-08-06.

[30] BaBar Collaboration (2012). "Evidence for an excess of B → D$^{(*)}$ τ$^-$ ντ decays". *Physical Review Letters* **109** (10): 101802. arXiv:1205.5442. Bibcode:2012PhRvL.109j1802L. doi:10.1103/PhysRevLett.109.101802.

[31] "BaBar data hint at cracks in the Standard Model". *e! Science News*. 18 June 2012. Retrieved 2013-08-06.

[32] J. Bagdonaite; et al. (2012). "A Stringent Limit on a Drifting Proton-to-Electron Mass Ratio from Alcohol in the Early Universe". *Science* **339** (6115): 46. Bibcode:2013Sci...339...46B. doi:10.1126/science.1224898.

[33] C. Moskowitz (13 December 2012). "Phew! Universe's Constant Has Stayed Constant". Space.com. Retrieved 2012-12-14.

[34] "Particle chameleon caught in the act of changing". CERN. 31 May 2010. Retrieved 2012-07-05.

[35] S. Weinberg (1979). "Baryon and Lepton Non-conserving Processes". *Physical Review Letters* **43** (21): 1566. Bibcode:1979PhRvL..43.1566W. doi:10.1103/PhysRevLett.43.1566.

[36] P. Minkowski (1977). "μ → e γ at a Rate of One Out of 10^9 Muon Decays?". *Physics Letters B* **67** (4): 421. Bibcode:1977PhLB...67..421M. doi:10.1016/0370-2693(77)90435-X.

[37] R. N. Mohapatra, G. Senjanovic (1980). "Neutrino Mass and Spontaneous Parity Nonconservation". *Physical Review Letters* **44** (14): 912–915. Bibcode:1980PhRvL..44..912M. doi:10.1103/PhysRevLett.44.912.

[38] M. Gell-Mann, P. Ramond and R. Slansky (1979). F. van Nieuwenhuizen and D. Z. Freedman, ed. *Supergravity*. North Holland. pp. 315–321. ISBN 0-444-85438-X.

[39] A. Blumhofer, M. Hutter (1997). "Family Structure from Periodic Solutions of an Improved Gap Equation". *Nuclear Physics* **B 484**: 80–96. Bibcode:1997NuPhB.484...80B. doi:10.1016/S0550-3213(96)00644-X.

[40] Salvio, Strumia (2014-03-17). "Agravity". *JHEP* *1406* *(2014)* *080*. arXiv:1403.4226. Bibcode:2014JHEP...06..080S. doi:10.1007/JHEP06(2014)080.

12.11   Further reading

- R. Oerter (2006). *The Theory of Almost Everything: The Standard Model, the Unsung Triumph of Modern Physics*. Plume.

- B.A. Schumm (2004). *Deep Down Things: The Breathtaking Beauty of Particle Physics*. Johns Hopkins University Press. ISBN 0-8018-7971-X.

- "The Standard Model of Particle Physics Interactive Graphic".

Introductory textbooks

- I. Aitchison, A. Hey (2003). *Gauge Theories in Particle Physics: A Practical Introduction*. Institute of Physics. ISBN 978-0-585-44550-2.

- W. Greiner, B. Müller (2000). *Gauge Theory of Weak Interactions*. Springer. ISBN 3-540-67672-4.

- G.D. Coughlan, J.E. Dodd, B.M. Gripaios (2006). *The Ideas of Particle Physics: An Introduction for Scientists*. Cambridge University Press.

- D.J. Griffiths (1987). *Introduction to Elementary Particles*. John Wiley & Sons. ISBN 0-471-60386-4.

- G.L. Kane (1987). *Modern Elementary Particle Physics*. Perseus Books. ISBN 0-201-11749-5.

Advanced textbooks

- T.P. Cheng, L.F. Li (2006). *Gauge theory of elementary particle physics*. Oxford University Press. ISBN 0-19-851961-3. Highlights the gauge theory aspects of the Standard Model.

- J.F. Donoghue, E. Golowich, B.R. Holstein (1994). *Dynamics of the Standard Model*. Cambridge University Press. ISBN 978-0-521-47652-2. Highlights dynamical and phenomenological aspects of the Standard Model.

- L. O'Raifeartaigh (1988). *Group structure of gauge theories*. Cambridge University Press. ISBN 0-521-34785-8.

- Nagashima Y. Elementary Particle Physics: Foundations of the Standard Model, Volume 2. (Wiley 2013) 920 рапуы

- Schwartz, M.D. Quantum Field Theory and the Standard Model (Cambridge University Press 2013) 952 pages

- Langacker P. The standard model and beyond. (CRC Press, 2010) 670 pages Highlights group-theoretical aspects of the Standard Model.

Journal articles

- E.S. Abers, B.W. Lee (1973). "Gauge theories". *Physics Reports* **9**: 1–141. Bibcode:1973PhR.....9....1A. doi:10.1016/0370-1573(73)90027-6.

- M. Baak; et al. (2012). "The Electroweak Fit of the Standard Model after the Discovery of a New Boson at the LHC". *The European Physical Journal C* **72** (11). arXiv:1209.2716. Bibcode:2012EPJC...72.2205B. doi:10.1140/epjc/s10052-012-2205-9.

- Y. Hayato; et al. (1999). "Search for Proton Decay through $p \rightarrow \nu K^+$ in a Large Water Cherenkov Detector". *Physical Review Letters* **83** (8): 1529. arXiv:hep-ex/9904020. Bibcode:1999PhRvL..83.1529H. doi:10.1103/PhysRevLett.83.1529.

- S.F. Novaes (2000). "Standard Model: An Introduction". arXiv:hep-ph/0001283 [hep-ph].

- D.P. Roy (1999). "Basic Constituents of Matter and their Interactions — A Progress Report". arXiv:hep-ph/9912523 [hep-ph].

- F. Wilczek (2004). "The Universe Is A Strange Place". *Nuclear Physics B - Proceedings Supplements* **134**: 3. arXiv:astro-ph/0401347. Bibcode:2004NuPhS.134....3W. doi:10.1016/j.nuclphysbps.2004.08.001.

12.12 External links

- "The Standard Model explained in Detail by CERN's John Ellis" omega tau podcast.

- "The Standard Model" The Standard Model on the CERN web site explains how the basic building blocks of matter interact, governed by four fundamental forces.

- "Standard Model" on YouTube

Chapter 13

Fundamental interaction

Fundamental interactions, also known as fundamental forces, are the interactions in physical systems that do not appear to be reducible to more basic interactions. There are four conventionally accepted fundamental interactions—gravitational, electromagnetic, strong nuclear, and weak nuclear. Each one is understood as the dynamics of a *field*. The gravitational force is modelled as a continuous classical field. The other three are each modelled as discrete quantum fields, and exhibit a measurable unit or *elementary particle*.

The two nuclear interactions produce strong forces at minuscule, subatomic distances. The strong nuclear interaction is responsible for the binding of atomic nuclei. The weak nuclear interaction also acts on the nucleus, mediating radioactive decay. Electromagnetism and gravity produce significant forces at macroscopic scales where the effects can be seen directly in every day life. Electrical and magnetic fields tend to cancel each other out when large collections of objects are considered, so over the largest distances (on the scale of planets and galaxies), gravity tends to be the dominant force.

Theoretical physicists working beyond the Standard Model seek to quantize the gravitational field toward predictions that particle physicists can experimentally confirm, thus yielding acceptance to a theory of quantum gravity (QG). (Phenomena suitable to model as a fifth force—perhaps an added gravitational effect—remain widely disputed.) Other theorists seek to unite the electroweak and strong fields within a Grand Unified Theory (GUT). While all four fundamental interactions are widely thought to align on a highly minuscule scale, particle accelerators cannot produce the massive energy levels required to experimentally probe at that Planck scale (which would experimentally confirm such theories.) Yet some theories, such as the string theory, seek both QG and GUT within one framework, unifying all four fundamental interactions along with mass generation within a theory of everything (ToE).

13.1 General relativity

In his 1687 theory, Isaac Newton postulated space as an infinite and unalterable physical structure existing before, within, and around all objects while their states and relations unfold at a constant pace everywhere, thus absolute space and time. Inferring that all objects bearing mass approach at a constant rate, but collide by impact proportional to their masses, Newton inferred that matter exhibits an attractive force. His law of universal gravitation mathematically stated it to span the entire universe instantly (despite absolute time), or, if not actually a force, to be instant interaction among all objects (despite absolute space.) As conventionally interpreted, Newton's theory of motion modelled a *central force* without a communicating medium.[2] Thus Newton's theory violated the first principle of mechanical philosophy, as stated by Descartes, *No action at a distance.* Conversely, during the 1820s, when explaining magnetism, Michael Faraday inferred a *field* filling space and transmitting that force. Faraday conjectured that ultimately, all forces unified into one.

In the early 1870s, James Clerk Maxwell unified electricity and magnetism as effects of an electromagnetic field whose third consequence was light, travelling at constant speed in a vacuum. The electromagnetic field theory contradicted predictions of Newton's theory of motion, unless physical states of the luminiferous aether—presumed to fill all space whether within matter or in a vacuum and to manifest the electromagnetic field—aligned all phenomena and thereby held valid the Newtonian principle relativity or invariance. Disfavouring hypotheses at unobservables, Albert Einstein discarded the aether, and aligned electrodynamics with relativity by denying absolute space and time, and stating relative space and time. The two phenomena altered in the vicinity of an object measured to be in motion—length contraction and time dilation for the object experienced to be in relative motion—Einstein's principle special relativity, published in 1905.

Special relativity was accepted as a theory too. It rendered Newton's theory of motion apparently untenable, especially since Newtonian physics postulated an object's mass to be constant. A consequence of special relativity is mass being a variant form of energy, condensed into an object. By the equivalence principle, published by Einstein in 1907, gravitation is indistinguishable from acceleration, perhaps two phenomena sharing a mechanism. That year, Hermann Minkowski modelled special relativity to a unification of space and time, 4D spacetime. Stretching the three spatial dimensions onto the single dimension of time's arrow, Einstein arrived at the general theory of relativity in 1915.[3] Einstein interpreted space as a substance, *Einstein-aether*, whose physical properties receive motion from an object and transmit it to other objects while modulating events unfolding. Equivalent to energy, mass contracts space, which dilates time—events unfold more slowly—establishing local tension. The object relieves it in the likeness of a free fall at light speed along the pathway of least resistance, a straight line's equivalent on the curved surface of 4D spacetime, a pathway termed *worldline*.

Einstein abolished *action at a distance* by theorizing a gravitational field—4D spacetime—that waves while transmitting motion across the universe at light speed. All objects always travel at light speed in 4D spacetime. At zero relative speed, an object is observed to travel none through space, but age most rapidly. That is, an object at relative rest in 3D space exhibits its constant energy to an observer by exhibiting top speed along 1D time flow. Conversely, at highest relative speed, an object traverses 3D space at light speed, yet is ageless, none of its constant energy available to internal motion as flow along 1D time. Whereas Newtonian inertia is an idealized case of an object either keeping rest or holding constant velocity by its hypothetical existence in a universe otherwise devoid of matter, Einsteinian inertia is indistinguishable from an object experiencing no acceleration by existing in a gravitational field possibly full of matter distributed uniformly. Conversely, even massless energy manifests gravitation—which is acceleration—on local objects by "curving" the surface of 4D spacetime. Physicists renounced belief that motion must be mediated by a *force*.

13.2 Standard Model

Main article: Standard Model
See also: Lambda-CDM model

The electromagnetic, strong, and weak interactions associate with elementary particles, whose behaviours are modelled in quantum mechanics (QM). For predictive success with QM's probabilistic outcomes, particle physics conventionally models QM events across a field set to special relativity, altogether relativistic quantum field theory (QFT).[4] Force particles, called gauge bosons—*force carriers* or *messenger particles* of underlying fields—interact with mat-

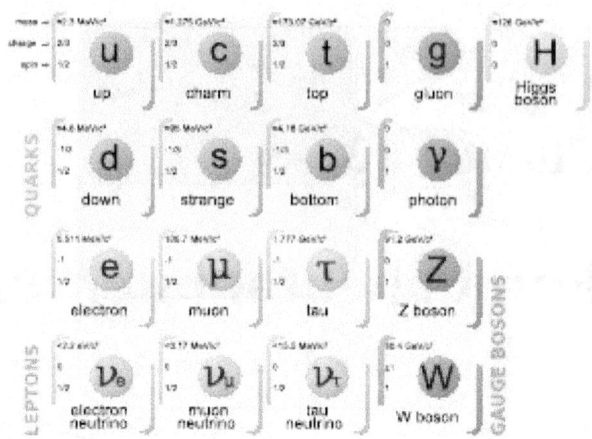

The Standard Model of elementary particles, with the fermions in the first three columns, the gauge bosons in the fourth column, and the Higgs boson in the fifth column

ter particles, called fermions. Everyday matter is atoms, composed of three fermion types: up-quarks and down-quarks constituting, as well as electrons orbiting, the atom's nucleus. Atoms interact, form molecules, and manifest further properties through electromagnetic interactions among their electrons absorbing and emitting photons, the electromagnetic field's force carrier, which if unimpeded traverse potentially infinite distance. Electromagnetism's QFT is quantum electrodynamics (QED).

The electromagnetic interaction was modelled with the weak interaction, whose force carriers are W and Z bosons, traversing the minuscule distance, in electroweak theory (EWT). Electroweak interaction would operate at such high temperatures as soon after the presumed Big Bang but, as the early universe cooled, split into electromagnetic and weak interactions. The strong interaction, whose force carrier is the gluon, traversing minuscule distance among quarks, is modeled in quantum chromodynamics (QCD). EWT, QCD, and the Higgs mechanism, whereby the Higgs field manifests Higgs bosons that interact with some quantum particles and thereby endow those particles with mass comprise particle physics' Standard Model (SM). Predictions are usually made using calculational approximation methods, although such perturbation theory is inadequate to model some experimental observations (for instance bound states and solitons.) Still, physicists widely accept the Standard Model as science's most experimentally confirmed theory.

Beyond the Standard Model, some theorists work to unite the electroweak and strong interactions within a Grand Unified Theory (GUT). Some attempts at GUTs hypothesize "shadow" particles, such that every known matter particle associates with an undiscovered force particle, and vice versa, altogether supersymmetry (SUSY). Other theorists seek to quantize the gravitational field by the modelling be-haviour of its hypothetical force carrier, the graviton and achieve quantum gravity (QG). One approach to QG is loop quantum gravity (LQG). Still other theorists seek both QG and GUT within one framework, reducing all four funda-mental interactions to a Theory of Everything (ToE). The most prevalent aim at a ToE is string theory, although to model matter particles, it added SUSY to force particles—and so, strictly speaking, became superstring theory. Mul-tiple, seemingly disparate superstring theories were unified on a backbone, M-theory. Theories beyond the Standard Model remain highly speculative, lacking great experimen-tal support.

13.3 Overview of the fundamental interactions

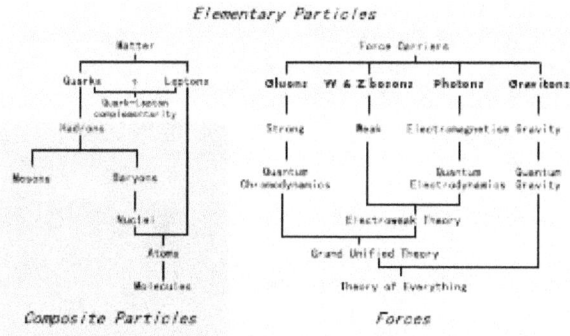

An overview of the various families of elementary and composite particles, and the theories describing their interactions. Fermions are on the left, and Bosons are on the right.

In the conceptual model of fundamental interactions, matter consists of fermions, which carry properties called charges and spin $\pm\frac{1}{2}$ (intrinsic angular momentum $\pm\frac{\hbar}{2}$, where $\hbar$ is the reduced Planck constant). They attract or repel each other by exchanging bosons.

The interaction of any pair of fermions in perturbation the-ory can then be modelled thus:

> Two fermions go in → *interaction* by boson ex-change → Two changed fermions go out.

The exchange of bosons always carries energy and momentum between the fermions, thereby changing their speed and direction. The exchange may also transport a charge between the fermions, changing the charges of the fermions in the process (e.g., turn them from one type of fermion to another). Since bosons carry one unit of angular momentum, the fermion's spin direction will flip from $+\frac{1}{2}$

to $-\frac{1}{2}$ (or vice versa) during such an exchange (in units of the reduced Planck's constant).

Because an interaction results in fermions attracting and re-pelling each other, an older term for "interaction" is force.

According to the present understanding, there are four fundamental interactions or forces: gravitation, electromagnetism, the weak interaction, and the strong interaction. Their magnitude and behaviour vary greatly, as described in the table below. Modern physics attempts to explain every observed physical phenomenon by these fundamental interactions. Moreover, reducing the number of different interaction types is seen as desirable. Two cases in point are the unification of:

- Electric and magnetic force into electromagnetism;

- The electromagnetic interaction and the weak interac-tion into the electroweak interaction; see below.

Both magnitude ("relative strength") and "range", as given in the table, are meaningful only within a rather complex theoretical framework. It should also be noted that the table below lists properties of a conceptual scheme that is still the subject of ongoing research.

The modern (perturbative) quantum mechanical view of the fundamental forces other than gravity is that particles of matter (fermions) do not directly interact with each other, but rather carry a charge, and exchange virtual particles (gauge bosons), which are the interaction carriers or force mediators. For example, photons mediate the interaction of electric charges, and gluons mediate the interaction of color charges.

13.4 The interactions

13.4.1 Gravity

Main article: Gravity

Gravitation is by far the weakest of the four interactions. The weakness of gravity can easily be demonstrated by sus-pending a pin using a simple magnet (such as a refrigerator magnet). The magnet is able to hold the pin against the gravitational pull of the entire Earth.

Yet gravitation is very important for macroscopic objects and over macroscopic distances for the following reasons. Gravitation:

- Is the only interaction that acts on all particles having mass, energy and/or momentum

- Has an infinite range, like electromagnetism but unlike strong and weak interaction

- Cannot be absorbed, transformed, or shielded against

- Always attracts and never repels

Even though electromagnetism is far stronger than gravitation, electrostatic attraction is not relevant for large celestial bodies, such as planets, stars, and galaxies, simply because such bodies contain equal numbers of protons and electrons and so have a net electric charge of zero. Nothing "cancels" gravity, since it is only attractive, unlike electric forces which can be attractive or repulsive. On the other hand, all objects having mass are subject to the gravitational force, which only attracts. Therefore, only gravitation matters on the large-scale structure of the universe.

The long range of gravitation makes it responsible for such large-scale phenomena as the structure of galaxies and black holes and it retards the expansion of the universe. Gravitation also explains astronomical phenomena on more modest scales, such as planetary orbits, as well as everyday experience: objects fall; heavy objects act as if they were glued to the ground, and animals can only jump so high.

Gravitation was the first interaction to be described mathematically. In ancient times, Aristotle hypothesized that objects of different masses fall at different rates. During the Scientific Revolution, Galileo Galilei experimentally determined that this was not the case — neglecting the friction due to air resistance, and buoyancy forces if an atmosphere is present (e.g. the case of a dropped air-filled balloon vs a water-filled balloon) all objects accelerate toward the Earth at the same rate. Isaac Newton's law of Universal Gravitation (1687) was a good approximation of the behaviour of gravitation. Our present-day understanding of gravitation stems from Albert Einstein's General Theory of Relativity of 1915, a more accurate (especially for cosmological masses and distances) description of gravitation in terms of the geometry of spacetime.

Merging general relativity and quantum mechanics (or quantum field theory) into a more general theory of quantum gravity is an area of active research. It is hypothesized that gravitation is mediated by a massless spin-2 particle called the graviton.

Although general relativity has been experimentally confirmed (at least for weak fields) on all but the smallest scales, there are rival theories of gravitation. Those taken seriously by [citation needed] the physics community all reduce to general relativity in some limit, and the focus of observational work is to establish limitations on what deviations from general relativity are possible.

Proposed extra dimensions could explain why the gravity force is so weak.[6]

13.4.2 Electroweak interaction

Main article: Electroweak interaction

Electromagnetism and weak interaction appear to be very different at everyday low energies. They can be modeled using two different theories. However, above unification energy, on the order of 100 GeV, they would merge into a single electroweak force.

Electroweak theory is very important for modern cosmology, particularly on how the universe evolved. This is because shortly after the Big Bang, the temperature was approximately above 10^{15} K. Electromagnetic force and weak force were merged into a combined electroweak force.

For contributions to the unification of the weak and electromagnetic interaction between elementary particles, Abdus Salam, Sheldon Glashow and Steven Weinberg were awarded the Nobel Prize in Physics in 1979.[7][8]

Electromagnetism

Main article: Electromagnetism

Electromagnetism is the force that acts between electrically charged particles. This phenomenon includes the electrostatic force acting between charged particles at rest, and the combined effect of electric and magnetic forces acting between charged particles moving relative to each other.

Electromagnetism is infinite-ranged like gravity, but vastly stronger, and therefore describes a number of macroscopic phenomena of everyday experience such as friction, rainbows, lightning, and all human-made devices using electric current, such as television, lasers, and computers. Electromagnetism fundamentally determines all macroscopic, and many atomic levels, properties of the chemical elements, including all chemical bonding.

In a four kilogram (~1 gallon) jug of water there are

$$4000 \text{ g } H_2O \cdot \frac{1 \text{ mol } H_2O}{18 \text{ g } H_2O} \cdot \frac{10 \text{ mol } e^-}{1 \text{ mol } H_2O} \cdot \frac{96{,}000 \text{ C}}{1 \text{ mol } e^-} = 2.1 \times 10^8 C$$

of total electron charge. Thus, if we place two such jugs a meter apart, the electrons in one of the jugs repel those in the other jug with a force of

$$\frac{1}{4\pi\varepsilon_0} \frac{(2.1\times10^6 C)^2}{(1m)^2} = 4.1 \times 10^{26} N.$$

This is larger than the planet Earth would weigh if weighed on another Earth. The atomic nuclei in one jug also repel those in the other with the same force. However, these repulsive forces are canceled by the attraction of the electrons in jug A with the nuclei in jug B and the attraction of the nuclei in jug A with the electrons in jug B, resulting in no net force. Electromagnetic forces are tremendously stronger than gravity but cancel out so that for large bodies gravity dominates.

Electrical and magnetic phenomena have been observed since ancient times, but it was only in the 19th century that it was discovered that electricity and magnetism are two aspects of the same fundamental interaction. By 1864, Maxwell's equations had rigorously quantified this unified interaction. Maxwell's theory, restated using vector calculus, is the classical theory of electromagnetism, suitable for most technological purposes.

The constant speed of light in a vacuum (customarily described with the letter "c") can be derived from Maxwell's equations, which are consistent with the theory of special relativity. Einstein's 1905 theory of special relativity, however, which flows from the observation that the speed of light is constant no matter how fast the observer is moving, showed that the theoretical result implied by Maxwell's equations has profound implications far beyond electromagnetism on the very nature of time and space.

In another work that departed from classical electromagnetism, Einstein also explained the photoelectric effect by hypothesizing that light was transmitted in quanta, which we now call photons. Starting around 1927, Paul Dirac combined quantum mechanics with the relativistic theory of electromagnetism. Further work in the 1940s, by Richard Feynman, Freeman Dyson, Julian Schwinger, and Sin-Itiro Tomonaga, completed this theory, which is now called quantum electrodynamics, the revised theory of electromagnetism. Quantum electrodynamics and quantum mechanics provide a theoretical basis for electromagnetic behavior such as quantum tunneling, in which a certain percentage of electrically charged particles move in ways that would be impossible under the classical electromagnetic theory, that is necessary for everyday electronic devices such as transistors to function.

Weak interaction

Main article: Weak interaction

The *weak interaction* or *weak nuclear force* is responsible for some nuclear phenomena such as beta decay. Electromagnetism and the weak force are now understood to be two aspects of a unified electroweak interaction — this discovery was the first step toward the unified theory known as the Standard Model. In the theory of the electroweak interaction, the carriers of the weak force are the massive gauge bosons called the W and Z bosons. The weak interaction is the only known interaction which does not conserve parity; it is left-right asymmetric. The weak interaction even violates CP symmetry but does conserve CPT.

13.4.3   Strong interaction

Main article: Strong interaction

The *strong interaction*, or *strong nuclear force*, is the most complicated interaction, mainly because of the way it varies with distance. At distances greater than 10 femtometers, the strong force is practically unobservable. Moreover, it holds only inside the atomic nucleus.

After the nucleus was discovered in 1908, it was clear that a new force was needed to overcome the electrostatic repulsion, a manifestation of electromagnetism, of the positively charged protons. Otherwise, the nucleus could not exist. Moreover, the force had to be strong enough to squeeze the protons into a volume that is 10^{-15} of that of the entire atom. From the short range of this force, Hideki Yukawa predicted that it was associated with a massive particle, whose mass is approximately 100 MeV.

The 1947 discovery of the pion ushered in the modern era of particle physics. Hundreds of hadrons were discovered from the 1940s to 1960s, and an extremely complicated theory of hadrons as strongly interacting particles was developed. Most notably:

- The pions were understood to be oscillations of vacuum condensates;

- Jun John Sakurai proposed the rho and omega vector bosons to be force carrying particles for approximate symmetries of isospin and hypercharge;

- Geoffrey Chew, Edward K. Burdett and Steven Frautschi grouped the heavier hadrons into families that could be understood as vibrational and rotational excitations of strings.

While each of these approaches offered deep insights, no approach led directly to a fundamental theory.

Murray Gell-Mann along with George Zweig first proposed fractionally charged quarks in 1961. Throughout the 1960s, different authors considered theories similar to the modern fundamental theory of quantum chromodynamics (QCD) as simple models for the interactions of quarks. The first to hypothesize the gluons of QCD were Moo-Young Han and Yoichiro Nambu, who introduced the quark color charge and hypothesized that it might be associated with a force-carrying field. At that time, however, it was difficult to see how such a model could permanently confine quarks. Han and Nambu also assigned each quark color an integer electrical charge, so that the quarks were fractionally charged only on average, and they did not expect the quarks in their model to be permanently confined.

In 1971, Murray Gell-Mann and Harald Fritzsch proposed that the Han/Nambu color gauge field was the correct theory of the short-distance interactions of fractionally charged quarks. A little later, David Gross, Frank Wilczek, and David Politzer discovered that this theory had the property of asymptotic freedom, allowing them to make contact with experimental evidence. They concluded that QCD was the complete theory of the strong interactions, correct at all distance scales. The discovery of asymptotic freedom led most physicists to accept QCD since it became clear that even the long-distance properties of the strong interactions could be consistent with experiment if the quarks are permanently confined.

Assuming that quarks are confined, Mikhail Shifman, Arkady Vainshtein, and Valentine Zakharov were able to compute the properties of many low-lying hadrons directly from QCD, with only a few extra parameters to describe the vacuum. In 1980, Kenneth G. Wilson published computer calculations based on the first principles of QCD, establishing, to a level of confidence tantamount to certainty, that QCD will confine quarks. Since then, QCD has been the established theory of the strong interactions.

QCD is a theory of fractionally charged quarks interacting by means of 8 photon-like particles called gluons. The gluons interact with each other, not just with the quarks, and at long distances the lines of force collimate into strings. In this way, the mathematical theory of QCD not only explains how quarks interact over short distances but also the string-like behavior, discovered by Chew and Frautschi, which they manifest over longer distances.

13.4.4 Beyond the Standard Model

Main article: Physics beyond the Standard Model
See also: Elementary particle § Beyond the Standard Model

Numerous theoretical efforts have been made to systematize the existing four fundamental interactions on the model of electroweak unification.

Grand Unified Theories (GUTs) are proposals to show that all of the fundamental interactions, other than gravity, arise from a single interaction with symmetries that break down at low energy levels. GUTs predict relationships among constants of nature that are unrelated in the SM. GUTs also predict gauge coupling unification for the relative strengths of the electromagnetic, weak, and strong forces, a prediction verified at the Large Electron–Positron Collider in 1991 for supersymmetric theories.

Theories of everything, which integrate GUTs with a quantum gravity theory face a greater barrier, because no quantum gravity theories, which include string theory, loop quantum gravity, and twistor theory, have secured wide acceptance. Some theories look for a graviton to complete the Standard Model list of force-carrying particles, while others, like loop quantum gravity, emphasize the possibility that time-space itself may have a quantum aspect to it.

Some theories beyond the Standard Model include a hypothetical fifth force, and the search for such a force is an ongoing line of experimental research in physics. In supersymmetric theories, there are particles that acquire their masses only through supersymmetry breaking effects and these particles, known as moduli can mediate new forces. Another reason to look for new forces is the recent discovery that the expansion of the universe is accelerating (also known as dark energy), giving rise to a need to explain a nonzero cosmological constant, and possibly to other modifications of general relativity. Fifth forces have also been suggested to explain phenomena such as CP violations, dark matter, and dark flow.

13.5 See also

- Standard Model
  - Strong interaction
  - Electroweak interaction
  - Weak interaction
  - Gravity
    - Quantum gravity
    - String Theory
    - Theory of Everything
- Grand Unified Theory
  - Gauge coupling unification
  - Unified Field Theory
- Quintessence, a hypothesized fifth force.
- *People*: Isaac Newton, James Clerk Maxwell, Albert Einstein, Richard Feynman, Sheldon Glashow, Abdus Salam, Steven Weinberg, Gerardus 't Hooft, David Gross, Edward Witten, Howard Georgi.

13.6   References

[1] http://www.pha.jhu.edu/~{}dfehling/particle.gif

[2] Newton's absolute space was a medium, but not one transmitting gravitation.

[3] Special relativity holds for objects at vast speed but of negligible mass, for instance elementary particles. Yet by yielding gravitation, which is a manner of acceleration, notable mass breaks inertia—that is, constant speed and direction—and thereby violates special relativity. Special relativity could approximately predict a massive object's motion during barely an instant, however, and thus is a temporally limited case of general relativity.

[4] Meinard Kuhlmann, "Physicists debate whether the world is made of particles or fields—or something else entirely", *Scientific American*, 24 Jul 2013.

[5] Approximate.   See Coupling constant for more exact strengths, depending on the particles and energies involved.

[6] CERN (20 January 2012). "Extra dimensions, gravitons, and tiny black holes".

[7] Bais, Sander (2005), *The Equations. Icons of knowledge*, ISBN 0-674-01967-9 p.84

[8] "The Nobel Prize in Physics 1979". The Nobel Foundation. Retrieved 2008-12-16.

Bibliography General:

- Davies, Paul (1986), *The Forces of Nature*, Cambridge Univ. Press 2nd ed.

- Feynman, Richard (1967), *The Character of Physical Law*, MIT Press, ISBN 0-262-56003-8

- Schumm, Bruce A. (2004), *Deep Down Things*, Johns Hopkins University Press While all interactions are discussed, discussion is especially thorough on the weak.

- Weinberg, Steven (1993), *The First Three Minutes: A Modern View of the Origin of the Universe*, Basic Books, ISBN 0-465-02437-8

- Weinberg, Steven (1994), *Dreams of a Final Theory*, Basic Books, ISBN 0-679-74408-8

Texts:

- Padmanabhan, T. (1998), *After The First Three Minutes: The Story of Our Universe*, Cambridge Univ. Press, ISBN 0-521-62972-1

- Perkins, Donald H. (2000), *Introduction to High Energy Physics*, Cambridge Univ. Press, ISBN 0-521-62196-8

- Riazuddin (December 29, 2009). "Non-standard interactions" (PDF). *NCP 5th Particle Physics Sypnoisis* (Islamabad: Riazuddin, Head of High-Energy Theory Group at National Center for Physics) 1 (1): 1–25. Retrieved March 19, 2011.

Chapter 14

Physics beyond the Standard Model

Physics beyond the Standard Model (BSM) refers to the theoretical developments needed to explain the deficiencies of the Standard Model, such as the origin of mass, the strong CP problem, neutrino oscillations, matter–antimatter asymmetry, and the nature of dark matter and dark energy.[1] Another problem lies within the mathematical framework of the Standard Model itself—the Standard Model is inconsistent with that of general relativity, to the point that one or both theories break down under certain conditions (for example within known spacetime singularities like the Big Bang and black hole event horizons).

Theories that lie beyond the Standard Model include various extensions of the standard model through supersymmetry, such as the Minimal Supersymmetric Standard Model (MSSM) and Next-to-Minimal Supersymmetric Standard Model (NMSSM), or entirely novel explanations, such as string theory, M-theory, and extra dimensions. As these theories tend to reproduce the entirety of current phenomena, the question of which theory is the right one, or at least the "best step" towards a Theory of Everything, can only be settled via experiments, and is one of the most active areas of research in both theoretical and experimental physics.

14.1 Problems with the Standard Model

Despite being the most successful theory of particle physics to date, the Standard Model is not perfect.[2] A large share of the published output of theoretical physicists consists of proposals for various forms of "Beyond the Standard Model" new physics proposals that would modify the Standard Model in ways subtle enough to be consistent with existing data, yet address its imperfections materially enough to predict non-Standard Model outcomes of new experiments that can be proposed.

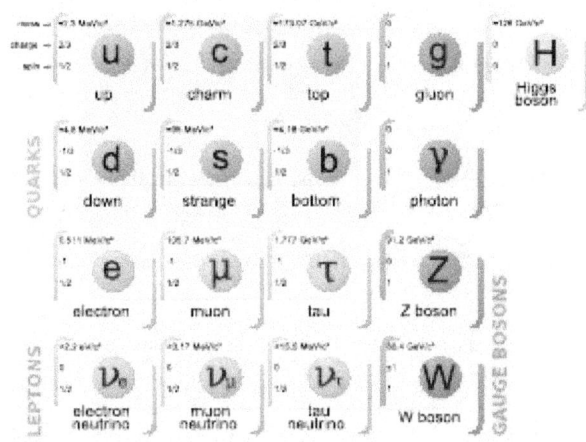

The Standard Model of elementary particles

14.1.1 Phenomena not explained

The Standard Model is inherently an incomplete theory. There are fundamental physical phenomena in nature that the Standard Model does not adequately explain:

- *Gravity.* The standard model does not explain gravity. The approach of simply adding a "graviton" (whose properties are the subject of considerable consensus among physicists if it exists) to the Standard Model does not recreate what is observed experimentally without other modifications, as yet undiscovered, to the Standard Model. Moreover, instead, the Standard Model is widely considered to be incompatible with the most successful theory of gravity to date, general relativity.[3]

- *Dark matter and dark energy.* Cosmological observations tell us the standard model explains about 5% of the energy present in the universe. About 26% should be dark matter, which would behave just like other matter, but which only interacts weakly (if at all) with the Standard Model fields. Yet, the Standard Model does not supply any fundamental particles that are good dark matter candidates. The rest (69%) should be dark energy, a constant energy density for the vacuum. Attempts to explain dark energy in terms of vacuum energy of the standard model lead to a mismatch of 120 orders of magnitude.[4]

- *Neutrino masses.* According to the standard model, neutrinos are massless particles. However, neutrino oscillation experiments have shown that neutrinos do have mass. Mass terms for the neutrinos can be added to the standard model by hand, but these lead to new theoretical problems. For example, the mass terms need to be extraordinarily small and it is not clear if the neutrino masses would arise in the same way that the masses of other fundamental particles do in the Standard Model.

- *Matter–antimatter asymmetry.* The universe is made out of mostly matter. However, the standard model predicts that matter and antimatter should have been created in (almost) equal amounts if the initial conditions of the universe did not involve disproportionate matter relative to antimatter. Yet, no mechanism sufficient to explain this asymmetry exists in the Standard Model.

14.1.2   Experimental results not explained

No experimental result is widely accepted as definitively contradicting the Standard Model at the "five sigma" level, widely considered to be the threshold of a "discovery" in particle physics. But because every experiment contains some degree of statistical and systemic uncertainty, and the theoretical predictions themselves are also almost never calculated exactly and are subject to uncertainties in measurements of the fundamental constants of the Standard Model (some of which are tiny and others of which are substantial), it is mathematically expected that some of the hundreds of experimental tests of the Standard Model will deviate to some extent from it, even if there were no "new physics" to be discovered.

At any given time there are a number of experimental results that are significantly different from the Standard Model expectation, although many of these have been found to be statistical flukes or experimental errors as more data has been collected. On the other hand, any "beyond the Standard Model" physics would necessarily first manifest experimentally as a statistically significant difference between an experiment and the theoretical prediction.

In each case, physicists seek to determine if a result is a mere statistical fluke or experimental error on the one hand, or a sign of new physics on the other. More statistically significant results cannot be mere statistical flukes but can

still result from experimental error or inaccurate estimates of experimental precision. Frequently, experiments are tailored to be more sensitive to experimental results that would distinguish the Standard Model from theoretical alternatives.

Some of the most notable examples include the following:

- *Muonic hydrogen* – the Standard Model makes precise theoretical predictions regarding the atomic radius size of ordinary hydrogen (a proton-electron system) and that of muonic hydrogen (a proton-muon system in which a muon is a "heavy" variant of an electron). However, the measured atomic radius of muonic hydrogen differs significantly from that of the radius predicted by the Standard Model using existing physical constant measurements by what appears to be as many as seven standard deviations.[5] Doubts about the accuracy of the error estimates in earlier experiments, which are still within 4% of each other in measuring a truly tiny distance, and a lack of a well motivated theory that could explain the discrepancy, have caused physicists to be hesitant to describe these results as contradicting the Standard Model despite the apparent statistical significance of the result and a lack of any clearly identified possible source of experimental error in the results.

- *Anomalous magnetic dipole moment of muon* – the experimentally measured value of muon's anomalous magnetic dipole moment ("muon g-2") is significantly different from the Standard Model prediction.[6]

- *BaBar data suggests possible flaws in the Standard Model* – results from a BaBar experiment may suggest a surplus over Standard Model predictions of a type of particle decay $(B \rightarrow D^{(*)}\tau^{-}\tau\nu)$. In this, an electron and positron collide, resulting in a B meson and an antimatter B meson, which then decays into a D meson and a tau lepton as well as a tau antineutrino. While the level of certainty of the excess (3.4 sigma in statistical language) is not enough to claim a break from the Standard Model, the results are a potential sign of something amiss and are likely to affect existing theories, including those attempting to deduce the properties of Higgs bosons.[7] In 2015, LHCb reported observing a 2.1 sigma excess in the same ratio of branching fractions.[8]

- Proton radius – the proton's charge radius measured using electron probes is different than when measured using muons.[9]

14.1.3 Theoretical predictions not observed

Observation at particle colliders of all of the fundamental particles predicted by the Standard Model has been confirmed. The Higgs boson is predicted by the Standard Model's explanation of the Higgs mechanism, which describes how the weak SU(2) gauge symmetry is broken and how fundamental particles obtain mass; it was the last particle predicted by the Standard Model to be observed. On July 4, 2012, CERN scientists using the Large Hadron Collider announced the discovery of a particle consistent with the Higgs boson, with a mass of about $126\,\text{GeV}/c^2$. A Higgs boson was confirmed to exist on March 14, 2013, although efforts to confirm that it has all of the properties predicted by the Standard Model are ongoing.[10]

A few hadrons (i.e. composite particles made of quarks) whose existence is predicted by the Standard Model, which can be produced only at very high energies in very low frequencies have not yet been definitively observed, and "glueballs"[11] (i.e. composite particles made of gluons) have also not yet been definitively observed. Some very low frequency particle decays predicted by the Standard Model have also not yet been definitively observed because insufficient data is available to make a statistically significant observation.

14.1.4 Theoretical problems

Some features of the standard model are added in an ad hoc way. These are not problems per se (i.e. the theory works fine with these ad hoc features), but they imply a lack of understanding. These ad hoc features have motivated theorists to look for more fundamental theories with fewer parameters. Some of the ad hoc features are:

- *Hierarchy problem* – the standard model introduces particle masses through a process known as spontaneous symmetry breaking caused by the Higgs field. Within the standard model, the mass of the Higgs gets some very large quantum corrections due to the presence of virtual particles (mostly virtual top quarks). These corrections are much larger than the actual mass of the Higgs. This means that the bare mass parameter of the Higgs in the standard model must be fine tuned in such a way that almost completely cancels the quantum corrections. This level of fine-tuning is deemed unnatural by many theorists.

- *Number of parameters* – the standard model depends on 19 numerical parameters. Their values are known from experiment, but the origin of the values is unknown. Some theorists have tried to find relations between different parameters, for example, between the masses of particles in different generations.

- *Quantum triviality* – suggests that it may not be possible to create a consistent quantum field theory involving elementary scalar Higgs particles.

- *Strong CP problem* – theoretically it can be argued that the standard model should contain a term that breaks CP symmetry—relating matter to antimatter—in the strong interaction sector. Experimentally, however, no such violation has been found, implying that the coefficient of this term is very close to zero. This fine tuning is also considered unnatural.

14.2 Grand unified theories

Main article: Grand Unified Theory

The standard model has three gauge symmetries; the colour SU(3), the weak isospin SU(2), and the hypercharge U(1) symmetry, corresponding to the three fundamental forces. Due to renormalization the coupling constants of each of these symmetries vary with the energy at which they are measured. Around 10^{16} GeV these couplings become approximately equal. This has led to speculation that above this energy the three gauge symmetries of the standard model are unified in one single gauge symmetry with a simple group gauge group, and just one coupling constant. Below this energy the symmetry is spontaneously broken to the standard model symmetries.[12] Popular choices for the unifying group are the special unitary group in five dimensions SU(5) and the special orthogonal group in ten dimensions SO(10).[13]

Theories that unify the standard model symmetries in this way are called Grand Unified Theories (or GUTs), and the energy scale at which the unified symmetry is broken is called the GUT scale. Generically, grand unified theories predict the creation of magnetic monopoles in the early universe,[14] and instability of the proton.[15] Neither of these have been observed, and this absence of observation puts limits on the possible GUTs.

14.3 Supersymmetry

Main article: Supersymmetry

Supersymmetry extends the Standard Model by adding another class of symmetries to the Lagrangian. These symmetries exchange fermionic particles with bosonic ones. Such a symmetry predicts the existence of *supersymmetric particles*, abbreviated as *sparticles*, which include the sleptons, squarks, neutralinos and charginos. Each particle in the Standard Model would have a superpartner whose spin differs by 1/2 from the ordinary particle. Due to the breaking of supersymmetry, the sparticles are much heavier than their ordinary counterparts; they are so heavy that existing particle colliders may not be powerful enough to produce them.

14.4　Neutrinos

In the standard model, neutrinos have exactly zero mass. This is a consequence of the standard model containing only left-handed neutrinos. With no suitable right-handed partner, it is impossible to add a renormalizable mass term to the standard model.[16] Measurements however indicated that neutrinos spontaneously change flavour, which implies that neutrinos have a mass. These measurements only give the relative masses of the different flavours. The best constraint on the absolute mass of the neutrinos comes from precision measurements of tritium decay, providing an upper limit 2 eV, which makes them at least five orders of magnitude lighter than the other particles in the standard model.[17] This necessitates an extension of the standard model, which not only needs to explain how neutrinos get their mass, but also why the mass is so small.[18]

One approach to add masses to the neutrinos, the so-called seesaw mechanism, is to add right-handed neutrinos and have these couple to left-handed neutrinos with a Dirac mass term. The right-handed neutrinos have to be sterile, meaning that they do not participate in any of the standard model interactions. Because they have no charges, the right-handed neutrinos can act as their own anti-particles, and have a Majorana mass term. Like the other Dirac masses in the standard model, the neutrino Dirac mass is expected to be generated through the Higgs mechanism, and is therefore unpredictable. The standard model fermion masses differ by many orders of magnitude; the Dirac neutrino mass has at least the same uncertainty. On the other hand, the Majorana mass for the right-handed neutrinos does not arise from the Higgs mechanism, and is therefore expected to be tied to some energy scale of new physics beyond the standard model, for example the Planck scale.[19] Therefore, any process involving right-handed neutrinos will be suppressed at low energies. The correction due to these suppressed processes effectively gives the left-handed neutrinos a mass that is inversely proportional to the right-handed Majorana mass, a mechanism known as the see-saw.[20] The presence of heavy right-handed neutrinos thereby explains both the small mass of the left-handed neutrinos and the absence of the right-handed neutrinos in observations. However, due to the uncertainty in the Dirac neutrino masses, the right-handed neutrino masses can lie anywhere. For example, they could be as light as keV and be dark matter,[21] they can have a mass in the LHC energy range[22][23] and lead to observable lepton number violation,[24] or they can be near the GUT scale, linking the right-handed neutrinos to the possibility of a grand unified theory.[25][26]

The mass terms mix neutrinos of different generations. This mixing is parameterized by the PMNS matrix, which is the neutrino analogue of the CKM quark mixing matrix. Unlike the quark mixing, which is almost minimal, the mixing of the neutrinos appears to be almost maximal. This has led to various speculations of symmetries between the various generations that could explain the mixing patterns.[27] The mixing matrix could also contain several complex phases that break CP invariance, although there has been no experimental probe of these. These phases could potentially create a surplus of leptons over anti-leptons in the early universe, a process known as leptogenesis. This asymmetry could then at a later stage be converted in an excess of baryons over anti-baryons, and explain the matter-antimatter asymmetry in the universe.[13]

The light neutrinos are disfavored as an explanation for the observation of dark matter, due to considerations of large-scale structure formation in the early universe. Simulations of structure formation show that they are too hot—i.e. their kinetic energy is large compared to their mass—while formation of structures similar to the galaxies in our universe requires cold dark matter. The simulations show that neutrinos can at best explain a few percent of the missing dark matter. However, the heavy sterile right-handed neutrinos *are* a possible candidate for a dark matter WIMP.[28]

14.5　Preon Models

Several preon models have been proposed to address the unsolved problem concerning the fact that there are three generations of quarks and leptons. Preon models generally postulate some additional new particles which are further postulated to be able to combine to form the quarks and leptons of the standard model. One of the earliest preon models was the Rishon model.[29][30][31]

To date, no preon model is widely accepted or fully verified.

14.6　Theories of everything

14.6.1　Theory of everything

Main article: Theory of everything

Theoretical physics continues to strive toward a theory of

everything, a theory that fully explains and links together all known physical phenomena, and predicts the outcome of any experiment that could be carried out in principle. In practical terms the immediate goal in this regard is to develop a theory which would unify the Standard Model with General Relativity in a theory of quantum gravity. Additional features, such as overcoming conceptual flaws in either theory or accurate prediction of particle masses, would be desired. The challenges in putting together such a theory are not just conceptual - they include the experimental aspects of the very high energies needed to probe exotic realms.

Several notable attempts in this direction are supersymmetry, string theory, and loop quantum gravity.

14.6.2 String theory

Main article: String theory

Extensions, revisions, replacements, and reorganizations of the Standard Model exist in attempt to correct for these and other issues. String theory is one such reinvention, and many theoretical physicists think that such theories are the next theoretical step toward a true Theory of Everything. Theories of quantum gravity such as loop quantum gravity and others are thought by some to be promising candidates to the mathematical unification of quantum field theory and general relativity, requiring less drastic changes to existing theories.[32] However recent work places stringent limits on the putative effects of quantum gravity on the speed of light, and disfavours some current models of quantum gravity.[33]

Among the numerous variants of string theory, M-theory, whose mathematical existence was first proposed at a String Conference in 1995, is believed by many to be a proper "ToE" candidate, notably by physicists Brian Greene and Stephen Hawking. Though a full mathematical description is not yet known, solutions to the theory exist for specific cases.[34] Recent works have also proposed alternate string models, some of which lack the various harder-to-test features of M-theory (e.g. the existence of Calabi–Yau manifolds, many extra dimensions, etc.) including works by well-published physicists such as Lisa Randall.[35][36]

14.7 See also

- Antimatter tests of Lorentz violation
- Beyond black holes
- Fundamental physical constants in the standard model
- Higgsless model

- Holographic principle
- Little Higgs
- Lorentz-violating neutrino oscillations
- Minimal Supersymmetric Standard Model
- Peccei–Quinn theory
- Preon
- Standard-Model Extension
- Supergravity
- Seesaw mechanism
- Supersymmetry
- Superfluid vacuum theory
- String theory
- Technicolor (physics)
- Theory of everything
- Unsolved problems in physics
- Unparticle physics

14.8 References

[1] Womersley, J. (February 2005). "Beyond the Standard Model" (PDF). *Symmetry Magazine*. Retrieved 2010-11-23.

[2] Lykken, J. D. (2010). "Beyond the Standard Model". *CERN Yellow Report*. CERN. pp. 101–109. arXiv:1005.1676. CERN-2010-002.

[3] Sushkov, A. O.; Kim, W. J.; Dalvit, D. A. R.; Lamoreaux, S. K. (2011). "New Experimental Limits on Non-Newtonian Forces in the Micrometer Range". *Physical Review Letters* **107** (17): 171101. arXiv:1108.2547. Bibcode:2011PhRvL.107q1101S. doi:10.1103/PhysRevLett.107.171101. It is remarkable that two of the greatest successes of 20th century physics, general relativity and the standard model, appear to be fundamentally incompatible. But see also Donoghue, John F. (2012). "The effective field theory treatment of quantum gravity". *AIP Conference Proceedings* **1473**: 73. arXiv:1209.3511. doi:10.1063/1.4756964. One can find thousands of statements in the literature to the effect that "general relativity and quantum mechanics are incompatible". These are completely outdated and no longer relevant. Effective field theory shows that general relativity and quantum mechanics work together perfectly normally over a range of scales and curvatures, including those relevant for the world that we see around us. However, effective field theories are only valid over some range of scales. General

relativity certainly does have problematic issues at extreme scales. There are important problems which the effective field theory does not solve because they are beyond its range of validity. However, this means that the issue of quantum gravity is not what we thought it to be. Rather than a fundamental incompatibility of quantum mechanics and gravity, we are in the more familiar situation of needing a more complete theory beyond the range of their combined applicability. The usual marriage of general relativity and quantum mechanics is fine at ordinary energies, but we now seek to uncover the modifications that must be present in more extreme conditions. This is the modern view of the problem of quantum gravity, and it represents progress over the outdated view of the past."

[4] Krauss, L. (2009). *A Universe from Nothing*. AAI Conference.

[5] Randolf Pohl; Ronald Gilman; Gerald A. Miller; Krzysztof Pachucki (2013). "Muonic hydrogen and the proton radius puzzle". *Annu. Rev. Nucl. Part. Sci.* 63. arXiv:1301.0905. Bibcode:2013ARNPS..63..175P. doi:10.1146/annurev-nucl-102212-170627. The recent determination of the proton radius using the measurement of the Lamb shift in the muonic hydrogen atom startled the physics world. The obtained value of 0.84087(39) fm differs by about 4% or 7 standard deviations from the CODATA value of 0.8775(51) fm. The latter is composed from the electronic hydrogenate atom value of 0.8758(77) fm and from a similar value with larger uncertainties determined by electron scattering.

[6] Thomas Blum, Achim Denig, Ivan Logashenko, Eduardo de Rafael, B. Lee Roberts, Thomas Teubner, Graziano Venanzoni. "The Muon (g-2) Theory Value: Present and Future". arXiv:1311.2198.

[7] Lees, J. P.; et al. (BaBar Collaboration) (1970). "Evidence for an excess of $B \rightarrow D^{(*)}\tau^-\bar{\nu}$ decays". *Physical Review Letters* 109 (10). arXiv:1205.5442. Bibcode:2012PhRvL.109j1802L. doi:10.1103/PhysRevLett.109.101802.

[8] . arXiv:1506.08614. Bibcode:2015PhRvL.115k1803A. doi:10.1103/PhysRevLett.115.111803. Missing or empty |title= (help)

[9] . arXiv:1502.05314. Bibcode:2015PrPNP..82...59C. doi:10.1016/j.ppnp.2015.01.002. Missing or empty |title= (help)

[10] O'Luanaigh, C. (14 March 2013). "New results indicate that new particle is a Higgs boson". CERN.

[11] Marco Frasca, "What is a Glueball?" (March 31, 2009) http://marcofrasca.wordpress.com/2009/03/31/what-is-a-glueball-2/

[12] Peskin, M. E.; Schroeder, D. V. (1995). *An introduction to quantum field theory*. Addison-Wesley. pp. 786–791. ISBN 978-0-201-50397-5.

[13] Buchmüller, W. (2002). "Neutrinos, Grand Unification and Leptogenesis". arXiv:hep-ph/0204288 [hep-ph].

[14] Milstead, D.; Weinberg, E.J. (2009). "Magnetic Monopoles" (PDF). Particle Data Group. Retrieved 2010-12-20.

[15] P., Nath; P. F., Perez (2006). "Proton stability in grand unified theories, in strings, and in branes". *Physics Reports* 441 (5–6): 191–317. arXiv:hep-ph/0601023. Bibcode:2007PhR...441..191N. doi:10.1016/j.physrep.2007.02.010.

[16] Peskin, M. E.; Schroeder, D. V. (1995). *An introduction to quantum field theory*. Addison-Wesley. pp. 713–715. ISBN 978-0-201-50397-5.

[17] Nakamura, K.; et al. (Particle Data Group) (2010). "Neutrino Properties". Particle Data Group. Retrieved 2010-12-20.

[18] Mohapatra, R. N.; Pal, P. B. (2007). *Massive neutrinos in physics and astrophysics*. Lecture Notes in Physics 72 (3rd ed.). World Scientific. ISBN 978-981-238-071-5.

[19] Senjanovic, G. (2011). "Probing the Origin of Neutrino Mass: from GUT to LHC". arXiv:1107.5322 [hep-ph].

[20] Grossman, Y. (2003). "TASI 2002 lectures on neutrinos". arXiv:hep-ph/0305245v1 [hep-ph].

[21] Dodelson, S.; Widrow, L. M. (1993). "Sterile neutrinos as dark matter". *Physical Review Letters* 72: 17. arXiv:hep-ph/9303287. Bibcode:1994PhRvL..72...17D. doi:10.1103/PhysRevLett.72.17.

[22] Minkowski, P. (1977). "$\mu \rightarrow e\,\gamma$ at a Rate of One Out of 10^9 Muon Decays?". *Physics Letters B* 67 (4): 421. Bibcode:1977PhLB...67..421M. doi:10.1016/0370-2693(77)90435-X.

[23] Mohapatra, R. N.; Senjanovic, G. (1980). "Neutrino mass and spontaneous parity nonconservation". *Physical Review Letters* 44 (14): 912. Bibcode:1980PhRvL..44..912M. doi:10.1103/PhysRevLett.44.912.

[24] Keung, W.-Y.; Senjanovic, G. (1983). "Majorana Neutrinos And The Production Of The Right-handed Charged Gauge Boson". *Physical Review Letters* 50 (19): 1427. Bibcode:1983PhRvL..50.1427K. doi:10.1103/PhysRevLett.50.1427.

[25] Gell-Mann, M.; Ramond, P.; Slansky, R. (1979). P. van Nieuwenhuizen; D. Freedman, eds. *Supergravity*. North Holland.

[26] Glashow, S. L. (1979). M. Levy, ed. *Proceedings of the 1979 Cargèse Summer Institute on Quarks and Leptons*. Plenum Press.

[27] Altarelli, G. (2007). "Lectures on Models of Neutrino Masses and Mixings". arXiv:0711.0161 [hep-ph].

[28] Murayama, H. (2007). "Physics Beyond the Standard Model and Dark Matter". arXiv:0704.2276 [hep-ph].

[29] Harari, H. (1979). "A Schematic Model of Quarks and Leptons". *Physics Letters B* **86** (1): 83–86. Bibcode:1979PhLB...86...83H. doi:10.1016/0370-2693(79)90626-9.

[30] Shupe, M. A. (1979). "A Composite Model of Leptons and Quarks". *Physics Letters B* **86** (1): 87–92. Bibcode:1979PhLB...86...87S. doi:10.1016/0370-2693(79)90627-0.

[31] Zenczykowski, P. (2008). "The Harari-Shupe preon model and nonrelativistic quantum phase space". *Physics Letters B* **660** (5): 567–572. arXiv:0803.0223. Bibcode:2008PhLB..660..567Z. doi:10.1016/j.physletb.2008.01.045.

[32] Smolin, L. (2001). *Three Roads to Quantum Gravity*. Basic Books. ISBN 0-465-07835-4.

[33] Abdo, A. A.; et al. (Fermi GBM/LAT Collaborations) (2009). "A limit on the variation of the speed of light arising from quantum gravity effects". *Nature* **462** (7271): 331–4. arXiv:0908.1832. Bibcode:2009Natur.462..331A. doi:10.1038/nature08574. PMID 19865083.

[34] Maldacena, J.; Strominger, A.; Witten, E. (1997). "Black hole entropy in M-Theory". *Journal of High Energy Physics* **1997** (12): 2. arXiv:hep-th/9711053. Bibcode:1997JHEP...12..002M. doi:10.1088/1126-6708/1997/12/002.

[35] Randall, L.; Sundrum, R. (1999). "Large Mass Hierarchy from a Small Extra Dimension". *Physical Review Letters* **83** (17): 3370. arXiv:hep-ph/9905221. Bibcode:1999PhRvL..83.3370R. doi:10.1103/PhysRevLett.83.3370.

[36] Randall, L.; Sundrum, R. (1999). "An Alternative to Compactification". *Physical Review Letters* **83** (23): 4690. arXiv:hep-th/9906064. Bibcode:1999PhRvL..83.4690R. doi:10.1103/PhysRevLett.83.4690.

- Open Questions
- Working group - schedule
- Les Houches Conference, Summer 2005

14.9   Further reading

- Lisa Randall (2005). *Warped Passages: Unraveling the Mysteries of the Universe's Hidden Dimensions*. HarperCollins. ISBN 0-06-053108-8.

14.10   External resources

- Standard Model Theory @ SLAC
- Scientific American Apr 2006
- LHC. Nature July 2007

Chapter 15

Modified Newtonian dynamics

"MOND" redirects here. For other uses, see Mond.

In physics, **modified Newtonian dynamics (MOND)** is a theory that proposes a modification of Newton's laws to account for observed properties of galaxies. Created in 1983 by Israeli physicist Mordehai Milgrom,[1] the theory's original motivation was to explain the fact that the velocities of stars in galaxies were observed to be larger than expected based on Newtonian mechanics. Milgrom noted that this discrepancy could be resolved if the gravitational force experienced by a star in the outer regions of a galaxy was proportional to the square of its centripetal acceleration (as opposed to the centripetal acceleration itself, as in Newton's Second Law), or alternatively if gravitational force came to vary inversely with radius (as opposed to the inverse square of the radius, as in Newton's Law of Gravity). In MOND, violation of Newton's Laws occurs at extremely small accelerations, characteristic of galaxies yet far below anything typically encountered in the Solar System or on Earth.

MOND is an example of a class of theories known as modified gravity, and is an alternative to the hypothesis that the dynamics of galaxies are determined by massive, invisible dark matter halos. Since Milgrom's original proposal, MOND has successfully predicted a variety of galactic phenomena that are difficult to understand from a dark matter perspective.[2] However, MOND and its generalisations do not adequately account for observed properties of galaxy clusters, and no satisfactory cosmological model has been constructed from the theory.

15.1 Overview

Several independent observations point to the fact that the visible mass in galaxies and galaxy clusters is insufficient to account for their dynamics, when analysed using Newton's laws. This discrepancy – known as the "missing mass problem" – was first identified for clusters by Swiss astronomer Fritz Zwicky in 1933 (who studied the Coma clus-

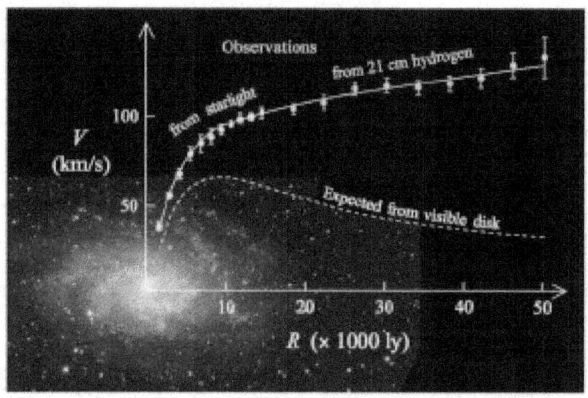

Comparison of the observed and expected rotation curves of the typical spiral galaxy M33[3]

ter),[4][5] and subsequently extended to include spiral galaxies by the 1939 work of Horace Babcock on Andromeda.[6] These early studies were augmented and brought to the attention of the astronomical community in the 1960s and 1970s by the work of Vera Rubin at the Carnegie Institute in Washington, who mapped in detail the rotation velocities of stars in a large sample of spirals. While Newton's Laws predict that stellar rotation velocities should decrease with distance from the galactic centre, Rubin and collaborators found instead that they remain almost constant[7] – the rotation curves are said to be "flat". This observation necessitates at least one of the following: 1) There exists in galaxies large quantities of unseen matter which boosts the stars' velocities beyond what would be expected on the basis of the visible mass alone, or 2) Newton's Laws do not apply to galaxies. The former leads to the dark matter hypothesis; the latter leads to MOND.

The basic premise of MOND is that while Newton's laws have been extensively tested in high-acceleration environments (in the Solar System and on Earth), they have not been verified for objects with extremely low acceleration, such as stars in the outer parts of galaxies. This led Milgrom to postulate a new effective gravitational force law (sometimes referred to as "Milgrom's law") that relates the

78

MOND was proposed by Mordehai Milgrom in 1983

true acceleration of an object to the acceleration that would be predicted for it on the basis of Newtonian mechanics.[1] This law, the keystone of MOND, is chosen to reduce to the Newtonian result at high acceleration but lead to different ("deep-MOND") behaviour at low acceleration:

Here **FN** is the Newtonian force, m is the object's (gravitational) mass, a is its acceleration, $\mu(x)$ is an as-yet unspecified function (known as the "interpolating function"), and a_0 is a new fundamental constant which marks the transition between the Newtonian and deep-MOND regimes. Agreement with Newtonian mechanics requires $\mu(x) \rightarrow 1$ for $x \gg 1$, and consistency with astronomical observations requires $\mu(x) \rightarrow x$ for $x \ll 1$. Beyond these limits, the interpolating function is not specified by the theory, although it is possible to weakly constrain it empirically.[8][9] Two common choices are:

$$\mu\left(\frac{a}{a_0}\right) = \left(1 + \frac{a_0}{a}\right)^{-1}$$

and

$$\mu\left(\frac{a}{a_0}\right) = \left(1 + \left(\frac{a_0}{a}\right)^2\right)^{-1/2}$$

Thus, in the deep-MOND regime ($a \ll a_0$):

$$F_N = ma^2/a_0$$

Applying this to an object of mass m in circular orbit around a point mass M (a crude approximation for a star in the outer regions of a galaxy), we find:

that is, the star's rotation velocity is independent of its distance r from the centre of the galaxy – the rotation curve is flat, as required. By fitting his law to rotation curve data, Milgrom found $a_0 \approx 1.2 \times 10^{-10}$ m s^{-2} to be optimal. This simple law is sufficient to make predictions for a broad range of galactic phenomena.

Milgrom's law can be interpreted in two different ways. One possibility is to treat it as a modification to the classical law of inertia (Newton's second law), so that the force on an object is not proportional to the particle's acceleration a but rather to $\mu(a/a_0)a$. In this case, the modified dynamics would apply not only to gravitational phenomena, but also those generated by other forces, for example electromagnetism.[10] Alternatively, Milgrom's law can be viewed as leaving Newton's Second Law intact and instead modifying the inverse-square law of gravity, so that the true gravitational force on an object of mass m due to another of mass M is roughly of the form $GMm/(\mu(a/a_0)r^2)$. In this interpretation, Milgrom's modification would apply exclusively to gravitational phenomena.

By itself, Milgrom's law is not a complete and self-contained physical theory, but rather an ad-hoc empirically motivated variant of one of the several equations that constitute classical mechanics. Its status within a coherent non-relativistic theory of MOND is akin to Kepler's Third Law within Newtonian mechanics; it provides a succinct description of observational facts, but must itself be explained by more fundamental concepts situated within the underlying theory. Several complete classical theories have been proposed (typically along "modified gravity" as opposed to "modified inertia" lines), which generally yield Milgrom's law exactly in situations of high symmetry and otherwise deviate from it slightly. A subset of these non-relativistic theories have been further embedded within relativistic theories, which are capable of making contact with non-classical phenomena (e.g., gravitational lensing) and cosmology.[11] Distinguishing both theoretically and observationally between these alternatives is a subject of current research.

The majority of astronomers, astrophysicists and cosmologists accept ΛCDM[12] (based on General Relativity, and hence Newtonian mechanics), and are committed to a dark matter solution of the missing-mass problem. MOND, by contrast, is actively studied by only a handful of researchers. The primary difference between supporters of ΛCDM and MOND is in the observations for which they demand a robust, quantitative explanation and those for which they are satisfied with a qualitative account, or are prepared to leave for future work. Proponents of MOND emphasize predictions made on galaxy scales (where MOND enjoys its most notable successes) and believe that a cosmological model consistent with galaxy dynamics has yet to be discovered; proponents of ΛCDM require high levels of cosmological accuracy (which concordance cosmology provides) and argue that a resolution of galaxy-scale issues will follow from a better understanding of the complicated baryonic astrophysics underlying galaxy formation.[2]

15.2 Observational evidence for MOND

Since MOND was specifically designed to produce flat rotation curves, these do not constitute evidence for the theory. Nevertheless, a broad range of astrophysical phenomena are neatly accounted for within the MOND framework.[11][13] Many of these came to light after the publication of Milgrom's original papers and are difficult to explain using the alternative dark matter hypothesis. The most prominent are the following:

1. In addition to demonstrating that rotation curves in MOND are flat, equation 2 provides a concrete relation between a galaxy's total baryonic mass (the sum of its mass in stars and gas) and its asymptotic rotation velocity. Observationally, this is known as the baryonic Tully–Fisher relation (BTFR),[14] and is found to conform quite closely to the MOND prediction.[15]

2. Milgrom's law fully specifies the rotation curve of a galaxy given only the distribution of its baryonic mass. In particular, MOND predicts a far stronger correlation between features in the baryonic mass distribution and features in the rotation curve than does the dark matter hypothesis (since dark matter dominates the galaxy's mass budget and is conventionally assumed not to closely track the distribution of baryons). Such a tight correlation is claimed to be observed in several spiral galaxies, a fact which has been referred to as "Renzo's rule".[11]

3. Since MOND modifies Newtonian dynamics in an acceleration-dependent way, it predicts a specific relationship between the acceleration of a star at any radius from the centre of a galaxy and the amount of unseen (dark matter) mass within that radius that would be inferred in a Newtonian analysis. This is known as the "mass discrepancy-acceleration relation", and has been measured observationally.[16][17] One aspect of the MOND prediction is that the mass of the inferred dark matter go to zero when the stellar centripetal acceleration becomes greater than a_0, where MOND reverts to Newtonian mechanics. In dark matter theory, it is a challenge to understand why this mass should correlate so closely with acceleration, and why there appears to be a critical acceleration above which dark matter is not required.[2]

4. Both MOND and dark matter halos stabilise disk galaxies, helping them retain their rotation-supported structure and preventing their transformation into elliptical galaxies. In MOND, this added stability is only available for regions of galaxies within the deep-MOND regime (i.e., with $a<a_0$), suggesting that spirals with $a>a_0$ in their central regions should be prone to instabilities and hence less likely to survive to the present day.[18] This may explain the "Freeman limit" to the observed central surface mass density of spiral galaxies, which is roughly a_0/G.[19] This scale must be put in by hand in dark matter-based galaxy formation models.[20]

5. Particularly massive galaxies are within the Newtonian regime ($a > a_0$) out to radii enclosing the vast majority of their baryonic mass. At these radii, MOND predicts that the rotation curve should fall as $1/r$, in accordance with Kepler's Laws. In contrast, from a dark matter perspective one would expect the halo to significantly boost the rotation velocity and cause it to asymptote to a constant value, as in less massive galaxies. Observations of high-mass ellipticals bear out the MOND prediction.[21][22]

6. In MOND, all gravitationally bound objects with $a < a_0$ – regardless of their origin – should exhibit a mass discrepancy when analysed using Newtonian mechanics, and should lie on the BTFR. Under the dark matter hypothesis, objects formed from baryonic material ejected during the merger or tidal interaction of two galaxies ("tidal dwarf galaxies") are expected to be devoid of dark matter and hence show no mass discrepancy. Three objects unambiguously identified as Tidal Dwarf Galaxies appear to have mass discrepancies in close agreement with the MOND prediction.[23][24][25]

7. Recent work has shown that many of the dwarf galaxies around the Milky Way and Andromeda are located preferentially in a single plane and have correlated motions. This suggests that they may have formed during a close encounter with another galaxy and hence be Tidal Dwarf Galaxies. If so, the presence of mass discrepancies in these systems constitutes further evidence for MOND. In addition, it has been claimed that a gravitational force stronger than Newton's (such as Milgrom's) is required for these galaxies to retain their orbits over time.[26]

15.3   Complete MOND theories

Milgrom's law requires incorporation into a complete theory if it is to satisfy conservation laws and provide a unique solution for the time evolution of any physical system. Each of the theories described here reduce to Milgrom's law in situations of high symmetry (and thus enjoy the successes described above), but produce different behaviour in detail.

15.3.1   Nonrelativistic

The first complete theory of MOND (dubbed AQUAL) was constructed in 1984 by Milgrom and Jacob Bekenstein.[27] AQUAL generates MONDian behaviour by modifying the gravitational term in the classical lagrangian from being quadratic in the gradient of the Newtonian potential to a more general function. (AQUAL is an acronym for AQUAdratic Lagrangian.) In formulae:

$$\mathcal{L}_{Newton} = -\frac{\|\nabla\phi\|^2}{8\pi G}$$

$$\mathcal{L}_{AQUAL} = -\frac{a_0^2 F(\|\nabla\phi\|^2/a_0^2)}{8\pi G}$$

where Φ is the standard Newtonian gravitational potential and F is a new dimensionless function. Applying the Euler–Lagrange equations in the standard way then leads to a nonlinear generalisation of the Newton–Poisson equation:

$$\nabla \cdot \left[\mu\left(\frac{\|\nabla\phi\|}{a_0}\right)\nabla\phi\right] = 4\pi G\rho$$

This can be solved given suitable boundary conditions and choice of F to yield Milgrom's law (up to a curl field correction which vanishes in situations of high symmetry).

An alternative way to modify the gravitational term in the lagrangian it is to introduce a distinction between the true (MONDian) acceleration field $\mathfrak{a}$ and the Newtonian acceleration field $\mathbf{aN}$. The Lagrangian may be constructed so that $\mathbf{aN}$ satisfies the usual Newton-Poisson equation, and is then used to find $\mathfrak{a}$ via an additional algebraic but non-linear step, which is chosen to satisfy Milgrom's law. This is called the "quasi-linear formulation of MOND", or QUMOND,[28] and is particularly useful for calculating the distribution of "phantom" dark matter that would be inferred from a Newtonian analysis of a given physical situation.[11]

Both AQUAL and QUMOND propose changes to the gravitational part of the classical matter action, and hence interpret Milgrom's law as a modification of Newtonian gravity as opposed to Newton's second law. The alternative is to turn the kinetic term of the action into a functional depending on the trajectory of the particle. Such "modified inertia" theories, however, are difficult to use because they are time-nonlocal, require energy and momentum to be non-trivially redefined to be conserved, and have predictions that depend on the entirety of a particle's orbit.[11]

15.3.2   Relativistic

In 2004, Jacob Bekenstein formulated TeVeS, the first complete relativistic theory with MONDian behaviour.[29] TeVeS is constructed from a local Lagrangian (and hence respects conservation laws), and employs a unit vector field, a dynamical and non-dynamical scalar field, a free function and a non-Einsteinian metric in order to yield AQUAL in the non-relativistic limit (low speeds and weak gravity). TeVeS has enjoyed some success in making contact with gravitational lensing and structure formation observations,[30] but faces problems when confronted with data on the anisotropy of the cosmic microwave background,[31] the lifetime of compact objects,[32] and the relationship between the lensing and matter overdensity potentials.[33]

Several alternative relativistic generalisations of MOND exist, including BIMOND and Generalised Einstein-Aether theories.[11]

15.4   The external field effect

In Newtonian mechanics, an object's acceleration can be found as the vector sum of the acceleration due to each of the individual forces acting on it. This means that a subsystem can be decoupled from the larger system in which it is embedded simply by referring the motion of its constituent particles to their centre of mass; in other words, the influence of the larger system is irrelevant for the internal dynamics of the subsystem. Since Milgrom's law is nonlinear in acceleration, MONDian subsystems cannot be decoupled from their environment in this way, and in certain situations this leads to behaviour with no Newtonian parallel. This is known as the "external field effect" (EFE).[1]

The external field effect is best described by classifying physical systems according to their relative values of a_{in} (the characteristic acceleration of one object within a subsystem due to the influence of another), a_{ex} (the acceleration of the entire subsystem due to forces exerted by objects outside of it), and a_0:

- $a_{in} > a_0$: Newtonian regime

- $a_{ex} < a_{in} < a_0$: Deep-MOND regime

- $a_{in} < a_0 < a_{ex}$: The external field is dominant and the behaviour of the system is Newtonian.

- $a_{in} < a_{ex} < a_0$: The external field is larger than the internal acceleration of the system, but both are smaller than the critical value. In this case, dynamics is Newtonian but the effective value of G is enhanced by a factor of a_0/a_{ex}.[34]

The external field effect implies a fundamental break with the strong equivalence principle (but not necessarily the weak equivalence principle). The effect was postulated by Milgrom in the first of his 1983 papers to explain why some open clusters were observed to have no mass discrepancy even though their internal accelerations were below a_0. It has since come to be recognised as a crucial element of the MOND paradigm.

The dependence in MOND of the internal dynamics of a system on its external environment (in principle, the rest of the universe) is strongly reminiscent of Mach's Principle, and may hint towards a more fundamental structure underlying Milgrom's law. In this regard, Milgrom has commented:[35]

> "It has been long suspected that local dynamics is strongly influenced by the universe at large, a-la Mach's Principle, but MOND seems to be the first to supply concrete evidence for such a connection. This may turn out to be the most fundamental implication of MOND, beyond its implied modification of Newtonian dynamics and General Relativity, and beyond the elimination of dark matter."

Indeed, the potential link between MONDian dynamics and the universe as a whole (that is, cosmology) is augmented by the observation that the value of a_0 (determined by fits to internal properties of galaxies) is within an order of magnitude of cH_0, where c is the speed of light and H_0 is the Hubble constant (a measure of the present-day expansion rate of the universe).[1] It is also close to the acceleration rate of the universe, and hence the cosmological constant. However, as yet no full theory has been constructed which manifests these connections in a natural way.

15.5   Responses and criticism

15.5.1   Attempts to explain MOND phenomenology using dark matter

While acknowledging that Milgrom's law provides a succinct and accurate description of a range of galactic phenomena, many physicists reject the idea that classical dynamics itself needs to be modified and attempt instead to explain the law's success by reference to the behaviour of dark matter. Some effort has gone towards establishing the presence of a characteristic acceleration scale as a natural consequence of the behaviour of cold dark matter halos,[36][37] although Milgrom has argued that such arguments explain only a small subset of MOND phenomena.[38] An alternative proposal is to modify the properties of dark matter (e.g., to make it interact strongly with itself or baryons) in order to induce the tight coupling between the baryonic and dark matter mass that the observations point to.[39] Finally, some researchers suggest that explaining the empirical success of Milgrom's law requires a more radical break with conventional assumptions about the nature of dark matter. One idea (dubbed "dipolar dark matter") is to make dark matter gravitationally polarisable by ordinary matter and have this polarisation enhance the gravitational attraction between baryons.[40]

15.5.2   Outstanding problems for MOND

The most serious problem facing Milgrom's law is that it cannot completely eliminate the need for dark matter in all astrophysical systems: galaxy clusters show a residual mass discrepancy even when analysed using MOND.[2] The fact that some form of unseen mass must exist in these systems detracts from the elegance of MOND as a solution to the missing mass problem, although the amount of extra mass required is 5 times less than in a Newtonian analysis, and there is no requirement that the missing mass be non-baryonic. It has been speculated that 2 eV neutrinos could account for the cluster observations in MOND while preserving the theory's successes at the galaxy scale.[41][42]

The 2006 observation of a pair of colliding galaxy clusters known as the "Bullet Cluster",[43] poses a significant challenge for all theories proposing a modified gravity solution to the missing mass problem, including MOND. Astronomers measured the distribution of stellar and gas mass in the clusters using visible and X-ray light, respectively, and in addition mapped the inferred dark matter density using gravitational lensing. In MOND, one would expect the missing mass (which is only apparent since it results from using Newtonian as opposed to MONDian dynamics) to be centred on the visible mass. In ΛCDM, on the other hand, one would expect the dark matter to be significantly offset from the visible mass because the halos of the two colliding clusters would pass through each other (assuming, as is conventional, that dark matter is collisionless), whilst the cluster gas would interact and end up at the centre. An offset is clearly seen in the observations. It has been suggested, however, that MOND-based models may be able to generate such an offset in strongly non-spherically-symmetric systems, such as the Bullet Cluster.[44]

Several other studies have noted observational difficulties with MOND. For example, it has been claimed that MOND offers a poor fit to the velocity dispersion profile of globular clusters and the temperature profile of galaxy clusters,[45][46] that different values of a_0 are required for agreement with different galaxies' rotation curves,[47] and that MOND is naturally unsuited to forming the basis of a theory of cosmology.[48]

Besides these observational issues, MOND and its generalisations are plagued by theoretical difficulties.[48][49] Several ad-hoc and inelegant additions to general relativity are required to create a theory with a non-Newtonian non-relativistic limit, the plethora of different versions of the theory offer diverging predictions in simple physical situations and thus make it difficult to test the framework conclusively, and some formulations (most prominently those based on modified inertia) have long suffered from poor compatibility with cherished physical principles such as conservation laws.

15.6 Proposals for testing MOND

Several observational and experimental tests have been proposed to help distinguish[50] between MOND and dark matter-based models:

- The detection of particles suitable for constituting cosmological dark matter would strongly suggest that ΛCDM is correct and no modification to Newton's laws is required.

- MOND predicts the existence of anomalous accelerations on the Earth at particular places and times of the year. These could be detected in a precision experiment.[51][52]

- It has been suggested that MOND could be tested in the Solar System using the LISA Pathfinder mission (launched in 2015). In particular, it may be possible to detect the anomalous tidal stresses predicted by MOND to exist at the Earth-Sun saddle-point of the Newtonian gravitational potential.[53] It may also be possible to measure MOND corrections to the perihelion precession of the planets in the Solar System,[54] or a purpose-built spacecraft.[55]

- One potential astrophysical test of MOND is to investigate whether isolated galaxies behave differently to otherwise-identical galaxies that are under the influence of a strong external field. Another is to search for non-Newtonian behaviour in the motion of binary star systems where the stars are sufficiently separated for their accelerations to be below a_0.[56]

- Testing modified gravity with black hole shadows - John Moffat calculates that the black hole shadow should appear larger by a factor of about ten in MOG as compared to general relativity[57]

15.7 See also

- MOND researchers:
  - Mordehai Milgrom
  - Jacob Bekenstein
  - Stacy McGaugh
  - Pavel Kroupa

- AQUAL

- TeVeS

- Modified Gravity

- Cold dark matter

- Dark matter

- Lambda-CDM

- Pioneer anomaly

- Tully-Fisher relation

- Universal rotation curve

15.8 References

[1] Milgrom, M. (1983). "A modification of the Newtonian dynamics as a possible alternative to the hidden mass hypothesis". *Astrophysical Journal* 270: 365–370. Bibcode:1983ApJ...270..365M. doi:10.1086/161130.. Milgrom, M. (1983). "A modification of the Newtonian dynamics - Implications for galaxies". *Astrophysical Journal* 270: 371–389. Bibcode:1983ApJ...270..371M. doi:10.1086/161131.. Milgrom, M. (1983). "A modification of the Newtonian dynamics - Implications for galaxy systems". *Astrophysical Journal* 270: 384. Bibcode:1983ApJ...270..384M. doi:10.1086/161132..

[2] McGaugh, S. (2014). "A Tale of Two Paradigms: the Mutual Incommensurability of LCDM and MOND". arXiv:1404.7525. Bibcode:2015CaJPh..93..250M. doi:10.1139/cjp-2014-0203.

[3] Data are from: E. Corbelli; P. Salucci (2000). "The extended rotation curve and the dark matter halo of M33". *Monthly Notices of the Royal Astronomical Society* **311** (2): 441–447. arXiv:astro-ph/9909252. Bibcode:2000MNRAS.311..441C. doi:10.1046/j.1365-8711.2000.03075.x.

[4] Zwicky, F. (1933). "Die Rotverschiebung von extragalaktischen Nebeln". *Helvetica Physica Acta* **6**: 110–127. Bibcode:1933AcHPh...6..110Z.

[5] Zwicky, F. (1937). "On the Masses of Nebulae and of Clusters of Nebulae". *The Astrophysical Journal* **86**: 217. Bibcode:1937ApJ....86..217Z. doi:10.1086/143864.

[6] Babcock, H, 1939, "The rotation of the Andromeda Nebula", Lick Observatory bulletin ; no. 498

[7] Rubin, Vera C.; Ford, W. Kent, Jr. (February 1970). "Rotation of the Andromeda Nebula from a Spectroscopic Survey of Emission Regions". *The Astrophysical Journal* **159**: 379–403. Bibcode:1970ApJ...159..379R. doi:10.1086/150317.

[8] G. Gentile, B. Famaey, W.J.G. de Blok (2011). "THINGS about MOND", Astron. Astrophys. **527**, A76. arXiv:1011.4148

[9] B. Famaey, J. Binney (2005), "Modified Newtonian Dynamics in the Milky Way", MNRAS, arXiv:astro-ph/0506723

[10] M. Milgrom, "MOND - Particularly as Modified Inertia", arXiv:1111.1611

[11] B. Famaey and S. McGaugh, "Modified Newtonian Dynamics (MOND): Observational Phenomenology and Relativistic Extensions", arXiv:1112.3960

[12] Pavel Kroupa – The vast polar structures around the Milky Way and Andromeda, YouTube, Nov. 18, 2013 Pavel Kroupa claims that the majority opinion is wrong and that empirical evidence rules out the ΛCDM model.

[13] M. Milgrom (2013), "MOND Laws of Galaxy Dynamics", arXiv:1212.2568

[14] S. S. McGaugh, J. M. Schombert, G. D. Bothun,2 and W. J. G. de Blok (2000), "The Baryonic Tully-Fisher Relation", arXiv:astro-ph/0003001

[15] S. McGaugh (2011), "The Baryonic Tully-Fisher Relation of Gas-Rich Galaxies as a Test of ΛCDM and MOND", ApJ, arXiv:1107.2934

[16] R. Sanders, "Mass discrepancies in galaxies: dark matter and alternatives", The Astronomy and Astrophysics Review 1990, Volume 2, Issue 1, pp 1-28

[17] S. McGaugh (2004), "The Mass Discrepancy-Acceleration Relation: Disk Mass and the Dark Matter Distribution", ApJ, arXiv:astro-ph/0403610

[18] M. A. Jimenez and X. Hernandez (2014), "Local galactic disk stability under MONDian gravity", arXiv:1406.0537

[19] McGaugh, S. (1998). "Testing the Hypothesis of Modified Dynamics with Low Surface Brightness Galaxies and Other Evidence". *Astrophys J* **499**: 66–81. arXiv:astro-ph/9801102. Bibcode:1998ApJ...499...66M. doi:10.1086/305629.

[20] McGaugh, S. (2005). "Balance of Dark and Luminous Mass in Rotating Galaxies". *Phys. Rev. Lett.* **95** (17): 171302. arXiv:astro-ph/0509305. Bibcode:2005PhRvL..95q1302M. doi:10.1103/physrevlett.95.171302.

[21] Romanowsky, A.J.; Douglas, N.G.; Arnaboldi, M.; Kuijken, K.; Merrifield, M.R.; Napolitano, N.R.; Capaccioli, M.; Freeman, K.C. (2003). "A Dearth of Dark Matter in Ordinary Elliptical Galaxies". *Science* **301**: 1696–1698. arXiv:astro-ph/0308518. Bibcode:2003Sci...301.1696R. doi:10.1126/science.1087441. PMID 12947033.</

[22] Milgrom, M.; Sanders, R.H. (2003). "Modified Newtonian Dynamics and the 'Dearth of Dark Matter in Ordinary Elliptical Galaxies'". *Astrophys J* **599**: 25–28. arXiv:astro-ph/0309617. Bibcode:2003ApJ...599L..25M. doi:10.1086/381138.

[23] F. Bournaud, P.-A. Duc, E. Brinks, M. Boquien, P. Amram, U. Lisenfeld, B. S. Koribalski, F. Walter, V. Charmandaris (2007), "Missing Mass in Collisional Debris from Galaxies", arXiv:0705.1356

[24] G. Gentile, B. Famaey, F. Combes, P. Kroupa, H. S. Zhao, O. Tiret (2007), "Tidal dwarf galaxies as a test of fundamental physics" arXiv:0706.1976]

[25] P. Kroupa (2012), "The dark matter crisis: falsification of the current standard model of cosmology", arXiv:1204.2546

[26] P. Kroupa (2014), "Lessons from the Local Group (and beyond) on dark matter", arXiv:1409.6302

[27] Jacob Bekenstein & M. Milgrom (1984). "Does the missing mass problem signal the breakdown of Newtonian gravity?". *Astrophys. J.* **286**: 7–14. Bibcode:1984ApJ...286....7B. doi:10.1086/162570.

[28] M. Milgrom (2010), "Quasi-linear formulation of MOND", arXiv:0911.5464

[29] Jacob D. Bekenstein (2004). "Relativistic gravitation theory for the MOND paradigm". *Phys. Rev.* **D70** (8): 83509. arXiv:astro-ph/0403694. Bibcode:2004PhRvD..70h3509B. doi:10.1103/PhysRevD.70.083509.

[30] T. Clifton, P. G. Ferreira, A. Padilla, C. Skordis (2011), "Modified Gravity and Cosmology", arXiv:1106.2476

[31] See Slosar, Melchiorri and Silk arXiv:astro-ph/0508048

[32] Seifert, M. D. (2007). "Stability of spherically symmetric solutions in modified theories of gravity", *Physical Review D* **76** (6): 064002, arXiv:gr-qc/0703060, Bibcode:2007PhRvD..76f4002S, doi:10.1103/PhysRevD.76.064002

[33] Zhang, P.; Liguori, M.; Bean, R.; Dodelson, S. (2007), "Probing Gravity at Cosmological Scales by Measurements which Test the Relationship between Gravitational Lensing and Matter Overdensity", *Physical Review Letters* 99 (14): 141302, arXiv:0704.1932, Bibcode:2007PhRvL..99n1302Z, doi:10.1103/PhysRevLett.99.141302

[34] S. McGaugh, The EFE in MOND

[35] M. Milgrom (2008), "The MOND paradigm", arXiv:0801.3133

[36] M. Kaplinghat and M. Turner (2012), "How Cold Dark Matter Theory Explains Milgrom's Law", arXiv:0107284

[37] M. H. Chan (2013), "Reconciliation of MOND and Dark Matter theory", arXiv:1310.6810

[38] M. Milgrom (2002), "Do Modified Newtonian Dynamics Follow from the Cold Dark Matter Paradigm?", arXiv:astro-ph/0110362

[39] J. Bullock (2014), Self-Interacting Dark Matter

[40] L. Blanchet, "Gravitational polarization and the phenomenology of MOND", Class. Quan- tum Grav., 24, 3529–3539, (2007), arXiv:astro-ph/0605637

[41] Angus, Garry W.; Shan, Huan Yuan; Zhao, Hong Sheng & Famaey, Benoit (2007). "On the Proof of Dark Matter, the Law of Gravity, and the Mass of Neutrinos". *The Astrophysical Journal Letters* 654 (1): L13–L16. arXiv:astro-ph/0609125. Bibcode:2007ApJ...654L..13A. doi:10.1086/510738.

[42] R.H. Sanders (2007). "Neutrinos as cluster dark matter". *Monthly Notices of the Royal Astronomical Society.* arXiv:astro-ph/0703590. Bibcode:2007MNRAS.380..331S. doi:10.1111/j.1365-2966.2007.12073.x.

[43] Clowe, Douglas; Bradač, Maruša; Gonzalez, Anthony H.; Markevitch, Maxim; Randall, Scott W.; Jones, Christine & Zaritsky, Dennis (2006). "A Direct Empirical Proof of the Existence of Dark Matter". *The Astrophysical Journal Letters* 648 (2): L109–L113. arXiv:astro-ph/0608407. Bibcode:2006ApJ...648L.109C. doi:10.1086/508162.

[44] G.W. Angus; B. Famaey & H. Zhao (September 2006). "Can MOND take a bullet? Analytical comparisons of three versions of MOND beyond spherical symmetry". *Mon.Not.Roy.Astron.Soc.* 371 (1): 138–146. arXiv:astro-ph/0606216v1. Bibcode:2006MNRAS.371..138A. doi:10.1111/j.1365-2966.2006.10668.x.

[45] Charles Seife (2004). *Alpha and Omega.* Penguin Books. pp. 100–101. ISBN 0-14-200446-4.

[46] Anthony Aguirre; Joop Schaye & Eliot Quataert (2001). "Problems for Modified Newtonian Dynamics in Clusters and the Lyα Forest?". *The Astrophysical Journal* 561 (2): 550–558. arXiv:astro-ph/0105184. Bibcode:2001ApJ...561..550A. doi:10.1086/323376.

[47] S. M. Kent, "Dark matter in spiral galaxies. II - Galaxies with H I rotation curves", 1987, AJ, 93, 816

[48] Scott, D.; White, M.; Cohn, J. D.; Pierpaoli, E. (2001). "Cosmological Difficulties with Modified Newtonian Dynamics (or: La Fin du MOND?)". arXiv:astro-ph/0104435. Bibcode:2001astro.ph..4435S.

[49] C. R. Contaldi, T. Wiseman, B. Withers (2008), "TeVeS gets caught on caustics", arXiv:0802.1215

[50] Wallin, John F.; Dixon, David S.; Page, Gary L. (23 May 2007). "Testing Gravity in the Outer Solar System: Results from Trans-Neptunian Objects". arXiv:0705.3408.

[51] A. Ignatiev (2014), "Testing MOND on Earth", arXiv:1408.3059

[52] V. A. De Lorenci, M. Faundez-Abans, J. P. Pereira (2010). "Testing the Newton second law in the regime of small accelerations" arXiv:1002.2766

[53] Christian Trenkel, Steve Kemble, Neil Bevis, Joao Magueijo (2010). "Testing MOND/TEVES with LISA Pathfinder" arXiv:1001.1303

[54] L. Blanchet (2011), "Testing MOND in the Solar System", arXiv:1105.5815

[55] V. Sahni, Y. Shtanov (2006), "APSIS - an Artificial Planetary System in Space to probe extra-dimensional gravity and MOND", arXiv:gr-qc/0606063

[56] X. Hernandez, M. A. Jimenez, and C. Allen (2012), "Wide binaries as a critical test of Classical Gravity", arXiv:1105.1873

[57] "Testing modified gravity with black hole shadows".

15.9 Further reading

Technical:

- Scholarpedia entry on the subject written by Mordehai Milgrom

- Modified Newtonian Dynamics: Observational Phenomenology and Relativistic Extensions, Famaey, McGaugh, 16 Dec 2011, updated 20 May 2012

- A Tale of Two Paradigms: the Mutual Incommensurability of LCDM and MOND, McGaugh, 29 Apr 2014, updated 17 May 2014

- MOND Theory, Milgrom, 30 Apr 2014, updated 31 Aug 2014

- Galaxies as simple dynamical systems: observational data disfavor dark matter and stochastic star formation, Kroupa, 18 Jun 2014

- Modified Gravity and Cosmology, Clifton, 13 Jun 2011, updated 12 Mar 2012

- Alternatives to Dark Matter and Dark Energy, Mannheim, 12 May 2005, updated 1 Aug 2005

Popular:

- MOND: time for a change of mind?, Milgrom, 26 Aug 2009

- "Dark matter" doubters not silenced yet, World Science, 2 Aug 2007

- Does Dark Matter Really Exist?, Milgrom, Scientific American, Aug 2002

- Dark matter critics focus on details, ignore big picture, Lee, 14 Nov 2012

15.10   External links

- The MOND pages, Stacy McGaugh

- Mordehai Milgrom's website

- "The Dark Matter Crisis" blog, Pavel Kroupa, Marcel Pawlowski

- Pavel Kroupa's website

- Hossenfelder, Sabine (1 Feb 2016). "The superfluid Universe". Retrieved 2 Feb 2016. Superfluid dark matter may provide a more natural way to arrive at the MOND equation.

Chapter 16

Scalar field

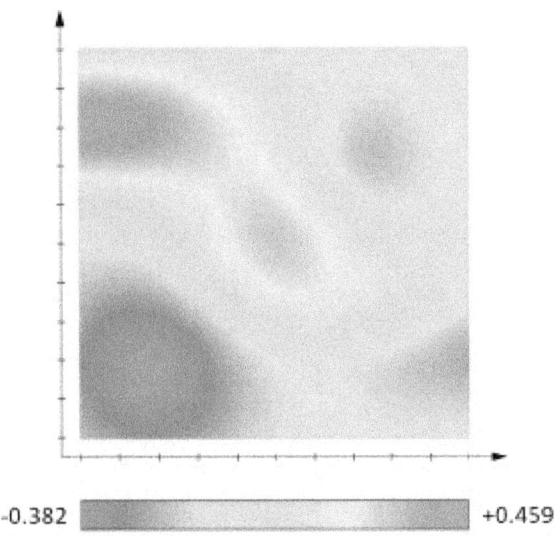

-0.382 +0.459

A scalar field such as temperature or pressure, where intensity of the field is represented by different hues of color.

In mathematics and physics, a **scalar field** associates a scalar value to every point in a space. The scalar may either be a mathematical number or a physical quantity. Scalar fields are required to be coordinate-independent, meaning that any two observers using the same units will agree on the value of the scalar field at the same absolute point in space (or spacetime) regardless of their respective points of origin. Examples used in physics include the temperature distribution throughout space, the pressure distribution in a fluid, and spin-zero quantum fields, such as the Higgs field. These fields are the subject of scalar field theory.

16.1 Definition

Mathematically, a scalar field on a region U is a real or complex-valued function or distribution on U.[1][2] The region U may be a set in some Euclidean space, Minkowski space, or more generally a subset of a manifold, and it is typical in mathematics to impose further conditions on the field, such that it be continuous or often continuously differentiable to some order. A scalar field is a tensor field of order zero,[3] and the term "scalar field" may be used to distinguish a function of this kind with a more general tensor field, density, or differential form.

The scalar field of $\sin(2\pi(xy + \sigma))$ oscillating as σ increases. Red represents positive values, purple represents negative values, and sky blue represents values close to zero.

Physically, a scalar field is additionally distinguished by having units of measurement associated with it. In this context, a scalar field should also be independent of the coordinate system used to describe the physical system—that is, any two observers using the same units must agree on the numerical value of a scalar field at any given point of physical space. Scalar fields are contrasted with other physical quantities such as vector fields, which associate a vector to every point of a region, as well as tensor fields and spinor fields. More subtly, scalar fields are often contrasted with pseudoscalar fields.

16.2   Uses in physics

In physics, scalar fields often describe the potential energy associated with a particular force. The force is a vector field, which can be obtained as the gradient of the potential energy scalar field. Examples include:

- Potential fields, such as the Newtonian gravitational potential, or the electric potential in electrostatics, are scalar fields which describe the more familiar forces.

- A temperature, humidity or pressure field, such as those used in meteorology.

16.2.1   Examples in quantum theory and relativity

- In quantum field theory, a scalar field is associated with spin-0 particles. The scalar field may be real or complex valued. Complex scalar fields represent charged particles. These include the charged Higgs field of the Standard Model, as well as the charged pions mediating the strong nuclear interaction.[4]

- In the Standard Model of elementary particles, a scalar Higgs field is used to give the leptons and massive vector bosons their mass, via a combination of the Yukawa interaction and the spontaneous symmetry breaking. This mechanism is known as the Higgs mechanism.[5] A candidate for the Higgs boson was first detected at CERN in 2012.

- In scalar theories of gravitation scalar fields are used to describe the gravitational field.

- scalar-tensor theories represent the gravitational interaction through both a tensor and a scalar. Such attempts are for example the Jordan theory [6] as a generalization of the Kaluza–Klein theory and the Brans–Dicke theory.[7]

  - Scalar fields like the Higgs field can be found within scalar-tensor theories, using as scalar field the Higgs field of the Standard Model.[8][9] This field interacts gravitationally and Yukawa-like (short-ranged) with the particles that get mass through it.[10]

- Scalar fields are found within superstring theories as dilaton fields, breaking the conformal symmetry of the string, though balancing the quantum anomalies of this tensor.[11]

- Scalar fields are supposed to cause the accelerated expansion of the universe (inflation [12]), helping to solve the horizon problem and giving a hypothetical reason for the non-vanishing cosmological constant of cosmology. Massless (i.e. long-ranged) scalar fields in this context are known as inflatons. Massive (i.e. short-ranged) scalar fields are proposed, too, using for example Higgs-like fields.[13]

16.3   Other kinds of fields

- Vector fields, which associate a vector to every point in space. Some examples of vector fields include the electromagnetic field and the Newtonian gravitational field.

- Tensor fields, which associate a tensor to every point in space. For example, in general relativity gravitation is associated with the tensor field called Einstein tensor. In Kaluza–Klein theory, spacetime is extended to five dimensions and its Riemann curvature tensor can be separated out into ordinary four-dimensional gravitation plus an extra set, which is equivalent to Maxwell's equations for the electromagnetic field, plus an extra scalar field known as the "dilaton". The dilaton scalar is also found among the massless bosonic fields in string theory.

16.4   See also

- Scalar field theory

- Vector-valued function

16.5   References

[1] Apostol, Tom (1969), *Calculus, Volume II* (2nd ed.), Wiley

[2] Hazewinkel, Michiel, ed. (2001), "Scalar", *Encyclopedia of Mathematics*, Springer, ISBN 978-1-55608-010-4

[3] Hazewinkel, Michiel, ed. (2001), "Scalar field", *Encyclopedia of Mathematics*, Springer, ISBN 978-1-55608-010-4

[4] Technically, pions are actually examples of pseudoscalar mesons, which fail to be invariant under spatial inversion, but are otherwise invariant under Lorentz transformations.

[5] P.W. Higgs (Oct 1964). "Broken Symmetries and the Masses of Gauge Bosons". *Phys. Rev. Lett* **13** (16): 508. Bibcode:1964PhRvL..13..508H. doi:10.1103/PhysRevLett.13.508.

[6] P. Jordan *Schwerkraft und Weltall*, Vieweg (Braunschweig) 1955.

[7] C. Brans and R. Dicke; *Phys. Rev. 124(3): 925*, 1961.

[8] A. Zee; *Phys. Rev. Lett. 42(7): 417*, 1979.

[9] H. Dehnen *et al.*; Int. J. of Theor. Phys. 31(1): 109, *1992*.

[10] H. Dehnen and H. Frommmert, *Int. J. of theor. Phys. 30(7): 987*, 1991.

[11] C.H. Brans; "The Roots of scalar-tensor theory", arXiv: gr-qc/0506063v1, June 2005.

[12] A. Guth; *Phys. Rev. D23: 347*, 1981.

[13] J.L. Cervantes-Cota and H. Dehnen; *Phys. Rev. D51, 395*, 1995.

Chapter 17

Equivalence principle

In the physics of general relativity, the **equivalence principle** is any of several related concepts dealing with the equivalence of gravitational and inertial mass, and to Albert Einstein's observation that the gravitational "force" as experienced locally while standing on a massive body (such as the Earth) is actually the same as the *pseudo-force* experienced by an observer in a non-inertial (accelerated) frame of reference.

17.1 Einstein's statement of the equality of inertial and gravitational mass

A little reflection will show that the law of the equality of the inertial and gravitational mass is equivalent to the assertion that the acceleration imparted to a body by a gravitational field is independent of the nature of the body. For Newton's equation of motion in a gravitational field, written out in full, it is:

(Inertial mass) · (Acceleration) = (Intensity of the gravitational field) · (Gravitational mass).

It is only when there is numerical equality between the inertial and gravitational mass that the acceleration is independent of the nature of the body.[1][2]

17.2 Development of gravitation theory

Something like the equivalence principle emerged in the early 17th century, when Galileo expressed experimentally that the acceleration of a test mass due to gravitation is independent of the amount of mass being accelerated.

During the Apollo 15 mission in 1971, astronaut David Scott showed that Galileo was right: acceleration is the same for all bodies subject to gravity on the Moon, even for a hammer and a feather.

Kepler, using Galileo's discoveries, showed knowledge of the equivalence principle by accurately describing what would occur if the moon were stopped in its orbit and dropped towards Earth. This can be deduced without knowing if or in what manner gravity decreases with distance, but requires assuming the equivalency between gravity and inertia.

If two stones were placed in any part of the world near each other, and beyond the sphere of influence of a third cognate body, these stones, like two magnetic needles, would come together in the intermediate point, each approaching the other by a space proportional to the comparative mass of the other. If the moon and earth were not retained in their orbits by their animal force or some other equivalent, the earth would mount to the moon by a fifty-fourth part of their distance, and the moon fall towards the earth through the other fifty-three parts, and they would there meet, assuming, however, that the substance of both is of the same density.

— Kepler, "Astronomia Nova", 1609[3]

The 1/54 ratio is Kepler's estimate of the Moon–Earth mass ratio, based on their diameters. The accuracy of his statement can be deduced by using Newton's inertia law F=ma and Galileo's gravitational observation that distance D = 1/2*a*t^2. Setting these accelerations equal for a mass is the equivalence principle. Noting the time to collision for each mass is the same gives Kepler's statement that $D_{moon}/DE_{arth}=ME_{arth}/M_{moon}$, without knowing the time to collision or how or if the acceleration force from gravity is a function of distance.

Newton's gravitational theory simplified and formalized Galileo's and Kepler's ideas by recognizing Kepler's "animal force or some other equivalent" beyond gravity and inertia were not needed, deducing from Kepler's planetary laws how gravity reduces with distance.

The equivalence principle was properly introduced by Albert Einstein in 1907, when he observed that the acceleration of bodies towards the center of the Earth at a rate of 1g ($g = 9.81$ m/s^2 being a standard reference of gravitational acceleration at the Earth's surface) is equivalent to the acceleration of an inertially moving body that would be observed on a rocket in free space being accelerated at a rate of 1g. Einstein stated it thus:

> we [...] assume the complete physical equivalence of a gravitational field and a corresponding acceleration of the reference system.
> — Einstein, 1907

That is, being on the surface of the Earth is equivalent to being inside a spaceship (far from any sources of gravity) that is being accelerated by its engines. The direction or vector of acceleration equivalence on the surface of the earth is "up" or directly opposite the center of the planet while the vector of acceleration in a spaceship is directly opposite from the mass ejected by its thrusters. From this principle, Einstein deduced that free-fall is actually inertial motion. Objects in free-fall do not experience being accelerated downward (e.g. toward the earth or other massive body) but rather weightlessness and no acceleration. In an inertial frame of reference bodies (and photons, or light) obey Newton's first law, moving at constant velocity in straight lines. Analogously, in a curved spacetime the world line of an inertial particle or pulse of light is *as straight as possible* (in space *and* time).[4] Such a world line is called a geodesic and from the point of view of the inertial frame is a straight line. This is why an accelerometer in free-fall doesn't register any acceleration; there isn't any.

As an example: an inertial body moving along a geodesic through space can be trapped into an orbit around a large gravitational mass without ever experiencing acceleration. This is possible because space is radically curved in close vicinity to a large gravitational mass. In such a situation the geodesic lines bend inward around the center of the mass and a free-floating (weightless) inertial body will simply follow those curved geodesics into an elliptical orbit. An accelerometer on-board would never record any acceleration.

By contrast, in Newtonian mechanics, gravity is assumed to be a force. This force draws objects having mass towards the center of any massive body. At the Earth's surface, the force of gravity is counteracted by the mechanical (physical) resistance of the Earth's surface. So in Newtonian physics, a person at rest on the surface of a (non-rotating) massive object is in an inertial frame of reference. These considerations suggest the following corollary to the equivalence principle, which Einstein formulated precisely in 1911:

> Whenever an observer detects the local presence of a force that acts on all objects in direct proportion to the inertial mass of each object, that observer is in an accelerated frame of reference.

Einstein also referred to two reference frames, K and K'. K is a uniform gravitational field, whereas K' has no gravitational field but is uniformly accelerated such that objects in the two frames experience identical forces:

> We arrive at a very satisfactory interpretation of this law of experience, if we assume that the systems K and K' are physically exactly equivalent, that is, if we assume that we may just as well regard the system K as being in a space free from gravitational fields, if we then regard K as uniformly accelerated. This assumption of exact physical equivalence makes it impossible for us to speak of the absolute acceleration of the system of reference, just as the usual theory of relativity forbids us to talk of the absolute velocity of a system; and it makes the equal falling of all bodies in a gravitational field seem a matter of course.
> — Einstein, 1911

This observation was the start of a process that culminated in general relativity. Einstein suggested that it should be elevated to the status of a general principle, which he called the "principle of equivalence" when constructing his theory of relativity:

As long as we restrict ourselves to purely mechanical processes in the realm where Newton's mechanics holds sway, we are certain of the equivalence of the systems K and K'. But this view of ours will not have any deeper significance unless the systems K and K' are equivalent with respect to all physical processes, that is, unless the laws of nature with respect to K are in entire agreement with those with respect to K'. By assuming this to be so, we arrive at a principle which, if it is really true, has great heuristic importance. For by theoretical consideration of processes which take place relatively to a system of reference with uniform acceleration, we obtain information as to the career of processes in a homogeneous gravitational field.

— Einstein, 1911

Einstein combined (postulated) the equivalence principle with special relativity to predict that clocks run at different rates in a gravitational potential, and light rays bend in a gravitational field, even before he developed the concept of curved spacetime.

So the original equivalence principle, as described by Einstein, concluded that free-fall and inertial motion were physically equivalent. This form of the equivalence principle can be stated as follows. An observer in a windowless room cannot distinguish between being on the surface of the Earth, and being in a spaceship in deep space accelerating at 1g. This is not strictly true, because massive bodies give rise to tidal effects (caused by variations in the strength and direction of the gravitational field) which are absent from an accelerating spaceship in deep space. The room, therefore, should be small enough that tidal effects can be neglected.

Although the equivalence principle guided the development of general relativity, it is not a founding principle of relativity but rather a simple consequence of the *geometrical* nature of the theory. In general relativity, objects in free-fall follow geodesics of spacetime, and what we perceive as the force of gravity is instead a result of our being unable to follow those geodesics of spacetime, because the mechanical resistance of matter prevents us from doing so.

Since Einstein developed general relativity, there was a need to develop a framework to test the theory against other possible theories of gravity compatible with special relativity. This was developed by Robert Dicke as part of his program to test general relativity. Two new principles were suggested, the so-called Einstein equivalence principle and the strong equivalence principle, each of which assumes the weak equivalence principle as a starting point. They only differ in whether or not they apply to gravitational experiments.

Another clarification needed is that the equivalence principle assumes a constant acceleration of 1g without considering the mechanics of generating 1g. If we do consider the mechanics of it, then we must assume the aforementioned windowless room has a fixed mass. Accelerating it at 1g means there is a constant force being applied, which = m*g where m is the mass of the windowless room along with its contents (including the observer). Now, if the observer jumps inside the room, an object lying freely on the floor will decrease in weight momentarily because the acceleration is going to decrease momentarily due to the observer pushing back against the floor in order to jump. The object will then gain weight while the observer is in the air and the resulting decreased mass of the windowless room allows greater acceleration; it will lose weight again when the observer lands and pushes once more against the floor; and it will finally return to its initial weight afterwards. To make all these affects equal those we would measure on a planet producing 1g, the windowless room must be assumed to have the same mass as that planet. Additionally, the windowless room must not cause its own gravity, otherwise the scenario changes even further. These are technicalities, clearly, but practical ones if we wish the experiment to demonstrate more or less precisely the equivalence of 1g gravity and 1g acceleration.

17.3   Modern usage

Three forms of the equivalence principle are in current use: weak (Galilean), Einsteinian, and strong.

17.3.1   The weak equivalence principle

The **weak equivalence principle**, also known as the **universality of free fall** or the **Galilean equivalence principle** can be stated in many ways. The strong EP includes (astronomic) bodies with gravitational binding energy[5] (e.g., 1.74 solar-mass pulsar PSR J1903+0327, 15.3% of whose separated mass is absent as gravitational binding energy[6]). The weak EP assumes falling bodies are bound by non-gravitational forces only. Either way:

> The trajectory of a point mass in a gravitational field depends only on its initial position and velocity, and is independent of its composition and *structure*.

> All test particles at the alike spacetime point, in a given gravitational field, will undergo the same acceleration, independent of their properties, including their rest mass.[7]

All local centers of mass free-fall (in vacuum), along identical (parallel-displaced, same speed) minimum action trajectories independent of all observable properties.

The vacuum world-line of a body immersed in a gravitational field is independent of all observable properties.

The local effects of motion in a curved space (gravitation) are indistinguishable from those of an accelerated observer in flat space, without exception.

Mass (measured with a balance) and weight (measured with a scale) are locally in identical ratio for all bodies (the opening page to Newton's *Philosophiæ Naturalis Principia Mathematica*, 1687).

Locality eliminates measurable tidal forces originating from a radial divergent gravitational field (e.g., the Earth) upon finite sized physical bodies. The "falling" equivalence principle embraces Galileo's, Newton's, and Einstein's conceptualization. The equivalence principle does not deny the existence of measurable effects caused by a *rotating* gravitating mass (frame dragging), or bear on the measurements of light deflection and gravitational time delay made by non-local observers.

Active, passive, and inertial masses

By definition of active and passive gravitational mass, the force on M_1 due to the gravitational field of M_0 is:

$$F_1 = \frac{M_0^{\text{act}} M_1^{\text{pass}}}{r^2}$$

Likewise the force on a second object of arbitrary mass$_2$ due to the gravitational field of mass$_0$ is:

$$F_2 = \frac{M_0^{\text{act}} M_2^{\text{pass}}}{r^2}$$

By definition of inertial mass:

$$F = m^{\text{inert}} a$$

If m_1 and m_2 are the same distance r from m_0 then, by the weak equivalence principle, they fall at the same rate (i.e. their accelerations are the same)

$$a_1 = \frac{F_1}{m_1^{\text{inert}}} = a_2 = \frac{F_2}{m_2^{\text{inert}}}$$

Hence:

$$\frac{M_0^{\text{act}} M_1^{\text{pass}}}{r^2 m_1^{\text{inert}}} = \frac{M_0^{\text{act}} M_2^{\text{pass}}}{r^2 m_2^{\text{inert}}}$$

Therefore:

$$\frac{M_1^{\text{pass}}}{m_1^{\text{inert}}} = \frac{M_2^{\text{pass}}}{m_2^{\text{inert}}}$$

In other words, passive gravitational mass must be proportional to inertial mass for all objects.

Furthermore, by Newton's third law of motion:

$$F_1 = \frac{M_0^{\text{act}} M_1^{\text{pass}}}{r^2}$$

must be equal and opposite to

$$F_0 = \frac{M_1^{\text{act}} M_0^{\text{pass}}}{r^2}$$

It follows that:

$$\frac{M_0^{\text{act}}}{M_0^{\text{pass}}} = \frac{M_1^{\text{act}}}{M_1^{\text{pass}}}$$

In other words, passive gravitational mass must be proportional to active gravitational mass for all objects.

The dimensionless Eötvös-parameter $\eta(A, B)$ is the difference of the ratios of gravitational and inertial masses divided by their average for the two sets of test masses "A" and "B."

$$\eta(A, B) = 2 \frac{\left(\frac{m_g}{m_i}\right)_A - \left(\frac{m_g}{m_i}\right)_B}{\left(\frac{m_g}{m_i}\right)_A + \left(\frac{m_g}{m_i}\right)_B}$$

Tests of the weak equivalence principle

Tests of the weak equivalence principle are those that verify the equivalence of gravitational mass and inertial mass. An obvious test is dropping different objects, ideally in a vacuum environment, e.g., inside Fallturm Bremen.

See:[12]

Experiments are still being performed at the University of Washington which have placed limits on the differential acceleration of objects towards the Earth, the Sun and towards dark matter in the galactic center. Future satellite experiments[31] – STEP (Satellite Test of the Equivalence Principle), Galileo Galilei, and MICROSCOPE (MICROSatellite à traînée Compensée pour l'Observation du Principe d'Équivalence) – will test the weak equivalence principle in space, to much higher accuracy.

With the first successful production of antimatter, in particular anti-hydrogen, a new approach to test the weak equivalence principle has been proposed. Experiments to compare the gravitational behavior of matter and antimatter are currently being developed.[32]

Proposals that may lead to a quantum theory of gravity such as string theory and loop quantum gravity predict violations of the weak equivalence principle because they contain many light scalar fields with long Compton wavelengths, which should generate fifth forces and variation of the fundamental constants. Heuristic arguments suggest that the magnitude of these equivalence principle violations could be in the 10^{-13} to 10^{-18} range.[33] Currently envisioned tests of the weak equivalence principle are approaching a degree of sensitivity such that *non-discovery* of a violation would be just as profound a result as discovery of a violation. Non-discovery of equivalence principle violation in this range would suggest that gravity is so fundamentally different from other forces as to require a major reevaluation of current attempts to unify gravity with the other forces of nature. A positive detection, on the other hand, would provide a major guidepost towards unification.[33]

17.3.2   The Einstein equivalence principle

What is now called the "Einstein equivalence principle" states that the weak equivalence principle holds, and that:[34]

> The outcome of any local non-gravitational experiment in a freely falling laboratory is independent of the velocity of the laboratory and its location in spacetime.

Here "local" has a very special meaning: not only must the experiment not look outside the laboratory, but it must also be small compared to variations in the gravitational field, tidal forces, so that the entire laboratory is freely falling. It also implies the absence of interactions with "external" fields *other than the gravitational field.*

The principle of relativity implies that the outcome of local experiments must be independent of the velocity of the apparatus, so the most important consequence of this principle is the Copernican idea that dimensionless physical values such as the fine-structure constant and electron-to-proton mass ratio must not depend on where in space or time we measure them. Many physicists believe that any Lorentz invariant theory that satisfies the weak equivalence principle also satisfies the Einstein equivalence principle.

Schiff's conjecture suggests that the weak equivalence principle actually implies the Einstein equivalence principle, but it has not been proven. Nonetheless, the two principles are tested with very different kinds of experiments. The Einstein equivalence principle has been criticized as imprecise, because there is no universally accepted way to distinguish gravitational from non-gravitational experiments (see for instance Hadley[35] and Durand[36]).

Tests of the Einstein equivalence principle

In addition to the tests of the weak equivalence principle, the Einstein equivalence principle can be tested by searching for variation of dimensionless constants and mass ratios. The present best limits on the variation of the fundamental constants have mainly been set by studying the naturally occurring Oklo natural nuclear fission reactor, where nuclear reactions similar to ones we observe today have been shown to have occurred underground approximately two billion years ago. These reactions are extremely sensitive to the values of the fundamental constants.

There have been a number of controversial attempts to constrain the variation of the strong interaction constant. There have been several suggestions that "constants" do vary on cosmological scales. The best known is the reported detection of variation (at the 10^{-5} level) of the fine-structure constant from measurements of distant quasars, see Webb et al.[37] Other researchers dispute these findings. Other tests of the Einstein equivalence principle are gravitational redshift experiments, such as the Pound–Rebka experiment which test the position independence of experiments.

17.3.3   The strong equivalence principle

The strong equivalence principle suggests the laws of gravitation are independent of velocity and location. In particular,

> The gravitational motion of a small test body de-

pends only on its initial position in spacetime and velocity, and not on its constitution.

and

The outcome of any local experiment (gravitational or not) in a freely falling laboratory is independent of the velocity of the laboratory and its location in spacetime.

The first part is a version of the weak equivalence principle that applies to objects that exert a gravitational force on themselves, such as stars, planets, black holes or Cavendish experiments. The second part is the Einstein equivalence principle (with the same definition of "local"), restated to allow gravitational experiments and self-gravitating bodies. The freely-falling object or laboratory, however, must still be small, so that tidal forces may be neglected (hence "local experiment").

This is the only form of the equivalence principle that applies to self-gravitating objects (such as stars), which have substantial internal gravitational interactions. It requires that the gravitational constant be the same everywhere in the universe and is incompatible with a fifth force. It is much more restrictive than the Einstein equivalence principle.

The strong equivalence principle suggests that gravity is entirely geometrical by nature (that is, the metric alone determines the effect of gravity) and does not have any extra fields associated with it. If an observer measures a patch of space to be flat, then the strong equivalence principle suggests that it is absolutely equivalent to any other patch of flat space elsewhere in the universe. Einstein's theory of general relativity (including the cosmological constant) is thought to be the only theory of gravity that satisfies the strong equivalence principle. A number of alternative theories, such as Brans–Dicke theory, satisfy only the Einstein equivalence principle.

Tests of the strong equivalence principle

The strong equivalence principle can be tested by searching for a variation of Newton's gravitational constant G over the life of the universe, or equivalently, variation in the masses of the fundamental particles. A number of independent constraints, from orbits in the solar system and studies of big bang nucleosynthesis have shown that G cannot have varied by more than 10%.

Thus, the strong equivalence principle can be tested by searching for fifth forces (deviations from the gravitational force-law predicted by general relativity). These experiments typically look for failures of the inverse-square law

(specifically Yukawa forces or failures of Birkhoff's theorem) behavior of gravity in the laboratory. The most accurate tests over short distances have been performed by the Eöt-Wash group. A future satellite experiment, SEE (Satellite Energy Exchange), will search for fifth forces in space and should be able to further constrain violations of the strong equivalence principle. Other limits, looking for much longer-range forces, have been placed by searching for the Nordtvedt effect, a "polarization" of solar system orbits that would be caused by gravitational self-energy accelerating at a different rate from normal matter. This effect has been sensitively tested by the Lunar Laser Ranging Experiment. Other tests include studying the deflection of radiation from distant radio sources by the sun, which can be accurately measured by very long baseline interferometry. Another sensitive test comes from measurements of the frequency shift of signals to and from the Cassini spacecraft. Together, these measurements have put tight limits on Brans–Dicke theory and other alternative theories of gravity.

In 2014, astronomers discovered a stellar triple system including a millisecond pulsar PSR J0337+1715 and two white dwarfs orbiting it. The system will provide them a chance to test the strong equivalence principle in a strong gravitational field.[38]

17.4 Challenges to the equivalence principle

One challenge to the equivalence principle is the Brans–Dicke theory. Self-creation cosmology is a modification of the Brans–Dicke theory. The Fredkin Finite Nature Hypothesis is an even more radical challenge to the equivalence principle and has even fewer supporters.

In August 2010, researchers from the University of New South Wales, Swinburne University of Technology, and Cambridge University published a paper titled "Evidence for spatial variation of the fine structure constant", whose tentative conclusion is that, "qualitatively, [the] results suggest a violation of the Einstein Equivalence Principle, and could infer a very large or infinite universe, within which our 'local' Hubble volume represents a tiny fraction."[39]

17.5 Explanations of the equivalence principle

Dutch physicist and string theorist Erik Verlinde has generated a self-contained, logical derivation of the equivalence principle based on the starting assumption of a holographic universe. Given this situation, gravity would not be a true fundamental force as is currently thought but instead an "emergent property" related to entropy. Verlinde's entropic gravity theory apparently leads naturally to the correct observed strength of dark energy; previous failures to explain its incredibly small magnitude have been called by such people as cosmologist Michael Turner (who is credited as having coined the term "dark energy") as "the greatest embarrassment in the history of theoretical physics".[40] However, it should be noted that these ideas are far from settled and still very controversial.

17.6 Experiments

- University of Washington[41]

- Lunar Laser Ranging[42]

- Galileo-Galilei satellite experiment[43]

- Satellite Test of the Equivalence Principle (STEP)[44]

- MICROSCOPE[45]

- Satellite Energy Exchange (SEE)[46]

- "...Physicists in Germany have used an atomic interferometer to perform the most accurate ever test of the equivalence principle at the level of atoms..."[47]

17.7 See also

- General Relativity

- General covariance

- Classical Mechanics

- Frame of reference

- Inertial frame of reference

- Mach's principle

- Equivalence principle (geometric)

- Brans–Dicke theory

- Gauge gravitation theory

- Self-creation cosmology

- Fredkin Finite Nature Hypothesis

- Tests of general relativity

- Unsolved problems in astronomy

- Unsolved problems in physics

17.8 Notes

[1] Einstein, Albert, *How I Constructed the Theory of Relativity*, translated by Masahiro Morikawa from the text recorded in Japanese by Jun Ishiwara, Association of Asia Pacific Physical Societies (AAPPS) Bulletin, Vol. 15, No. 2, pp. 17-19, April 2005. Einstein recalls events of 1907 in a talk in Japan on 14 December 1922.

[2] Einstein, Albert (2003). *The Meaning of Relativity*. Routledge. p. 59. ISBN 9781134449798.

[3] http://quotes.yourdictionary.com/orbits/quote/71225/

[4] Macdonald, Alan (September 15, 2012). "General Relativity in a Nutshell" (PDF). Luther College. p. 32. Retrieved February 8, 2013.

[5] Wagner, Todd A.; Schlamminger, Stephan; Gundlach, Jens H.; Adelberger, Eric G.; "Torsion-balance tests of the weak equivalence principle", *Classical Quantum Gravity* 29, 184002 (2012); http://arXiv.org/abs/1207.2442

[6] Champion, David J.; Ransom, Scott M.; Lazarus, Patrick; Camilo, Fernando; et al.; *Science* 320(5881), 1309 (2008), http://arXiv.org/abs/0805.2396

[7] Wesson, Paul S. (2006). *Five-dimensional Physics*. World Scientific. p. 82. ISBN 981-256-661-9.

[8] Devreese, Jozef T.; Vanden Berghe, Guido (2008). *'Magic Is No Magic': The Wonderful World of Simon Stevin*. p. 154. ISBN 9781845643911.

[9] Roll, Peter G.; Krotkov, Robert; Dicke, Robert H.; *The equivalence of inertial and passive gravitational mass*, Annals of Physics, Volume 26, Issue 3, 20 February 1964, pp. 442-517

[10] http://www.youtube.com/watch?v=MJyUDpm9Kvk

[11] Schlamminger, Stephan; Choi, Ki-Young; Wagner, Todd A.; Gundlach, Jens H.; Adelberger, Eric G. (2008). "Test of the Equivalence Principle Using a Rotating Torsion Balance". *Physical Review Letters* **100** (4). arXiv:0712.0607. Bibcode:2008PhRvL.100d1101S. doi:10.1103/PhysRevLett.100.041101.

[12] Ciufolini, Ignazio; Wheeler, John A.; "Gravitation and Inertia", Princeton, New Jersey: Princeton University Press, 1995, pp. 117-119

[13] Philoponus, John; "Corollaries on Place and Void", translated by David Furley, Ithaca, New York: Cornell University Press, 1987

[14] Stevin, Simon; *De Beghinselen der Weeghconst ["Principles of the Art of Weighing"]*, Leyden, 1586; Dijksterhuis, Eduard J.; "The Principal Works of Simon Stevin", Amsterdam, 1955

[15] Galilei, Galileo; "Discorsi e Dimostrazioni Matematiche Intorno a Due Nuove Scienze", Leida: Appresso gli Elsevirii, 1638; "Discourses and Mathematical Demonstrations Concerning Two New Sciences", Leiden: Elsevier Press, 1638

[16] Newton, Isaac; "Philosophiae Naturalis Principia Mathematica" [Mathematical Principles of Natural Philosophy and his System of the World], translated by Andrew Motte, revised by Florian Cajori, Berkeley, California: University of California Press, 1934; Newton, Isaac; "The Principia: Mathematical Principles of Natural Philosophy", translated by I. Bernard Cohen and Anne Whitman, with the assistance of Julia Budenz, Berkeley, California: University of California Press, 1999

[17] Bessel, Friedrich W.; "Versuche Uber die Kraft, mit welcher die Erde Körper von verschiedner Beschaffenhelt anzieht", *Annalen der Physik und Chemie*, Berlin: J. Ch. Poggendorff, 25 401–408 (1832)

[18] Southerns, Leonard; "A Determination of the Ratio of Mass to Weight for a Radioactive Substance", *Proceedings of the Royal Society of London*, 84 325–344 (1910), doi:10.1098/rspa.1910.0078

[19] Zeeman, Pieter; "Some experiments on gravitation: The ratio of mass to weight for crystals and radioactive substances", *Proceedings of the Koninklijke Nederlandse Akademie van Wetenschappen*, Amsterdam 20(4) 542–553 (1918)

[20] Eötvös, Loránd; *Mathematische and naturnissenschaftliche Berichte aus Ungarn* 8 65 (1889); *Annalen der Physik* (Leipzig) 68 11 (1922); *Physical Review D* 61(2) 022001 (1999)

[21] Potter, Harold H.; "Some Experiments on the Proportionality of Mass and Weight", *Proceedings of the Royal Society of London* 104 588–610 (1923), doi:10.1098/rspa.1923.0130

[22] Renner, János; "Kísérleti vizsgálatok a tömegvonzás és tehetetlenség arányosságáról", *Mathematikai és Természettudományi Értesítő* 53 569 (1935), Budapest

[23] Braginski, Vladimir Borisovich; Panov, Vladimir Ivanovich; Журнал Экспериментальной и Теоретической Физики (*Zhurnal Éksperimental'noĭ i Teoreticheskoĭ Fiziki, Journal of Experimental and Theoretical Physics*) 61 873 (1971)

[24] Shapiro, Irwin I.; Counselman, III; Charles, C.; King, Robert W. (1976). "Verification of the principle of equivalence for massive bodies". *Physical Review Letters* 36: 555–558. Bibcode:1976PhRvL..36..555S. doi:10.1103/physrevlett.36.555.

[25] Keiser, George M.; Faller, James E.; *Bulletin of the American Physical Society* 24 579 (1979)

[26] Niebauer, Timothy M.; McHugh, Martin P.; Faller, James E. (1987). "Galilean test for the fifth force". *Physical Review Letters* 59: 609–612. Bibcode:1987PhRvL..59..609N. doi:10.1103/physrevlett.59.609.

[27] Stubbs, Christopher W.; Adelberger, Eric G.; Heckel, Blayne R.; Rogers, Warren F.; Swanson, H. Erik; Watanabe, R.; Gundlach, Jens H.; Raab, Frederick J. (1989). "Limits on Composition-Dependent Interactions Using a Laboratory Source: Is There a "Fifth Force" Coupled to Isospin?". *Physical Review Letters* 62: 609. Bibcode:1989PhRvL..62..609S. doi:10.1103/physrevlett.62.609.

[28] Adelberger, Eric G.; Stubbs, Christopher W.; Heckel, Blayne R.; Su, Y.; Swanson, H. Erik; Smith, G. L.; Gundlach, Jens H.; Rogers, Warren F. (1990). "Testing the equivalence principle in the field of the Earth: Particle physics at masses below 1 μeV?". *Physical Review D* 42: 3267–3292. Bibcode:1990PhRvD..42.3267A. doi:10.1103/physrevd.42.3267.

[29] Baeßler, Stefan; et al.: *Classical Quantum Gravity* 18(13) 2393 (2001); Baeßler, Stefan; Heckel, Blayne R.; Adelberger, Eric G.; Gundlach, Jens H.; Schmidt, Ulrich; Swanson, H. Erik; "Improved Test of the Equivalence Principle for Gravitational Self-Energy", *Physical Review Letters* 83(18) 3585 (1999)

[30] Reasenberg, Robert D.; Patla, Biju R.; Phillips, James D.; Thapa, Rajesh; "Design and characteristics of a WEP test in a sounding-rocket payload", *Classical Quantum Gravity* 27, 095005 (2010); http://www.cfa.harvard.edu/PAG/6-%2520Presentations/Reasenberg_Q2C3_web.pdf

[31] Dittus, Hansjörg; Lämmerzahl, Claus. "Experimental Tests of the Equivalence Principle and Newton's Law in Space" (PDF). *Gravitation and Cosmology: 2nd Mexican Meeting on Mathematical and Experimental Physics*, AIP Conference Proceedings 758: 95. Bibcode:2005AIPC..758...95D. doi:10.1063/1.1900510.

[32] "Testing the Weak Equivalence Principle with an antimatter beam at CERN". *Journal of Physics: Conference Series* 631 (1): 012047. 2015. Bibcode:2015JPhCS.631a2047K. doi:10.1088/1742-6596/631/1/012047.

[33] Overduin, James; Everitt, Francis; Mester, John; Worden, Paul (2009). "The Science Case for STEP". *Advances in Space Research* 43 (10): 1532. arXiv:0902.2247. Bibcode:2009AdSpR..43.1532O. doi:10.1016/j.asr.2009.02.012.

[34] Haugen, Mark P.; Lämmerzahl, Claus (2001). *Principles of Equivalence: Their Role in Gravitation Physics and Experiments that Test Them.* Springer. arXiv:gr-qc/0103067. ISBN 978-3-540-41236-6.

[35] Hadley, Mark J. (1997). "The Logic of Quantum Mechanics Derived from Classical General Relativity". *Foundations of Physics Letters* 10: 43–60. arXiv:quant-ph/9706018. Bibcode:1997FoPhL..10...43H. doi:10.1007/BF02764119.

[36] Durand, Stéphane; "An amusing analogy: modelling quantum-type behaviours with wormhole-based time travel", *Journal of Optics B: Quantum and Semiclassical Optics*, vol. 4, no. 4, doi:10.1088/1464-4266/4/4/319

[37] Webb, John K.; Murphy, Michael T.; Flambaum, Victor V.; Dzuba, Vladimir A.; Barrow, John D.; Churchill, Chris W.; Prochaska, Jason X.; Wolfe, Arthur M. (2000). "Further Evidence for Cosmological Evolution of the Fine Structure Constant". *Physical Review Letters* **87** (9): 091301. arXiv:astro-ph/0012539. Bibcode:2001PhRvL..87i1301W. doi:10.1103/PhysRevLett.87.091301. PMID 11531558.

[38] Ransom, Scott M.; et al. (2014). "A millisecond pulsar in a stellar triple system". *Nature*. arXiv:1401.0535. Bibcode:2014Natur.505..520R. doi:10.1038/nature12917. Retrieved 8 January 2014.

[39] Webb, John K.; King, Julian A.; Murphy, Michael T.; Flambaum, Victor V.; Carswell, Robert F.; Bainbridge, Matthew B. (2010). "Evidence for spatial variation of the fine structure constant". arXiv:1008.3907 [astro-ph.CO].

[40] Wright, Karen (1 March 2001). "Very Dark Energy". Discover Magazine. Retrieved 26 February 2013.

[41] Eöt-Wash group

[42] http://funphysics.jpl.nasa.gov/technical/grp/lunar-laser.html

[43] http://eotvos.dm.unipi.it/nobili/

[44] http://einstein.stanford.edu/STEP/

[45] http://smsc.cnes.fr/MICROSCOPE/index.htm

[46] http://www.phys.utk.edu/see/

[47] 16 November 2004, physicsweb: Equivalence principle passes atomic test

17.9 References

- Dicke, Robert H.; "New Research on Old Gravitation", *Science* **129**, 3349 (1959). This paper is the first to make the distinction between the strong and weak equivalence principles.

- Dicke, Robert H.; "Mach's Principle and Equivalence", in *Evidence for gravitational theories: proceedings of course 20 of the International School of Physics "Enrico Fermi"*, ed. C. Møller (Academic Press, New York, 1962). This article outlines the approach to precisely testing general relativity advocated by Dicke and pursued from 1959 onwards.

- Einstein, Albert; "Über das Relativitätsprinzip und die aus demselben gezogene Folgerungen", *Jahrbuch der Radioaktivitaet und Elektronik* **4** (1907); translated "On the relativity principle and the conclusions drawn from it", in *The collected papers of Albert Einstein. Vol. 2 : The Swiss years: writings, 1900–1909* (Princeton University Press, Princeton, New Jersey, 1989), Anna Beck translator. This is Einstein's first statement of the equivalence principle.

- Einstein, Albert; "Über den Einfluß der Schwerkraft auf die Ausbreitung des Lichtes", *Annalen der Physik* **35** (1911); translated "On the Influence of Gravitation on the Propagation of Light" in *The collected papers of Albert Einstein. Vol. 3 : The Swiss years: writings, 1909–1911* (Princeton University Press, Princeton, New Jersey, 1994), Anna Beck translator, and in *The Principle of Relativity*, (Dover, 1924), pp 99–108, W. Perrett and G. B. Jeffery translators, ISBN 0-486-60081-5. The two Einstein papers are discussed online at The Genesis of General Relativity.

- Brans, Carl H.; "The roots of scalar-tensor theory: an approximate history", arXiv:gr-qc/0506063. Discusses the history of attempts to construct gravity theories with a scalar field and the relation to the equivalence principle and Mach's principle.

- Misner, Charles W.; Thorne, Kip S.; and Wheeler, John A.; *Gravitation*, New York: W. H. Freeman and Company, 1973, Chapter 16 discusses the equivalence principle.

- Ohanian, Hans; and Ruffini, Remo; *Gravitation and Spacetime 2nd edition*, New York: Norton, 1994, ISBN 0-393-96501-5 Chapter 1 discusses the equivalence principle, but incorrectly, according to modern usage, states that the strong equivalence principle is wrong.

- Uzan, Jean-Philippe; "The fundamental constants and their variation: Observational status and theoretical motivations", *Reviews of Modern Physics* **75**, 403 (2003). arXiv:hep-ph/0205340 This technical article reviews the best constraints on the variation of the fundamental constants.

- Will, Clifford M.; *Theory and experiment in gravitational physics*, Cambridge, UK: Cambridge University Press, 1993. This is the standard technical reference for tests of general relativity.

- Will, Clifford M.; *Was Einstein Right?: Putting General Relativity to the Test*, Basic Books (1993). This is a popular account of tests of general relativity.

- Will, Clifford M.; *The Confrontation between General Relativity and Experiment*, Living Reviews in Relativity (2006). An online, technical review, covering much of the material in *Theory and experiment in gravitational physics*. The Einstein and strong variants of the equivalence principles are discussed in sections 2.1 and 3.1, respectively.

- Friedman, Michael; *Foundations of Space-Time Theories*, Princeton, New Jersey: Princeton University Press, 1983. Chapter V discusses the equivalence principle.

17.10   External links

- Equivalence Principle at NASA, including tests

- Introducing The Einstein Principle of Equivalence from Syracuse University

- The Equivalence Principle at MathPages

- The Einstein Equivalence Principle at Living Reviews on General Relativity

Chapter 18

Alternatives to general relativity

Alternatives to general relativity are physical theories that attempt to describe the phenomena of gravitation in competition to Einstein's theory of general relativity.

There have been many different attempts at constructing an ideal theory of gravity. These attempts can be split into four broad categories:

- Straightforward alternatives to general relativity (GR), such as the Cartan, Brans–Dicke and Rosen bimetric theories.

- Those that attempt to construct a quantized gravity theory such as loop quantum gravity.

- Those that attempt to unify gravity and other forces such as Kaluza–Klein.

- Those that attempt to do several at once, such as M-theory.

This article deals only with straightforward alternatives to GR. For quantized gravity theories, see the article quantum gravity. For the unification of gravity and other forces, see the article classical unified field theories. For those theories that attempt to do several at once, see the article theory of everything.

18.1 Motivations

Motivations for developing new theories of gravity have changed over the years, with the first one to explain planetary orbits (Newton) and more complicated orbits (e.g. Lagrange). Then came unsuccessful attempts to combine gravity and either wave or corpuscular theories of gravity. The whole landscape of physics was changed with the discovery of Lorentz transformations, and this led to attempts to reconcile it with gravity. At the same time, experimental physicists started testing the foundations of gravity and relativity – Lorentz invariance, the gravitational deflection of light, the Eötvös experiment. These considerations led to and past the development of general relativity.

After that, motivations differ. Two major concerns were the development of quantum theory and the discovery of the strong and weak nuclear forces. Attempts to quantize and unify gravity are outside the scope of this article, and so far none has been completely successful.

After general relativity (GR), attempts were made either to improve on theories developed before GR, or to improve GR itself. Many different strategies were attempted, for example the addition of spin to GR, combining a GR-like metric with a space-time that is static with respect to the expansion of the universe, getting extra freedom by adding another parameter. At least one theory was motivated by the desire to develop an alternative to GR that is completely free from singularities.

Experimental tests improved along with the theories. Many of the different strategies that were developed soon after GR were abandoned, and there was a push to develop more general forms of the theories that survived, so that a theory would be ready the moment any test showed a disagreement with GR.

By the 1980s, the increasing accuracy of experimental tests had all led to confirmation of GR, no competitors were left except for those that included GR as a special case. Further, shortly after that, theorists switched to string theory which was starting to look promising, but has since lost popularity. In the mid-1980s a few experiments were suggesting that gravity was being modified by the addition of a fifth force (or, in one case, of a fifth, sixth and seventh force) acting on the scale of meters. Subsequent experiments eliminated these.

Motivations for the more recent alternative theories are almost all cosmological, associated with or replacing such constructs as "inflation", "dark matter" and "dark energy". Investigation of the Pioneer anomaly has caused renewed public interest in alternatives to General Relativity.

18.2 Notation in this article

Main article: Mathematics of general relativity

c is the speed of light, G is the gravitational constant. "Geometric variables" are not used.

Latin indexes go from 1 to 3, Greek indexes go from 0 to 3. The Einstein summation convention is used.

$\eta_{\mu\nu}$ is the Minkowski metric. $g_{\mu\nu}$ is a tensor, usually the metric tensor. These have signature $(-,+,+,+)$.

Partial differentiation is written $\partial_\mu\phi$ or $\phi_{,\mu}$. Covariant differentiation is written $\nabla_\mu\phi$ or $\phi_{;\mu}$.

18.3 Classification of theories

Theories of gravity can be classified, loosely, into several categories. Most of the theories described here have:

- an 'action' (see the principle of least action, a variational principle based on the concept of action)

- a Lagrangian density

- a metric

If a theory has a Lagrangian density for gravity, say L , then the gravitational part of the action S is the integral of that.

$$S = \int L \sqrt{-g}\, \mathrm{d}^4 x$$

In this equation it is usual, though not essential, to have $g = -1$ at spatial infinity when using Cartesian coordinates. For example, the Einstein–Hilbert action uses

$$L \propto R$$

where R is the scalar curvature, a measure of the curvature of space.

Almost every theory described in this article has an action. It is the only known way to guarantee that the necessary conservation laws of energy, momentum and angular momentum are incorporated automatically; although it is easy to construct an action where those conservation laws are violated. The original 1983 version of MOND did not have an action.

A few theories have an action but not a Lagrangian density. A good example is Whitehead (1922), the action there is termed non-local.

A theory of gravity is a "metric theory" if and only if it can be given a mathematical representation in which two conditions hold:

Condition 1: There exists a symmetric metric tensor $g_{\mu\nu}$ of signature $(-,+,+,+)$, which governs proper-length and proper-time measurements in the usual manner of special and general relativity:

$$d\tau^2 = -g_{\mu\nu}\, dx^\mu\, dx^\nu$$

where there is a summation over indices μ and ν .

Condition 2: Stressed matter and fields being acted upon by gravity respond in accordance with the equation:

$$0 = \nabla_\nu T^{\mu\nu} = T^{\mu\nu}{}_{,\nu} + \Gamma^\mu_{\sigma\nu} T^{\sigma\nu} + \Gamma^\nu_{\sigma\nu} T^{\mu\sigma}$$

where $T^{\mu\nu}$ is the stress–energy tensor for all matter and non-gravitational fields, and where ∇_ν is the covariant derivative with respect to the metric and $\Gamma^\alpha_{\sigma\nu}$ is the Christoffel symbol. The stress–energy tensor should also satisfy an energy condition.

Metric theories include (from simplest to most complex):

- Scalar field theories (includes Conformally flat theories & Stratified theories with conformally flat space slices)

  - Bergman
  - Coleman
  - Einstein (1912)
  - Einstein–Fokker theory
  - Lee–Lightman–Ni
  - Littlewood
  - Ni
  - Nordström's theory of gravitation (first metric theory of gravity to be developed)
  - Page–Tupper
  - Papapetrou
  - Rosen (1971)
  - Whitrow–Morduch
  - Yilmaz theory of gravitation (attempted to eliminate event horizons from the theory.)

- Quasilinear theories (includes Linear fixed gauge)

  - Bollini–Giambiagi–Tiomno
  - Deser–Laurent
  - Whitehead's theory of gravity (intended to use only retarded potentials)

- Tensor theories

  - Einstein's GR

  - Fourth order gravity (allows the Lagrangian to depend on second-order contractions of the Riemann curvature tensor)

  - f(R) gravity (allows the Lagrangian to depend on higher powers of the Ricci scalar)

  - Gauss–Bonnet gravity

  - Lovelock theory of gravity (allows the Lagrangian to depend on higher-order contractions of the Riemann curvature tensor)

  - Infinite derivative theorem gravity

- Scalar-tensor theories

  - Bekenstein

  - Bergmann-Wagoner

  - Brans–Dicke theory (the most well-known alternative to GR, intended to be better at applying Mach's principle)

  - Jordan

  - Nordtvedt

  - Thiry

  - Chameleon

  - Pressuron

- Vector-tensor theories

  - Hellings–Nordtvedt

  - Will–Nordtvedt

- Bimetric theories

  - Lightman–Lee

  - Rastall

  - Rosen (1975)

- Other metric theories

(see section Modern theories below)

Non-metric theories include

- Belinfante–Swihart

- Einstein–Cartan theory (intended to handle spin-orbital angular momentum interchange)

- Kustaanheimo (1967)

- Teleparallelism

- Gauge theory gravity

A word here about Mach's principle is appropriate because a few of these theories rely on Mach's principle (e.g. Whitehead (1922)), and many mention it in passing (e.g. Einstein–Grossmann (1913), Brans–Dicke (1961)). Mach's principle can be thought of a half-way-house between Newton and Einstein. It goes this way:[1]

- Newton: Absolute space and time.

- Mach: The reference frame comes from the distribution of matter in the universe.

- Einstein: There is no reference frame.

So far, all the experimental evidence points to Mach's principle being wrong, but it has not entirely been ruled out.

18.4   Early theories, 1686 to 1916

Main articles: History of gravitational theory and History of general relativity

Newton (1686)

Main articles: Newton's law of universal gravitation and Gauss's law for gravity

In Newton's (1686) theory (rewritten using more modern mathematics) the density of mass ρ generates a scalar field, the gravitational potential ϕ in joules per kilogram, by

$$\frac{\partial^2 \phi}{\partial x^j \partial x^j} = 4\pi G\rho .$$

Using the Nabla operator ∇ for the gradient and divergence (partial derivatives), this can be conveniently written as:

$$\nabla^2 \phi = 4\pi G\rho .$$

This scalar field governs the motion of a free-falling particle by:

$$\frac{d^2 x^j}{dt^2} = -\frac{\partial \phi}{\partial x^j} .$$

At distance, r, from an isolated mass, M, the scalar field is

$$\phi = -GM/r .$$

The theory of Newton, and Lagrange's improvement on the calculation (applying the variational principle), completely fails to take into account relativistic effects of course, and so can be rejected as a viable theory of gravity. Even so, Newton's theory is thought to be exactly correct in the limit of weak gravitational fields and low speeds and all other theories of gravity need to reproduce Newton's theory in the appropriate limits.

Mechanical explanations (1650–1900)

To explain Newton's theory, some mechanical explanations of gravitation (incl. Le Sage's theory) were created between 1650 and 1900, but they were overthrown because most of them lead to an unacceptable amount of drag, which is not observed. Other models are violating the energy conservation law and are incompatible with modern thermodynamics.

Electrostatic models (1870–1900)

At the end of the 19th century, many tried to combine Newton's force law with the established laws of electrodynamics, like those of Weber, Carl Friedrich Gauss, Bernhard Riemann and James Clerk Maxwell. Those models were used to explain the perihelion advance of Mercury. In 1890, Lévy succeeded in doing so by combining the laws of Weber and Riemann, whereby the speed of gravity is equal to the speed of light in his theory. And in another attempt, Paul Gerber (1898) even succeeded in deriving the correct formula for the Perihelion shift (which was identical to that formula later used by Einstein). However, because the basic laws of Weber and others were wrong (for example, Weber's law was superseded by Maxwell's theory), those hypothesis were rejected.[2] In 1900, Hendrik Lorentz tried to explain gravity on the basis of his Lorentz ether theory and the Maxwell equations. He assumed, like Ottaviano Fabrizio Mossotti and Johann Karl Friedrich Zöllner, that the attraction of opposite charged particles is stronger than the repulsion of equal charged particles. The resulting net force is exactly what is known as universal gravitation, in which the speed of gravity is that of light. But Lorentz calculated that the value for the perihelion advance of Mercury was much too low.[3]

Lorentz-invariant models (1905–1910)

Based on the principle of relativity, Henri Poincaré (1905, 1906), Hermann Minkowski (1908), and Arnold Sommerfeld (1910) tried to modify Newton's theory and to establish a Lorentz invariant gravitational law, in which the speed of gravity is that of light. However, as in Lorentz's model, the value for the perihelion advance of Mercury was much too low.[4]

Einstein (1908, 1912)

Einstein's two part publication in 1912 (and before in 1908) is really only important for historical reasons. By then he knew of the gravitational redshift and the deflection of light. He had realized that Lorentz transformations are not generally applicable, but retained them. The theory states that the speed of light is constant in free space but varies in the presence of matter. The theory was only expected to hold when the source of the gravitational field is stationary. It includes the principle of least action:

$$\delta \int d\tau = 0$$

$$d\tau^2 = -\eta_{\mu\nu} dx^\mu dx^\nu$$

where $\eta_{\mu\nu}$ is the Minkowski metric, and there is a summation from 1 to 4 over indices μ and ν .

Einstein and Grossmann (1913) includes Riemannian geometry and tensor calculus.

$$\delta \int d\tau = 0$$

$$d\tau^2 = -g_{\mu\nu} dx^\mu dx^\nu$$

The equations of electrodynamics exactly match those of GR. The equation

$$T^{\mu\nu} = \rho \frac{dx^\mu}{d\tau} \frac{dx^\nu}{d\tau}$$

is not in GR. It expresses the stress–energy tensor as a function of the matter density.

Abraham (1912)

While this was going on, Abraham was developing an alternative model of gravity in which the speed of light depends on the gravitational field strength and so is variable almost everywhere. Abraham's 1914 review of gravitation models is said to be excellent, but his own model was poor.

Nordström (1912)

The first approach of Nordström (1912) was to retain the Minkowski metric and a constant value of c but to let mass depend on the gravitational field strength ϕ . Allowing this field strength to satisfy

$$\Box \phi = \rho$$

where ρ is rest mass energy and $\Box$ is the d'Alembertian,

$$m = m_0 \exp(\phi/c^2)$$

and

$$-\frac{\partial \phi}{\partial x^\mu} = \dot{u}_\mu + \frac{u_\mu}{c^2 \dot{\phi}}$$

where u is the four-velocity and the dot is a differential with respect to time.

The second approach of Nordström (1913) is remembered as the first logically consistent relativistic field theory of gravitation ever formulated. From (note, notation of Pais (1982) not Nordström):

$$\delta \int \psi d\tau = 0$$

$$d\tau^2 = -\eta_{\mu\nu} dx^\mu dx^\nu$$

where ψ is a scalar field,

$$-\frac{\partial T^{\mu\nu}}{\partial x^\nu} = T \frac{1}{\psi} \frac{\partial \psi}{\partial x_\mu}$$

This theory is Lorentz invariant, satisfies the conservation laws, correctly reduces to the Newtonian limit and satisfies the weak equivalence principle.

Einstein and Fokker (1914)

This theory is Einstein's first treatment of gravitation in which general covariance is strictly obeyed. Writing:

$$\delta \int ds = 0$$

$$ds^2 = g_{\mu\nu} dx^\mu dx^\nu$$

$$g_{\mu\nu} = \psi^2 \eta_{\mu\nu}$$

they relate Einstein-Grossmann (1913) to Nordström (1913). They also state:

$$T \propto R.$$

That is, the trace of the stress energy tensor is proportional to the curvature of space.

Einstein (1916, 1917)

This theory is what we now know of as General Relativity. Discarding the Minkowski metric entirely, Einstein gets:

$$\delta \int ds = 0$$

$$ds^2 = g_{\mu\nu} dx^\mu dx^\nu$$

$$R_{\mu\nu} = \frac{8\pi G}{c^4} \left(T_{\mu\nu} - \frac{1}{2} g_{\mu\nu} T \right)$$

which can also be written

$$T^{\mu\nu} = \frac{c^4}{8\pi G} \left(R^{\mu\nu} - \frac{1}{2} g^{\mu\nu} R \right).$$

Five days before Einstein presented the last equation above, Hilbert had submitted a paper containing an almost identical equation. See relativity priority dispute. Hilbert was the first to correctly state the Einstein–Hilbert action for GR, which is:

$$S = \frac{c^4}{16\pi G} \int R\sqrt{-g} d^4x + S_m$$

where G is Newton's gravitational constant, $R = R_\mu{}^\mu$ is the Ricci curvature of space, $g = \det(g_{\mu\nu})$ and S_m is the action due to mass.

GR is a tensor theory, the equations all contain tensors. Nordström's theories, on the other hand, are scalar theories because the gravitational field is a scalar. Later in this article you will see scalar-tensor theories that contain a scalar field in addition to the tensors of GR, and other variants containing vector fields as well have been developed recently.

18.5   Theories from 1917 to the 1980s

This section includes alternatives to GR published after GR but before the observations of galaxy rotation that led to the hypothesis of "dark matter".

Those considered here include (see Will (1981),[5] Lang (2002)[6]):

Listed by date (the hyperlinks take you further down this article)

Whitehead (1922), Cartan (1922, 1923), Fierz & Pauli (1939), Birkhov (1943), Milne (1948), Thiry (1948), Papapetrou (1954a, 1954b), Littlewood (1953), Jordan

(1955), Bergman (1956), Belinfante & Swihart (1957), Yilmaz (1958, 1973), Brans & Dicke (1961), Whitrow & Morduch (1960, 1965), Kustaanheimo (1966), Kustaanheimo & Nuotio (1967), Deser & Laurent (1968), Page & Tupper (1968), Bergmann (1968), Bollini-Giambiagi-Tiomno (1970), Nordtveldt (1970), Wagoner (1970), Rosen (1971, 1975, 1975), Wei-Tou Ni (1972, 1973), Will & Nordtveldt (1972), Hellings & Nordtveldt (1973), Lightman & Lee (1973), Lee, Lightman & Ni (1974), Bekenstein (1977), Barker (1978), Rastall (1979)

These theories are presented here without a cosmological constant or added scalar or vector potential unless specifically noted, for the simple reason that the need for one or both of these was not recognised before the supernova observations by the Supernova Cosmology Project and High-Z Supernova Search Team. How to add a cosmological constant or quintessence to a theory is discussed under Modern Theories (see also here).

18.5.1 Scalar field theories

See also: Scalar theories of gravitation

The scalar field theories of Nordström (1912, 1913) have already been discussed. Those of Littlewood (1953), Bergman (1956), Yilmaz (1958), Whitrow and Morduch (1960, 1965) and Page and Tupper (1968) follow the general formula give by Page and Tupper.

According to Page and Tupper (1968), who discuss all these except Nordström (1913), the general scalar field theory comes from the principle of least action:

$$\delta \int f\left(\frac{\phi}{c^2}\right) ds = 0$$

where the scalar field is,

$$\phi = GM/r$$

and c may or may not depend on ϕ.

In Nordström (1912),

$$f(\phi/c^2) = \exp(-\phi/c^2), \qquad c = c_\infty$$

In Littlewood (1953) and Bergmann (1956),

$$f(\phi/c^2) = \exp(-\phi/c^2 - (\phi/c^2)^2/2), \qquad c = c_\infty$$

In Whitrow and Morduch (1960),

$$f(\phi/c^2) = 1, \qquad c^2 = c_\infty^2 - 2\phi$$

In Whitrow and Morduch (1965),

$$f(\phi/c^2) = \exp(-\phi/c^2), \qquad c^2 = c_\infty^2 - 2\phi$$

In Page and Tupper (1968),

$$f(\phi/c^2) = \phi/c^2 + \alpha(\phi/c^2)^2$$

$$c_\infty^2/c^2 = 1 + 4(\phi/c_\infty^2) + (15 + 2\alpha)(\phi/c_\infty^2)^2$$

Page and Tupper (1968) matches Yilmaz (1958) (see also Yilmaz theory of gravitation) to second order when $\alpha = -7/2$.

The gravitational deflection of light has to be zero when c is constant. Given that variable c and zero deflection of light are both in conflict with experiment, the prospect for a successful scalar theory of gravity looks very unlikely. Further, if the parameters of a scalar theory are adjusted so that the deflection of light is correct then the gravitational redshift is likely to be wrong.

Ni (1972) summarised some theories and also created two more. In the first, a pre-existing special relativity spacetime and universal time coordinate acts with matter and non-gravitational fields to generate a scalar field. This scalar field acts together with all the rest to generate the metric.

The action is:

$$S = \frac{1}{16\pi G} \int d^4x \sqrt{-g} L_\phi + S_m$$

$$L_\phi = \phi R - 2g^{\mu\nu} \partial_\mu \phi \partial_\nu \phi$$

Misner et al. (1973) gives this without the ϕR term. S_m is the matter action.

$$\Box \phi = 4\pi T^{\mu\nu} \left[\eta_{\mu\nu} e^{-2\phi} + \left(e^{2\phi} + e^{-2\phi} \right) \partial_\mu t \partial_\nu t \right]$$

t is the universal time coordinate. This theory is self-consistent and complete. But the motion of the solar system through the universe leads to serious disagreement with experiment.

In the second theory of Ni (1972) there are two arbitrary functions $f(\phi)$ and $k(\phi)$ that are related to the metric by:

$$ds^2 = e^{-2f(\phi)} dt^2 - e^{2f(\phi)} \left[dx^2 + dy^2 + dz^2 \right]$$

$$\eta^{\mu\nu} \partial_\mu \partial_\nu \phi = 4\pi \rho^* k(\phi)$$

Ni (1972) quotes Rosen (1971) as having two scalar fields ϕ and ψ that are related to the metric by:

$$ds^2 = \phi^2 dt^2 - \psi^2 \left[dx^2 + dy^2 + dz^2 \right]$$

In Papapetrou (1954a) the gravitational part of the Lagrangian is:

$$L_\phi = e^\phi \left(\tfrac{1}{2} e^{-\phi} \partial_\alpha \phi \partial_\alpha \phi + \tfrac{3}{2} e^\phi \partial_0 \phi \partial_0 \phi \right)$$

In Papapetrou (1954b) there is a second scalar field χ . The gravitational part of the Lagrangian is now:

$$L_\phi = e^{\frac{1}{2}(3\phi+\chi)} \left(-\tfrac{1}{2} e^{-\phi} \partial_\alpha \phi \partial_\alpha \phi \right.$$

$$\left. - e^{-\phi} \partial_\alpha \phi \partial_\chi \phi + \tfrac{3}{2} e^{-\chi} \partial_0 \phi \partial_0 \phi \right)$$

18.5.2 Bimetric theories

See also: Bimetric theory

Bimetric theories contain both the normal tensor metric and the Minkowski metric (or a metric of constant curvature), and may contain other scalar or vector fields.

Rosen (1973, 1975) Bimetric Theory The action is:

$$S = \frac{1}{64\pi G} \int$$

$$d^4 x \sqrt{-\eta} \eta^{\mu\nu} g^{\alpha\beta} g^{\gamma\delta} (g_{\alpha\gamma|\mu} g_{\alpha\delta|\nu} - \tfrac{1}{2} g_{\alpha\beta|\mu} g_{\gamma\delta|\nu}) + S_m$$

where the vertical line "|" denotes covariant derivative with respect to η . The field equations may be written in the form:

$$\Box_\eta g_{\mu\nu} - g^{\alpha\beta} \eta^{\gamma\delta} g_{\mu\alpha|\gamma} g_{\nu\beta|\delta} = -16\pi G \sqrt{g/\eta} (T_{\mu\nu} - \tfrac{1}{2} g_{\mu\nu} T)$$

Lightman-Lee (1973) developed a metric theory based on the non-metric theory of Belinfante and Swihart (1957a, 1957b). The result is known as BSLL theory. Given a tensor field $B_{\mu\nu}$, $B = B_{\mu\nu} \eta^{\mu\nu}$, and two constants a and f the action is:

$$S = \frac{1}{16\pi G} \int d^4 x \sqrt{-\eta} (a B^{\mu\nu|\alpha} B_{\mu\nu|\alpha} + f B_{,\alpha} B^{,\alpha}) + S_m$$

and the stress–energy tensor comes from:

$$a \Box_\eta B^{\mu\nu} + f \eta^{\mu\nu} \Box_\eta B = -4\pi G \sqrt{g/\eta} T^{\alpha\beta} (\partial g_{\alpha\beta} / \partial B_\mu \nu)$$

In Rastall (1979), the metric is an algebraic function of the Minkowski metric and a Vector field.[7] The Action is:

$$S = \frac{1}{16\pi G} \int d^4 x \sqrt{-g} F(N) K^{\mu;\nu} K_{\mu;\nu} + S_m$$

where

$$F(N) = -N/(2+N) \ \text{ and } \ N = g^{\mu\nu} K_\mu K_\nu$$

(see Will (1981) for the field equation for $T^{\mu\nu}$ and K_μ).

18.5.3 Quasilinear theories

In Whitehead (1922), the physical metric g is constructed (by Synge) algebraically from the Minkowski metric η and matter variables, so it doesn't even have a scalar field. The construction is:

$$g_{\mu\nu}(x^\alpha) = \eta_{\mu\nu} - 2 \int_{\Sigma-} \frac{y_\mu^- y_\nu^-}{(w^-)^3} [\sqrt{-g} \rho u^\alpha d\Sigma_\alpha]^-$$

where the superscript (-) indicates quantities evaluated along the past η light cone of the field point x^α and

$$(y^\mu)^- = x^\mu - (x^\mu)^- \ , \ (y^\mu)^- (y_\mu)^- = 0,$$
$$w^- = (y^\mu)^- (u_\mu)^- \ , \ (u_\mu) = dx^\mu / d\sigma,$$
$$d\sigma^2 = \eta_{\mu\nu} dx^\mu dx^\nu$$

Nevertheless, the metric construction (from a non-metric theory) using the "length contraction" ansatz[8] is criticised.[9]

Deser and Laurent (1968) and Bollini-Giambiagi-Tiomno (1970) are Linear Fixed Gauge (LFG) theories. Taking an approach from quantum field theory, combine a Minkowski spacetime with the gauge invariant action of a spin-two tensor field (i.e. graviton) $h_{\mu\nu}$ to define

$$g_{\mu\nu} = \eta_{\mu\nu} + h_{\mu\nu}$$

The action is:

$$S = \frac{1}{16\pi G} \int d^4 x \sqrt{-\eta} [2h^{\mu\nu}_{|\nu} h^{|\lambda}_{\mu\lambda} - 2h^{\mu\nu}_{|\nu} h^\lambda_{\lambda|\mu} + h^\nu_{\nu|\mu} h^{\lambda|\mu}_\lambda$$

$$- h^{\mu\nu|\lambda} h_{\mu\nu|\lambda}] + S_m$$

The Bianchi identity associated with this partial gauge invariance is wrong. LFG theories seek to remedy this by breaking the gauge invariance of the gravitational action through the introduction of auxiliary gravitational fields that couple to $h_{\mu\nu}$.

A cosmological constant can be introduced into a quasi-linear theory by the simple expedient of changing the Minkowski background to a de Sitter or anti-de Sitter space-time, as suggested by G. Temple in 1923. Temple's suggestions on how to do this were criticized by C. B. Rayner in 1955.[10]

18.5.4   Tensor theories

Einstein's general relativity is the simplest plausible theory of gravity that can be based on just one symmetric tensor field (the metric tensor). Others include: Gauss–Bonnet gravity, f(R) gravity, and Lovelock theory of gravity.

18.5.5   Scalar-tensor theories

See also: Scalar-tensor theory, Brans–Dicke theory, Dilaton, Chameleon_particle, and Pressuron

These all contain at least one free parameter, as opposed to GR which has no free parameters.

Although not normally considered a Scalar-Tensor theory of gravity, the 5 by 5 metric of Kaluza–Klein reduces to a 4 by 4 metric and a single scalar. So if the 5th element is treated as a scalar gravitational field instead of an electromagnetic field then Kaluza–Klein can be considered the progenitor of Scalar-Tensor theories of gravity. This was recognised by Thiry (1948).

Scalar-Tensor theories include Thiry (1948), Jordan (1955), Brans and Dicke (1961), Bergman (1968), Nordtveldt (1970), Wagoner (1970), Bekenstein (1977) and Barker (1978).

The action S is based on the integral of the Lagrangian L_ϕ

$$S = \frac{1}{16\pi G} \int d^4x \sqrt{-g} L_\phi + S_m$$

$$L_\phi = \phi R - \frac{\omega(\phi)}{\phi} g^{\mu\nu} \partial_\mu \phi \partial_\nu \phi + 2\phi\lambda(\phi)$$

$$S_m = \int d^4x \sqrt{g} G_N L_m$$

$$T^{\mu\nu} \stackrel{\text{def}}{=} \frac{2}{\sqrt{g}} \frac{\delta S_m}{\delta g_{\mu\nu}}$$

where $\omega(\phi)$ is a different dimensionless function for each different scalar-tensor theory. The function $\lambda(\phi)$ plays the same role as the cosmological constant in GR. G_N is a dimensionless normalization constant that fixes the present-day value of G . An arbitrary potential can be added for the scalar.

The full version is retained in Bergman (1968) and Wagoner (1970). Special cases are:

Nordtvedt (1970), $\lambda = 0$

Since λ was thought to be zero at the time anyway, this would not have been considered a significant difference. The role of the cosmological constant in more modern work is discussed under Cosmological constant.

Brans–Dicke (1961), ω is constant

Bekenstein (1977) Variable Mass Theory Starting with parameters r and q , found from a cosmological solution, $\phi = [1 - qf(\phi)]f(\phi)^{-r}$ determines function f then

$$\omega(\phi) = -\tfrac{3}{2} - \tfrac{1}{4}f(\phi)[(1-6q)qf(\phi)-1][r+(1-r)qf(\phi)]^{-2}$$

Barker (1978) Constant G Theory

$$\omega(\phi) = (4 - 3\phi)/(2\phi - 2)$$

Adjustment of $\omega(\phi)$ allows Scalar Tensor Theories to tend to GR in the limit of $\omega \to \infty$ in the current epoch. However, there could be significant differences from GR in the early universe.

So long as GR is confirmed by experiment, general Scalar-Tensor theories (including Brans–Dicke) can never be ruled out entirely, but as experiments continue to confirm GR more precisely and the parameters have to be fine-tuned so that the predictions more closely match those of GR.

18.5.6   Vector-tensor theories

Before we start, Will (2001) has said: "Many alternative metric theories developed during the 1970s and 1980s could be viewed as "straw-man" theories, invented to prove that such theories exist or to illustrate particular properties. Few of these could be regarded as well-motivated theories from the point of view, say, of field theory or particle physics. Examples are the vector-tensor theories studied by Will, Nordtvedt and Hellings."

Hellings and Nordtvedt (1973) and Will and Nordtvedt (1972) are both vector-tensor theories. In addition to the metric tensor there is a timelike vector field K_μ . The gravitational action is:

$$S = \frac{1}{16\pi G} \int d^4x \sqrt{-g}[R + \omega K_\mu K^\mu R + \eta K^\mu K^\nu R_{\mu\nu}$$

$$- \epsilon F_{\mu\nu} F^{\mu\nu} + \tau K_{\mu;\nu} K^{\mu;\nu}] + S_m$$

where ω , η , ϵ and τ are constants and

$$F_{\mu\nu} = K_{\nu;\mu} - K_{\mu;\nu}$$

See Will (1981) for the field equations for $T^{\mu\nu}$ and K_μ .

Will and Nordtvedt (1972) is a special case where

$$\omega = \eta = \epsilon = 0 \; ; \tau = 1$$

Hellings and Nordtvedt (1973) is a special case where

$$\tau = 0 \; ; \epsilon = 1 \; ; \eta = -2\omega$$

These vector-tensor theories are semi-conservative, which means that they satisfy the laws of conservation of momentum and angular momentum but can have preferred frame effects. When $\omega = \eta = \epsilon = \tau = 0$ they reduce to GR so, so long as GR is confirmed by experiment, general vector-tensor theories can never be ruled out.

18.5.7   Other metric theories

Others metric theories have been proposed; that of Bekenstein (2004) is discussed under Modern Theories.

18.5.8   Non-metric theories

See also: Einstein–Cartan theory and Cartan connection

Cartan's theory is particularly interesting both because it is a non-metric theory and because it is so old. The status of Cartan's theory is uncertain. Will (1981) claims that all non-metric theories are eliminated by Einstein's Equivalence Principle (EEP). Will (2001) tempers that by explaining experimental criteria for testing non-metric theories against EEP. Misner et al. (1973) claims that Cartan's theory is the only non-metric theory to survive all experimental tests up to that date and Turyshev (2006) lists Cartan's theory among the few that have survived all experimental tests up to that date. The following is a quick sketch of Cartan's theory as restated by Trautman (1972).

Cartan (1922, 1923) suggested a simple generalization of Einstein's theory of gravitation. He proposed a model of space time with a metric tensor and a linear "connection" compatible with the metric but not necessarily symmetric. The torsion tensor of the connection is related to the density of intrinsic angular momentum. Independently of Cartan, similar ideas were put forward by Sciama, by Kibble in the years 1958 to 1966, culminating in a 1976 review by Hehl et al.

The original description is in terms of differential forms, but for the present article that is replaced by the more familiar language of tensors (risking loss of accuracy). As in GR, the Lagrangian is made up of a massless and a mass part. The Lagrangian for the massless part is:

$$L = \frac{1}{32\pi G}\Omega_\nu^\mu g^{\nu\xi} x^\eta x^\zeta \varepsilon_{\xi\mu\eta\zeta}$$

$$\Omega_\nu^\mu = d\omega_\nu^\mu + \omega_\xi^\eta$$

$$\nabla x^\mu = -\omega_\nu^\mu x^\nu$$

The ω_ν^μ is the linear connection. $\varepsilon_{\xi\mu\eta\zeta}$ is the completely antisymmetric pseudo-tensor (Levi-Civita symbol) with $\varepsilon_{0123} = \sqrt{-g}$, and $g^{\nu\xi}$ is the metric tensor as usual. By assuming that the linear connection is metric, it is possible to remove the unwanted freedom inherent in the non-metric theory. The stress–energy tensor is calculated from:

$$T^{\mu\nu} = \frac{1}{16\pi G}(g^{\mu\nu}\eta_\eta^\xi - g^{\xi\mu}\eta_\eta^\nu - g^{\xi\nu}\eta_\eta^\mu)\Omega_\xi^\eta$$

The space curvature is not Riemannian, but on a Riemannian space-time the Lagrangian would reduce to the Lagrangian of GR.

Some equations of the non-metric theory of Belinfante and Swihart (1957a, 1957b) have already been discussed in the section on bimetric theories.

A distinctively non-metric theory is given by gauge theory gravity, which replaces the metric in its field equations with a pair of gauge fields in flat spacetime. On the one hand, the theory is quite conservative because it is substantially equivalent to Einstein–Cartan theory (or general relativity in the limit of vanishing spin), differing mostly in the nature of its global solutions. On the other hand, it is radical because it replaces differential geometry with geometric algebra.

18.6   Modern theories 1980s to present

This section includes alternatives to GR published after the observations of galaxy rotation that led to the hypothesis of "dark matter".

There is no known reliable list of comparison of these theories.

Those considered here include: Beckenstein (2004), Moffat (1995), Moffat (2002), Moffat (2005a, b).

These theories are presented with a cosmological constant or added scalar or vector potential.

18.6.1 Motivations

Motivations for the more recent alternatives to GR are almost all cosmological, associated with or replacing such constructs as "inflation", "dark matter" and "dark energy". The basic idea is that gravity agrees with GR at the present epoch but may have been quite different in the early universe.

There was a slow dawning realisation in the physics world that there were several problems inherent in the then big bang scenario, two of these were the horizon problem and the observation that at early times when quarks were first forming there was not enough space on the universe to contain even one quark. Inflation theory was developed to overcome these. Another alternative was constructing an alternative to GR in which the speed of light was larger in the early universe.

The discovery of unexpected rotation curves for galaxies took everyone by surprise. Could there be more mass in the universe than we are aware of, or is the theory of gravity itself wrong? The consensus now is that the missing mass is "cold dark matter", but that consensus was only reached after trying alternatives to general relativity and some physicists still believe that alternative models of gravity might hold the answer.

The discovery of the accelerated expansion of the universe by the supernova surveys led to the rapid reinstatement of Einstein's cosmological constant, and quintessence arrived as an alternative to the cosmological constant. At least one new alternative to GR attempted to explain the supernova surveys' results in a completely different way.

Another observation that sparked recent interest in alternatives to General Relativity is the Pioneer anomaly. It was quickly discovered that alternatives to GR could explain this anomaly. This is now believed to be accounted for by non-uniform thermal radiation.

18.6.2 Cosmological constant and quintessence

(also see Cosmological constant, Einstein–Hilbert action, Quintessence (physics))

The cosmological constant Λ is a very old idea, going back to Einstein in 1917. The success of the Friedmann model of the universe in which $\Lambda = 0$ led to the general acceptance that it is zero, but the use of a non-zero value came back with a vengeance when data from supernovae indicated that the expansion of the universe is accelerating

First, let's see how it influences the equations of Newtonian gravity and General Relativity.

In Newtonian gravity, the addition of the cosmological constant changes the Newton-Poisson equation from:

$$\nabla^2 \phi = 4\pi \rho \, G;$$

to

$$\nabla^2 \phi - \Lambda \phi = 4\pi \rho \, G;$$

In GR, it changes the Einstein–Hilbert action from

$$S = \frac{1}{16\pi G} \int R\sqrt{-g}\, d^4x \, + S_m$$

to

$$S = \frac{1}{16\pi G} \int (R - 2\Lambda)\sqrt{-g}\, d^4x \, + S_m$$

which changes the field equation

$$T^{\mu\nu} = \frac{1}{8\pi G} \left(R^{\mu\nu} - \frac{1}{2} g^{\mu\nu} R \right)$$

to

$$T^{\mu\nu} = \frac{1}{8\pi G} \left(R^{\mu\nu} - \frac{1}{2} g^{\mu\nu} R + g^{\mu\nu} \Lambda \right)$$

In alternative theories of gravity, a cosmological constant can be added to the action in exactly the same way.

The cosmological constant is not the only way to get an accelerated expansion of the universe in alternatives to GR. We've already seen how the scalar potential $\lambda(\phi)$ can be added to scalar tensor theories. This can also be done in every alternative the GR that contains a scalar field ϕ by adding the term $\lambda(\phi)$ inside the Lagrangian for the gravitational part of the action, the L_ϕ part of

$$S = \frac{1}{16\pi G} \int d^4x \, \sqrt{-g} L_\phi + S_m$$

Because $\lambda(\phi)$ is an arbitrary function of the scalar field, it can be set to give an acceleration that is large in the early universe and small at the present epoch. This is known as quintessence.

A similar method can be used in alternatives to GR that use vector fields, including Rastall (1979) and vector-tensor theories. A term proportional to

$$K^\mu K^\nu g_{\mu\nu}$$

is added to the Lagrangian for the gravitational part of the action.

18.6.3   Relativistic MOND

(see Modified Newtonian dynamics, Tensor-vector-scalar gravity, and Bekenstein (2004) for more details).

The original theory of MOND by Milgrom was developed in 1983 as an alternative to "dark matter". Departures from Newton's law of gravitation are governed by an acceleration scale, not a distance scale. MOND successfully explains the Tully-Fisher observation that the luminosity of a galaxy should scale as the fourth power of the rotation speed. It also explains why the rotation discrepancy in dwarf galaxies is particularly large.

There were several problems with MOND in the beginning.

1. It did not include relativistic effects

2. It violated the conservation of energy, momentum and angular momentum

3. It was inconsistent in that it gives different galactic orbits for gas and for stars

4. It did not state how to calculate gravitational lensing from galaxy clusters.

By 1984, problems 2 and 3 had been solved by introducing a Lagrangian (AQUAL). A relativistic version of this based on scalar-tensor theory was rejected because it allowed waves in the scalar field to propagate faster than light. The Lagrangian of the non-relativistic form is:

$$L = -\frac{a_0^2}{8\pi G} f\left[\frac{|\nabla\phi|^2}{a_0^2}\right] - \rho\phi$$

The relativistic version of this has:

$$L = -\frac{a_0^2}{8\pi G} \tilde{f}\left(l_0^2 g^{\mu\nu} \,\partial_\mu\phi\,\partial_\nu\phi\right)$$

with a nonstandard mass action. Here f and $\tilde{f}$ are arbitrary functions selected to give Newtonian and MOND behaviour in the correct limits, and $l_0 = c^2/a_0$ is the MOND length scale.

By 1988, a second scalar field (PCC) fixed problems with the earlier scalar-tensor version but is in conflict with the

perihelion precession of Mercury and gravitational lensing by galaxies and clusters.

By 1997, MOND had been successfully incorporated in a stratified relativistic theory [Sanders], but as this is a preferred frame theory it has problems of its own.

Bekenstein (2004) introduced a tensor-vector-scalar model (TeVeS). This has two scalar fields ϕ and σ and vector field U_α. The action is split into parts for gravity, scalars, vector and mass.

$$S = S_g + S_s + S_v + S_m$$

The gravity part is the same as in GR.

$$S_s = -\frac{1}{2}\int\left[\sigma^2 h^{\alpha\beta}\phi_{,\alpha}\phi_{,\beta} + \frac{1}{2}Gl_0^{-2}\sigma^4 F(kG\sigma^2)\right]\sqrt{-g}\,d^4x$$

$$S_v = -\frac{K}{32\pi G}\int\left[g^{\alpha\beta}g^{\mu\nu}U_{[\alpha,\mu]}U_{[\beta,\nu]}\right.$$

$$\left. -\frac{2\lambda}{K}\left(g^{\mu\nu}U_\mu U_\nu + 1\right)\right]\sqrt{-g}\,d^4x$$

$$S_m = \int L\left(\tilde{g}_{\mu\nu}, f^\alpha, f^\alpha_{|\mu}, \cdots\right)\sqrt{-g}\,d^4x$$

where

$$h^{\alpha\beta} = g^{\alpha\beta} - U^\alpha U^\beta$$

$$\tilde{g}^{\alpha\beta} = e^{2\phi}g^{\alpha\beta} + 2U^\alpha U^\beta \sinh(2\phi)$$

k, K are constants, square brackets in indices $U_{[\alpha,\mu]}$ represent anti-symmetrization, λ is a Lagrange multiplier (calculated elsewhere), and L is a Lagrangian translated from flat spacetime onto the metric $\tilde{g}^{\alpha\beta}$. Note that G need not equal the observed gravitational constant G_{Newton}

F is an arbitrary function, and

$$F(\mu) = \frac{3}{4}\frac{\mu^2(\mu-2)^2}{1-\mu}$$

is given as an example with the right asymptotic behaviour; note how it becomes undefined when $\mu = 1$

The PPN parameters of this theory are calculated in,[11] which shows that all its parameters are equal to GR's, except for

$$\alpha_1 = \frac{4G}{K}\left((2K-1)e^{-4\phi_0} - e^{4\phi_0} + 8\right) - 8$$

$$\alpha_2 = \frac{6G}{2-K} - \frac{2G(K+4)e^{4\phi_0}}{(2-K)^2} - 1$$

both of which expressed in geometric units where $c = G_{Newtonian} = 1$; so

$$G^{-1} = \frac{2}{2-K} + \frac{k}{4\pi}.$$

The parameter ϕ_0 measures the value of the scalar field ϕ at infinity, and is given by

$$\frac{K}{2-K} = e^{-4\phi_0} - 1.$$

Milgrom[12] proposed a "bimetric MOND" or "BIMOND" theory, with action

$$S - S_M - \hat{S}_M = -\frac{c^4}{16\pi G} \int \left[\beta g^{1/2} R + \alpha \hat{g}^{1/2} \right.$$

$$\left. R - 2(g\hat{g})^{1/4} f(\kappa) l_0^{-2} \mathcal{M}\left(l_0^m \Upsilon^{(m)} \right) \right] d^4 x$$

with S_M and $\hat{S}_M$ the (noninteracting) matter actions attached to the two metrics, Υ a tensor derived from the difference in the metrics' connections, $\kappa = (g/\hat{g})^{\frac{1}{4}}$ the ratio between the two metric traces, and α, β are free parameters. $\mathcal{M}$ is a function which depends on some contractions of the Υ tensors.

Assuming that $\mathcal{M}$ depends only on the scalar contraction of Υ, Milgrom obtained as a nonrelativistic limit his bipotential version of MOND with action

$$S - S_M = -\frac{1}{8\pi G} \int \left[\beta \left(\nabla \phi \right)^2 + \alpha \left(\nabla \hat{\phi} \right)^2 \right.$$

$$\left. - a_0^2 \mathcal{M}\left(\left(\nabla \phi - \nabla \hat{\phi} \right)^2 a_0^{-2} \right) \right] d^4 x$$

$$S_M = \rho(v^2/2 - \phi)$$

Here $\mathcal{M}(z)$ should scale as $z^{-1/4}$ in the deep-MOND limit and as z in the Newtonian limit.

18.6.4 Moffat's theories

J. W. Moffat (1995) developed a non-symmetric gravitation theory (NGT). This is not a metric theory. It was first claimed that it does not contain a black hole horizon, but Burko and Ori (1995) have found that NGT can contain black holes. Later, Moffat claimed that it has also been applied to explain rotation curves of galaxies without invoking "dark matter". Damour, Deser & MaCarthy (1993) have criticised NGT, saying that it has unacceptable asymptotic behaviour.

The mathematics is not difficult but is intertwined so the following is only a brief sketch. Starting with a non-symmetric tensor $g_{\mu\nu}$, the Lagrangian density is split into

$$L = L_R + L_M$$

where L_M is the same as for matter in GR.

$$L_R = \sqrt{-g} \left[R(W) - 2\lambda - \frac{1}{4}\mu^2 g^{\mu\nu} g_{[\mu\nu]} \right] - \frac{1}{6} g^{\mu\nu} W_\mu W_\nu$$

where $R(W)$ is a curvature term analogous to but not equal to the Ricci curvature in GR, λ and μ^2 are cosmological constants, $g_{[\nu\mu]}$ is the antisymmetric part of $g_{\nu\mu}$. W_μ is a connection, and is a bit difficult to explain because it's defined recursively. However, $W_\mu \approx -2g_{[\mu\nu]}^{,\nu}$

Moffat's (2002) theory is a scalar-tensor bimetric gravity theory (BGT) and is one of the many theories of gravity in which the speed of light is faster in the early universe. These theories were motivated partly be the desire to avoid the "horizon problem" without invoking inflation. It has a variable G. The theory also attempts to explain the dimming of supernovae from a perspective other than the acceleration of the universe and so runs the risk of predicting an age for the universe that is too small.

Moffat's (2005a) metric-skew-tensor-gravity (MSTG) theory is able to predict rotation curves for galaxies without either dark matter or MOND, and claims that it can also explain gravitational lensing of galaxy clusters without dark matter. It has variable G, increasing to a final constant value about a million years after the big bang.

The theory seems to contain an asymmetric tensor $A_{\mu\nu}$ field and a source current J_μ vector. The action is split into:

$$S = S_G + S_F + S_{FM} + S_M$$

Both the gravity and mass terms match those of GR with cosmological constant. The skew field action and the skew field matter coupling are:

$$S_F = \int d^4 x \sqrt{-g} \left(\frac{1}{12} F_{\mu\nu\rho} F^{\mu\nu\rho} - \frac{1}{4}\mu^2 A_{\mu\nu} A^{\mu\nu} \right)$$

$$S_{FM} = \int d^4 x \, \epsilon^{\alpha\beta\mu\nu} A_{\alpha\beta} \partial_\mu J_\nu$$

where

$$F_{\mu\nu\rho} = \partial_\mu A_{\nu\rho} + \partial_\rho A_{\mu\nu}$$

and $\epsilon^{\alpha\beta\mu\nu}$ is the Levi-Civita symbol. The skew field coupling is a Pauli coupling and is gauge invariant for any source current. The source current looks like a matter fermion field associated with baryon and lepton number.

Moffat (2005b) Scalar-tensor-vector gravity (SVTG) theory.

The theory contains a tensor, vector and three scalar fields. But the equations are quite straightforward. The action is split into: $S = S_G + S_K + S_S + S_M$ with terms for gravity, vector field K_μ, scalar fields G, ω & μ, and mass. S_G is the standard gravity term with the exception that G is moved inside the integral.

$$S_K = -\int d^4x \sqrt{-g}\,\omega\left(\frac{1}{4}B_{\mu\nu}B^{\mu\nu} + V(K)\right)$$

where $B_{\mu\nu} = \partial_\mu K_\nu - \partial_\nu K_\mu$

$$S_S = -\int d^4x \sqrt{-g}\,\frac{1}{G^3}\left(\frac{1}{2}g^{\mu\nu}\nabla_\mu G\,\nabla_\nu G - V(G)\right)$$

$$+ \frac{1}{G}\left(\frac{1}{2}g^{\mu\nu}\nabla_\mu\omega\,\nabla_\nu\omega - V(\omega)\right) + \frac{1}{\mu^2 G}\left(\frac{1}{2} - \right.$$

$$\left. g^{\mu\nu}\nabla_\mu\mu\,\nabla_\nu\mu - V(\mu)\right)$$

The potential function for the vector field is chosen to be:

$$V(K) = -\frac{1}{2}\mu^2\phi^\mu\phi_\mu - \frac{1}{4}g(\phi^\mu\phi_\mu)^2$$

where g is a coupling constant. The functions assumed for the scalar potentials are not stated.

18.6.5 Infinite derivative theorem

In order to remove ghosts in the modified propagator, as well as to obtain asymptotic freedom, Biswas, Mazumdar and Siegel (2005) considered an infinite set of higher derivative terms

$$S = \int d^4x \sqrt{-g}\left(\frac{R}{2} + RF_1(\Box)R + R^{\mu\nu}F_2(\Box)R_{\mu\nu}\right.$$

$$\left. + C_{\mu\nu\lambda\sigma}F_3(\Box)C^{\mu\nu\lambda\sigma}\right)$$

where the $F_i(\Box)$ are functions of the D'Alembertian operator and $C^{\mu\nu\lambda\sigma}$ is the Weyl tensor.[13][14] This avoids a singularity near the origin, while recovering the 1/r fall of the GR potential at large distances.[15]

18.7 Testing of alternatives to general relativity

Main article: Tests of general relativity

Any putative alternative to general relativity would need to meet a variety of tests for it to become accepted. For in-depth coverage of these tests, see Misner et al. (1973) Ch.39, Will (1981) Table 2.1, and Ni (1972). Most such tests can be categorized as in the following subsections.

18.7.1 Self-consistency

Self-consistency among non-metric theories includes eliminating theories allowing tachyons, ghost poles and higher order poles, and those that have problems with behaviour at infinity.

Among metric theories, self-consistency is best illustrated by describing several theories that fail this test. The classic example is the spin-two field theory of Fierz and Pauli (1939); the field equations imply that gravitating bodies move in straight lines, whereas the equations of motion insist that gravity deflects bodies away from straight line motion. Yilmaz (1971, 1973) contains a tensor gravitational field used to construct a metric; it is mathematically inconsistent because the functional dependence of the metric on the tensor field is not well defined.

18.7.2 Completeness

To be complete, a theory of gravity must be capable of analysing the outcome of every experiment of interest. It must therefore mesh with electromagnetism and all other physics. For instance, any theory that cannot predict from first principles the movement of planets or the behaviour of atomic clocks is incomplete.

Many early theories are incomplete in that it is unclear whether the density ρ used by the theory should be calculated from the stress–energy tensor T as $\rho = T_{\mu\nu}u^\mu u^\nu$ or as $\rho = T_{\mu\nu}\delta^{\mu\nu}$, where u is the four-velocity, and δ is the Kronecker delta.

The theories of Thirry (1948) and Jordan (1955) are incomplete unless Jordan's parameter η is set to -1, in which case they match the theory of Brans–Dicke (1961) and so are worthy of further consideration.

Milne (1948) is incomplete because it makes no gravitational red-shift prediction.

The theories of Whitrow and Morduch (1960, 1965), Kustaanheimo (1966) and Kustaanheimo and Nuotio (1967) are either incomplete or inconsistent. The incorporation of Maxwell's equations is incomplete unless it is assumed that they are imposed on the flat background space-time, and when that is done they are inconsistent, because they predict zero gravitational redshift when the wave version of light (Maxwell theory) is used, and nonzero redshift when the particle version (photon) is used. Another more obvious example is Newtonian gravity with Maxwell's equations; light as photons is deflected by gravitational fields (by twice that of GR) but light as waves is not.

18.7.3 Classical tests

Main article: Tests of general relativity

There are three "classical" tests (dating back to the 1910s or earlier) of the ability of gravity theories to handle relativistic effects; they are:

- gravitational redshift

- gravitational lensing (generally tested around the Sun)

- anomalous perihelion advance of the planets (see Tests of General Relativity)

Each theory should reproduce the observed results in these areas, which have to date always aligned with the predictions of general relativity.

In 1964, Irwin I. Shapiro found a fourth test, called the Shapiro delay. It is usually regarded as a "classical" test as well.

18.7.4 Agreement with Newtonian mechanics and special relativity

As an example of disagreement with Newtonian experiments, Birkhoff (1943) theory predicts relativistic effects fairly reliably but demands that sound waves travel at the speed of light. This was the consequence of an assumption made to simplify handling the collision of masses.

18.7.5 The Einstein equivalence principle (EEP)

Main article: Equivalence principle

The EEP has three components.

The first is the uniqueness of free fall, also known as the Weak Equivalence Principle (WEP). This is satisfied if inertial mass is equal to gravitational mass. η is a parameter used to test the maximum allowable violation of the WEP. The first tests of the WEP were done by Eötvös before 1900 and limited η to less than 5×10^{-9}. Modern tests have reduced that to less than 5×10^{-13}.

The second is Lorentz invariance. In the absence of gravitational effects the speed of light is constant. The test parameter for this is δ. The first tests of Lorentz invariance were done by Michelson and Morley before 1890 and limited δ to less than 5×10^{-3}. Modern tests have reduced this to less than 1×10^{-21}.

The third is local position invariance, which includes spatial and temporal invariance. The outcome of any local non-gravitational experiment is independent of where and when it is performed. Spatial local position invariance is tested using gravitational redshift measurements. The test parameter for this is α. Upper limits on this found by Pound and Rebka in 1960 limited α to less than 0.1. Modern tests have reduced this to less than 1×10^{-4}.

Schiff's conjecture states that any complete, self-consistent theory of gravity that embodies the WEP necessarily embodies EEP. This is likely to be true if the theory has full energy conservation.

Metric theories satisfy the Einstein Equivalence Principle. Extremely few non-metric theories satisfy this. For example, the non-metric theory of Belinfante & Swihart (1957) is eliminated by the *THeμ* formalism for testing EEP. Gauge theory gravity is a notable exception, where the strong equivalence principle is essentially the minimal coupling of the gauge covariant derivative.

18.7.6 Parametric post-Newtonian (PPN) formalism

Main article: Parameterized post-Newtonian formalism

See also Tests of general relativity, Misner et al. (1973) and Will (1981) for more information.

Work on developing a standardized rather than ad-hoc set of tests for evaluating alternative gravitation models began with Eddington in 1922 and resulted in a standard set of PPN numbers in Nordtvedt and Will (1972) and Will and Nordtvedt (1972). Each parameter measures a different aspect of how much a theory departs from Newtonian gravity. Because we are talking about deviation from Newtonian theory here, these only measure weak-field effects. The effects of strong gravitational fields are examined later.

These ten are called : γ , β , η , α_1 , α_2 , α_3 , ζ_1 , ζ_2 , ζ_3 , ζ_4

γ is a measure of space curvature, being zero for Newtonian gravity and one for GR.

β is a measure of nonlinearity in the addition of gravitational fields, one for GR.

η is a check for preferred location effects.

α_1 , α_2 , α_3 measure the extent and nature of "preferred-frame effects". Any theory of gravity with at least one α nonzero is called a preferred-frame theory.

ζ_1 , ζ_2 , ζ_3 , ζ_4 , α_3 measure the extent and nature of breakdowns in global conservation laws. A theory of gravity possesses 4 conservation laws for energy-momentum and 6 for angular momentum only if all five are zero.

18.7.7 Strong gravity and gravitational waves

Main article: Tests of general relativity

PPN is only a measure of weak field effects. Strong gravity effects can be seen in compact objects such as white dwarfs, neutron stars, and black holes. Experimental tests such as the stability of white dwarfs, spin-down rate of pulsars, orbits of binary pulsars and the existence of a black hole horizon can be used as tests of alternative to GR.

GR predicts that gravitational waves travel at the speed of light. Many alternatives to GR say that gravitational waves travel faster than light. If true, this could result in failure of causality.

18.7.8 Cosmological tests

Many of these have been developed recently. For those theories that aim to replace dark matter, the galaxy rotation curve, the Tully-Fisher relation, the faster rotation rate of dwarf galaxies, and the gravitational lensing due to galactic clusters act as constraints.

For those theories that aim to replace inflation, the size of ripples in the spectrum of the cosmic microwave background radiation is the strictest test.

For those theories that incorporate or aim to replace dark energy, the supernova brightness results and the age of the universe can be used as tests.

Another test is the flatness of the universe. With GR, the combination of baryonic matter, dark matter and dark energy add up to make the universe exactly flat. As the accuracy of experimental tests improve, alternatives to GR that aim to replace dark matter or dark energy will have to explain why.

18.8   Results of testing theories

18.8.1   PPN parameters for a range of theories

(See Will (1981) and Ni (1972) for more details. Misner et al. (1973) gives a table for translating parameters from the notation of Ni to that of Will)

General Relativity is now more than 100 years old, during which one alternative theory of gravity after another has failed to agree with ever more accurate observations. One illustrative example is Parameterized post-Newtonian formalism (PPN).

The following table lists PPN values for a large number of theories. If the value in a cell matches that in the column heading then the full formula is too complicated to include here.

† The theory is incomplete, and ζ_4 can take one of two values. The value closest to zero is listed.

All experimental tests agree with GR so far, and so PPN analysis immediately eliminates all the scalar field theories in the table.

A full list of PPN parameters is not available for Whitehead (1922), Deser-Laurent (1968), Bollini-Giambiagi-Tiomino (1970), but in these three cases $\beta = \xi$, which is in strong conflict with GR and experimental results. In particular, these theories predict incorrect amplitudes for the Earth's tides. (A minor modification of Whitehead's theory avoids this problem. However, the modification predicts the Nordtvedt effect, which has been experimentally constrained.)

18.8.2   Theories that fail other tests

The stratified theories of Ni (1973), Lee Lightman and Ni (1974) are non-starters because they all fail to explain the perihelion advance of Mercury.

The bimetric theories of Lightman and Lee (1973), Rosen (1975), Rastall (1979) all fail some of the tests associated with strong gravitational fields.

The scalar-tensor theories include GR as a special case, but only agree with the PPN values of GR when they are equal to GR to within experimental error. As experimental tests get more accurate, the deviation of the scalar-tensor theories from GR is being squashed to zero.

The same is true of vector-tensor theories, the deviation of the vector-tensor theories from GR is being squashed to zero. Further, vector-tensor theories are semi-conservative; they have a nonzero value for α_2 which can have a measurable effect on the Earth's tides.

Non-metric theories, such as Belinfante and Swihart (1957a, 1957b), usually fail to agree with experimental tests of Einstein's equivalence principle.

And that leaves, as a likely valid alternative to GR, nothing except possibly Cartan (1922).

That was the situation until cosmological discoveries pushed the development of modern alternatives.

18.9 Footnotes

[1] this isn't exactly the way Mach originally stated it, see other variants in Mach principle

[2] Zenneck, J. (1903). "Gravitation". *Encyklopädie der mathematischen Wissenschaften mit Einschluss ihrer Anwendungen* (in German) **5**: 25–67. doi:10.1007/978-3-663-16016-8_2. ISBN 978-3-663-15445-7.

[3] Lorentz, H.A. (1900). "Considerations on Gravitation". *Proc. Acad. Amsterdam* **2**: 559–574.

[4] Walter, S. (2007). Renn, J., ed. "Breaking in the 4-vectors: the four-dimensional movement in gravitation, 1905–1910". *The Genesis of General Relativity* (Berlin: Springer) **3**: 193–252.

[5] A later edition is Will (1993). See also Ni (1972)

[6] Although an important source for this article, the presentations of Turyshev (2006) and Lang (2002) contain many errors of fact

[7] Will (1981) lists this as bimetric but I don't see why it isn't just a vector field theory

[8] http://arxiv.org/pdf/0704.1574v2.pdf – Retarded electric and magnetic fields of a moving charge: Feynman's derivation of Liénard-Wiechert potentials revisited

[9] http://gsjournal.net/Science-Journals/Research%20Papers-Relativity%20Theory/Download/4217 – On the Multiple Interpretations of Gravity

[10] Gary Gibbons; Will (2008). "On the Multiple Deaths of Whitehead's Theory of Gravity". *Stud.Hist.Philos.Mod.Phys.* **39**: 41–61. arXiv:gr-qc/0611006. doi:10.1016/j.shpsb.2007.04.004. Cf. Ronny Desmet and Michel Weber (edited by), Whitehead. The Algebra of Metaphysics. Applied Process Metaphysics Summer Institute Memorandum, Louvain-la-Neuve, Éditions Chromatika, 2010.

[11] Sagi, Eva (July 2009). "Preferred frame parameters in the tensor-vector-scalar theory of gravity and its generalization". *Physical Review D* **80** (4): 044032. arXiv:0905.4001. Bibcode:2009PhRvD..80d4032S. doi:10.1103/PhysRevD.80.044032.

[12] Milgrom, M (2009). "Bimetric MOND gravity". *Physical Review D* **80** (12): 123536. arXiv:0912.0790. Bibcode:2009PhRvD..80l3536M. doi:10.1103/PhysRevD.80.123536.

[13] "Bouncing Universes in String-inspired Gravity". arXiv:hep-th/0508194. Bibcode:2006JCAP...03..009B. doi:10.1088/1475-7516/2006/03/009.

[14] Biswas, Tirthabir; Conroy, Aindriú; Koshelev, Alexey S.; Mazumdar, Anupam (2013). "Generalized ghost-free quadratic curvature gravity". *Classical and Quantum Gravity* **31**: 015022. arXiv:1308.2319. Bibcode:2014CQGra..31a5022B. doi:10.1088/0264-9381/31/1/015022.

[15] "Towards singularity and ghost free theories of gravity".

18.10 References

- Barker, B. M. (1978). "General scalar-tensor theory of gravity with constant G". *The Astrophysical Journal* **219**: 5. Bibcode:1978ApJ...219....5B. doi:10.1086/155749.

- Bekenstein, Jacob (1977). "Are particle rest masses variable? Theory and constraints from solar system experiments". *Physical Review D* **15** (6): 1458–1468. Bibcode:1977PhRvD..15.1458B. doi:10.1103/PhysRevD.15.1458.

- Bekenstein, J. D. (2004) Revised gravitation theory for the modified Newtonian dynamics paradigm. Phys. Rev. D 70, 083509

- Belinfante, F. J. and Swihart, J. C. (1957a) Phenomenological linear theory of gravitation Part I, Ann. Phys. 1, 168

- Belinfante, F. J. and Swihart, J. C. (1957b) Phenomenological linear theory of gravitation Part II, Ann. Phys. 2, 196

- Bergman, O. (1956) Scalar field theory as a theory of gravitation, Amer. J. Phys. 24, 39

- Bergmann, P. G. (1968) Comments on the scalar-tensor theory, Int. J. Theor. Phys. 1, 25-36

- Birkhoff, G. D. (1943) Matter, electricity and gravitation in flat space-time. Proc. Nat Acad. Sci. U.S. 29, 231-239

- Bollini, C. G., Giambiagi, J. J., and Tiomno, J. (1970) A linear theory of gravitation, Nuovo Com. Lett. 3, 65-70

- Burko, L.M. and Ori, A. (1995) On the Formation of Black Holes in Nonsymmetric Gravity, Phys. Rev. Lett. 75, 2455–2459

- Brans, C. and Dicke, R. H. (1961) Mach's principle and a relativistic theory of gravitation. Phys. Rev. 124, 925-935

- Carroll, Sean. Video lecture discussion on the possibilities and constraints to revision of the General Theory of Relativity. Dark Energy or Worse: Was Einstein Wrong?

- Cartan, É. (1922) Sur une généralisation de la notion de courbure de Riemann et les espaces à torsion. Acad. Sci. Paris, Comptes Rend. 174, 593-595

- Cartan, É. (1923) Sur les variétés à connexion affine et la théorie de la relativité généralisée. Annales Scientifiques de l'École Normale Superieure Sér. 3, 40, 325-412. http://archive.numdam.org/article/ASENS_1923_3_40__325_0.pdf

- Damour; Deser; McCarthy (1993). "Nonsymmetric Gravity has Unacceptable Global Asymptotics". arXiv:gr-qc/9312030 [gr-qc].

- Deser, S. and Laurent, B. E. (1968) Gravitation without self-interaction, Annals of Physics 50, 76-101

- Einstein, A. (1912a) Lichtgeschwindigkeit und Statik des Gravitationsfeldes. Annalen der Physik 38, 355-369

- Einstein, A. (1912b) Zur Theorie des statischen Gravitationsfeldes. Annalen der Physik 38, 443

- Einstein, A. and Grossmann, M. (1913), Z. Math Physik 62, 225

- Einstein, A. and Fokker, A. D. (1914) Die Nordströmsche Gravitationstheorie vom Standpunkt des absoluten Differentkalküls. Annalen der Physik 44, 321-328

- Einstein, A. (1916) Annalen der Physik 49, 769

- Einstein, A. (1917) Über die Spezielle und die Allgemeinen Relativitätstheorie, Gemeinverständlich, Vieweg, Braunschweig

- Fierz, M. and Pauli, W. (1939) On relativistic wave equations for particles of arbitrary spin in an electromagnetic field. Proc. Royal Soc. London 173, 211-232

- J.Foukzon, S.A.Podosenov, A.A.Potapov, E.Menkova, Bimetric Theory of Gravitational-Inertial Field in Riemannian and in Finsler-Lagrange Approximation 2010.http://arxiv.org/abs/1007.3290

- Hellings, Ronald; Nordtvedt, Kenneth (1973). "Vector-Metric Theory of Gravity". Physical Review D 7 (12): 3593–3602. Bibcode:1973PhRvD...7.3593H. doi:10.1103/PhysRevD.7.3593.

- Jordan, P.(1955) Schwerkraft und Weltall, Vieweg, Braunschweig

- Kustaanheimo, P. (1966) Route dependence of the gravitational redshift. Phys. Lett. 23, 75-77

- Kustaanheimo, P. E. and Nuotio, V. S. (1967) Publ. Astron. Obs. Helsinki No. 128

- Lang, R. (2002) Experimental foundations of general relativity, http://www.mppmu.mpg.de/~{}rlang/talks/melbourne2002.ppt

- Lee, D.; Lightman, A.; Ni, W. (1974). "Conservation laws and variational principles in metric theories of gravity". Physical Review D 10 (6): 1685–1700. Bibcode:1974PhRvD..10.1685L. doi:10.1103/PhysRevD.10.1685.

- Lightman, Alan; Lee, David (1973). "New Two-Metric Theory of Gravity with Prior Geometry". Physical Review D 8 (10): 3293–3302. Bibcode:1973PhRvD...8.3293L. doi:10.1103/PhysRevD.8.3293.

- Littlewood, D. E. (1953) Proceedings of the Cambridge Philosophical Society 49, 90-96

- Milne E. A. (1948) Kinematic Relativity, Clarendon Press, Oxford

- Misner, C. W., Thorne, K. S. and Wheeler, J. A. (1973) Gravitation, W. H. Freeman & Co.

- Moffat (1995). "Nonsymmetric Gravitational Theory". Physics Letters B 355 (3–4): 447–452. arXiv:gr-qc/9411006. Bibcode:1995PhLB..355..447M. doi:10.1016/0370-2693(95)00670-G.

- Moffat (2003). "Bimetric Gravity Theory, Varying Speed of Light and the Dimming of Supernovae". International Journal of Modern Physics D [Gravitation; Astrophysics and Cosmology] 12 (2): 281. arXiv:gr-qc/0202012. Bibcode:2003IJMPD..12..281M. doi:10.1142/S0218271803002366.

- Moffat (2005). "Gravitational Theory, Galaxy Rotation Curves and Cosmology without Dark Matter". Journal of Cosmology and Astroparticle Physics 2005 (5): 003–003. arXiv:astro-ph/0412195. Bibcode:2005JCAP...05..003M. doi:10.1088/1475-7516/2005/05/003.

- Moffat (2006). "Scalar-Tensor-Vector Gravity Theory". Journal of Cosmology and Astroparticle Physics 2006 (3): 004–004. arXiv:gr-qc/0506021. Bibcode:2006JCAP...03..004M. doi:10.1088/1475-7516/2006/03/004.

- Newton, I. (1686) Philosophiæ Naturalis Principia Mathematica

- Ni, Wei-Tou (1972). "Theoretical Frameworks for Testing Relativistic Gravity.IV. a Compendium of

Metric Theories of Gravity and Their POST Newtonian Limits". *The Astrophysical Journal* **176**: 769. Bibcode:1972ApJ...176..769N. doi:10.1086/151677.

- Ni, Wei-Tou (1973). "A New Theory of Gravity". *Physical Review D* **7** (10): 2880–2883. Bibcode:1973PhRvD...7.2880N. doi:10.1103/PhysRevD.7.2880.

- Nordtvedt, Jr., K. (1970) Post-Newtonian metric for a general class of scalar-tensor gravitational theories with observational consequences, The Astrophysical Journal 161, 1059

- Nordtvedt, Jr., K. and Will C. M. (1972) Conservation laws and preferred frames in relativistic gravity II, The Astrophysical Journal 177, 775

- Nordström, G. (1912), Relativitätsprinzip und Gravitation. *Phys. Zeitschr.* 13, 1126

- Nordström, G. (1913), Zur Theorie der Gravitation vom Standpunkt des Relativitätsprinzips, *Annalen der Physik* 42, 533

- Pais, A. (1982) *Subtle is the Lord*, Clarendon Press

- Page, C. and Tupper, B. O. J. (1968) Scalar gravitational theories with variable velocity of light, *Mon. Not. R. Astr. Soc.* 138, 67-72

- Papapetrou, A. (1954a) Zs Phys., 139, 518

- Papapetrou, A. (1954b) Math. Nach., 12, 129 & Math. Nach., 12, 143

- Poincaré, H. (1908) *Science and Method*

- Rastall, P. (1979) The Newtonian theory of gravitation and its generalization, *Canadian Journal of Physics* 57, 944-973

- Rosen, N. (1971) Theory of gravitation, *Physical Review D* 3, 2317

- Rosen, N. (1973) A bimetric theory of gravitation, *General Relativity and Gravitation* 4, 435-447.

- Rosen, N. (1975) A bimetric theory of gravitation II, *General Relativity and Gravitation* 6, 259-268

- Seljak, Uros, et al. (2010) Study Validates General Relativity on Cosmic Scale, abstract appears in physorg.com

- Thiry, Y. (1948) Les équations de la théorie unitaire de Kaluza, *Comptes Rendus Acad. Sci* (Paris) 226, 216

- Trautman, A. (1972) On the Einstein–Cartan equations I, Bulletin de l'Academie Polonaise des Sciences 20, 185-190

- Turyshev, S. G. (2006) Testing gravity in the solar system, http://star-www.st-and.ac.uk/~{ }hz4/workshop/workshopppt/turyshev.pdf

- Wagoner, Robert V. (1970). "Scalar-Tensor Theory and Gravitational Waves". *Physical Review D* **1** (12): 3209–3216. Bibcode:1970PhRvD...1.3209W. doi:10.1103/PhysRevD.1.3209.

- Whitehead, A.N. (1922) *The Principles of Relativity*, Cambridge Univ. Press

- Whitrow, G. J. and Morduch, G. E. (1960) General relativity and Lorentz-invariant theories of gravitation, *Nature* 188, 790-794

- Whitrow, G. J. and Morduch, G. E. (1965) Relativistic theories of gravitation, *Vistas in Astronomy* 6, 1-67

- Will, C. M. (1981, 1993) *Theory and Experiment in Gravitational Physics*, Cambridge Univ. Press

- Will, C. M. (2006) The Confrontation between General Relativity and Experiment, *Living Rev. Relativity* 9 (3), http://www.livingreviews.org/lrr-2006-3

- Will, C. M. and Nordtvedt Jr., K. (1972) Conservation laws and preferred frames in relativistic gravity I, *The Astrophysical Journal* 177, 757

- Yilmaz, H. (1958) New approach to general relativity, *Phys. Rev.* 111, 1417

- Yilmaz, H. (1973) New approach to relativity and gravitation, *Annals of Physics* 81, 179-200

Chapter 19

Phantom energy

Phantom energy is a hypothetical form of dark energy that is even more potent than the cosmological constant at increasing the expansion of the universe (i.e., it satisfies the equation of state with $w < -1$).The remarkable feature is that phantom energy possesses a negative kinetic energy.If it exists, it could cause the expansion of the universe to accelerate so quickly that a scenario known as the Big Rip would occur. If this is true, the expansion of the universe reaches an infinite degree in finite time, causing expansion to accelerate without bounds. This acceleration will pass the speed of light (since it involves expansion of the universe itself, not particles moving within it), causing more and more objects to leave our observable universe faster than its expansion, as light and information emitted from distant stars and other cosmic sources cannot "catch up" with the expansion. As the observable universe expands, objects will be unable to interact with each other via fundamental forces, and eventually the expansion will prevent any action of forces between any particles, even within atoms, "ripping apart" the universe. This characterizes the Big Rip as a possible end to the universe.

One application of phantom energy in 2007 was to a cyclic model of the universe.[1]

19.1   External links

- Robert R. Caldwell et al.: Phantom Energy and Cosmic Doomsday

19.2   References

[1] Lauris Baum and Paul Frampton (2007). "Turnaround In Cyclic Cosmology". *Phys. Rev. Lett.* **98** (7): 071301. arXiv:hep-th/0610213. Bibcode:2007PhRvL..98g1301B. doi:10.1103/PhysRevLett.98.071301. PMID 17359014.

Chapter 20

Self-energy

In most theoretical physics such as quantum field theory, a particle's **self-energy** Σ represents the contribution to the particle's energy, or effective mass, due to interactions between the particle and the system it is part of. For example, in electrostatics the self-energy of a given charge distribution is the energy required to assemble the distribution by bringing in the constituent charges from infinity, where the electric force goes to zero. In a condensed matter context relevant to electrons moving in a material, the self-energy represents the potential felt by the electron due to the surrounding medium's interactions with it: for example, the fact that electrons repel each other means that a moving electron polarizes (causes to displace) the electrons in its vicinity and this in turn changes the potential the moving electron feels; these and other effects are included in the self-energy. In basic terms, the self-energy is the energy that a particle has as a result of changes that it itself causes in its environment.

20.1 Characteristics

Mathematically, this energy is equal to the so-called on-the-mass-shell value of the proper self-energy *operator* (or proper mass *operator*) in the momentum-energy representation (more precisely, to $\hbar$ times this value). In this, or other representations (such as the space-time representation), the self-energy is pictorially (and economically) represented by means of Feynman diagrams, such as the one shown below. In this particular diagram, the three arrowed straight lines represent particles, or particle propagators, and the wavy line a particle-particle interaction; removing (or *amputating*) the left-most and the right-most straight lines in the diagram shown below (these so-called *external* lines correspond to prescribed values for, for instance, momentum and energy, or four-momentum), one retains a contribution to the self-energy operator (in, for instance, the momentum-energy representation). Using a small number of simple rules, each Feynman diagram can be readily expressed in its corresponding algebraic form.

In general, the on-the-mass-shell value of the self-energy operator in the momentum-energy representation is complex. In such cases, it is the real part of this self-energy that is identified with the physical self-energy (referred to above as particle's **self-energy**); the inverse of the imaginary part is a measure for the lifetime of the particle under investigation. For clarity, elementary excitations, or dressed particles (see quasi-particle), in interacting systems are distinct from stable particles in vacuum; their state functions consist of complicated superpositions of the eigenstates of the underlying many-particle system, which only, if at all, momentarily behave like those specific to isolated particles; the above-mentioned lifetime is the time over which a dressed particle behaves as if it were a single particle with well-defined momentum and energy.

The self-energy operator (often denoted by Σ, and less frequently by M) is related to the bare and dressed propagators (often denoted by G_0 and G respectively) via the Dyson equation (named after Freeman John Dyson):

$$G = G_0 + G_0 \Sigma G.$$

Multiplying on the left by the inverse G_0^{-1} of the operator G_0 and on the right by G^{-1} yields

$$\Sigma = G_0^{-1} - G^{-1}.$$

The photon and gluon do not get a mass through renormalization because gauge symmetry protects them from getting a mass. This is a consequence of the Ward identity. The W-boson and the Z-boson get their masses through the Higgs mechanism; they do undergo mass renormalization through the renormalization of the electroweak theory.

Neutral particles with internal quantum numbers can mix with each other through virtual pair production. The primary example of this phenomenon is the mixing of neutral kaons. Under appropriate simplifying assumptions this can be described without quantum field theory.

In chemistry, the self-energy or *Born energy* of an ion is the energy associated with the field of the ion itself.

In solid state and condensed-matter physics self-energies and a myriad related quasiparticle properties are calculated by Green's function methods and Green's function (many-body theory) of **interacting low-energy excitations** on the basis of electronic band structure calculations.

20.2   See also

- Quantum field theory

- QED vacuum

- Renormalization

- GW approximation

- Wheeler–Feynman absorber theory

20.3   References

- A. L. Fetter, and J. D. Walecka, *Quantum Theory of Many-Particle Systems* (McGraw-Hill, New York, 1971); (Dover, New York, 2003)

- J. W. Negele, and H. Orland, *Quantum Many-Particle Systems* (Westview Press, Boulder, 1998)

- A. A. Abrikosov, L. P. Gorkov and I. E. Dzyaloshinski (1963): *Methods of Quantum Field Theory in Statistical Physics* Englewood Cliffs: Prentice-Hall.

- Alexei M. Tsvelik (2007). *Quantum Field Theory in Condensed Matter Physics* (2nd ed.). Cambridge University Press. ISBN 0-521-52980-8.

- A. N. Vasil'ev *The Field Theoretic Renormalization Group in Critical Behavior Theory and Stochastic Dynamics* (Routledge Chapman & Hall 2004); ISBN 0-415-31002-4; ISBN 978-0-415-31002-4

Chapter 21

Spacetime

For other uses of this term, see Spacetime (disambiguation).

In physics, **spacetime** is any mathematical model that combines space and time into a single interwoven continuum. Since 300 BCE, the spacetime of our universe has historically been interpreted from a Euclidean space perspective, which regards space as consisting of three dimensions, and time as consisting of one dimension, the "fourth dimension". By combining space and time into a single manifold called Minkowski space in 1905, physicists have significantly simplified a large number of physical theories, as well as described in a more uniform way the workings of the universe at both the supergalactic and subatomic levels.

21.1   Explanation

In non-relativistic classical mechanics, the use of Euclidean space instead of spacetime is appropriate, because time is treated as universal with a constant rate of passage that is independent of the state of motion of an observer. In relativistic contexts, time cannot be separated from the three dimensions of space, because the observed rate at which time passes for an object depends on the object's velocity relative to the observer and also on the strength of gravitational fields, which can slow the passage of time for an object as seen by an observer outside the field.

In cosmology, the concept of spacetime combines space and time to a single abstract universe. Mathematically it is a manifold consisting of "events" which are described by some type of coordinate system. Typically **three spatial dimensions** (length, width, height), and one **temporal dimension** (time) are required. Dimensions are independent components of a coordinate grid needed to locate a point in a certain defined "space". For example, on the globe the latitude and longitude are two independent coordinates which together uniquely determine a location. In spacetime, a coordinate grid that spans the 3+1 dimensions locates

events (rather than just points in space), i.e., time is added as another dimension to the coordinate grid. This way the coordinates specify *where* and *when* events occur. However, the unified nature of spacetime and the freedom of coordinate choice it allows imply that to express the temporal coordinate in one coordinate system requires both temporal and spatial coordinates in another coordinate system. Unlike in normal spatial coordinates, there are still restrictions for how measurements can be made spatially and temporally (see Spacetime intervals). These restrictions correspond roughly to a particular mathematical model which differs from Euclidean space in its manifest symmetry.

Until the beginning of the 20th century, time was believed to be independent of motion, progressing at a fixed rate in all reference frames; however, following its prediction by special relativity, later experiments confirmed that time slows at higher speeds of the reference frame relative to another reference frame. Such slowing, called time dilation, is explained in special relativity theory. Many experiments have confirmed time dilation, such as the relativistic decay of muons from cosmic ray showers and the slowing of atomic clocks aboard a Space Shuttle relative to synchronized Earth-bound inertial clocks.[1] The duration of time can therefore vary according to events and reference frames.

When dimensions are understood as mere components of the grid system, rather than physical attributes of space, it is easier to understand the alternate dimensional views as being simply the result of coordinate transformations.

The term *spacetime* has taken on a generalized meaning beyond treating spacetime events with the normal 3+1 dimensions. It is really the combination of space and time. Other proposed spacetime theories include additional dimensions—normally spatial but there exist some speculative theories that include additional temporal dimensions and even some that include dimensions that are neither temporal nor spatial (e.g., superspace). How many dimensions are needed to describe the universe is still an open question. Speculative theories such as string theory predict 10 or 26 dimensions (with M-theory predicting 11 dimensions: 10

spatial and 1 temporal), but the existence of more than four dimensions would only appear to make a difference at the subatomic level.[2]

21.2   Spacetime in literature

Incas regarded space and time as a single concept, referred to as *pacha* (Quechua: *pacha*, Aymara: *pacha*).[3][4] The peoples of the Andes maintain a similar understanding.[5]

The idea of a unified spacetime is stated by Edgar Allan Poe in his essay on cosmology titled *Eureka* (1848) that "Space and duration are one". In 1895, in his novel *The Time Machine*, H. G. Wells wrote, "There is no difference between time and any of the three dimensions of space except that our consciousness moves along it", and that "any real body must have extension in four directions: it must have Length, Breadth, Thickness, and Duration".

Marcel Proust, in his novel *Swann's Way* (published 1913), describes the village church of his childhood's Combray as "a building which occupied, so to speak, four dimensions of space—the name of the fourth being Time".

21.2.1   Mathematical concept

In Encyclopedie, published in 1754, under the term *dimension* Jean le Rond d'Alembert speculated that duration (time) might be considered a fourth dimension if the idea was not too novel.[6]

Another early venture was by Joseph Louis Lagrange in his *Theory of Analytic Functions* (1797, 1813). He said, "One may view mechanics as a geometry of four dimensions, and mechanical analysis as an extension of geometric analysis".[7]

The ancient idea of the cosmos gradually was described mathematically with differential equations, differential geometry, and abstract algebra. These mathematical articulations blossomed in the nineteenth century as electrical technology stimulated men like Michael Faraday and James Clerk Maxwell to describe the reciprocal relations of electric and magnetic fields. Daniel Siegel phrased Maxwell's role in relativity as follows:

> [...] the idea of the propagation of forces at the velocity of light through the electromagnetic field as described by Maxwell's equations—rather than instantaneously at a distance—formed the necessary basis for relativity theory.[8]

Maxwell used vortex models in his papers on On Physical Lines of Force, but ultimately gave up on any substance but the electromagnetic field. Pierre Duhem wrote:

> [Maxwell] was not able to create the theory that he envisaged except by giving up the use of any model, and by extending by means of analogy the abstract system of electrodynamics to displacement currents.[9]

In Siegel's estimation, "this very abstract view of the electromagnetic fields, involving no visualizable picture of what is going on out there in the field, is Maxwell's legacy."[10] Describing the behaviour of electric fields and magnetic fields led Maxwell to view the combination as an electromagnetic field. These fields have a value at every point of spacetime. It is the intermingling of electric and magnetic manifestations, described by Maxwell's equations, that give spacetime its structure. In particular, the rate of motion of an observer determines the electric and magnetic profiles of the electromagnetic field. The propagation of the field is determined by the electromagnetic wave equation, which requires spacetime for description.

Spacetime was described as an affine space with quadratic form in Minkowski space of 1908.[11] In his 1914 textbook *The Theory of Relativity*, Ludwik Silberstein used biquaternions to represent events in Minkowski space. He also exhibited the Lorentz transformations between observers of differing velocities as biquaternion mappings. Biquaternions were described in 1853 by W. R. Hamilton, so while the physical interpretation was new, the mathematics was well known in English literature, making relativity an instance of applied mathematics.

The first inkling of general relativity in spacetime was articulated by W. K. Clifford. Description of the effect of gravitation on space and time was found to be most easily visualized as a "warp" or stretching in the geometrical fabric of space and time, in a smooth and continuous way that changed smoothly from point-to-point along the spacetime fabric. In 1947 James Jeans provided a concise summary of the development of spacetime theory in his book *The Growth of Physical Science*.[12]

21.3   Basic concepts

The basic elements of spacetime are events. In any given spacetime, an event is a unique position at a unique time. Because events are spacetime points, an example of an event in classical relativistic physics is (x, y, z, t), the location of an elementary (point-like) particle at a particular time. A spacetime itself can be viewed as the union of all events in the same way that a line is the union of all of its points, formally organized into a manifold, a space which can be described at small scales using coordinate systems.

A spacetime is independent of any observer.[13] However, in describing physical phenomena (which occur at certain moments of time in a given region of space), each observer chooses a convenient metrical coordinate system. Events are specified by four real numbers in any such coordinate system. The trajectories of elementary (point-like) particles through space and time are thus a continuum of events called the world line of the particle. Extended or composite objects (consisting of many elementary particles) are thus a union of many world lines twisted together by virtue of their interactions through spacetime into a "world-braid".

However, in physics, it is common to treat an extended object as a "particle" or "field" with its own unique (e.g., center of mass) position at any given time, so that the world line of a particle or light beam is the path that this particle or beam takes in the spacetime and represents the history of the particle or beam. The world line of the orbit of the Earth (in such a description) is depicted in two spatial dimensions x and y (the plane of the Earth's orbit) and a time dimension orthogonal to x and y. The orbit of the Earth is an ellipse in space alone, but its world line is a helix in spacetime.[14]

The unification of space and time is exemplified by the common practice of selecting a metric (the measure that specifies the interval between two events in spacetime) such that all four dimensions are measured in terms of units of distance: representing an event as $(x_0, x_1, x_2, x_3) = (ct, x, y, z)$ (in the Lorentz metric) or $(x_1, x_2, x_3, x_4) = (x, y, z, ict)$ (in the original Minkowski metric) where c is the speed of light.[15] The metrical descriptions of Minkowski Space and spacelike, lightlike, and timelike intervals given below follow this convention, as do the conventional formulations of the Lorentz transformation.

21.3.1 Spacetime intervals in flat space

In a Euclidean space, the separation between two points is measured by the distance between the two points. The distance is purely spatial, and is always positive. In spacetime, the displacement four-vector ΔR is given by the space displacement vector Δr and the time difference Δt between the events. The *spacetime interval*, also called *invariant interval*, between the two events, s^2,[16] is defined as:

$$s^2 = \Delta r^2 - c^2 \Delta t^2 \text{ (spacetime interval)},$$

where c is the speed of light. The choice of signs for s^2 above follows the space-like convention $(-+++)$.[17] Spacetime intervals may be classified into three distinct types, based on whether the temporal separation ($c^2 \Delta t^2$) or the spatial separation (Δr^2) of the two events is greater: timelike, light-like or space-like.

Certain types of world lines are called geodesics of the spacetime – straight lines in the case of Minkowski space and their closest equivalent in the curved spacetime of general relativity. In the case of purely time-like paths, geodesics are (locally) the paths of greatest separation (spacetime interval) as measured along the path between two events, whereas in Euclidean space and Riemannian manifolds, geodesics are paths of shortest distance between two points.[18][19] The concept of geodesics becomes central in general relativity, since geodesic motion may be thought of as "pure motion" (inertial motion) in spacetime, that is, free from any external influences.

Time-like interval

$$c^2 \Delta t^2 > \Delta r^2$$
$$s^2 < 0$$

For two events separated by a time-like interval, enough time passes between them that there could be a cause–effect relationship between the two events. For a particle traveling through space at less than the speed of light, any two events which occur to or by the particle must be separated by a time-like interval. Event pairs with time-like separation define a negative spacetime interval ($s^2 < 0$) and may be said to occur in each other's future or past. There exists a reference frame such that the two events are observed to occur in the same spatial location, but there is no reference frame in which the two events can occur at the same time.

The measure of a time-like spacetime interval is described by the proper time interval, $\Delta \tau$:

$$\Delta \tau = \sqrt{\Delta t^2 - \frac{\Delta r^2}{c^2}} \text{ (proper time interval)}.$$

The proper time interval would be measured by an observer with a clock traveling between the two events in an inertial reference frame, when the observer's path intersects each event as that event occurs. (The proper time interval defines a real number, since the interior of the square root is positive.)

Light-like interval

$$c^2 \Delta t^2 = \Delta r^2$$
$$s^2 = 0$$

In a light-like interval, the spatial distance between two events is exactly balanced by the time between the two events. The events define a spacetime interval of zero (

$s^2 = 0$). Light-like intervals are also known as "null" intervals.

Events which occur to or are initiated by a photon along its path (i.e., while traveling at c , the speed of light) all have light-like separation. Given one event, all those events which follow at light-like intervals define the propagation of a light cone, and all the events which preceded from a light-like interval define a second (graphically inverted, which is to say "*pastward*") light cone.

Space-like interval

$$c^2 \Delta t^2 < \Delta r^2$$
$$s^2 > 0$$

When a space-like interval separates two events, not enough time passes between their occurrences for there to exist a causal relationship crossing the spatial distance between the two events at the speed of light or slower. Generally, the events are considered not to occur in each other's future or past. There exists a reference frame such that the two events are observed to occur at the same time, but there is no reference frame in which the two events can occur in the same spatial location.

For these space-like event pairs with a positive spacetime interval ($s^2 > 0$), the measurement of space-like separation is the proper distance, $\Delta \sigma$:

$$\Delta \sigma = \sqrt{s^2} = \sqrt{\Delta r^2 - c^2 \Delta t^2} \text{ (proper distance)}.$$

Like the proper time of time-like intervals, the proper distance of space-like spacetime intervals is a real number value.

21.3.2　Interval as area

The interval has been presented as the area of an oriented rectangle formed by two events and isotropic lines through them. Time-like or space-like separations correspond to oppositely oriented rectangles, one type considered to have rectangles of negative area. The case of two events separated by light corresponds to the rectangle degenerating to the segment between the events and zero area.[20] The transformations leaving interval-length invariant are the area-preserving squeeze mappings.

The parameters traditionally used rely on quadrature of the hyperbola, which is the natural logarithm. This transcendental function is essential in mathematical analysis as its inverse unites circular functions and hyperbolic functions:

The exponential function, e^t, t a real number, used in the hyperbola (e^t, e^{-t}), generates hyperbolic sectors and the hyperbolic angle parameter. The functions cosh and sinh, used with rapidity as hyperbolic angle, provide the common representation of squeeze in the form $\begin{pmatrix} \cosh \phi & \sinh \phi \\ \sinh \phi & \cosh \phi \end{pmatrix}$, or as the split-complex unit $e^{j\phi} = \cosh \phi + j \sinh \phi$.

21.4　Mathematics of spacetimes

For physical reasons, a spacetime continuum is mathematically defined as a four-dimensional, smooth, connected Lorentzian manifold (M, g) . This means the smooth Lorentz metric g has signature $(3, 1)$. The metric determines the geometry of spacetime, as well as determining the geodesics of particles and light beams. About each point (event) on this manifold, coordinate charts are used to represent observers in reference frames. Usually, Cartesian coordinates (x, y, z, t) are used. Moreover, for simplicity's sake, units of measurement are usually chosen such that the speed of light c is equal to 1.

A reference frame (observer) can be identified with one of these coordinate charts; any such observer can describe any event p . Another reference frame may be identified by a second coordinate chart about p . Two observers (one in each reference frame) may describe the same event p but obtain different descriptions.

Usually, many overlapping coordinate charts are needed to cover a manifold. Given two coordinate charts, one containing p (representing an observer) and another containing q (representing another observer), the intersection of the charts represents the region of spacetime in which both observers can measure physical quantities and hence compare results. The relation between the two sets of measurements is given by a non-singular coordinate transformation on this intersection. The idea of coordinate charts as local observers who can perform measurements in their vicinity also makes good physical sense, as this is how one actually collects physical data—locally.

For example, two observers, one of whom is on Earth, but the other one who is on a fast rocket to Jupiter, may observe a comet crashing into Jupiter (this is the event p). In general, they will disagree about the exact location and timing of this impact, i.e., they will have different 4-tuples (x, y, z, t) (as they are using different coordinate systems). Although their kinematic descriptions will differ, dynamical (physical) laws, such as momentum conservation and the first law of thermodynamics, will still hold. In fact, relativity theory requires more than this in the sense that it stipulates these (and all other physical) laws must take the same form in all coordinate systems. This introduces tensors into relativity, by which all physical quantities are represented.

Geodesics are said to be time-like, null, or space-like if the tangent vector to one point of the geodesic is of this nature. Paths of particles and light beams in spacetime are represented by time-like and null (light-like) geodesics, respectively.

21.4.1 Topology

Main article: Spacetime topology

The assumptions contained in the definition of a spacetime are usually justified by the following considerations.

The connectedness assumption serves two main purposes. First, different observers making measurements (represented by coordinate charts) should be able to compare their observations on the non-empty intersection of the charts. If the connectedness assumption were dropped, this would not be possible. Second, for a manifold, the properties of connectedness and path-connectedness are equivalent, and one requires the existence of paths (in particular, geodesics) in the spacetime to represent the motion of particles and radiation.

Every spacetime is paracompact. This property, allied with the smoothness of the spacetime, gives rise to a smooth linear connection, an important structure in general relativity. Some important theorems on constructing spacetimes from compact and non-compact manifolds include the following:

- A compact manifold can be turned into a spacetime if, and only if, its Euler characteristic is 0. (Proof idea: the existence of a Lorentzian metric is shown to be equivalent to the existence of a nonvanishing vector field.)

- Any non-compact 4-manifold can be turned into a spacetime.[21]

21.4.2 Spacetime symmetries

Main article: Spacetime symmetries

Often in relativity, spacetimes that have some form of symmetry are studied. As well as helping to classify spacetimes, these symmetries usually serve as a simplifying assumption in specialized work. Some of the most popular ones include:

- Axisymmetric spacetimes

- Spherically symmetric spacetimes

- Static spacetimes

- Stationary spacetimes

21.4.3 Causal structure

Main article: Causal structure
See also: Causality (physics) and Causality

The causal structure of a spacetime describes causal relationships between pairs of points in the spacetime based on the existence of certain types of curves joining the points.

21.5 Spacetime in special relativity

Main article: Minkowski space

The geometry of spacetime in special relativity is described by the Minkowski metric on R^4. This spacetime is called Minkowski space. The Minkowski metric is usually denoted by η and can be written as a four-by-four matrix:

$$\eta_{ab} = \text{diag}(1, -1, -1, -1)$$

where the Landau–Lifshitz time-like convention is being used. A basic assumption of relativity is that coordinate transformations must leave spacetime intervals invariant. Intervals are invariant under Lorentz transformations. This invariance property leads to the use of four-vectors (and other tensors) in describing physics.

Strictly speaking, one can also consider events in Newtonian physics as a single spacetime. This is Galilean–Newtonian relativity, and the coordinate systems are related by Galilean transformations. However, since these preserve spatial and temporal distances independently, such a spacetime can be decomposed into spatial coordinates plus temporal coordinates, which is not possible in the general case.

21.6 Spacetime in general relativity

In general relativity, it is assumed that spacetime is curved by the presence of matter (energy), this curvature being represented by the Riemann tensor. In special relativity, the Riemann tensor is identically zero, and so this concept of "non-curvedness" is sometimes expressed by the statement *Minkowski spacetime is flat.*

The earlier discussed notions of time-like, light-like and space-like intervals in special relativity can similarly be used to classify one-dimensional curves through curved spacetime. A time-like curve can be understood as one where the interval between any two infinitesimally close events on the curve is time-like, and likewise for light-like and space-like curves. Technically the three types of curves are usually defined in terms of whether the tangent vector at each point on the curve is time-like, light-like or space-like. The world line of a slower-than-light object will always be a time-like curve, the world line of a massless particle such as a photon will be a light-like curve, and a space-like curve could be the world line of a hypothetical tachyon. In the local neighborhood of any event, time-like curves that pass through the event will remain inside that event's past and future light cones, light-like curves that pass through the event will be on the surface of the light cones, and space-like curves that pass through the event will be outside the light cones. One can also define the notion of a three-dimensional "space-like hypersurface", a continuous three-dimensional "slice" through the four-dimensional property with the property that every curve that is contained entirely within this hypersurface is a space-like curve.[22]

Many spacetime continua have physical interpretations which most physicists would consider bizarre or unsettling. For example, a compact spacetime has closed timelike curves, which violate our usual ideas of causality (that is, future events could affect past ones). For this reason, mathematical physicists usually consider only restricted subsets of all the possible spacetimes. One way to do this is to study "realistic" solutions of the equations of general relativity. Another way is to add some additional "physically reasonable" but still fairly general geometric restrictions and try to prove interesting things about the resulting spacetimes. The latter approach has led to some important results, most notably the Penrose–Hawking singularity theorems.

21.7 Quantized spacetime

Main article: Quantum spacetime

In general relativity, spacetime is assumed to be smooth and continuous—and not just in the mathematical sense. In the theory of quantum mechanics, there is an inherent discreteness present in physics. In attempting to reconcile these two theories, it is sometimes postulated that spacetime should be quantized at the very smallest scales. Current theory is focused on the nature of spacetime at the Planck scale. Causal sets, loop quantum gravity, string theory, causal dynamical triangulation, and black hole thermodynamics all predict a quantized spacetime with agreement on the order of magnitude. Loop quantum gravity makes precise predic-

tions about the geometry of spacetime at the Planck scale.

Spin networks provide a language to describe quantum geometry of space. Spin foam does the same job on spacetime. A spin network is a one-dimensional graph, together with labels on its vertices and edges which encodes aspects of a spatial geometry.

21.8 See also

- Anthropic_principle § Applications of the principle §§ Spacetime
- Basic introduction to the mathematics of curved spacetime
- Four-vector
- Frame-dragging
- Global spacetime structure
- Hole argument
- List of mathematical topics in relativity
- Local spacetime structure
- Lorentz invariance
- Manifold
- Mathematics of general relativity
- Metric space
- Philosophy of space and time
- Relativity of simultaneity
- Strip photography
- World manifold

21.9 References

[1] Ashby, Neil (2003). "Relativity in the Global Positioning System" (PDF). *Living Reviews in Relativity* 6: 16. Bibcode:2003LRR.....6....1A. doi:10.12942/lrr-2003-1.

[2] Kopeikin, Sergei; Efroimsky, Michael; Kaplan, George (2011). *Relativistic Celestial Mechanics of the Solar System*. John Wiley & Sons. p. 157. ISBN 3527634576. Retrieved 2016-02-28. Extract of page 157

[3] Atuq Eusebio Manga Qespi, Instituto de lingüística y Cultura Amerindia de la Universidad de Valencia. *Pacha: un concepto andino de espacio y tiempo*. Revista española de Antropología Americana, 24, p. 155–189. Edit. Complutense, Madrid. 1994

[4] Paul Richard Steele, Catherine J. Allen, *Handbook of Inca mythology*, p. 86, (ISBN 1-57607-354-8)

[5] Shirley Ardener, University of Oxford, *Women and space: ground rules and social maps*, p. 36 (ISBN 0-85496-728-1)

[6] Jean d'Alembert (1754) Dimension from ARTFL Encyclopedie project

[7] R.C. Archibald (1914) *Time as a fourth dimension Bulletin of the American Mathematical Society 20:409.*

[8] Daniel M. Siegel (2014) "Maxwell's contributions to electricity and magnetism", chapter 10 in *James Clerk Maxwell: Perspectives on his Life and Work*, Raymond Flood, Mark McCartney, Andrew Whitaker, editors, Oxford University Press ISBN 978-0-19-966437-5

[9] Pierre Duhem (1954) *The Aim and Structure of Physical Theory*, page 98, Princeton University Press

[10] Siegel 2014 p 191

[11] Minkowski, Hermann (1909), "Raum und Zeit", *Physikalische Zeitschrift* **10**: 75–88

   • Various English translations on Wikisource: Space and Time.

[12] James Jeans (1947) The Growth of Physical Science, "Space-time", pp. 205–301, link from Internet Archive

[13] Matolcsi, Tamás (1994). *Spacetime Without Reference Frames*. Budapest: Akadémiai Kiadó.

[14] Ellis, G. F. R.; Williams, Ruth M. (2000). *Flat and curved space–times* (2nd ed.). Oxford University Press. p. 9. ISBN 0-19-850657-0.

[15] Petkov, Vesselin (2010). *Minkowski Spacetime: A Hundred Years Later*. Springer. p. 70. ISBN 90-481-3474-9. Retrieved 2016-02-28., Section 3.4, p. 70

[16] Note that the term *spacetime interval* is applied by several authors to the quantity s^2 and not to s. The reason that the quantity s^2 is used and not s is that s^2 can be positive, zero or negative, and is a more generally convenient and useful quantity than the Minkowski norm with a timelike/null/spacelike distinguisher: the pair $(\sqrt{|s^2|}, \operatorname{sgn}(s^2))$. Despite the notation, it should not be regarded as the square of a number, but as a symbol. The cost for this convenience is that this "interval" is quadratic in linear separation along a straight line.

[17] More generally the spacetime interval in flat space can be written as $s^2 = g_{\alpha\beta}\Delta x^\alpha \Delta x^\beta$ with metric tensor g independent of spacetime position.

[18] This characterization is not universal: both the arcs between two points of a great circle on a sphere are geodesics.

[19] Berry, Michael V. (1989). *Principles of Cosmology and Gravitation*. CRC Press. p. 58. ISBN 0-85274-037-9. Retrieved 2016-02-28. Extract of page 58, caption of Fig. 25

[20] I. M. Yaglom (1979) *A Simple Non-Euclidean Geometry and its Physical Basis*, page 178, Springer, ISBN 0387-90332-1, MR 520230

[21] Geroch, Robert; Horowitz, Gary T. (1979). "Chapter 5. Global structure of spacetimes". In Hawking, S.W.; Israel, W. *General Relativity: An Einstein Centenary Survey*. Cambridge University Press. p. 219. ISBN 0521299284.

[22] See "Quantum Spacetime and the Problem of Time in Quantum Gravity" by Leszek M. Sokolowski, where on this page he writes "Each of these hypersurfaces is spacelike, in the sense that every curve, which entirely lies on one of such hypersurfaces, is a spacelike curve." More commonly a spacelike hypersurface is defined technically as a surface such that the normal vector at every point is time-like, but the definition above may be somewhat more intuitive.

21.10   Further reading

• Albert Einstein on Space-Time 13th edition Encyclopedia Britannica Historical: Albert Einstein's 1926 article

• Ehrenfest, Paul (1920) "How do the fundamental laws of physics make manifest that Space has 3 dimensions?" *Annalen der Physik 366*: 440.

• George F. Ellis and Ruth M. Williams (1992) *Flat and curved space–times*. Oxford Univ. Press. ISBN 0-19-851164-7

• Encyclopedia of Space-time and gravitation Scholarpedia Expert articles

21.11   External links

• http://universaltheory.org

• Barrow, John D.; Tipler, Frank J. (1988). *The Anthropic Cosmological Principle*. Oxford University Press. ISBN 978-0-19-282147-8. LCCN 87028148.

• Isenberg, J. A. (1981). "Wheeler–Einstein–Mach spacetimes". *Phys. Rev. D* **24** (2): 251–256. Bibcode:1981PhRvD..24..251I. doi:10.1103/PhysRevD.24.251.

• Kant, Immanuel (1929) "Thoughts on the true estimation of living forces" in J. Handyside, trans., *Kant's Inaugural Dissertation and Early Writings on Space*. Univ. of Chicago Press.

• Lorentz, H. A., Einstein, Albert, Minkowski, Hermann, and Weyl, Hermann (1952) *The Principle of Relativity: A Collection of Original Memoirs*. Dover.

- Lucas, John Randolph (1973) *A Treatise on Time and Space*. London: Methuen.

- Penrose, Roger (2004). *The Road to Reality*. Oxford: Oxford University Press. ISBN 0-679-45443-8. Chpts. 17–18.

- Poe, Edgar A. (1848). *Eureka; An Essay on the Material and Spiritual Universe*. Hesperus Press Limited. ISBN 1-84391-009-8.

- Robb, A. A. (1936). *Geometry of Time and Space*. University Press.

- Erwin Schrödinger (1950) *Space–time structure*. Cambridge Univ. Press.

- Schutz, J. W. (1997). *Independent axioms for Minkowski Space–time*. Addison-Wesley Longman. ISBN 0-582-31760-6.

- Tangherlini, F. R. (1963). "Schwarzschild Field in n Dimensions and the Dimensionality of Space Problem". *Nuovo Cimento* **14** (27): 636.

- Taylor, E. F.; Wheeler, John A. (1963). *Spacetime Physics*. W. H. Freeman. ISBN 0-7167-2327-1.

- Wells, H.G. (2004). *The Time Machine*. New York: Pocket Books. ISBN 0-671-57554-6. (pp. 5–6)

- Stanford Encyclopedia of Philosophy: "Space and Time: Inertial Frames" by Robert DiSalle.

Chapter 22

Neutrino

Not to be confused with Neutralino.

For other uses, see Neutrino (disambiguation).

A **neutrino** (/nuːˈtriːnoʊ/ or /njuːˈtriːnoʊ/) (denoted by the Greek letter ν) is a lepton, an elementary particle with half-integer spin, that interacts only via the weak subatomic force[4] and gravity. The mass of the neutrino is tiny compared to other subatomic particles. Neutrinos, named as such because they are electrically neutral, are not affected by the electromagnetic or strong forces. The weak force is a very short-range interaction, and gravity is extremely weak on the subatomic scale. Thus, neutrinos typically pass through normal matter unimpeded and undetected.

Neutrinos come in three flavors: electron neutrinos (ν e), muon neutrinos (ν μ), and tau neutrinos (ν τ), associated with the electron, muon, and tau, respectively. Each neutrino also has a corresponding antiparticle, called an *antineutrino*, which also has no electric charge and half-integer spin. Neutrinos are produced such that there is no overall change in lepton number; that is, electron neutrinos are produced together with positrons (anti-electrons), and electron antineutrinos are produced with electrons.[5][6][7]

Neutrinos can be created in several ways, including in certain types of radioactive decay, in nuclear reactions such as those that take place in the Sun, in nuclear reactors, when cosmic rays hit atoms, and in supernovae. The majority of neutrinos in the vicinity of the Earth are from nuclear reactions in the Sun. About 65 billion (6.5×10^{10}) solar neutrinos per second pass through every square centimeter perpendicular to the direction of the Sun in the region of the Earth.[8]

Neutrinos oscillate between different flavors in flight. That is, an electron neutrino produced in a beta decay reaction may arrive in a detector as a muon or tau neutrino. This oscillation requires that the different neutrino flavors have different masses, and although the value of the masses is not known, experiments have shown that these masses are tiny. From cosmological measurements, it has been calculated that the sum of the three neutrino masses must be less than one millionth that of the electron.[9]

There are several active research areas involving the neutrino. Large neutrino detectors near nuclear reactors or in neutrino beams from particle accelerators attempt to measure the neutrino masses and determine the precise values for the magnitude and rates of oscillations between neutrino flavors. These experiments are also searching for the existence of CP violation in the neutrino sector; that is, whether or not the laws of physics treat neutrinos and antineutrinos differently. Many are looking for evidence of a sterile neutrino, a fourth neutrino flavor that does not interact with matter like the three known neutrino flavors. There are also experiments searching for neutrinoless double-beta decay, which, if it exists, would require that the neutrino and antineutrino are really the same particle. Then there are solar and cosmic neutrino experiments, which use neutrinos from space to understand the universe around us. Neutrinos are also the only identified candidate for dark matter, specifically hot dark matter.[10]

22.1 History

22.1.1 Pauli's proposal

The neutrino[nb 1] was postulated first by Wolfgang Pauli in 1930 to explain how beta decay could conserve energy, momentum, and angular momentum (spin). In contrast to Niels Bohr, who proposed a statistical version of the conservation laws to explain the observed continuous energy spectra in beta decay, Pauli hypothesized an undetected particle that he called a "neutron", using the same *-on* ending employed for naming both the proton and the electron. He considered that the new particle was emitted from the nucleus together with the electron or beta particle in the process of beta decay.[11][nb 2]

James Chadwick discovered a much more massive nuclear particle in 1932 and also named it a neutron, leaving two kinds of particles with the same name. Pauli earlier (in

1930) had used the term "neutron" for both the neutral particle that conserved energy in beta decay, and a presumed neutral particle in the nucleus, and initially did not consider these two neutral particles as distinct from each other.[11] The word "neutrino" entered the scientific vocabulary through Enrico Fermi, who used it during a conference in Paris in July 1932 and at the Solvay Conference in October 1933, where Pauli also employed it. The name (the Italian equivalent of "little neutral one") was jokingly coined by Edoardo Amaldi during a conversation with Fermi at the Institute of physics of via Panisperna in Rome, in order to distinguish this light neutral particle from Chadwick's neutron.[12]

In Fermi's theory of beta decay, Chadwick's large neutral particle could decay to a proton, electron, and the smaller neutral particle (flavored as an electron antineutrino):

$$n0 \rightarrow p+ + e- + \nu_e$$

Fermi's paper, written in 1934, unified Pauli's neutrino with Paul Dirac's positron and Werner Heisenberg's neutron–proton model and gave a solid theoretical basis for future experimental work. However, the journal Nature rejected Fermi's paper, saying that the theory was "too remote from reality". He submitted the paper to an Italian journal, which accepted it, but the general lack of interest in his theory at that early date caused him to switch to experimental physics.[13]:24 [14]

However by 1934 there was experimental evidence against Bohr's idea that energy conservation is invalid for beta decay. At the Solvay conference of that year, measurements of the energy spectra of beta particles (electrons) were reported, showing that there is a strict limit on the energy of electrons from each type of beta decay. Such a limit is not expected if the conservation of energy is invalid, in which case any amount of energy would be statistically available in at least a few decays. The natural explanation of the beta decay spectrum as first measured in 1934 was that only a limited (and conserved) amount of energy was available, and a new particle was sometimes taking a varying fraction of this limited energy, leaving the rest for the beta particle. Pauli made use of the occasion to publicly emphasize that the still-undetected "neutrino" must be an actual particle.[13]:25

22.1.2 Direct detection

In 1942, Wang Ganchang first proposed the use of beta capture to experimentally detect neutrinos.[15] In the 20 July 1956 issue of *Science*, Clyde Cowan, Frederick Reines, F. B. Harrison, H. W. Kruse, and A. D. McGuire published

Clyde Cowan conducting the neutrino experiment c. 1956

confirmation that they had detected the neutrino,[16][17] a result that was rewarded almost forty years later with the 1995 Nobel Prize.[18]

In this experiment, now known as the Cowan–Reines neutrino experiment, antineutrinos created in a nuclear reactor by beta decay reacted with protons to produce neutrons and positrons:

$$\nu_e + p+ \rightarrow n0 + e+$$

The positron quickly finds an electron, and they annihilate each other. The two resulting gamma rays (γ) are detectable. The neutron can be detected by its capture on an appropriate nucleus, releasing a gamma ray. The coincidence of both events – positron annihilation and neutron capture – gives a unique signature of an antineutrino interaction.

22.1.3 Neutrino flavor

The antineutrino discovered by Cowan and Reines is the antiparticle of the electron neutrino. In 1962, Leon M. Lederman, Melvin Schwartz and Jack Steinberger showed that more than one type of neutrino exists by first detecting interactions of the muon neutrino (already hypothesised with the name *neutretto*),[19] which earned them the 1988 Nobel Prize in Physics. When the third type of lepton, the tau, was discovered in 1975 at the Stanford Linear Accelerator Center, it too was expected to have an associated neutrino (the tau neutrino). First evidence for this third neutrino type came from the observation of missing energy and momentum in tau decays analogous to the beta decay leading to the discovery of the electron neutrino. The first detection of tau neutrino interactions was announced in summer of 2000 by the DONUT collaboration at Fermilab; its existence had already been inferred by both theoretical consistency and experimental data from the Large Electron–Positron Collider.[20]

22.1.4 Solar neutrino problem

Main article: Solar neutrino problem

Starting in the late 1960s, several experiments found that the number of electron neutrinos arriving from the Sun was between one third and one half the number predicted by the Standard Solar Model. This discrepancy, which became known as the solar neutrino problem, remained unresolved for some thirty years. It was resolved by discovery of neutrino oscillation and mass. (The Standard Model of particle physics had assumed that neutrinos are massless and cannot change flavor. However, if neutrinos had mass, they could change flavor, or *oscillate* between flavors).

22.1.5 Oscillation

Main article: Neutrino oscillation

A practical method for investigating neutrino oscillations was first suggested by Bruno Pontecorvo in 1957 using an analogy with kaon oscillations; over the subsequent 10 years he developed the mathematical formalism and the modern formulation of vacuum oscillations. In 1985 Stanislav Mikheyev and Alexei Smirnov (expanding on 1978 work by Lincoln Wolfenstein) noted that flavor oscillations can be modified when neutrinos propagate through matter. This so-called Mikheyev–Smirnov–Wolfenstein effect (MSW effect) is important to understand because many neutrinos emitted by fusion in the Sun pass through the dense matter in the solar core (where essentially all solar fusion takes place) on their way to detectors on Earth.

Starting in 1998, experiments began to show that solar and atmospheric neutrinos change flavors (see Super-Kamiokande and Sudbury Neutrino Observatory). This resolved the solar neutrino problem: the electron neutrinos produced in the Sun had partly changed into other flavors which the experiments could not detect.

Although individual experiments, such as the set of solar neutrino experiments, are consistent with non-oscillatory mechanisms of neutrino flavor conversion, taken altogether, neutrino experiments imply the existence of neutrino os-

cillations. Especially relevant in this context are the reactor experiment KamLAND and the accelerator experiments such as MINOS. The KamLAND experiment has indeed identified oscillations as the neutrino flavor conversion mechanism involved in the solar electron neutrinos. Similarly MINOS confirms the oscillation of atmospheric neutrinos and gives a better determination of the mass squared splitting.[21] Takaaki Kajita of Japan and Arthur B. McDonald of Canada received the 2015 Nobel Prize for Physics for their landmark finding, theoretical and experimental, that neutrinos can change flavors.

22.1.6 Supernova neutrinos

See also: Supernova Early Warning System

Raymond Davis, Jr. and Masatoshi Koshiba were jointly awarded the 2002 Nobel Prize in Physics; Davis for his pioneer work on cosmic neutrinos and Koshiba for the first real time observation of supernova neutrinos. The detection of solar neutrinos, and of neutrinos of the SN 1987A supernova in 1987 marked the beginning of neutrino astronomy.[22] In an average supernova, approximately 10^{57} (an Octodecillion) neutrinos are released.[22]

22.2 Properties and reactions

The neutrino has half-integer spin ($\frac{1}{2}$) and is therefore a fermion. Neutrinos interact primarily through the weak force. The discovery of neutrino flavor oscillations implies that neutrinos have mass. The existence of a neutrino mass strongly suggests the existence of a tiny neutrino magnetic moment[23] of the order of 10^{-19} μB, allowing the possibility that neutrinos may interact electromagnetically as well. An experiment done by C. S. Wu at Columbia University showed that neutrinos always have left-handed chirality.[24] It is very hard to uniquely identify neutrino interactions among the natural background of radioactivity. For this reason, in early experiments a special reaction channel was chosen to facilitate the identification: the interaction of an antineutrino with one of the hydrogen nuclei in the water molecules. A hydrogen nucleus is a single proton, so simultaneous nuclear interactions, which would occur within a heavier nucleus, don't need to be considered for the detection experiment. Within a cubic metre of water placed right outside a nuclear reactor, only relatively few such interactions can be recorded, but the setup is now used for measuring the reactor's plutonium production rate.

22.2.1 Mikheyev–Smirnov–Wolfenstein effect

Main article: Mikheyev–Smirnov–Wolfenstein effect

Neutrinos traveling through matter, in general, undergo a process analogous to light traveling through a transparent material. This process is not directly observable because it does not produce ionizing radiation, but gives rise to the MSW effect. Only a small fraction of the neutrino's energy is transferred to the material.

22.2.2 Nuclear reactions

Neutrinos can interact with a nucleus, changing it to another nucleus. This process is used in radiochemical neutrino detectors. In this case, the energy levels and spin states within the target nucleus have to be taken into account to estimate the probability for an interaction. In general the interaction probability increases with the number of neutrons and protons within a nucleus.

22.2.3 Induced fission

Very much like neutrons do in nuclear reactors, neutrinos can induce fission reactions within heavy nuclei.[25] So far, this reaction has not been measured in a laboratory, but is predicted to happen within stars and supernovae. The process affects the abundance of isotopes seen in the universe.[26] Neutrino fission of deuterium nuclei has been observed in the Sudbury Neutrino Observatory, which uses a heavy water detector.

22.2.4 No self interaction

Observations of the cosmic microwave background suggest that neutrinos do not interact with themselves.[27]

22.2.5 Types

There are three known types (*flavors*) of neutrinos: electron neutrino ν
e, muon neutrino ν
μ and tau neutrino ν
τ, named after their partner leptons in the Standard Model (see table at right). The current best measurement of the number of neutrino types comes from observing the decay of the Z boson. This particle can decay into any light neutrino and its antineutrino, and the more types of light neutrinos[nb 3] available, the shorter the lifetime of the Z boson. Measurements of the Z lifetime have shown that the number of light neutrino types is 3.[23] The correspondence between the six quarks in the Standard Model and the six leptons, among them the three neutrinos, suggests to physicists' intuition that there should be exactly three types of neutrino. However, actual proof that there are only three kinds of neutrinos remains an elusive goal of particle physics.

The possibility of *sterile* neutrinos—relatively light neutrinos which do not participate in the weak interaction but which could be created through flavor oscillation (see below)—is unaffected by these Z-boson-based measurements, and the existence of such particles is in fact hinted by experimental data from the LSND experiment. However, the currently running MiniBooNE experiment suggested, until recently, that sterile neutrinos are not required to explain the experimental data,[28] although the latest research into this area is on-going and anomalies in the MiniBooNE data may allow for exotic neutrino types, including sterile neutrinos.[29] A recent re-analysis of reference electron spectra data from the Institut Laue-Langevin[30] has also hinted at a fourth, sterile neutrino.[31]

Recently analyzed data from the Wilkinson Microwave Anisotropy Probe of the cosmic background radiation is compatible with either three or four types of neutrinos. It is hoped that the addition of two more years of data from the probe will resolve this uncertainty.[32]

22.2.6 Antineutrinos

Antineutrinos, the antiparticles of neutrinos, are neutral particles produced in nuclear beta decay. These are emitted during beta particle emissions, in which a neutron decays into a proton, electron, and antineutrino. They have a spin of ½, and are part of the lepton family of particles. All antineutrinos observed thus far possess right-handed helicity (i.e. only one of the two possible spin states has ever been seen), while neutrinos are left-handed. Antineutrinos, like neutrinos, interact with other matter only through the gravitational and weak forces, making them very difficult to detect experimentally. Neutrino oscillation experiments indicate that antineutrinos have mass, but beta decay experiments constrain that mass to be very small. A neutrino–antineutrino interaction has been suggested in attempts to form a composite photon with the neutrino theory of light.

Because antineutrinos and neutrinos are neutral particles, it is possible that they are the same particle. Particles that have this property are known as Majorana particles. Majorana neutrinos have the property that the neutrino and antineutrino could be distinguished only by chirality; what experiments observe as a difference between the neutrino and antineutrino could simply be due to one particle with two possible chiralities. If neutrinos are indeed Majorana particles, neutrinoless double beta decay, as well as a range of other lepton number violating phenomena, would be allowed. Several experiments have been and are being conducted to search for this process.

Researchers around the world have begun to investigate the possibility of using antineutrinos for reactor monitoring in the context of preventing the proliferation of nuclear weapons.[33][34][35]

Antineutrinos were first detected as a result of their interaction with protons in a large tank of water. This was installed next to a nuclear reactor as a controllable source of the antineutrinos. (See: Cowan–Reines neutrino experiment)

Only antineutrinos, not neutrinos, take part in the Glashow resonance.

22.2.7 Flavor oscillations

Main article: Neutrino oscillation

Neutrinos are most often created or detected with a well defined flavor (electron, muon, tau). However, in a phenomenon known as neutrino flavor oscillation, neutrinos are able to oscillate among the three available flavors while they propagate through space. Specifically, this occurs because the neutrino flavor eigenstates are not the same as the neutrino mass eigenstates (simply called 1, 2, 3). This allows for a neutrino that was produced as an electron neutrino at a given location to have a calculable probability to be detected as either a muon or tau neutrino after it has traveled to another location. This quantum mechanical effect was first hinted by the discrepancy between the number of electron neutrinos detected from the Sun's core failing to match the expected numbers, dubbed as the "solar neutrino problem". In the Standard Model the existence of flavor oscillations implies nonzero differences between the neutrino masses, because the amount of mixing between neutrino flavors at a given time depends on the differences between their squared masses. There are other possibilities in which neutrino can oscillate even if they are massless. If Lorentz symmetry is not an exact symmetry, neutrinos can experience Lorentz-violating oscillations.[36]

It is possible that the neutrino and antineutrino are in fact the same particle, a hypothesis first proposed by the Italian physicist Ettore Majorana. The neutrino could transform into an antineutrino (and vice versa) by flipping the orientation of its spin state.[37]

This change in spin would require the neutrino and antineutrino to have nonzero mass, and therefore travel slower than light, because such a spin flip, caused only by a change in point of view, can take place only if inertial frames of reference exist that move faster than the particle: such a particle has a spin of one orientation when seen from a frame which moves slower than the particle, but the opposite spin when observed from a frame that moves faster than the particle.

On July 19, 2013 the results from the T2K experiment presented at the European Physical Society Conference on High Energy Physics in Stockholm, Sweden, confirmed neutrino oscillation theory.[38][39]

22.2.8 Speed

Main article: Measurements of neutrino speed

Before neutrinos were found to oscillate, they were generally assumed to be massless, propagating at the speed of light. According to the theory of special relativity, the question of neutrino velocity is closely related to their mass. If neutrinos are massless, they must travel at the speed of light. However, if they have mass, they cannot reach the speed of light.

Also some Lorentz-violating variants of quantum gravity might allow faster-than-light neutrinos. A comprehensive framework for Lorentz violations is the Standard-Model Extension (SME).

In the early 1980s, first measurements of neutrino speed were done using pulsed pion beams (produced by pulsed proton beams hitting a target). The pions decayed producing neutrinos, and the neutrino interactions observed within a time window in a detector at a distance were consistent with the speed of light. This measurement was repeated in 2007 using the MINOS detectors, which found the speed of 3 GeV neutrinos to be, at the 99% confidence level, in the range between $0.999976\,c$ and $1.000126\,c$. The central value of $1.000051c$ is higher than the speed of light but is also consistent with a velocity of exactly c or even slightly less. This measurement set an upper bound on the mass of the muon neutrino of 50 MeV at 99% confidence.[40][41] After the detectors for the project were upgraded in 2012, MINOS refined their initial result and found agreement with the speed of light, with the difference in the arrival time of neutrinos and light of -0.0006% ($\pm0.0012\%$).[42]

A similar observation was made, on a much larger scale, with supernova 1987A (SN 1987A). 10-MeV antineutrinos from the supernova were detected within a time window that was consistent with the speed of light for the neutrinos. Currently, the question of whether or not neutrinos have mass cannot be decided; their speed is (as yet) indistinguishable from the speed of light.[43][44]

In September 2011, the OPERA collaboration released calculations showing velocities of 17-GeV and 28-GeV neutrinos exceeding the speed of light in their experiments (see

Faster-than-light neutrino anomaly). In November 2011, OPERA repeated its experiment with changes so that the speed could be determined individually for each detected neutrino. The results showed the same faster-than-light speed. However, in February 2012, reports came out that the results may have been caused by a loose fiber optic cable attached to one of the atomic clocks which measured the departure and arrival times of the neutrinos. An independent recreation of the experiment in the same laboratory by ICARUS found no discernible difference between the speed of a neutrino and the speed of light.[45]

In June 2012, CERN announced that new measurements conducted by all four Gran Sasso experiments (OPERA, ICARUS, Borexino and LVD) found agreement between the speed of light and the speed of neutrinos, finally refuting the initial OPERA claim.[46]

22.2.9 Mass

The Standard Model of particle physics assumed that neutrinos are massless. However the experimentally established phenomenon of neutrino oscillation, which mixes neutrino flavour states with neutrino mass states (analogously to CKM mixing), requires neutrinos to have nonzero masses.[47] Massive neutrinos were originally conceived by Bruno Pontecorvo in the 1950s. Enhancing the basic framework to accommodate their mass is straightforward by adding a right-handed Lagrangian. This can be done in two ways. If, like other fundamental Standard Model particles, mass is generated by the Dirac mechanism, then the framework would require an SU(2) singlet. This particle would have no other Standard Model interactions (apart from the Yukawa interactions with the neutral component of the Higgs doublet), so is called a sterile neutrino. Or, mass can be generated by the Majorana mechanism, which would require the neutrino and antineutrino to be the same particle.

The strongest upper limit on the masses of neutrinos comes from cosmology: the Big Bang model predicts that there is a fixed ratio between the number of neutrinos and the number of photons in the cosmic microwave background. If the total energy of all three types of neutrinos exceeded an average of 50 eV per neutrino, there would be so much mass in the universe that it would collapse.[48] This limit can be circumvented by assuming that the neutrino is unstable; however, there are limits within the Standard Model that make this difficult. A much more stringent constraint comes from a careful analysis of cosmological data, such as the cosmic microwave background radiation, galaxy surveys, and the Lyman-alpha forest. These indicate that the summed masses of the three neutrinos must be less than 0.3 eV.[49]

In 1998, research results at the Super-Kamiokande neutrino detector determined that neutrinos can oscillate from one flavor to another, which requires that they must have a nonzero mass.[50] While this shows that neutrinos have mass, the absolute neutrino mass scale is still not known. This is because neutrino oscillations are sensitive only to the difference in the squares of the masses.[51] The best estimate of the difference in the squares of the masses of mass eigenstates 1 and 2 was published by KamLAND in 2005: Δm^2

$21 = 0.000079$ eV^2.[52] In 2006, the MINOS experiment measured oscillations from an intense muon neutrino beam, determining the difference in the squares of the masses between neutrino mass eigenstates 2 and 3. The initial results indicate $|\Delta m^2$

$32| = 0.0027$ eV^2, consistent with previous results from Super-Kamiokande.[53] Since $|\Delta m^2$

$32|$ is the difference of two squared masses, at least one of them has to have a value which is at least the square root of this value. Thus, there exists at least one neutrino mass eigenstate with a mass of at least 0.04 eV.[54]

In 2009, lensing data of a galaxy cluster were analyzed to predict a neutrino mass of about 1.5 eV.[55] This surprisingly high value requires that the three neutrino masses be nearly equal, with neutrino oscillations on the order of meV. The masses lie below the Mainz-Troitsk upper bound of 2.2 eV for the electron antineutrino.[56] The latter will be tested in 2015 in the KATRIN experiment, that searches for a mass between 0.2 eV and 2 eV.

A number of efforts are under way to directly determine the absolute neutrino mass scale in laboratory experiments. The methods applied involve nuclear beta decay (KATRIN and MARE).

On 31 May 2010, OPERA researchers observed the first tau neutrino candidate event in a muon neutrino beam, the first time this transformation in neutrinos had been observed, providing further evidence that they have mass.[57]

In July 2010 the 3-D MegaZ DR7 galaxy survey reported that they had measured a limit of the combined mass of the three neutrino varieties to be less than 0.28 eV.[58] A tighter upper bound yet for this sum of masses, 0.23 eV, was reported in March 2013 by the Planck collaboration,[59] whereas a February 2014 result estimates the sum as 0.320 ± 0.081 eV based on discrepancies between the cosmological consequences implied by Planck's detailed measurements of the Cosmic Microwave Background and predictions arising from observing other phenomena, combined with the assumption that neutrinos are responsible for the observed weaker gravitational lensing than would be expected from massless neutrinos.[60]

If the neutrino is a Majorana particle, the mass may be calculated by finding the half life of neutrinoless double-beta decay of certain nuclei. As of 2015, the lowest upper limit on the Majorana mass of the neutrino has been set by KamLAND-Zen: 0.12–0.25 eV.[61]

The Nobel prize in Physics 2015 was awarded to both Takaaki Kajita and Arthur B. McDonald for their experimental discovery of neutrino oscillations, which demonstrates that neutrinos have mass.[62][63]

22.2.10   Size

Standard Model neutrinos are fundamental point-like particles. An effective size can be defined using their electroweak cross section (apparent size in electroweak interaction). The average electroweak characteristic size is $r^2 = n \times 10^{-33}$ cm^2 ($n \times 1$ nanobarn), where $n = 3.2$ for electron neutrino, $n = 1.7$ for muon neutrino and $n = 1.0$ for tau neutrino; it depends on no other properties than mass.[64] However, this is best understood as being relevant only to probability of scattering. Since the neutrino does not interact electromagnetically, and is defined quantum mechanically by a wavefunction, it does not have a size in the same sense as everyday objects.[65] Furthermore, processes that produce neutrinos impart such high energies to them that they travel at almost the speed of light. Nevertheless, neutrinos are fermions, and thus obey the Pauli exclusion principle, i.e. that increasing their density forces them into progressively higher momentum states.

22.2.11   Chirality

Experimental results show that (nearly) all produced and observed neutrinos have left-handed helicities (spins antiparallel to momenta), and all antineutrinos have right-handed helicities, within the margin of error. In the massless limit, it means that only one of two possible chiralities is observed for either particle. These are the only chiralities included in the Standard Model of particle interactions.

It is possible that their counterparts (right-handed neutrinos and left-handed antineutrinos) simply do not exist. If they do, their properties are substantially different from observable neutrinos and antineutrinos. It is theorized that they are either very heavy (on the order of GUT scale—see *Seesaw mechanism*), do not participate in weak interaction (so-called sterile neutrinos), or both.

The existence of nonzero neutrino masses somewhat complicates the situation. Neutrinos are produced in weak interactions as chirality eigenstates. However, chirality of a massive particle is not a constant of motion; helicity is, but the chirality operator does not share eigenstates with the helicity operator. Free neutrinos propagate as mixtures of left- and right-handed helicity states, with mixing ampli-

tudes on the order of mv/E. This does not significantly affect the experiments, because neutrinos involved are nearly always ultrarelativistic, and thus mixing amplitudes are vanishingly small. For example, most solar neutrinos have energies on the order of 100 keV–1 MeV, so the fraction of neutrinos with "wrong" helicity among them cannot exceed 10^{-10}.[66][67]

22.3   Sources

22.3.1   Artificial

Reactor neutrinos

Nuclear reactors are the major source of human-generated neutrinos. Antineutrinos are made in the beta-decay of neutron-rich daughter fragments in the fission process. Generally, the four main isotopes contributing to the antineutrino flux are 235U, 238U, 239Pu and 241Pu (i.e. via the antineutrinos emitted during beta-minus decay of their respective fission fragments). The average nuclear fission releases about 200 MeV of energy, of which roughly 4.5% (or about 9 MeV)[68] is radiated away as antineutrinos. For a typical nuclear reactor with a thermal power of 4000 MW, meaning that the core produces this much heat, and an electrical power generation of 1300 MW, the total power production from fissioning atoms is actually 4185 MW, of which 185 MW is radiated away as antineutrino radiation and never appears in the engineering. This is to say, 185 MW of fission energy is *lost* from this reactor and does not appear as heat available to run turbines, since antineutrinos penetrate all building materials practically without interaction.[nb 4]

The antineutrino energy spectrum depends on the degree to which the fuel is burned (plutonium-239 fission antineutrinos on average have slightly more energy than those from uranium-235 fission), but in general, the *detectable* antineutrinos from fission have a peak energy between about 3.5 and 4 MeV, with a maximum energy of about 10 MeV.[69] There is no established experimental method to measure the flux of low-energy antineutrinos. Only antineutrinos with an energy above threshold of 1.8 MeV can be uniquely identified (see *neutrino detection* below). An estimated 3% of all antineutrinos from a nuclear reactor carry an energy above this threshold. Thus, an average nuclear power plant may generate over 10^{20} antineutrinos per second above this threshold, but also a much larger number (97%/3% = ~30 times this number) below the energy threshold, which cannot be seen with present detector technology.

Accelerator neutrinos

Some particle accelerators have been used to make neutrino beams. The technique is to collide protons with a
fixed target, producing charged pions or kaons. These unstable particles are then magnetically focused into a long
tunnel where they decay while in flight. Because of the
relativistic boost of the decaying particle, the neutrinos are
produced as a beam rather than isotropically. Efforts to
construct an accelerator facility where neutrinos are produced through muon decays are ongoing.[70] Such a setup is
generally known as a neutrino factory.

Nuclear bombs

Nuclear bombs also produce very large quantities of neutrinos. Fred Reines and Clyde Cowan considered the detection of neutrinos from a bomb prior to their search for reactor neutrinos; a fission reactor was recommended as a better
alternative by Los Alamos physics division leader J.M.B.
Kellogg.[71] Fission bombs produce antineutrinos (from the
fission process), and fusion bombs produce both neutrinos
(from the fusion process) and antineutrinos (from the initiating fission explosion).

22.3.2   Geologic

Main article: Geoneutrino

Neutrinos are part of the natural background radiation. In
particular, the decay chains of 238U and 232Th isotopes,
as well as40K, include beta decays which emit antineutrinos. These so-called geoneutrinos can provide valuable information on the Earth's interior. A first indication for
geoneutrinos was found by the KamLAND experiment in
2005. KamLAND's main background in the geoneutrino
measurement are the antineutrinos coming from reactors.
Several future experiments aim at improving the geoneutrino measurement and these will necessarily have to be far
away from reactors.

22.3.3   Atmospheric

Atmospheric neutrinos result from the interaction of cosmic
rays with atomic nuclei in the Earth's atmosphere, creating
showers of particles, many of which are unstable and produce neutrinos when they decay. A collaboration of particle
physicists from Tata Institute of Fundamental Research (India), Osaka City University (Japan) and Durham University
(UK) recorded the first cosmic ray neutrino interaction in
an underground laboratory in Kolar Gold Fields in India in
1965.[72]

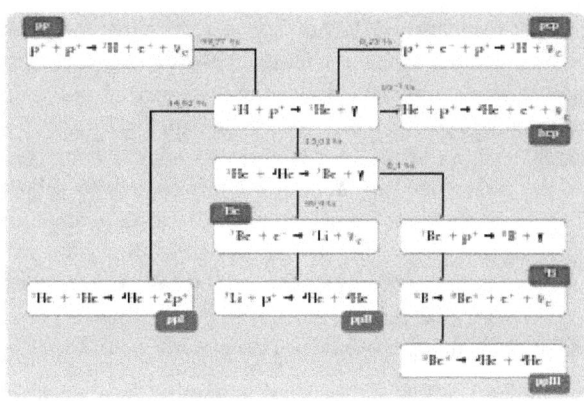

Solar neutrinos (proton–proton chain) in the Standard Solar Model

22.3.4   Solar

Solar neutrinos originate from the nuclear fusion powering
the Sun and other stars. The details of the operation of the
Sun are explained by the Standard Solar Model. In short:
when four protons fuse to become one helium nucleus, two
of them have to convert into neutrons, and each such conversion releases one electron neutrino.

The Sun sends enormous numbers of neutrinos in all directions. Each second, about 65 billion (6.5×10^{10}) solar neutrinos pass through every square centimeter on the part of
the Earth that faces the Sun.[8] Since neutrinos are insignif-
icantly absorbed by the mass of the Earth, the surface area
on the side of the Earth opposite the Sun receives about the
same number of neutrinos as the side facing the Sun.

22.3.5   Supernovae

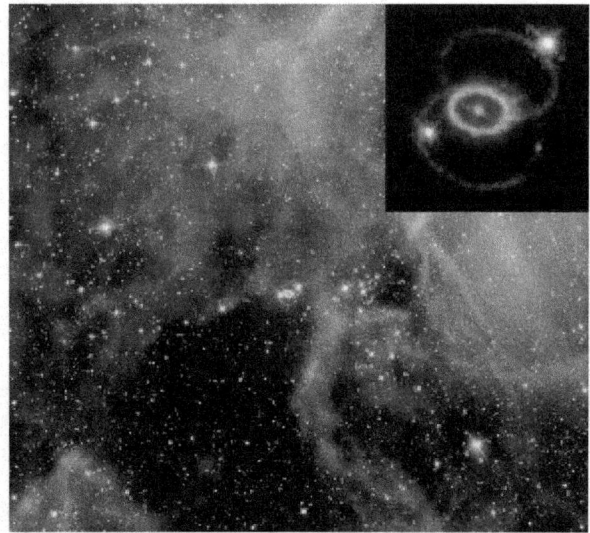

SN 1987A

In 1966 Colgate and White[73] calculated that neutrinos carry away most of the gravitational energy released by the collapse of massive stars, events now categorized as Type Ib and Ic and Type II supernovae. When such stars collapse, matter densities at the core become so high (10^{17} kg/m^3) that the degeneracy of electrons is not enough to prevent protons and electrons from combining to form a neutron and an electron neutrino. A second and more important neutrino source is the thermal energy (100 billion kelvins) of the newly formed neutron core, which is dissipated via the formation of neutrino–antineutrino pairs of all flavors.[74]

Colgate and White's theory of supernova neutrino production was confirmed in 1987, when neutrinos from supernova 1987A were detected. The water-based detectors Kamiokande II and IMB detected 11 and 8 antineutrinos of thermal origin,[74] respectively, while the scintillator-based Baksan detector found 5 neutrinos (lepton number = 1) of either thermal or electron-capture origin, in a burst lasting less than 13 seconds. The neutrino signal from the supernova arrived at earth several hours before the arrival of the first electromagnetic radiation, as expected from the evident fact that the latter emerges along with the shock wave. The exceptionally feeble interaction with normal matter allowed the neutrinos to pass through the churning mass of the exploding star, while the electromagnetic photons were slowed.

Because neutrinos interact so little with matter, it is thought that a supernova's neutrino emissions carry information about the innermost regions of the explosion. Much of the *visible* light comes from the decay of radioactive elements produced by the supernova shock wave, and even light from the explosion itself is scattered by dense and turbulent gases, and thus delayed. The neutrino burst is expected to reach Earth before any electromagnetic waves, including visible light, gamma rays or radio waves. The exact time delay depends on the velocity of the shock wave and on the thickness of the outer layer of the star. For a Type II supernova, astronomers expect the neutrino flood to be released seconds after the stellar core collapse, while the first electromagnetic signal may emerge hours later, after the explosion shock wave has had time to reach the surface of the star. The SNEWS project uses a network of neutrino detectors to monitor the sky for candidate supernova events; the neutrino signal will provide a useful advance warning of a star exploding in the Milky Way.

Although neutrinos pass through the outer gases of a supernova without scattering, they provide information about the deeper supernova core with evidence that here, even neutrinos scatter to a significant extent. In a supernova core the densities are those of a neutron star (which is expected to be formed in this type of supernova),[75] becoming large enough to influence the duration of the neutrino signal by delaying some neutrinos. The length of the neutrino signal

from SN 1987A, some 13 seconds, was far longer than it would take in theory for neutrinos to pass directly through the neutrino-generating core of a supernova, expected to be only 32 kilometers in diameter SN 1987A. The number of neutrinos counted was also consistent with a total neutrino energy of 2.2 x 10^{46} joules, which was estimated to be nearly all of the total energy of the supernova.[76]

22.3.6 Supernova remnants

The energy of supernova neutrinos ranges from a few to several tens of MeV. However, the sites where cosmic rays are accelerated are expected to produce neutrinos that are at least one million times more energetic, produced from turbulent gaseous environments left over by supernova explosions: the supernova remnants. The origin of the cosmic rays was attributed to supernovas by Walter Baade and Fritz Zwicky; this hypothesis was refined by Vitaly L. Ginzburg and Sergei I. Syrovatsky who attributed the origin to supernova remnants, and supported their claim by the crucial remark, that the cosmic ray losses of the Milky Way is compensated, if the efficiency of acceleration in supernova remnants is about 10 percent. Ginzburg and Syrovatskii's hypothesis is supported by the specific mechanism of "shock wave acceleration" happening in supernova remnants, which is consistent with the original theoretical picture drawn by Enrico Fermi, and is receiving support from observational data. The very-high-energy neutrinos are still to be seen, but this branch of neutrino astronomy is just in its infancy. The main existing or forthcoming experiments that aim at observing very-high-energy neutrinos from our galaxy are Baikal, AMANDA, IceCube, ANTARES, NEMO and Nestor. Related information is provided by very-high-energy gamma ray observatories, such as VERITAS, HESS and MAGIC. Indeed, the collisions of cosmic rays are supposed to produce charged pions, whose decay give the neutrinos, and also neutral pions, whose decay give gamma rays: the environment of a supernova remnant is transparent to both types of radiation.

Still-higher-energy neutrinos, resulting from the interactions of extragalactic cosmic rays, could be observed with the Pierre Auger Observatory or with the dedicated experiment named ANITA.

22.3.7 Big Bang

Main article: Cosmic neutrino background

It is thought that, just like the cosmic microwave background radiation left over from the Big Bang, there is a background of low-energy neutrinos in our Universe. In the 1980s it was proposed that these may be the explanation for the dark matter thought to exist in the universe. Neutrinos have one important advantage over most other dark matter candidates: it is known that they exist. However, this idea also has serious problems.

From particle experiments, it is known that neutrinos are very light. This means that they easily move at speeds close to the speed of light. For this reason, dark matter made from neutrinos is termed "hot dark matter". The problem is that being fast moving, the neutrinos would tend to have spread out evenly in the universe before cosmological expansion made them cold enough to congregate in clumps. This would cause the part of dark matter made of neutrinos to be smeared out and unable to cause the large galactic structures that we see.

Further, these same galaxies and groups of galaxies appear to be surrounded by dark matter that is not fast enough to escape from those galaxies. Presumably this matter provided the gravitational nucleus for formation. This implies that neutrinos cannot make up a significant part of the total amount of dark matter.

From cosmological arguments, relic background neutrinos are estimated to have density of 56 of each type per cubic centimeter and temperature 1.9 K (1.7×10^{-4} eV) if they are massless, much colder if their mass exceeds 0.001 eV. Although their density is quite high, they have not yet been observed in the laboratory, as their energy is below thresholds of most detection methods, and due to extremely low neutrino interaction cross-sections at sub-eV energies. In contrast, boron-8 solar neutrinos—which are emitted with a higher energy—have been detected definitively despite having a space density that is lower than that of relic neutrinos by some 6 orders of magnitude.

22.4   Detection

Main article: Neutrino detector

Neutrinos cannot be detected directly, because they do not ionize the materials they are passing through (they do not carry electric charge and other proposed effects, like the MSW effect, do not produce traceable radiation). A unique reaction to identify antineutrinos, sometimes referred to as inverse beta decay, as applied by Reines and Cowan (see below), requires a very large detector in order to detect a significant number of neutrinos. All detection methods require the neutrinos to carry a minimum threshold energy. So far, there is no detection method for low-energy neutrinos, in the sense that potential neutrino interactions (for example by the MSW effect) cannot be uniquely distinguished from other causes. Neutrino detectors are often built underground in order to isolate the detector from cosmic rays and

other background radiation.

Antineutrinos were first detected in the 1950s near a nuclear reactor. Reines and Cowan used two targets containing a solution of cadmium chloride in water. Two scintillation detectors were placed next to the cadmium targets. Antineutrinos with an energy above the threshold of 1.8 MeV caused charged current interactions with the protons in the water, producing positrons and neutrons. This is very much like β+ decay, where energy is used to convert a proton into a neutron, a positron (e+) and an electron neutrino (ν_e) is emitted:

From known β+ decay:

$$\text{Energy} + p \rightarrow n + e+ + \nu_e$$

In the Cowan and Reines experiment, instead of an outgoing neutrino, you have an incoming antineutrino ($\bar{\nu}_e$) from a nuclear reactor:

$$\text{Energy (>1.8 MeV)} + p + \nu_e \rightarrow n + e+$$

The resulting positron annihilation with electrons in the detector material created photons with an energy of about 0.5 MeV. Pairs of photons in coincidence could be detected by the two scintillation detectors above and below the target. The neutrons were captured by cadmium nuclei resulting in gamma rays of about 8 MeV that were detected a few microseconds after the photons from a positron annihilation event.

Since then, various detection methods have been used. Super Kamiokande is a large volume of water surrounded by photomultiplier tubes that watch for the Cherenkov radiation emitted when an incoming neutrino creates an electron or muon in the water. The Sudbury Neutrino Observatory is similar, but uses heavy water as the detecting medium, which uses the same effects, but also allows the additional reaction any-flavor neutrino photo-dissociation of deuterium, resulting in a free neutron which is then detected from gamma radiation after chlorine-capture. Other detectors have consisted of large volumes of chlorine or gallium which are periodically checked for excesses of argon or germanium, respectively, which are created by electron-neutrinos interacting with the original substance. MINOS uses a solid plastic scintillator coupled to photomultiplier tubes, while Borexino uses a liquid pseudocumene scintillator also watched by photomultiplier tubes and the proposed NOνA detector will use liquid scintillator watched by avalanche photodiodes. The IceCube Neutrino Observatory uses 1 km^3 of the Antarctic ice sheet near the south pole with photomultiplier tubes distributed throughout the volume.

22.5 Motivation for scientific interest

Neutrinos' low mass and neutral charge mean they interact exceedingly weakly with other particles and fields. This feature of weak interaction interests scientists because it means neutrinos can be used to probe environments that other radiation (such as light or radio waves) cannot penetrate.

Using neutrinos as a probe was first proposed in the mid-20th century as a way to detect conditions at the core of the Sun. The solar core cannot be imaged directly because electromagnetic radiation (such as light) is diffused by the great amount and density of matter surrounding the core. On the other hand, neutrinos pass through the Sun with few interactions. Whereas photons emitted from the solar core may require 40,000 years to diffuse to the outer layers of the Sun, neutrinos generated in stellar fusion reactions at the core cross this distance practically unimpeded at nearly the speed of light.[77][78]

Neutrinos are also useful for probing astrophysical sources beyond the Solar System because they are the only known particles that are not significantly attenuated by their travel through the interstellar medium. Optical photons can be obscured or diffused by dust, gas, and background radiation. High-energy cosmic rays, in the form of swift protons and atomic nuclei, are unable to travel more than about 100 megaparsecs due to the Greisen–Zatsepin–Kuzmin limit (GZK cutoff). Neutrinos, in contrast, can travel even greater distances barely attenuated.

The galactic core of the Milky Way is fully obscured by dense gas and numerous bright objects. Neutrinos produced in the galactic core might be measurable by Earth-based neutrino telescopes.[13]

Another important use of the neutrino is in the observation of supernovae, the explosions that end the lives of highly massive stars. The core collapse phase of a supernova is an extremely dense and energetic event. It is so dense that no known particles are able to escape the advancing core front except for neutrinos. Consequently, supernovae are known to release approximately 99% of their radiant energy in a short (10-second) burst of neutrinos.[79] These neutrinos are a very useful probe for core collapse studies.

The rest mass of the neutrino is an important test of cosmological and astrophysical theories (see *Dark matter*). The neutrino's significance in probing cosmological phenomena is as great as any other method, and is thus a major focus of study in astrophysical communities.[80]

The study of neutrinos is important in particle physics because neutrinos typically have the lowest mass, and hence are examples of the lowest-energy particles theorized in extensions of the Standard Model of particle physics.

In November 2012 American scientists used a particle accelerator to send a coherent neutrino message through 780 feet of rock. This marks the first use of neutrinos for communication, and future research may permit binary neutrino messages to be sent immense distances through even the densest materials, such as the Earth's core.[81]

22.6 See also

- List of neutrino experiments

22.7 Notes

[1] More specifically, the electron neutrino. Two other types were discovered later – see Neutrino flavor below.

[2] Niels Bohr was notably opposed to this interpretation of beta decay and was ready to accept that energy, momentum and angular momentum were not conserved quantities.

[3] In this context, "light neutrino" means neutrinos with less than half the mass of the Z boson.

[4] Typically about one third of the heat which is deposited in a reactor core is available to be converted to electricity, and a 4000 MW reactor would produce only 2700 MW of actual heat, with the rest being converted to its 1300 MW of electric power production.

22.8 References

[1] "Astronomers Accurately Measure the Mass of Neutrinos for the First Time". *scitechdaily.com*. Image credit:NASA, ESA, and J. Lotz, M. Mountain, A. Koekemoer, and the HFF Team (STScI). February 10, 2014. Archived from the original on May 7, 2014. Retrieved May 7, 2014.

[2] Foley, James A. (February 10, 2014). "Mass of Neutrinos Accurately Calculated for First Time, Physicists Report". *natureworldnews.com*. Image credit: . via Wikimedia Commons. Archived from the original on May 7, 2014. Retrieved May 7, 2014.

[3] Battye, Richard A.; Moss, Adam (2014). "Evidence for Massive Neutrinos from Cosmic Microwave Background and Lensing Observations". *Physical Review Letters* **112** (5): 051303. arXiv:1308.5870v2. Bibcode:2014PhRvL.112e1303B. doi:10.1103/PhysRevLett.112.051303. PMID 24580586.

[4] "Neutrino". *Glossary for the Research Perspectives of the Max Planck Society*. Max Planck Gesellschaft. Retrieved 2012-03-27.

[5] "Solar Neutrinos" (PDF). *Philip Armitage*. JILA, University of Colorado, Boulder. 2003. Retrieved 24 April 2016.

[6] "Neutrinos". particlecentral.com. Retrieved 24 April 2016.

[7] "Conservation of lepton number". HyperPhysics, Georgia State University. Retrieved 24 April 2016.

[8] Bahcall, John N.; Serenelli, Aldo M.; Basu, Sarbani (2005). "New Solar Opacities, Abundances, Helioseismology, and Neutrino Fluxes". *The Astrophysical Journal* 621 (1): L85–8. arXiv:astro-ph/0412440. Bibcode:2005ApJ...621L..85B. doi:10.1086/428929.

[9] Olive, K. A. "Sum of Neutrino Masses" (PDF). *Chinese Physics C*.

[10] Dodelson, Scott; Widrow, Lawrence M. (1994). "Sterile neutrinos as dark matter". *Physical Review Letters* 72 (17): 17–20. arXiv:hep-ph/9303287. Bibcode:1994PhRvL..72...17D. doi:10.1103/PhysRevLett.72.17.

[11] Brown, Laurie M. (1978). "The idea of the neutrino". *Physics Today* 31 (9): 23–8. Bibcode:1978PhT....31i..23B. doi:10.1063/1.2995181.

[12] E. Amaldi (1984). "From the discovery of the neutron to the discovery of nuclear fission". *Phys. Rep.* 111 (1–4): 306.

[13] F. Close (2012). *Neutrino*. Oxford University Press. ISBN 978-0199695997.

[14] E. Fermi (1934). "Versuch einer Theorie der β-Strahlen. I". *Zeitschrift für Physik A* 88 (3–4): 161–177. Bibcode:1934ZPhy...88..161F. doi:10.1007/BF01351864. Translated in F. L. Wilson (1968). "Fermi's Theory of Beta Decay" (PDF). *American Journal of Physics* 36 (12): 1150. Bibcode:1968AmJPh..36.1150W. doi:10.1119/1.1974382.

[15] K.-C. Wang (1942). "A Suggestion on the Detection of the Neutrino". *Physical Review* 61 (1–2): 97. Bibcode:1942PhRv...61...97W. doi:10.1103/PhysRev.61.97.

[16] C. L. Cowan Jr.; F. Reines; F. B. Harrison; H. W. Kruse; et al. (1956). "Detection of the Free Neutrino: a Confirmation". *Science* 124 (3212): 103–4. Bibcode:1956Sci...124..103C. doi:10.1126/science.124.3212.103. PMID 17796274.

[17] K. Winter (2000). *Neutrino physics*. Cambridge University Press. p. 38ff. ISBN 978-0-521-65003-8. This source reproduces the 1956 paper.

[18] "The Nobel Prize in Physics 1995". The Nobel Foundation. Retrieved 29 June 2010.

[19] I. V. Anicin (2005). "The Neutrino – Its Past, Present and Future". arXiv:physics/0503172.

[20] "Physicists Find First Direct Evidence for Tau Neutrino at Fermilab" (Press release). Fermilab. 20 July 2000. In 1989, experimenters at CERN found proof that the tau neutrino is the third and last light neutrino of the Standard Model, but a direct observation was not yet feasible.

[21] M. Maltoni; T. Schwetz; M. Tórtola; J. W. F. Valle (2004). "Status of global fits to neutrino oscillations". *New Journal of Physics* 6 (1): 122. arXiv:hep-ph/0405172. Bibcode:2004NJPh....6..122M. doi:10.1088/1367-2630/6/1/122.

[22] Pagliaroli, G.; Vissani, F.; Costantini, M. L.; Ianni, A. (2009). "Improved analysis of SN1987A antineutrino events". *Astroparticle Physics* 31 (3): 163–176. arXiv:0810.0466. Bibcode:2009APh....31..163P. doi:10.1016/j.astropartphys.2008.12.010.

[23] Particle Data Group; Eidelman, S.; Hayes, K. G.; Olive, K. A.; Aguilar-Benitez, M.; Amsler, C.; Asner, D.; Babu, K. S.; Barnett, R. M.; Beringer, J.; Burchat, P. R.; Carone, C. D.; Caso, S.; Conforto, G.; Dahl, O.; d'Ambrosio, G.; Doser, M.; Feng, J. L.; Gherghetta, T.; Gibbons, L.; Goodman, M.; Grab, C.; Groom, D. E.; Gurtu, A.; Hagiwara, K.; Hernández-Rey, J. J.; Hikasa, K.; Honscheid, K.; Jawahery, H.; et al. (2004). "Review of Particle Physics". *Physics Letters B* 592: 1–5. arXiv:astro-ph/0406663. Bibcode:2004PhLB..592....1P. doi:10.1016/j.physletb.2004.06.001.

[24] S.M. Caroll (25 March 2009). "Ada Lovelace Day: Chien-Shiung Wu". *Discover Magazine*. Retrieved 2011-09-23.

[25] Kolbe, E.; Langanke, K.; Fuller, G. M. (2004). "Neutrino-Induced Fission of Neutron-Rich Nuclei". *Physical Review Letters* 92 (11): 111101. arXiv:astro-ph/0308350. Bibcode:2004PhRvL..92k1101K. doi:10.1103/PhysRevLett.92.111101. PMID 15089120.

[26] Kelić, A.; Zinner, N.; Kolbe, E.; Langanke, K.; Schmidt, K.-H. (2005). "Cross sections and fragment distributions from neutrino-induced fission on r-process nuclei". *Physics Letters B* 616 (1–2): 48–58. arXiv:hep-ex/0312045. Bibcode:2005PhLB..616...48K. doi:10.1016/j.physletb.2005.04.074.

[27] Hall, Shannon (1 December 2015). "Astronomers Indirectly Spot Neutrinos Released Just 1 Second after the Birth of the Universe". *scientificamerican.com* (Nature America, Inc.). Retrieved 18 December 2015.

[28] Karagiorgi, G.; Aguilar-Arevalo, A.; Conrad, J. M.; Shaevitz, M. H.; Whisnant, K.; Sorel, M.; Barger, V. (2007). "LeptonicCPviolation studies at MiniBooNE in the (3+2) sterile neutrino oscillation hypothesis". *Physical Review D* 75: 013011. arXiv:hep-ph/0609177. Bibcode:2007PhRvD..75a3011K. doi:10.1103/PhysRevD.75.013011.

[29] M. Alpert (2007). "Dimensional Shortcuts". *Scientific American*. Retrieved 2009-10-31.

[30] Mueller, Th. A.; Lhuillier, D.; Fallot, M.; Letourneau, A.; Cormon, S.; Fechner, M.; Giot, L.; Lasserre, T.; Martino, J.; Mention, G.; Porta, A.; Yermia, F. (2011). "Improved predictions of reactor antineutrino spectra". *Physical Review C* **83** (5): 054615. arXiv:1101.2663. Bibcode:2011PhRvC..83e4615M. doi:10.1103/PhysRevC.83.054615.

[31] Mention, G.; Fechner, M.; Lasserre, Th.; Mueller, Th. A.; Lhuillier, D.; Cribier, M.; Letourneau, A. (2011). "Reactor antineutrino anomaly". *Physical Review D* **83** (7): 073006. arXiv:1101.2755. Bibcode:2011PhRvD..83g3006M. doi:10.1103/PhysRevD.83.073006.

[32] R. Cowen (2 February 2010). "Ancient Dawn's Early Light Refines the Age of the Universe". *Science News*. Retrieved 2010-02-03.

[33] "LLNL/SNL Applied Antineutrino Physics Project. LLNL-WEB-204112". 2006.

[34] "Applied Antineutrino Physics 2007 workshop". 2007.

[35] "New Tool To Monitor Nuclear Reactors Developed". ScienceDaily. 13 March 2008. Retrieved 2008-03-16.

[36] Alan Kostelecký, V.; Mewes, Matthew (2004). "Lorentz andCPTviolation in neutrinos". *Physical Review D* **69**: 016005. arXiv:hep-ph/0309025. Bibcode:2004PhRvD..69a6005A. doi:10.1103/PhysRevD.69.016005.

[37] C. Giunti; C.W. Kim (2007). *Fundamentals of neutrino physics and astrophysics*. Oxford University Press. p. 255. ISBN 0-19-850871-9.

[38] "Neutrino shape-shift points to new physics" *Physics News*, 19 July 2013.

[39] "Neutrino 'flavour' flip confirmed" *BBC News*, 19 July 2013.

[40] Adamson, P.; Andreopoulos, C.; Arms, K. E.; Armstrong, R.; Auty, D. J.; Avvakumov, S.; Ayres, D. S.; Baller, B.; Barish, B.; Barnes, P. D.; Barr, G.; Barrett, W. L.; Beall, E.; Becker, B. R.; Belias, A.; Bergfeld, T.; Bernstein, R. H.; Bhattacharya, D.; Bishai, M.; Blake, A.; Bock, B.; Bock, G. J.; Boehm, J.; Boehnlein, D. J.; Bogert, D.; Border, P. M.; Bower, C.; Buckley-Geer, E.; Cabrera, A.; et al. (2007). "Measurement of neutrino velocity with the MINOS detectors and NuMI neutrino beam". *Physical Review D* **76** (7): 072005. arXiv:0706.0437. Bibcode:2007PhRvD..76g2005A. doi:10.1103/PhysRevD.76.072005.

[41] D. Overbye (22 September 2011). "Tiny neutrinos may have broken cosmic speed limit". *New York Times*. That group found, although with less precision, that the neutrino speeds were consistent with the speed of light.

[42] Hesla, Leah (June 8, 2012). "MINOS reports new measurement of neutrino velocity". Fermilab today. Retrieved April 2, 2015.

[43] Stodolsky, Leo (1988). "The speed of light and the speed of neutrinos". *Physics Letters B* **201** (3): 353–354. Bibcode:1988PhLB..201..353S. doi:10.1016/0370-2693(88)91154-9.

[44] Andrew Cohen; Sheldon Glashow (28 October 2011). "New Constraints on Neutrino Velocities". *Phys. Rev. Lett.* **107** (18): 181803. arXiv:1109.6562. Bibcode:2011PhRvL.107r1803C. doi:10.1103/PhysRevLett.107.181803.

[45] Antonello, M.; Aprili, P.; Baiboussinov, B.; Baldo Ceolin, M.; Benetti, P.; Calligarich, E.; Canci, N.; Centro, S.; Cesana, A.; Cieślik, K.; Cline, D.B.; Cocco, A.G.; Dabrowska, A.; Dequal, D.; Dermenev, A.; Dolfini, R.; Farnese, C.; Fava, A.; Ferrari, A.; Fiorillo, G.; Gibin, D.; Gigli Berzolari, A.; Gninenko, S.; Guglielmi, A.; Haranczyk, M.; Holeczek, J.; Ivashkin, A.; Kisiel, J.; Kochanek, I.; et al. (2012). "Measurement of the neutrino velocity with the ICARUS detector at the CNGS beam". *Physics Letters B* **713**: 17–22. arXiv:1203.3433. Bibcode:2012PhLB..713...17I. doi:10.1016/j.physletb.2012.05.033.

[46] "Neutrinos sent from CERN to Gran Sasso respect the cosmic speed limit, experiments confirm" (Press release). CERN. June 8, 2012. Retrieved April 2, 2015.

[47] Schechter, J.; Valle, J. W. F. (1980). "Neutrino masses in SU(2) ⊗ U(1) theories". *Physical Review D* **22** (9): 2227–2235. Bibcode:1980PhRvD..22.2227S. doi:10.1103/PhysRevD.22.2227.

[48] Hut, P.; Olive, K.A. (1979). "A cosmological upper limit on the mass of heavy neutrinos". *Physics Letters B* **87** (1–2): 144–6. Bibcode:1979PhLB...87..144H. doi:10.1016/0370-2693(79)90039-X.

[49] Goobar, Ariel; Hannestad, Steen; Mörtsell, Edvard; Tu, Huitzu (2006). "The neutrino mass bound from WMAP 3 year data, the baryon acoustic peak, the SNLS supernovae and the Lyman-α forest". *Journal of Cosmology and Astroparticle Physics* **2006** (6): 019. arXiv:astro-ph/0602155. Bibcode:2006JCAP...06..019G. doi:10.1088/1475-7516/2006/06/019.

[50] Fukuda, Y.; Hayakawa, T.; Ichihara, E.; Inoue, K.; Ishihara, K.; Ishino, H.; Itow, Y.; Kajita, T.; Kameda, J.; Kasuga, S.; Kobayashi, K.; Kobayashi, Y.; Koshio, Y.; Martens, K.; Miura, M.; Nakahata, M.; Nakayama, S.; Okada, A.; Oketa, M.; Okumura, K.; Ota, M.; Sakurai, N.; Shiozawa, M.; Suzuki, Y.; Takeuchi, Y.; Totsuka, Y.; Yamada, S.; Earl, M.; Habig, A.; et al. (1998). "Measurements of the Solar Neutrino Flux from Super-Kamiokande's First 300 Days". *Physical Review Letters* **81** (6): 1158–1162. arXiv:hep-ex/9805021. Bibcode:1998PhRvL..81.1158F. doi:10.1103/PhysRevLett.81.1158.

[51] Mohapatra, R N; Antusch, S; Babu, K S; Barenboim, G; Chen, M-C; De Gouvêa, A; De Holanda, P; Dutta, B; Grossman, Y; Joshipura, A; Kayser, B; Kersten, J; Keum, Y Y; King, S F; Langacker, P; Lindner, M;

Loinaz, W; Masina, I; Mocioiu, I; Mohanty, S; Murayama, H; Pascoli, S; Petcov, S T; Pilaftsis, A; Ramond, P; Ratz, M; Rodejohann, W; Shrock, R; Takeuchi, T; et al. (2007). "Theory of neutrinos: A white paper". *Reports on Progress in Physics* 70 (11): 1757–1867. arXiv:hep-ph/0510213. Bibcode:2007RPPh...70.1757M. doi:10.1088/0034-4885/70/11/R02.

[52] Araki, T.; Eguchi, K.; Enomoto, S.; Furuno, K.; Ichimura, K.; Ikeda, H.; Inoue, K.; Ishihara, K.; Iwamoto, T.; Kawashima, T.; Kishimoto, Y.; Koga, M.; Koseki, Y.; Maeda, T.; Mitsui, T.; Motoki, M.; Nakajima, K.; Ogawa, H.; Owada, K.; Ricol, J.-S.; Shimizu, I.; Shirai, J.; Suekane, F.; Suzuki, A.; Tada, K.; Tajima, O.; Tamae, K.; Tsuda, Y.; Watanabe, H.; et al. (2005). "Measurement of Neutrino Oscillation with KamLAND: Evidence of Spectral Distortion". *Physical Review Letters* 94 (8): 081801. arXiv:hep-ex/0406035. Bibcode:2005PhRvL..94h1801A. doi:10.1103/PhysRevLett.94.081801. PMID 15783875.

[53] "MINOS experiment sheds light on mystery of neutrino disappearance" (Press release). Fermilab. 30 March 2006. Retrieved 2007-11-25.

[54] Amsler, C.; Doser, M.; Antonelli, M.; Asner, D.M.; Babu, K.S.; Baer, H.; Band, H.R.; Barnett, R.M.; Bergren, E.; Beringer, J.; Bernardi, G.; Bertl, W.; Bichsel, H.; Biebel, O.; Bloch, P.; Blucher, E.; Blusk, S.; Cahn, R.N.; Carena, M.; Caso, C.; Ceccucci, A.; Chakraborty, D.; Chen, M.-C.; Chivukula, R.S.; Cowan, G.; Dahl, O.; d'Ambrosio, G.; Damour, T.; De Gouvêa, A.; et al. (2008). "Review of Particle Physics". *Physics Letters B* 667: 1–6. Bibcode:2008PhLB..667....1P. doi:10.1016/j.physletb.2008.07.018.

[55] Nieuwenhuizen, Th. M. (2009). "Do non-relativistic neutrinos constitute the dark matter?". *EPL* 86 (5): 59001. arXiv:0812.4552. Bibcode:2009EL......8659001N. doi:10.1209/0295-5075/86/59001.

[56] "The most sensitive analysis on the neutrino mass [...] is compatible with a neutrino mass of zero. Considering its uncertainties this value corresponds to an upper limit on the electron neutrino mass of $m < 2.2$ eV/c^2 (95% Confidence Level)" The Mainz Neutrino Mass Experiment

[57] Agafonova, N.; Aleksandrov, A.; Altinok, O.; Ambrosio, M.; Anokhina, A.; Aoki, S.; Ariga, A.; Ariga, T.; Autiero, D.; Badertscher, A.; Bagulya, A.; Bendhabi, A.; Bertolin, A.; Besnier, M.; Bick, D.; Boyarkin, V.; Bozza, C.; Brugière, T.; Brugnera, R.; Brunet, F.; Brunetti, G.; Buontempo, S.; Cazes, A.; Chaussard, L.; Chernyavsky, M.; Chiarella, V.; Chon-Sen, N.; Chukanov, A.; Ciesielski, R.; et al. (2010). "Observation of a first $\nu\tau$ candidate event in the OPERA experiment in the CNGS beam". *Physics Letters B* 691 (3): 138–45. arXiv:1006.1623. Bibcode:2010PhLB..691..138A. doi:10.1016/j.physletb.2010.06.022.

[58] Thomas, Shaun A.; Abdalla, Filipe B.; Lahav, Ofer (2010). "Upper Bound of 0.28 eV on Neutrino Masses from the Largest Photometric Redshift Survey". *Physical Review Letters* 105 (3): 031301. arXiv:0911.5291. Bibcode:2010PhRvL.105c1301T. doi:10.1103/PhysRevLett.105.031301. PMID 20867754.

[59] Planck Collaboration, P. A. R.; Ade, P. A. R.; Aghanim, N.; Armitage-Caplan, C.; Arnaud, M.; Ashdown, M.; Atrio-Barandela, F.; Aumont, J.; Baccigalupi, C.; Banday, A. J.; Barreiro, R. B.; Bartlett, J. G.; Battaner, E.; Benabed, K.; Benoît, A.; Benoit-Lévy, A.; Bernard, J.-P.; Bersanelli, M.; Bielewicz, P.; Bobin, J.; Bock, J. J.; Bonaldi, A.; Bond, J. R.; Borrill, J.; Bouchet, F. R.; Bridges, M.; Bucher, M.; Burigana, C.; Butler, R. C.; et al. (2013). "Planck 2013 results. XVI. Cosmological parameters". *Astronomy & Astrophysics* 1303: 5076. arXiv:1303.5076. Bibcode:2014A&A...571A..16P. doi:10.1051/0004-6361/201321591.

[60] Battye, Richard A.; Moss, Adam (2014). "Evidence for Massive Neutrinos from Cosmic Microwave Background and Lensing Observations". *Physical Review Letters* 112 (5): 051303. arXiv:1308.5870. Bibcode:2014PhRvL.112e1303B. doi:10.1103/PhysRevLett.112.051303. PMID 24580586.

[61] A. Gando (KamLAND-Zen Collaboration); et al. (Feb 7, 2013). "Limit on Neutrinoless $\beta\beta$ Decay of Xe136 from the First Phase of KamLAND-Zen and Comparison with the Positive Claim in Ge76". *Phys. Rev. Lett. 110, 062502* 110. Bibcode:2013PhRvL.110f2502G. doi:10.1103/PhysRevLett.110.062502.

[62] Press Release (The Royal Swedish Academy of Sciences) 2015 Oct 6

[63] Day, Charles (2015-10-07). "Takaaki Kajita and Arthur McDonald share 2015 Physics Nobel". *Physics Today*. doi:10.1063/PT.5.7208. ISSN 0031-9228.

[64] Lucio, J. L.; Rosado, A.; Zepeda, A. (1985). "Characteristic size for the neutrino". *Physical Review D* 31 (5): 1091–1096. Bibcode:1985PhRvD..31.1091L. doi:10.1103/PhysRevD.31.1091. PMID 9955801.

[65] Choi, Charles Q. (2 June 2009). "Particles Larger Than Galaxies Fill the Universe?". *National Geographic News*.

[66] B. Kayser (2005). "Neutrino mass, mixing, and flavor change" (PDF). Particle Data Group. Retrieved 2007-11-25.

[67] S.M. Bilenky; C. Giunti (2001). "Lepton Numbers in the framework of Neutrino Mixing". *International Journal of Modern Physics A* 16 (24): 3931–3949. arXiv:hep-ph/0102320. Bibcode:2001IJMPA..16.3931B. doi:10.1142/S0217751X01004967.

[68] "Nuclear Fission and Fusion, and Nuclear Interactions". NLP National Physical Laboratory. 2008. Retrieved 2009-06-25.

[69] A. Bernstein; Wang, Y.; Gratta, G.; West, T. (2002). "Nuclear reactor safeguards and monitoring with antineutrino detectors". *Journal of Applied Physics* **91** (7): 4672. arXiv:nucl-ex/0108001. Bibcode:2002JAP....91.4672B. doi:10.1063/1.1452775.

[70] A. Bandyopadhyay et al. (ISS Physics Working Group); et al. (2007). "Physics at a future Neutrino Factory and super-beam facility". *Reports on Progress in Physics* **72** (10): 6201. arXiv:0710.4947. Bibcode:2009RPPh...72j6201B. doi:10.1088/0034-4885/72/10/106201.

[71] F. Reines; C. Cowan, Jr. (1997). "The Reines-Cowan Experiments: Detecting the Poltergeist" (PDF). *Los Alamos Science* **25**: 3.

[72] M. R. Krishnaswamy; et al. (6 July 1971). "The Kolar Gold Fields Neutrino Experiment. II. Atmospheric Muons at a Depth of 7000 hg cm-2 (Kolar)". *Proceedings of the Royal Society of London. Series A, Mathematical and Physical Sciences* **323** (1555): 511–522. Bibcode:1971RSPSA.323..511K. doi:10.1098/rspa.1971.0120. JSTOR 78071.

[73] S. A. Colgate & R. H. White (1966). "The Hydrodynamic Behavior of Supernova Explosions". *The Astrophysical Journal* **143**: 626. Bibcode:1966ApJ...143..626C. doi:10.1086/148549.

[74] A.K. Mann (1997). *Shadow of a star: The neutrino story of Supernova 1987A*. W. H. Freeman. p. 122. ISBN 0-7167-3097-9.

[75] Products of the 1987A supernova

[76] Diameter of neutrino-generating core, and total neutrino power of SN 1987A

[77] J.N. Bahcall (1989). *Neutrino Astrophysics*. Cambridge University Press. ISBN 0-521-37975-X.

[78] D.R. David Jr. (2003). "Nobel Lecture: A half-century with solar neutrinos". *Reviews of Modern Physics* **75** (3): 10. Bibcode:2003RvMP...75..985D. doi:10.1103/RevModPhys.75.985.

[79] "Physics – Supernova Starting Gun: Neutrinos". Focus.aps.org. 2009-07-17. Retrieved 2012-04-05.

[80] G.B. Gelmini; A. Kusenko; T.J. Weiler (May 2010). "Through Neutrino Eyes". *Scientific American* **302** (5): 38–45. Bibcode:2010SciAm.302e..38G. doi:10.1038/scientificamerican0510-38.

[81] Stancil, D. D.; Adamson, P.; Alania, M.; Aliaga, L.; et al. (2012). "Demonstration of Communication Using Neutrinos" (PDF). *Modern Physics Letters A* **27** (12): 1250077. arXiv:1203.2847. Bibcode:2012MPLA...2750077S. doi:10.1142/S0217732312500770. Lay summary – *Popular Science* (March 15, 2012).

22.9 Bibliography

- Adam, T.; (OPERA collaboration); et al. (2011). "Measurement of the neutrino velocity with the OPERA detector in the CNGS beam". arXiv:1109.4897 [hep-ex].

- Alberico, W. M.; Bilenky, S. M. (2004). "Neutrino Oscillations, Masses And Mixing". *Physics of Particles and Nuclei* **35**: 297–323. arXiv:hep-ph/0306239. Bibcode:2003hep.ph....6239A.

- Bahcall, J. N. (1989). *Neutrino Astrophysics*. Cambridge University Press. ISBN 0-521-35113-8.

- Bumfiel, G. (1 October 2001). "The Milky Way's Hidden Black Hole". *Scientific American*. Retrieved 2010-04-23.

- Close, F. (2010). *Neutrino*. Oxford University Press. ISBN 978-0-19-957459-9.

- Griffiths, D. J. (1987). *Introduction to Elementary Particles*. John Wiley & Sons. ISBN 0-471-60386-4.

- Perkins, D. H. (1999). *Introduction to High Energy Physics*. Cambridge University Press. ISBN 0-521-62196-8.

- Povh, B. (1995). *Particles and Nuclei: An Introduction to the Physical Concepts*. Springer-Verlag. ISBN 0-387-59439-6.

- Riazuddin (2005). "Neutrinos" (PDF). National Center for Physics.

- Schopper, H. F. (1966). *Weak interactions and nuclear beta decay*. North-Holland.

- Tammann, G. A.; Thielemann, F. K.; Trautmann, D. (2003). "Opening new windows in observing the Universe". Europhysics News. Retrieved 2006-06-08.

- Tipler, P.; Llewellyn, R. (2002). *Modern Physics* (4th ed.). W. H. Freeman. ISBN 0-7167-4345-0.

- Tomonaga, S.-I. (1997). *The Story of Spin*. University of Chicago Press.

- Zuber, K. (2003). *Neutrino Physics*. IOP Publishing. ISBN 978-0-7503-0750-5.

22.10 External links

- "What's a Neutrino?", Dave Casper (University of California, Irvine)

- Neutrino unbound: On-line review and e-archive on Neutrino Physics and Astrophysics

- Nova: The Ghost Particle: Documentary on US public television from WGBH

- Universe submerged in a sea of chilled neutrinos, *New Scientist*, 5 March 2008

- The neutrino oscillation industry

- Search for neutrinoless double beta decay with enriched 76Ge in Gran Sasso 1990–2003

- Cosmic Weight Gain: A Wispy Particle Bulks Up by George Johnson

- Neutrino 'ghost particle' sized up by astronomers BBC News 22 June 2010

- Merrifield, Michael; Copeland, Ed; Bowley, Roger (2010). "Neutrinos". *Sixty Symbols*. Brady Haran for the University of Nottingham.

- The Neutrino with Dr. Clyde L. Cowan (Lecture on Project Poltergeist by Clyde Cowan)

- Nuclear Reactor as the Source of Antineutrinos

- Pauli's letter (December 1930), the hypothesis of the neutrino (online and analyzed, for English version click 'Télécharger')

Chapter 23

Baryon

Not to be confused with Baryonyx.

A **baryon** is a composite subatomic particle made up of three quarks (as distinct from mesons, which are composed of one quark and one antiquark). Baryons and mesons belong to the hadron family of particles, which are the quark-based particles. The name "baryon" comes from the Greek word for "heavy" (βαρύς, *barys*), because, at the time of their naming, most known elementary particles had lower masses than the baryons.

As quark-based particles, baryons participate in the strong interaction, whereas leptons, which are not quark-based, do not. The most familiar baryons are the protons and neutrons that make up most of the mass of the visible matter in the universe. Electrons (the other major component of the atom) are leptons.

Each baryon has a corresponding antiparticle (antibaryon) where quarks are replaced by their corresponding antiquarks. For example, a proton is made of two up quarks and one down quark; and its corresponding antiparticle, the antiproton, is made of two up antiquarks and one down antiquark.

23.1 Background

Baryons are strongly interacting fermions, that is, they experience the strong nuclear force and are described by Fermi–Dirac statistics, which apply to all particles obeying the Pauli exclusion principle. This is in contrast to the bosons, which do not obey the exclusion principle.

Baryons, along with mesons, are hadrons, meaning they are particles composed of quarks. Quarks have baryon numbers of $B = 1/3$ and antiquarks have baryon number of $B = -1/3$. The term "baryon" usually refers to *triquarks*—baryons made of three quarks ($B = 1/3 + 1/3 + 1/3 = 1$).

Other exotic baryons have been proposed, such as pentaquarks—baryons made of four quarks and one anti-

quark ($B = 1/3 + 1/3 + 1/3 + 1/3 - 1/3 = 1$), but their existence is not generally accepted. In theory, heptaquarks (5 quarks, 2 antiquarks), nonaquarks (6 quarks, 3 antiquarks), etc. could also exist. Until recently, it was believed that some experiments showed the existence of pentaquarks—baryons made of four quarks and one antiquark.[1][2] The particle physics community as a whole did not view their existence as likely in 2006,[3] and in 2008, considered evidence to be overwhelmingly against the existence of the reported pentaquarks.[4] However, in July 2015, the LHCb experiment observed two resonances consistent with pentaquark states in the Λ0

b → J/ψK−

p decay, with a combined statistical significance of 15σ.[5][6]

23.2 Baryonic matter

Nearly all matter that may be encountered or experienced in everyday life is baryonic matter, which includes atoms of any sort, and provides those with the quality of mass. Non-baryonic matter, as implied by the name, is any sort of matter that is not composed primarily of baryons. Those might include neutrinos or free electrons, dark matter, such as supersymmetric particles, axions, or black holes.

The very existence of baryons is also a significant issue in cosmology because it is assumed that the Big Bang produced a state with equal amounts of baryons and antibaryons. The process by which baryons came to outnumber their antiparticles is called baryogenesis.

23.3 Baryogenesis

Main article: Baryogenesis

Experiments are consistent with the number of quarks in the universe being a constant and, to be more specific, the number of baryons being a constant ; in technical language, the total baryon number appears to be *conserved*. Within the prevailing Standard Model of particle physics, the number of baryons may change in multiples of three due to the action of sphalerons, although this is rare and has not been observed under experiment. Some grand unified theories of particle physics also predict that a single proton can decay, changing the baryon number by one; however, this has not yet been observed under experiment. The excess of baryons over antibaryons in the present universe is thought to be due to non-conservation of baryon number in the very early universe, though this is not well understood.

23.4 Properties

23.4.1 Isospin and charge

Main article: Isospin

The concept of isospin was first proposed by Werner

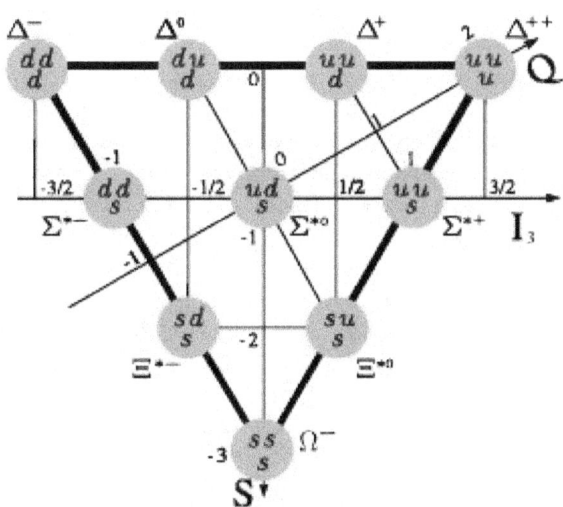

Combinations of three u, d or s quarks forming baryons with a spin-3/2 form the uds baryon decuplet

Heisenberg in 1932 to explain the similarities between protons and neutrons under the strong interaction.[7] Although they had different electric charges, their masses were so similar that physicists believed they were the same particle. The different electric charges were explained as being the result of some unknown excitation similar to spin. This unknown excitation was later dubbed *isospin* by Eugene Wigner in 1937.[8]

This belief lasted until Murray Gell-Mann proposed the quark model in 1964 (containing originally only the u, d, and s quarks).[9] The success of the isospin model is now understood to be the result of the similar masses of the u

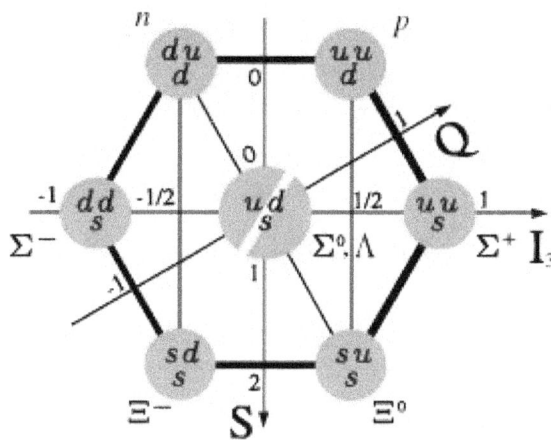

Combinations of three u, d or s quarks forming baryons with a spin-1/2 form the uds baryon octet

and d quarks. Since the u and d quarks have similar masses, particles made of the same number then also have similar masses. The exact specific u and d quark composition determines the charge, as u quarks carry charge $+2/3$ while d quarks carry charge $-1/3$. For example, the four Deltas all have different charges (Δ++ (uuu), Δ+ (uud), Δ0 (udd), Δ– (ddd)), but have similar masses (~1,232 MeV/c^2) as they are each made of a combination of three u and d quarks. Under the isospin model, they were considered to be a single particle in different charged states.

The mathematics of isospin was modeled after that of spin. Isospin projections varied in increments of 1 just like those of spin, and to each projection was associated a "charged state". Since the "Delta particle" had four "charged states", it was said to be of isospin $I = 3/2$. Its "charged states" Δ++, Δ+, Δ0, and Δ–, corresponded to the isospin projections $I_3 = +3/2$, $I_3 = +1/2$, $I_3 = -1/2$, and $I_3 = -3/2$, respectively. Another example is the "nucleon particle". As there were two nucleon "charged states", it was said to be of isospin 1/2. The positive nucleon N+ (proton) was identified with $I_3 = +1/2$ and the neutral nucleon N0 (neutron) with $I_3 = -1/2$.[10] It was later noted that the isospin projections were related to the up and down quark content of particles by the relation:

$$I_3 = \frac{1}{2}[(n_u - n_{\bar{u}}) - (n_d - n_{\bar{d}})],$$

where the n's are the number of up and down quarks and antiquarks.

In the "isospin picture", the four Deltas and the two nucleons were thought to be the different states of two particles. However, in the quark model, Deltas are different states of nucleons (the N^{++} or N^- are forbidden by Pauli's exclusion principle). Isospin, although conveying an inaccurate picture of things, is still used to classify baryons, leading to unnatural and often confusing nomenclature.

23.4.2 Flavour quantum numbers

Main article: Flavour (particle physics) § Flavour quantum numbers

The strangeness flavour quantum number S (not to be confused with spin) was noticed to go up and down along with particle mass. The higher the mass, the lower the strangeness (the more s quarks). Particles could be described with isospin projections (related to charge) and strangeness (mass) (see the uds octet and decuplet figures on the right). As other quarks were discovered, new quantum numbers were made to have similar description of udc and udb octets and decuplets. Since only the u and d mass are similar, this description of particle mass and charge in terms of isospin and flavour quantum numbers works well only for octet and decuplet made of one u, one d, and one other quark, and breaks down for the other octets and decuplets (for example, ucb octet and decuplet). If the quarks all had the same mass, their behaviour would be called *symmetric*, as they would all behave in exactly the same way with respect to the strong interaction. Since quarks do not have the same mass, they do not interact in the same way (exactly like an electron placed in an electric field will accelerate more than a proton placed in the same field because of its lighter mass), and the symmetry is said to be broken.

It was noted that charge (Q) was related to the isospin projection (I_3), the baryon number (B) and flavour quantum numbers (S, C, B', T) by the Gell-Mann–Nishijima formula:[10]

$$Q = I_3 + \frac{1}{2}(B + S + C + B' + T),$$

where S, C, B', and T represent the strangeness, charm, bottomness and topness flavour quantum numbers, respectively. They are related to the number of strange, charm, bottom, and top quarks and antiquark according to the relations:

$$S = -(n_s - n_{\bar{s}}),$$

$$C = +(n_c - n_{\bar{c}}),$$

$$B' = -(n_b - n_{\bar{b}}),$$

$$T = +(n_t - n_{\bar{t}}),$$

meaning that the Gell-Mann–Nishijima formula is equivalent to the expression of charge in terms of quark content:

$$Q = \frac{2}{3}[(n_u - n_{\bar{u}}) + (n_c - n_{\bar{c}}) + (n_t - n_{\bar{t}})] - \frac{1}{3}[(n_d - n_{\bar{d}}) + (n_s - n_{\bar{s}}) + (n_b - n_{\bar{b}})].$$

23.4.3 Spin, orbital angular momentum, and total angular momentum

Main articles: Spin (physics), Angular momentum operator, Quantum numbers and Clebsch–Gordan coefficients

Spin (quantum number S) is a vector quantity that represents the "intrinsic" angular momentum of a particle. It comes in increments of 1/2 ℏ (pronounced "h-bar"). The ℏ is often dropped because it is the "fundamental" unit of spin, and it is implied that "spin 1" means "spin 1 ℏ". In some systems of natural units, ℏ is chosen to be 1, and therefore does not appear anywhere.

Quarks are fermionic particles of spin 1/2 ($S = 1/2$). Because spin projections vary in increments of 1 (that is 1 ℏ), a single quark has a spin vector of length 1/2, and has two spin projections ($S_z = +1/2$ and $S_z = -1/2$). Two quarks can have their spins aligned, in which case the two spin vectors add to make a vector of length $S = 1$ and three spin projections ($S_z = +1$, $S_z = 0$, and $S_z = -1$). If two quarks have unaligned spins, the spin vectors add up to make a vector of length $S = 0$ and has only one spin projection ($S_z = 0$), etc. Since baryons are made of three quarks, their spin vectors can add to make a vector of length $S = 3/2$, which has four spin projections ($S_z = +3/2$, $S_z = +1/2$, $S_z = -1/2$, and $S_z = -3/2$), or a vector of length $S = 1/2$ with two spin projections ($S_z = +1/2$, and $S_z = -1/2$).[11]

There is another quantity of angular momentum, called the orbital angular momentum, (azimuthal quantum number L), that comes in increments of 1 ℏ, which represent the angular moment due to quarks orbiting around each other. The total angular momentum (total angular momentum quantum number J) of a particle is therefore the combination of intrinsic angular momentum (spin) and orbital angular momentum. It can take any value from $J = |L - S|$ to $J = |L + S|$, in increments of 1.

Particle physicists are most interested in baryons with no orbital angular momentum ($L = 0$), as they correspond to ground states—states of minimal energy. Therefore, the two groups of baryons most studied are the $S = 1/2$; $L = 0$ and $S = 3/2$; $L = 0$, which corresponds to $J = 1/2^+$ and $J =$

$3/2^+$, respectively, although they are not the only ones. It is also possible to obtain $J = 3/2^+$ particles from $S = 1/2$ and $L = 2$, as well as $S = 3/2$ and $L = 2$. This phenomenon of having multiple particles in the same total angular momentum configuration is called *degeneracy*. How to distinguish between these degenerate baryons is an active area of research in baryon spectroscopy.[12][13]

23.4.4   Parity

Main article: Parity (physics)

If the universe were reflected in a mirror, most of the laws of physics would be identical—things would behave the same way regardless of what we call "left" and what we call "right". This concept of mirror reflection is called *intrinsic parity* or *parity* (P). Gravity, the electromagnetic force, and the strong interaction all behave in the same way regardless of whether or not the universe is reflected in a mirror, and thus are said to conserve parity (P-symmetry). However, the weak interaction *does* distinguish "left" from "right", a phenomenon called parity violation (P-violation).

Based on this, one might think that, if the wavefunction for each particle (in more precise terms, the quantum field for each particle type) were simultaneously mirror-reversed, then the new set of wavefunctions would perfectly satisfy the laws of physics (apart from the weak interaction). It turns out that this is not quite true: In order for the equations to be satisfied, the wavefunctions of certain types of particles have to be multiplied by −1, in addition to being mirror-reversed. Such particle types are said to have *negative* or *odd* parity ($P = -1$, or alternatively $P = -$), while the other particles are said to have *positive* or *even* parity ($P = +1$, or alternatively $P = +$).

For baryons, the parity is related to the orbital angular momentum by the relation:[14]

$$P = (-1)^L.$$

As a consequence, baryons with no orbital angular momentum ($L = 0$) all have even parity ($P = +$).

23.5   Nomenclature

Baryons are classified into groups according to their isospin (I) values and quark (q) content. There are six groups of baryons—nucleon (N), Delta (Δ), Lambda (Λ), Sigma (Σ), Xi (Ξ), and Omega (Ω). The rules for classification are defined by the Particle Data Group. These rules consider the up (u), down (d) and strange (s) quarks to be *light* and the

charm (c), bottom (b), and top (t) quarks to be *heavy*. The rules cover all the particles that can be made from three of each of the six quarks, even though baryons made of t quarks are not expected to exist because of the t quark's short lifetime. The rules do not cover pentaquarks.[15]

- Baryons with three u and/or d quarks are N's ($I = 1/2$) or Δ's ($I = 3/2$).

- Baryons with two u and/or d quarks are Λ's ($I = 0$) or Σ's ($I = 1$). If the third quark is heavy, its identity is given by a subscript.

- Baryons with one u or d quark are Ξ's ($I = 1/2$). One or two subscripts are used if one or both of the remaining quarks are heavy.

- Baryons with no u or d quarks are Ω's ($I = 0$), and subscripts indicate any heavy quark content.

- Baryons that decay strongly have their masses as part of their names. For example, Σ^0 does not decay strongly, but $\Delta^{++}(1232)$ does.

It is also a widespread (but not universal) practice to follow some additional rules when distinguishing between some states that would otherwise have the same symbol.[10]

- Baryons in total angular momentum $J = 3/2$ configuration that have the same symbols as their $J = 1/2$ counterparts are denoted by an asterisk (*).

- Two baryons can be made of three different quarks in $J = 1/2$ configuration. In this case, a prime (′) is used to distinguish between them.

  - *Exception*: When two of the three quarks are one up and one down quark, one baryon is dubbed Λ while the other is dubbed Σ.

Quarks carry charge, so knowing the charge of a particle indirectly gives the quark content. For example, the rules above say that a Λ^+_c contains a c quark and some combination of two u and/or d quarks. The c quark has a charge of ($Q = +2/3$), therefore the other two must be a u quark ($Q = +2/3$), and a d quark ($Q = -1/3$) to have the correct total charge ($Q = +1$).

23.6   See also

- Eightfold way
- List of baryons
- List of particles

- Meson

- Timeline of particle discoveries

23.7 Notes

[1] H. Muir (2003)

[2] K. Carter (2003)

[3] W.-M. Yao *et al.* (2006): Particle listings – Θ^+

[4] C. Amsler *et al.* (2008): Pentaquarks

[5] LHCb (14 July 2015). "Observation of particles composed of five quarks, pentaquark-charmonium states, seen in $\Lambda_b^0 \rightarrow J/\psi pK^-$ decays.". *CERN website.* Retrieved 2015-07-14.

[6] R. Aaij et al. (LHCb collaboration) (2015). "Observation of $J/\psi p$ resonances consistent with pentaquark states in $\Lambda 0$ $b \rightarrow J/\psi K^-$
p decays". *Physical Review Letters* **115** (7). arXiv:1507.03414.   Bibcode:2015PhRvL.115g2001A. doi:10.1103/PhysRevLett.115.072001.

[7] W. Heisenberg (1932)

[8] E. Wigner (1937)

[9] M. Gell-Mann (1964)

[10] S.S.M. Wong (1998a)

[11] R. Shankar (1994)

[12] H. Garcilazo *et al.* (2007)

[13] D.M. Manley (2005)

[14] S.S.M. Wong (1998b)

[15] C. Amsler *et al.* (2008): Naming scheme for hadrons

23.8 References

- C. Amsler *et al.* (Particle Data Group) (2008). "Review of Particle Physics". *Physics Letters B* **667** (1): 1–1340. Bibcode:2008PhLB..667....1P. doi:10.1016/j.physletb.2008.07.018.

- H. Garcilazo, J. Vijande, and A. Valcarce (2007). "Faddeev study of heavy-baryon spectroscopy". *Journal of Physics G* **34** (5): 961–976. doi:10.1088/0954-3899/34/5/014.

- K. Carter (2006). "The rise and fall of the pentaquark". Fermilab and SLAC. Retrieved 2008-05-27.

- W.-M. Yao *et al.*(Particle Data Group) (2006). "Review of Particle Physics". *Journal of Physics G* **33**: 1–1232. arXiv:astro-ph/0601168. Bibcode:2006JPhG...33....1Y. doi:10.1088/0954-3899/33/1/001.

- D.M. Manley (2005). "Status of baryon spectroscopy". *Journal of Physics: Conference Series* **5**: 230–237. Bibcode:2005JPhCS...9..230M. doi:10.1088/1742-6596/9/1/043.

- H. Muir (2003). "Pentaquark discovery confounds sceptics". New Scientist. Retrieved 2008-05-27.

- S.S.M. Wong (1998a). "Chapter 2—Nucleon Structure". *Introductory Nuclear Physics* (2nd ed.). New York (NY): John Wiley & Sons. pp. 21–56. ISBN 0-471-23973-9.

- S.S.M. Wong (1998b). "Chapter 3—The Deuteron". *Introductory Nuclear Physics* (2nd ed.). New York (NY): John Wiley & Sons. pp. 57–104. ISBN 0-471-23973-9.

- R. Shankar (1994). *Principles of Quantum Mechanics* (2nd ed.). New York (NY): Plenum Press. ISBN 0-306-44790-8.

- E. Wigner (1937). "On the Consequences of the Symmetry of the Nuclear Hamiltonian on the Spectroscopy of Nuclei". *Physical Review* **51** (2): 106–119. Bibcode:1937PhRv...51..106W. doi:10.1103/PhysRev.51.106.

- M. Gell-Mann (1964). "A Schematic of Baryons and Mesons". *Physics Letters* **8** (3): 214–215. Bibcode:1964PhL......8..214G. doi:10.1016/S0031-9163(64)92001-3.

- W. Heisenberg (1932). "Über den Bau der Atomkerne I". *Zeitschrift für Physik* (in German) **77**: 1–11. Bibcode:1932ZPhy...77....1H. doi:10.1007/BF01342433.

- W. Heisenberg (1932). "Über den Bau der Atomkerne II". *Zeitschrift für Physik* (in German) **78** (3–4): 156–164. Bibcode:1932ZPhy...78..156H. doi:10.1007/BF01337585.

- W. Heisenberg (1932). "Über den Bau der Atomkerne III". *Zeitschrift für Physik* (in German) **80** (9–10): 587–596. Bibcode:1933ZPhy...80..587H. doi:10.1007/BF01335696.

23.9   External links

- Particle Data Group—Review of Particle Physics (2008).

- Georgia State University—HyperPhysics

- Baryons made thinkable, an interactive visualisation allowing physical properties to be compared

Chapter 24

Hypercharge

In particle physics, the **hypercharge** (from **hyperonic** + **charge**) Y of a particle is related to the strong interaction, and is distinct from the similarly named weak hypercharge, which has an analogous role in the electroweak interaction. The concept of hypercharge combines and unifies isospin and flavour into a single charge operator.

24.1 Definition

Hypercharge in particle physics is a quantum number relating the strong interactions of the SU(3) model. Isospin is defined in the SU(2) model while the SU(3) model defines hypercharge.

SU(3) weight diagrams (see below) are 2-dimensional with the coordinates referring to two quantum numbers, I_z, which is the z-component of isospin and Y, which is the hypercharge (the sum of strangeness (S), charm (C), bottomness (B'), topness (T), and baryon number (B)). Mathematically, hypercharge is

$$Y = S + C + B' + T + B$$

and conservation of hypercharge implies a conservation of flavour. Strong interactions conserve hypercharge, but weak interactions do not.

24.2 Relation with electric charge and isospin

Main article: Gell-Mann–Nishijima formula

The Gell-Mann–Nishijima formula relates isospin and electric charge

$$Q = I_3 + \frac{1}{2}Y,$$

where I_3 is the third component of isospin and Q is the particle's charge.

Isospin creates multiplets of particles whose average charge is related to the hypercharge by:

$$Y = 2\bar{Q}.$$

since the hypercharge is the same for all members of a multiplet, and the average of the I_3 values is 0.

24.3 SU(3) model in relation to hypercharge

The SU(2) model has multiplets characterized by a quantum number J, which is the total angular momentum. Each multiplet consists of $2J + 1$ substates with equally spaced values of J_z, forming a symmetric arrangement seen in atomic spectra and isospin. This formalises the observation that certain strong baryon decays were not observed, leading to the prediction of the mass, strangeness and charge of the $\Omega-$ baryon.

The SU(3) has *supermultiplets* containing SU(2) multiplets. SU(3) now needs 2 numbers to specify all its sub-states which are denoted by λ_1 and λ_2.

$(\lambda_1 + 1)$ specifies the number of points in the topmost side of the hexagon while $(\lambda_2 + 1)$ specifies the number of points on the bottom side.

24.4 Examples

- The nucleon group (protons with $Q = +1$ and neutrons with $Q = 0$) have an average charge of $+1/2$, so they both have hypercharge $Y = 1$ (baryon number $B = +1$, $S = C = B' = T = 0$). From the Gell-Mann–Nishijima formula we know that proton has isospin $I_3 = +1/2$, while neutron has $I_3 = -1/2$.

- This also works for quarks: for the *up* quark, with a charge of +2/3, and an I_3 of +1/2, we deduce a hypercharge of 1/3, due to its baryon number (since you need 3 quarks to make a baryon, a quark has baryon number of 1/3).

- For a *strange* quark, with charge −1/3, a baryon number of 1/3 and strangeness of −1 we get a hypercharge $Y = -2/3$, so we deduce an $I_3 = 0$. That means that a *strange* quark makes a singlet of its own (same happens with *charm*, *bottom* and *top* quarks), while *up* and *down* constitute an isospin doublet.

24.5   Practical obsolescence

Hypercharge was a concept developed in the 1960s, to organize groups of particles in the *"particle zoo"* and to develop *ad hoc* conservation laws based on their observed transformations. With the advent of the quark model, it is now obvious that (if one only includes the up, down and strange quarks out of the total 6 quarks in the Standard Model), hypercharge Y is the following combination of the numbers of up (n_u), down (n_d), strange quarks (n_s), charm quarks (n_c), top quarks (n_t) and bottom quarks (n_b):

$$Y = \frac{1}{3}(n_u + n_d - 2n_s + 4n_c + 4n_t - 2n_b).$$

In modern descriptions of hadron interaction, it has become more obvious to draw Feynman diagrams that trace through individual quarks composing the interacting baryons and mesons, rather than counting hypercharge quantum numbers. Weak hypercharge, however, remains of practical use in various theories of the electroweak interaction.

24.6   References

- Henry Semat, John R. Albright (1984). *Introduction to atomic and nuclear physics*. Chapman and Hall. ISBN 0-412-15670-9.

Chapter 25

Gravitational constant

Further information: Gravity of Earth and Standard gravity
The **gravitational constant** (also known as "universal

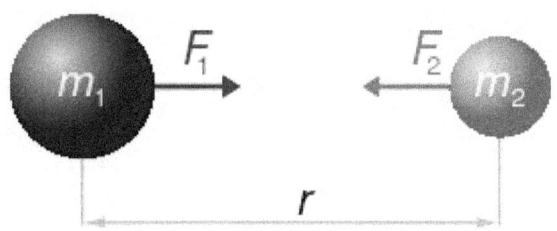

$$F_1 = F_2 = G\frac{m_1 \times m_2}{r^2}$$

The gravitational constant G is a key quantity in Newton's law of universal gravitation.

gravitational constant", or as "Newton's constant"), denoted by the letter G, is an empirical physical constant involved in the calculation of gravitational effects in Sir Isaac Newton's law of universal gravitation and in Albert Einstein's general theory of relativity. Its value is approximately $6.674{\times}10^{-11}$ N·m^2/kg^2.[1]

25.1 Law of gravitation

According to Newton's law of universal gravitation, the attractive force (F) between two bodies is directly proportional to the product of their masses (m_1 and m_2), and inversely proportional to the square of the distance, r, (inverse-square law) between them:

$$F = G\frac{m_1 m_2}{r^2}$$

The constant of proportionality, G, is the gravitational constant.

In the general theory of relativity, the Einstein field equations,[2][3]

$$R_{\mu\nu} - \tfrac{1}{2}R\,g_{\mu\nu} = \frac{8\pi G}{c^4}T_{\mu\nu}$$

Newton's constant appears in the proportionality between spacetime curvature and energy.

Colloquially, the gravitational constant is also called "Big G", for disambiguation with "small g" (g), which is the local gravitational field of Earth (equivalent to the free-fall acceleration).[4] The two quantities are related by $g = \frac{GM_E}{r_E^2}$ (where M_E is the mass of the Earth and r_E is the Earth radius).

25.2 Value and dimensions

The gravitational constant is a physical constant that is difficult to measure with high accuracy.[5] The reason is that the gravitational force is extremely weak compared with other fundamental forces.[6]

In SI units, the 2014 CODATA-recommended value of the gravitational constant (with standard uncertainty in parentheses) is:[1][7]

$$G = 6.67408(31) \times 10^{-11}\ \mathrm{m^3\ kg^{-1}\ s^{-2}}$$

This corresponds to a relative standard uncertainty of $4.7{\times}10^{-5}$.

The dimensions assigned to the gravitational constant are force times length squared divided by mass squared; this is equivalent to length cubed, divided by mass and by time squared:

$$[G] = \frac{[F][L]^2}{[M]^2} = \frac{[L]^3}{[M][T]^2}$$

153

In SI base units, this amounts to meters cubed per kilogram per second squared:

$$\mathrm{N\,m^2\,kg^{-2}} = \mathrm{m^3\,kg^{-1}\,s^{-2}}$$

The gravitational constant is taken as the basis of the Planck units: it is equal to the cube of the Planck length divided by the product of the Planck mass and the square of Planck time:

$$G = \frac{l_\mathrm{P}^3}{m_\mathrm{P} t_\mathrm{P}^2}.$$

In other words, in Planck units, G has the numerical value of 1.

In cgs, G can be written as:

$$G \approx 6.674 \times 10^{-8}\ \mathrm{cm^3\,g^{-1}\,s^{-2}}.$$

G can also be given as:

$$G \approx 0.8650\ \mathrm{cm^3\,g^{-1}\,hr^{-2}}.$$

In astrophysics, it is convenient to measure distances in parsecs (pc), velocities in kilometers per second (km/s) and masses in solar units $M\odot$. In these units, the gravitational constant is:

$$G \approx 4.302 \times 10^{-3}\ \mathrm{pc}\ M_\odot^{-1}\ \mathrm{(km/s)^2}.$$

In orbital mechanics, the period P of an object in circular orbit around a spherical object obeys

$$GM = \frac{3\pi V}{P^2}$$

where V is the volume inside the radius of the orbit. It follows that

$$P^2 = \frac{3\pi}{G}\frac{V}{M} \approx 10.896\ \mathrm{hr^2\,g\,cm^{-3}}\frac{V}{M}.$$

This way of expressing G shows the relationship between the average density of a planet and the period of a satellite orbiting just above its surface.

For elliptical orbits, applying Kepler's 3rd law, expressed in units characteristic of Earth's orbit:

$$G = 4\pi^2\ \mathrm{AU^3\,yr^{-2}}\ M_\odot^{-1}$$

where distance is measured in astronomical units (AU), time in years, and mass in solar masses ($M\odot$).

25.3 History of measurement

The gravitational constant appears in Newton's law of universal gravitation, but it was not measured until seventy-one years after Newton's death by Henry Cavendish with his Cavendish experiment, performed in 1798 (*Philosophical Transactions* 1798). Cavendish measured G implicitly, using a torsion balance invented by the geologist Rev. John Michell. He used a horizontal torsion beam with lead balls whose inertia (in relation to the torsion constant) he could tell by timing the beam's oscillation. Their faint attraction to other balls placed alongside the beam was detectable by the deflection it caused. Cavendish's aim was not actually to measure the gravitational constant, but rather to measure Earth's density relative to water, through the precise knowledge of the gravitational interaction. In modern units, the density that Cavendish calculated implied a value for G of $6.754 \times 10^{-11}\ \mathrm{m^3\,kg^{-1}\,s^{-2}}$.[8]

The accuracy of the measured value of G has increased only modestly since the original Cavendish experiment. G is quite difficult to measure, because gravity is much weaker than other fundamental forces, and an experimental apparatus cannot be separated from the gravitational influence of other bodies. Furthermore, gravity has no established relation to other fundamental forces, so it does not appear possible to calculate it indirectly from other constants that can be measured more accurately, as is done in some other areas of physics. Published values of G have varied rather broadly, and some recent measurements of high precision are, in fact, mutually exclusive.[5][9] This led to the 2010 CODATA value by NIST having 20% increased uncertainty than in 2006.[10] For the 2014 update, CODATA reduced the uncertainty to less than half the 2010 value.

In the January 2007 issue of *Science*, Fixler et al. described a new measurement of the gravitational constant by atom interferometry, reporting a value of $G = 6.693(34) \times 10^{-11}$ $\mathrm{m^3\,kg^{-1}\,s^{-2}}$.[11] An improved cold atom measurement by Rosi et al. was published in 2014 of $G = 6.67191(99) \times 10^{-11}$ $\mathrm{m^3\,kg^{-1}\,s^{-2}}$.[12]

A controversial 2015 study of some previous measurements of G, by Anderson et al., suggested that most of the mutually exclusive values can be explained by a periodic variation.[13] The variation was measured as having a period of 5.9 years, similar to that observed in length of day (LOD) measurements, hinting at a common physical cause which is not necessarily a variation in G. A response was produced by some of the original authors of the G measurements used in Anderson et al.[14] This response notes that Anderson et al. not only omitted measurements, they also used the time of publication not the time the experiments were performed. A plot with estimated time of measurement from contacting original authors seriously degrades the length of day correlation. Also taking the data collected over a decade by Karagioz and Izmailov shows no correlation with length of day measurements.[14][15] As such the variations in G most likely arise from systematic measurement errors which have not properly been accounted for.

Under the assumption that the physics of type Ia supernovae are universal, analysis of observations of 580 type Ia supernovae has shown that the gravitational constant has varied by less than one part in ten billion per year over the last nine billion years.[16]

25.4 The *GM* product

Main article: Standard gravitational parameter

The quantity *GM*—the product of the gravitational constant and the mass of a given astronomical body such as the Sun or Earth—is known as the standard gravitational parameter and is denoted μ. Depending on the body concerned, it may also be called the geocentric or heliocentric gravitational constant, among other names.

This quantity gives a convenient simplification of various gravity-related formulas. Also, for celestial bodies such as Earth and the Sun, the value of the product *GM* is known much more accurately than each factor independently. Indeed, the limited accuracy available for G limits the accuracy of determination of such masses in the first place.

For Earth, using $M\oplus$ as the symbol for the mass of Earth, we have

$$\mu = GM_\oplus = (398\ 600.4415 \pm 0.0008)\ \text{km}^3\ \text{s}^{-2}.\ \text{[17]}$$

For Sun, we have

$$\mu = GM_\odot = (1.327\ 124\ 400 \times 10^{11})\ \text{km}^3\ \text{s}^{-2}.$$

Calculations in celestial mechanics can also be carried out using the unit of solar mass rather than the standard SI unit kilogram. In this case we use the Gaussian gravitational constant k, where

$$k = 0.017\ 202\ 098\ 95\ A^{\frac{3}{2}}\ D^{-1}\ S^{-\frac{1}{2}}$$

and

A is the astronomical unit;

D is the mean solar day;

S is the solar mass.

If instead of mean solar day we use the sidereal year as our time unit, the value of ks is very close to 2π ($k = 6.28315$).

The standard gravitational parameter *GM* appears as above in Newton's law of universal gravitation, as well as in formulas for the deflection of light caused by gravitational lensing, in Kepler's laws of planetary motion, and in the formula for escape velocity.

25.5 See also

- Dirac large numbers hypothesis
- Accelerating universe
- Gravity expressed in terms of orbital period
- Lunar Laser Ranging experiment
- Cosmological constant
- Gravitational coupling constant
- Strong gravitational constant

25.6 Notes

[1] Mohr, Peter J.; Newell, David B.; Taylor, Barry N. (2015-07-21). "CODATA Recommended Values of the Fundamental Physical Constants: 2014". arXiv:1507.07956 [physics.atom-ph].

[2] Grøn, Øyvind; Hervik, Sigbjorn (2007). *Einstein's General Theory of Relativity: With Modern Applications in Cosmology* (illustrated ed.). Springer Science & Business Media. p. 180. ISBN 978-0-387-69200-5. Extract of page 180

[3] Einstein, Albert (1916). "The Foundation of the General Theory of Relativity". *Annalen der Physik* **354** (7): 769. Bibcode:1916AnP...354..769E. doi:10.1002/andp.19163540702. Archived from the original (PDF) on 2012-02-06.

[4] Gundlach, Jens H.; Merkowitz, Stephen M. (2002-12-23). "University of Washington Big G Measurement". *Astrophysics Science Division*. Goddard Space Flight Center. Since Cavendish first measured Newton's Gravitational constant 200 years ago, "Big G" remains one of the most elusive constants in physics. Fundamentals of Physics 8th ed., Halliday/Resnick/Walker, ISBN 978-0-470-04618-0 p. 336.

[5] George T. Gillies (1997), "The Newtonian gravitational constant: recent measurements and related studies", *Reports on Progress in Physics* **60** (2): 151–225. Bibcode:1997RPPh...60..151G. doi:10.1088/0034-4885/60/2/001. A lengthy, detailed review. See Figure 1 and Table 2 in particular.

[6] For example, the gravitational force between an electron and proton one meter apart is approximately 10^{-67} N, whereas the electromagnetic force between the same two particles is approximately 10^{-28} N. The electromagnetic force in this example is some 39 orders of magnitude (i.e. 10^{39}) greater than the force of gravity—roughly the same ratio as the mass of the Sun to a microgram.

[7] "Newtonian constant of gravitation G". CODATA, NIST.

[8] Brush, Stephen G.; Holton, Gerald James (2001), *Physics, the human adventure: from Copernicus to Einstein and beyond*, New Brunswick, N.J: Rutgers University Press, p. 137, ISBN 0-8135-2908-5

[9] Peter J. Mohr; Barry N. Taylor (January 2005), "CODATA recommended values of the fundamental physical constants: 2002" (PDF), *Reviews of Modern Physics* 77 (1): 1–107, Bibcode:2005RvMP...77....1M, doi:10.1103/RevModPhys.77.1, retrieved 2006-07-01. Section Q (pp. 42–47) describes the mutually inconsistent measurement experiments from which the CODATA value for G was derived.

[10] "CODATA recommended values of the fundamental physical constants: 2010" (PDF). *Rev Mod Phys* 84: 1527–1605. 13 November 2012. arXiv:1203.5425. Bibcode:2012RvMP...84.1527M. doi:10.1103/RevModPhys.84.1527.

[11] J. B. Fixler; G. T. Foster; J. M. McGuirk; M. A. Kasevich (2007-01-05), "Atom Interferometer Measurement of the Newtonian Constant of Gravity", *Science* 315 (5808): 74–77, Bibcode:2007Sci...315...74F, doi:10.1126/science.1135459, PMID 17204644

[12] Schlamminger, Stephan (18 June 2014). "Fundamental constants: A cool way to measure big G". *Nature* 510: 478–480. Bibcode:2014Natur.510..478S. doi:10.1038/nature13507.

[13] J.D. Anderson; G. Schubert; V. Trimble; M.R. Feldman (April 2015), "Measurements of Newton's gravitational constant and the length of day" (PDF), *EPL* 110: 10002, arXiv:1504.06604, Bibcode:2015EL....11010002A, doi:10.1209/0295-5075/110/10002

[14] Schlamminger, S.; Gundlach, J. H.; Newman, R. D. (2015). "Recent measurements of the gravitational constant as a function of time". *Physical Review D* 91 (12). arXiv:1505.01774. Bibcode:2015PhRvD..91l1101S. doi:10.1103/PhysRevD.91.121101. ISSN 1550-7998.

[15] Karagioz, O. V.; Izmailov, V. P. (1996). "Measurement of the gravitational constant with a torsion balance". *Measurement Techniques* 39 (10): 979–987. doi:10.1007/BF02377461. ISSN 0543-1972.

[16] J. Mould; S. A. Uddin (2014-04-10), "Constraining a Possible Variation of G with Type Ia Supernovae", *Publications of the Astronomical Society of Australia* 31: e015, arXiv:1402.1534, Bibcode:2014PASA...31...15M, doi:10.1017/pasa.2014.9

[17] Ries, J.C.; Eanes, R.J.; Shum, C.K.; Watkins, M.M. (20 March 1992). "Progress in the determination of the gravitational coefficient of the Earth". *Geophysical Research Letters* 19 (6): 529–531. Bibcode:1992GeoRL..19..529R. doi:10.1029/92GL00259. Retrieved 5 February 2016.

25.7 References

- E. Myles Standish. "Report of the IAU WGAS Subgroup on Numerical Standards". In *Highlights of Astronomy*, I. Appenzeller, ed. Dordrecht: Kluwer Academic Publishers, 1995. *(Complete report available online: PostScript; PDF. Tables from the report also available: Astrodynamic Constants and Parameters)*

- Jens H. Gundlach; Stephen M. Merkowitz (2000), "Measurement of Newton's Constant Using a Torsion Balance with Angular Acceleration Feedback", *Physical Review Letters* 85 (14): 2869–2872, arXiv:gr-qc/0006043, Bibcode:2000PhRvL..85.2869G, doi:10.1103/PhysRevLett.85.2869, PMID 11005956

25.8 External links

- Newtonian constant of gravitation G at the National Institute of Standards and Technology References on Constants, Units, and Uncertainty

- The Controversy over Newton's Gravitational Constant — additional commentary on measurement problems

Chapter 26

Supersymmetry

In particle physics, **Supersymmetry (SUSY)** is a proposed type of spacetime symmetry that relates two basic classes of elementary particles: bosons, which have an integer-valued spin, and fermions, which have a half-integer spin.[1] Each particle from one group is associated with a particle from the other, known as its superpartner, the spin of which differs by a half-integer. In a theory with perfectly "unbroken" supersymmetry, each pair of superpartners would share the same mass and internal quantum numbers besides spin. For example, there would be a "selectron" (superpartner electron), a bosonic version of the electron with the same mass as the electron, that would be easy to find in a laboratory. Thus, since no superpartners have been observed, if supersymmetry exists it must be a spontaneously broken symmetry so that superpartners may differ in mass.[2][3] Spontaneously-broken supersymmetry could solve many mysterious problems in particle physics including the hierarchy problem. The simplest realization of spontaneously-broken supersymmetry, the so-called Minimal Supersymmetric Standard Model, is one of the best studied candidates for physics beyond the Standard Model.

There is only indirect evidence and motivation for the existence of supersymmetry. Direct confirmation would entail production of superpartners in collider experiments, such as the Large Hadron Collider (LHC). The first run of the LHC found no evidence for supersymmetry (all results were consistent with the Standard Model), and thus set limits on superpartner masses in supersymmetric theories. While some remain enthusiastic about supersymmetry,[4] this first run at the LHC led some physicists to explore other ideas.[5] The LHC resumed its search for supersymmetry and other new physics in its second run.

26.1 Motivations

There are numerous phenomenological motivations for supersymmetry close to the electroweak scale, as well as technical motivations for supersymmetry at any scale.

26.1.1 The hierarchy problem

Supersymmetry close to the electroweak scale ameliorates the hierarchy problem that afflicts the Standard Model. In the Standard Model, the electroweak scale receives enormous Planck-scale quantum corrections. The observed hierarchy between the electroweak scale and the Planck scale must be achieved with extraordinary fine tuning. In a supersymmetric theory, on the other hand, Planck-scale quantum corrections cancel between partners and superpartners (owing to a minus sign associated with fermionic loops). The hierarchy between the electroweak scale and the Planck scale is achieved in a natural manner, without miraculous fine-tuning.

26.1.2 Gauge coupling unification

The idea that the gauge symmetry groups unify at high-energy is called Grand unification theory. In the Standard Model, however, the weak, strong and electromagnetic couplings fail to unify at high energy. In a supersymmetry theory, the running of the gauge couplings are modified, and precise high-energy unification of the gauge couplings is achieved. The modified running also provides a natural mechanism for radiative electroweak symmetry breaking.

26.1.3 Dark matter

TeV-scale supersymmetry (augmented with a discrete symmetry) typically provides a candidate dark matter particle at a mass scale consistent with thermal relic abundance calculations.[6][7]

26.1.4   Other technical motivations

Supersymmetry is also motivated by solutions to several theoretical problems, for generally providing many desirable mathematical properties, and for ensuring sensible behavior at high energies. Supersymmetric quantum field theory is often much easier to analyze, as many more problems become exactly solvable. When supersymmetry is imposed as a *local* symmetry, Einstein's theory of general relativity is included automatically, and the result is said to be a theory of supergravity. It is also a necessary feature of the most popular candidate for a theory of everything, superstring theory.

Another theoretically appealing property of supersymmetry is that it offers the only "loophole" to the Coleman–Mandula theorem, which prohibits spacetime and internal symmetries from being combined in any nontrivial way, for quantum field theories like the Standard Model with very general assumptions. The Haag-Lopuszanski-Sohnius theorem demonstrates that supersymmetry is the only way spacetime and internal symmetries can be combined consistently.[8]

26.2   History

A supersymmetry relating mesons and baryons was first proposed, in the context of hadronic physics, by Hironari Miyazawa during 1966. This supersymmetry did not involve spacetime, that is, it concerned internal symmetry, and was broken badly. Miyazawa's work was largely ignored at the time.[9][10][11][12]

J. L. Gervais and B. Sakita (during 1971),[13] Yu. A. Golfand and E. P. Likhtman (also during 1971), and D.V. Volkov and V.P. Akulov (1972),[14] independently rediscovered supersymmetry in the context of quantum field theory, a radically new type of symmetry of spacetime and fundamental fields, which establishes a relationship between elementary particles of different quantum nature, bosons and fermions, and unifies spacetime and internal symmetries of microscopic phenomena. Supersymmetry with a consistent Lie-algebraic graded structure on which the Gervais–Sakita rediscovery was based directly first arose during 1971[15] in the context of an early version of string theory by Pierre Ramond, John H. Schwarz and André Neveu.

Finally, Julius Wess and Bruno Zumino (during 1974)[16] identified the characteristic renormalization features of four-dimensional supersymmetric field theories, which identified them as remarkable QFTs, and they and Abdus Salam and their fellow researchers introduced early particle physics applications. The mathematical structure of supersymmetry (Graded Lie superalgebras) has subse-

quently been applied successfully to other topics of physics, ranging from nuclear physics,[17][18] critical phenomena,[19] quantum mechanics to statistical physics. It remains a vital part of many proposed theories of physics.

The first realistic supersymmetric version of the Standard Model was proposed during 1977 by Pierre Fayet and is known as the Minimal Supersymmetric Standard Model or MSSM for short. It was proposed to solve, amongst other things, the hierarchy problem.

26.3   Applications

26.3.1   Extension of possible symmetry groups

One reason that physicists explored supersymmetry is because it offers an extension to the more familiar symmetries of quantum field theory. These symmetries are grouped into the Poincaré group and internal symmetries and the Coleman–Mandula theorem showed that under certain assumptions, the symmetries of the S-matrix must be a direct product of the Poincaré group with a compact internal symmetry group or if there is not any mass gap, the conformal group with a compact internal symmetry group. During 1971 Golfand and Likhtman were the first to show that the Poincaré algebra can be extended through introduction of four anticommuting spinor generators (in four dimensions), which later became known as supercharges. During 1975 the Haag-Lopuszanski-Sohnius theorem analyzed all possible superalgebras in the general form, including those with an extended number of the supergenerators and central charges. This extended super-Poincaré algebra paved the way for obtaining a very large and important class of supersymmetric field theories.

The supersymmetry algebra

Main article: Supersymmetry algebra

Traditional symmetries of physics are generated by objects that transform by the tensor representations of the Poincaré group and internal symmetries. Supersymmetries, however, are generated by objects that transform by the spinor representations. According to the spin-statistics theorem, bosonic fields commute while fermionic fields anticommute. Combining the two kinds of fields into a single algebra requires the introduction of a Z_2-grading under which the bosons are the even elements and the fermions are the odd elements. Such an algebra is called a Lie superalgebra.

The simplest supersymmetric extension of the Poincaré algebra is the Super-Poincaré algebra. Expressed in terms of two Weyl spinors, has the following anti-commutation relation:

$$\{Q_\alpha, \bar{Q}\beta\} = 2(\sigma^\mu)_{\alpha\beta}P_\mu$$

and all other anti-commutation relations between the Qs and commutation relations between the Qs and Ps vanish. In the above expression $P_\mu = -i\partial_\mu$ are the generators of translation and σ^μ are the Pauli matrices.

There are representations of a Lie superalgebra that are analogous to representations of a Lie algebra. Each Lie algebra has an associated Lie group and a Lie superalgebra can sometimes be extended into representations of a Lie supergroup.

26.3.2 The Supersymmetric Standard Model

Main article: Minimal Supersymmetric Standard Model

Incorporating supersymmetry into the Standard Model requires doubling the number of particles since there is no way that any of the particles in the Standard Model can be superpartners of each other. With the addition of new particles, there are many possible new interactions. The simplest possible supersymmetric model consistent with the Standard Model is the Minimal Supersymmetric Standard Model (MSSM) which can include the necessary additional new particles that are able to be superpartners of those in the Standard Model.

One of the main motivations for SUSY comes from the quadratically divergent contributions to the Higgs mass squared. The quantum mechanical interactions of the Higgs boson causes a large renormalization of the Higgs mass and unless there is an accidental cancellation, the natural size of the Higgs mass is the greatest scale possible. This problem is known as the hierarchy problem. Supersymmetry reduces the size of the quantum corrections by having automatic cancellations between fermionic and bosonic Higgs interactions. If supersymmetry is restored at the weak scale, then the Higgs mass is related to supersymmetry breaking which can be induced from small non-perturbative effects explaining the vastly different scales in the weak interactions and gravitational interactions.

In many supersymmetric Standard Models there is a heavy stable particle (such as neutralino) which could serve as a weakly interacting massive particle (WIMP) dark matter candidate. The existence of a supersymmetric dark matter candidate is related closely to R-parity.

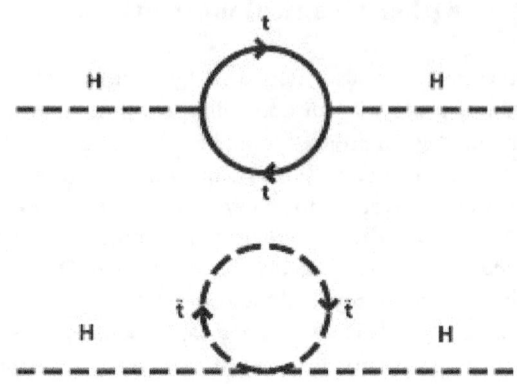

Cancellation of the Higgs boson quadratic mass renormalization between fermionic top quark loop and scalar stop squark tadpole Feynman diagrams in a supersymmetric extension of the Standard Model

The standard paradigm for incorporating supersymmetry into a realistic theory is to have the underlying dynamics of the theory be supersymmetric, but the ground state of the theory does not respect the symmetry and supersymmetry is broken spontaneously. The supersymmetry break can not be done permanently by the particles of the MSSM as they currently appear. This means that there is a new sector of the theory that is responsible for the breaking. The only constraint on this new sector is that it must break supersymmetry permanently and must give superparticles TeV scale masses. There are many models that can do this and most of their details do not matter. In order to parameterize the relevant features of supersymmetry breaking, arbitrary soft SUSY breaking terms are added to the theory which temporarily break SUSY explicitly but could never arise from a complete theory of supersymmetry breaking.

Gauge-coupling unification

Main article: Minimal Supersymmetric Standard Model § Gauge-coupling unification

One piece of evidence for supersymmetry existing is gauge coupling unification. The renormalization group evolution of the three gauge coupling constants of the Standard Model is somewhat sensitive to the present particle content of the theory. These coupling constants do not quite meet together at a common energy scale if we run the renormalization group using the Standard Model.[20] With the addition of minimal SUSY joint convergence of the coupling constants is projected at approximately 10^{16} GeV.[20]

26.3.3   Supersymmetric quantum mechanics

Main article: Supersymmetric quantum mechanics

Supersymmetric quantum mechanics adds the SUSY super-algebra to quantum mechanics as opposed to quantum field theory. Supersymmetric quantum mechanics often becomes relevant when studying the dynamics of supersymmetric solitons, and due to the simplified nature of having fields which are only functions of time (rather than space-time), a great deal of progress has been made in this subject and it is now studied in its own right.

SUSY quantum mechanics involves pairs of Hamiltonians which share a particular mathematical relationship, which are called *partner Hamiltonians*. (The potential energy terms which occur in the Hamiltonians are then known as *partner potentials*.) An introductory theorem shows that for every eigenstate of one Hamiltonian, its partner Hamiltonian has a corresponding eigenstate with the same energy. This fact can be exploited to deduce many properties of the eigenstate spectrum. It is analogous to the original description of SUSY, which referred to bosons and fermions. We can imagine a "bosonic Hamiltonian", whose eigenstates are the various bosons of our theory. The SUSY partner of this Hamiltonian would be "fermionic", and its eigenstates would be the theory's fermions. Each boson would have a fermionic partner of equal energy.

26.3.4   Supersymmetry: Applications   to condensed matter physics

SUSY concepts have provided useful extensions to the WKB approximation. Additionally, SUSY has been applied to disorder averaged systems both quantum and non-quantum (through statistical mechanics), the Fokker-Planck equation being an example of a non-quantum theory. The 'supersymmetry' in all these systems arises from the fact that one is modelling one particle and as such the 'statistics' don't matter. The use of the supersymmetry method provides a mathematical rigorous alternative to the replica trick, but only in non-interacting systems, which attempts to address the so-called 'problem of the denominator' under disorder averaging. For more on the applications of supersymmetry in condensed matter physics see the book[21]

26.3.5   Supersymmetry in optics

Integrated optics was recently found[22] to provide a fertile ground on which certain ramifications of SUSY can be explored in readily-accessible laboratory settings. Making use

of the analogous mathematical structure of the quantum-mechanical Schrödinger equation and the wave equation governing the evolution of light in one-dimensional settings, one may interpret the refractive index distribution of a structure as a potential landscape in which optical wave packets propagate. In this manner, a new class of functional optical structures with possible applications in phase matching, mode conversion[23] and space-division multiplexing becomes possible. SUSY transformations have been also proposed as a way to address inverse scattering problems in optics and as a one-dimensional transformation optics [24]

26.3.6   Mathematics

SUSY is also sometimes studied mathematically for its intrinsic properties. This is because it describes complex fields satisfying a property known as holomorphy, which allows holomorphic quantities to be exactly computed. This makes supersymmetric models useful "toy models" of more realistic theories. A prime example of this has been the demonstration of S-duality in four-dimensional gauge theories[25] that interchanges particles and monopoles.

The proof of the Atiyah-Singer index theorem is much simplified by the use of supersymmetric quantum mechanics.

26.4   General supersymmetry

Supersymmetry appears in many related contexts of theoretical physics. It is possible to have multiple supersymmetries and also have supersymmetric extra dimensions.

26.4.1   Extended supersymmetry

Main article: Extended supersymmetry

It is possible to have more than one kind of supersymmetry transformation. Theories with more than one supersymmetry transformation are known as extended supersymmetric theories. The more supersymmetry a theory has, the more constrained are the field content and interactions. Typically the number of copies of a supersymmetry is a power of 2, i.e. 1, 2, 4, 8. In four dimensions, a spinor has four degrees of freedom and thus the minimal number of supersymmetry generators is four in four dimensions and having eight copies of supersymmetry means that there are 32 supersymmetry generators.

The maximal number of supersymmetry generators possible is 32. Theories with more than 32 supersymmetry generators automatically have massless fields with spin greater than 2. It is not known how to make massless fields with spin greater than two interact, so the maximal number of supersymmetry generators considered is 32. This is due to the Weinberg-Witten theorem. This corresponds to an $N = 8$ supersymmetry theory. Theories with 32 supersymmetries automatically have a graviton.

For four dimensions there are the following theories, with the corresponding multiplets[26] (CPT adds a copy, whenever they are not invariant under such symmetry)

- $N = 1$

Chiral multiplet: $(0,\frac{1}{2})$ Vector multiplet: $(\frac{1}{2},1)$ Gravitino multiplet: $(1,\frac{3}{2})$ Graviton multiplet: $(\frac{3}{2},2)$

- $N = 2$

hypermultiplet: $(-\frac{1}{2},0^2,\frac{1}{2})$ vector multiplet: $(0,\frac{1}{2}^2,1)$ supergravity multiplet: $(1,\frac{3}{2}^2,2)$

- $N = 4$

Vector multiplet: $(-1,-\frac{1}{2}^4,0^6,\frac{1}{2}^4,1)$ Supergravity multiplet: $(0,\frac{1}{2}^4,1^6,\frac{3}{2}^4,2)$

- $N = 8$

Supergravity multiplet: $(-2,-\frac{3}{2}^8,-1^{28},-\frac{1}{2}^{56},0^{70},\frac{1}{2}^{56},1^{28},\frac{3}{2}^8,2)$

26.4.2 Supersymmetry in alternate numbers of dimensions

It is possible to have supersymmetry in dimensions other than four. Because the properties of spinors change drastically between different dimensions, each dimension has its characteristic. In d dimensions, the size of spinors is approximately $2^{d/2}$ or $2^{(d-1)/2}$. Since the maximum number of supersymmetries is 32, the greatest number of dimensions in which a supersymmetric theory can exist is eleven.

26.5 Supersymmetry in quantum gravity

Supersymmetry is part of a larger enterprise of theoretical physics to unify everything we know about the universe into a single consistent set of physical principles, known as the quest for a Theory of Everything (TOE). A significant part of this larger enterprise is the quest for a theory of quantum gravity, which would unify the classical theory of general relativity and the Standard Model, which explains the other three basic forces in physics (electromagnetism, the strong interaction, and the weak interaction), and provides a palette of fundamental particles upon which all four forces act. Two of the most active methods of forming a theory of quantum gravity are string theory and loop quantum gravity (LQG), although in theory, supersymmetry could be a component of other theories as well.

For string theory to be consistent, supersymmetry seems to be required at some level (although it may be a strongly broken symmetry). In particle theory, supersymmetry is recognized as a way to stabilize the hierarchy between the unification scale and the electroweak scale (or the Higgs boson mass), and can also provide a natural dark matter candidate. String theory also requires extra spatial dimensions which have to be compactified as in Kaluza–Klein theory.

Loop quantum gravity (LQG) predicts no additional spatial dimensions, nor anything else about particle physics. These theories can be formulated in three spatial dimensions and one dimension of time, although in some LQG theories dimensionality is an emergent property of the theory, rather than a fundamental assumption of the theory. Also, LQG is a theory of quantum gravity which does not require supersymmetry. Lee Smolin, one of the originators of LQG, has proposed that a loop quantum gravity theory incorporating either supersymmetry or extra dimensions, or both, be called "loop quantum gravity II".

If experimental evidence confirms supersymmetry in the form of supersymmetric particles such as the neutralino that is often believed to be the lightest superpartner, some people believe this would be a major boost to string theory. Since supersymmetry is a required component of string theory, any discovered supersymmetry would be consistent with string theory. If the Large Hadron Collider and other major particle physics experiments fail to detect supersymmetric partners or evidence of extra dimensions, many versions of string theory which had predicted certain low mass superpartners to existing particles may need to be significantly revised. The failure of experiments to discover either supersymmetric partners or extra spatial dimensions, as of 2013, has encouraged loop quantum gravity researchers.

26.6 Current status

Supersymmetric models are constrained by a variety of experiments, including measurements of low-energy observables – for example, the anomalous magnetic moment of the muon at Brookhaven; the WMAP dark matter density measurement and direct detection experiments – for example, XENON−100 and LUX; and by particle collider experiments, including B-physics, Higgs phenomenology and direct searches for superpartners (sparticles), at the Large Electron–Positron Collider, Tevatron and the LHC.

Historically, the tightest limits were from direct production at colliders. The first mass limits for squarks and gluinos were made at CERN by the UA1 experiment and the UA2 experiment at the Super Proton Synchrotron. LEP later set very strong limits.,[27] which in 2006 were extended by the D0 experiment at the Tevatron.[28][29] From 2003, WMAP's and Planck's dark matter density measurements have strongly constrained supersymmetry models, which, if they explain dark matter, have to be tuned to invoke a particular mechanism to sufficiently reduce the neutralino density.

Prior to the beginning of the LHC, in 2009 fits of available data to CMSSM and NUHM1 indicated that squarks and gluinos were most likely to have masses in the 500 to 800 GeV range, though values as high as 2.5 TeV were allowed with low probabilities. Neutralinos and sleptons were expected to be quite light, with the lightest neutralino and the lightest stau most likely to be found between 100 to 150 GeV.[30]

The first run of the LHC found no evidence for supersymmetry, and, as a result, surpassed existing experimental limits from the Large Electron–Positron Collider and Tevatron and partially excluded the aforementioned expected ranges.[31]

During 2011 and 2012, the LHC discovered a Higgs boson with a mass of about 125 GeV, and with couplings to fermions and bosons which are consistent with the Standard Model. The MSSM predicts that the mass of the lightest Higgs boson should not be much higher than the mass of the Z boson, and, in the absence of fine tuning (with the supersymmetry breaking scale on the order of 1 TeV), should not exceed 130 GeV. Furthermore, for values of the MSSM parameter $tan\ \beta \leq 3$, it predicts a Higgs mass below 114 GeV over most of the parameter space.[32] This region of Higgs mass was excluded by LEP by 2000. The LHC result is somewhat problematic for the minimal supersymmetric model, as the value of 125 GeV is relatively large for the model and can only be achieved with large radiative loop corrections from top squarks, which many theorists consider to be "unnatural" (see naturalness and fine tuning).[33] On the other hand, the lightest Higgs boson in the MSSM is Standard Model-like, which is consistent with measurements of the Higgs boson couplings at the LHC.

26.7   See also

- Supersymmetric gauge theory
- Wess–Zumino model
- Minimal Supersymmetric Standard Model
- Supersymmetry as a quantum group
- Quantum group
- Supercharge
- Superfield
- Supergeometry
- Supergravity
- Supergroup
- Superspace
- Superpartner

26.8   References

[1] Haber, Howie. "SUPERSYMMETRY, PART I (THEORY)" (PDF). *Reviews, Tables and Plots*. Particle Data Group (PDG). Retrieved 8 July 2015.

[2] Martin, Stephen P. (1997). "A Supersymmetry Primer". arXiv:hep-ph/9709356.

[3] Dine, Michael (2007). *Supersymmetry and String Theory: Beyond the Standard Model*. p. 169.

[4] Ellis, John. "The Physics Landscape after the Higgs Discovery at the LHC". *arXiv*. Invited plenary talk at SILAFAE 2014. Retrieved 8 July 2015.

[5] Wolchover, Natalie (November 20, 2012). "Supersymmetry Fails Test, Forcing Physics to Seek New Ideas". *Quanta Magazine*.

[6] Jonathan Feng: Supersymmetric Dark Matter *(pdf)*, University of California, Irvine, 11 May 2007

[7] Torsten Bringmann: The WIMP "Miracle" *(pdf)* University of Hamburg

[8] R. Haag, J. T. Lopuszanski and M. Sohnius, "All Possible Generators Of Supersymmetries Of The S Matrix", Nucl. Phys. B 88 (1975) 257

[9] H. Miyazawa (1966). "Baryon Number Changing Currents". *Prog. Theor. Phys.* **36** (6): 1266–1276. Bibcode:1966PThPh..36.1266M. doi:10.1143/PTP.36.1266.

[10] H. Miyazawa (1968). "Spinor Currents and Symmetries of Baryons and Mesons". *Phys. Rev.* **170** (5): 1586–1590. Bibcode:1968PhRv..170.1586M. doi:10.1103/PhysRev.170.1586.

[11] Michio Kaku, *Quantum Field Theory*, ISBN 0-19-509158-2, pg 663.

[12] Peter Freund, *Introduction to Supersymmetry*, ISBN 0-521-35675-X, pages 26-27, 138.

[13] Gervais, J. -L.; Sakita, B. (1971). "Field theory interpretation of supergauges in dual models". *Nuclear Physics B* **34** (2): 632–639. Bibcode:1971NuPhB..34..632G. doi:10.1016/0550-3213(71)90351-8.

[14] D.V. Volkov, V.P. Akulov, Pisma Zh.Eksp.Teor.Fiz. 16 (1972) 621; Phys.Lett. B46 (1973) 109; V.P. Akulov, D.V. Volkov, Teor.Mat.Fiz. 18 (1974) 39

[15] Ramond, P. (1971). "Dual Theory for Free Fermions". *Physical Review D* **3** (10): 2415–2418. Bibcode:1971PhRvD...3.2415R. doi:10.1103/PhysRevD.3.2415.

[16] Wess, J.; Zumino, B. (1974). "Supergauge transformations in four dimensions". *Nuclear Physics B* **70**: 39–50. Bibcode:1974NuPhB..70...39W. doi:10.1016/0550-3213(74)90355-1.

[17] http://users.physik.fu-berlin.de/~{}kleinert/kleinert/?p=supersym suggested here

[18] Iachello, F. (1980). "Dynamical Supersymmetries in Nuclei". *Physical Review Letters* **44** (12): 772–775. Bibcode:1980PhRvL..44..772I. doi:10.1103/PhysRevLett.44.772.

[19] Friedan, D.; Qiu, Z.; Shenker, S. (1984). "Conformal Invariance, Unitarity, and Critical Exponents in Two Dimensions". *Physical Review Letters* **52** (18): 1575–1578. Bibcode:1984PhRvL..52.1575F. doi:10.1103/PhysRevLett.52.1575.

[20] Gordon L. Kane, *The Dawn of Physics Beyond the Standard Model*, Scientific American, June 2003, page 60 and *The frontiers of physics*, special edition, Vol 15, #3, page 8

[21] *Supersymmetry in Disorder and Chaos*, Konstantin Efetov, Cambridge university press, 1997.

[22] Miri, M.-A.; Heinrich, M.; El-Ganainy, R.; Christodoulides, D. N. (2013). "Superymmetric optical structures". *Physical Review Letters* (APS) **110** (23): 233902. arXiv:1304.6646. Bibcode:2013PhRvL.110w3902M. doi:10.1103/PhysRevLett.110.233902. PMID 25167493. Retrieved April 2014.

[23] Heinrich, M.; Miri, M.-A.; Stützer, S.; El-Ganainy, R.; Nolte, S.; Szameit, A.; Christodoulides, D. N. (2014). "Superymmetric mode converters". *Nature Communications* (NPG) **5**: 3698. arXiv:1401.5734. Bibcode:2014NatCo...5E3698H. doi:10.1038/ncomms4698. PMID 24739256. Retrieved April 2014.

[24] Miri, M.-A.; Heinrich, Matthias; Christodoulides, D. N. (2014). "SUSY-inspired one-dimensional transformation optics". *Optica* (OSA) **1** (2): 89. arXiv:1408.0832. doi:10.1364/OPTICA.1.000089. Retrieved August 2014.

[25] Krasnitz, Michael (2002). *Correlation functions in supersymmetric gauge theories from supergravity fluctuafluctuations hHKtions* (PDF). Princeton University Department of Physics: Princeton University Department of Physics. p. 91.

[26] Polchinski,J. *String theory. Vol. 2: Superstring theory and beyond*, Appendix B

[27] LEPSUSYWG, ALEPH, DELPHI, L3 and OPAL experiments, charginos, large m0 LEPSUSYWG/01-03.1

[28] The D0-Collaboration (2009). "Search for associated production of charginos and neutralinos in the trilepton final state using 2.3 fb⁻¹ of data". arXiv:0901.0646. Bibcode:2009PhLB..680...34D. doi:10.1016/j.physletb.2009.08.011.

[29] The D0 Collaboration (2006). "Search for squarks and gluinos in events with jets and missing transverse energy using 2.1 fb-1 of pp⁻ collision data at s=1.96 TeV". arXiv:0712.3805. Bibcode:2008PhLB..660..449D. doi:10.1016/j.physletb.2008.01.042.

[30] O. Buchmueller; et al. (2009). "Likelihood Functions for Supersymmetric Observables in Frequentist Analyses of the CMSSM and NUHM1". *The European Physical Journal C* **64** (3): 391–415. arXiv:0907.5568. Bibcode:2009EPJC...64..391B. doi:10.1140/epjc/s10052-009-1159-z.

[31] Roszkowski, Leszek; Sessolo, Enrico Maria; Williams, Andrew J. (11 August 2014). "What next for the CMSSM and the NUHM: improved prospects for superpartner and dark matter detection". *Journal of High Energy Physics* **2014** (8). arXiv:1405.4289. Bibcode:2014JHEP...08..067R. doi:10.1007/JHEP08(2014)067.

[32] Marcela Carena and Howard E. Haber; Haber (1970). "Higgs Boson Theory and Phenomenology". *Progress in Particle and Nuclear Physics* **50**: 63–152. arXiv:hep-ph/0208209v3. Bibcode:2003PrPNP..50...63C. doi:10.1016/S0146-6410(02)00177-1.

[33] Patrick Draper; et al. (December 2011). "Implications of a 125 GeV Higgs for the MSSM and Low-Scale SUSY Breaking". *Physical Review D* **85** (9): 095007. arXiv:1112.3068. Bibcode:2012PhRvD..85i5007D. doi:10.1103/PhysRevD.85.095007.

26.9 Further reading

- Supersymmetry and Supergravity page in String Theory Wiki lists more books and reviews.

26.9.1 Theoretical introductions, free and online

- S. Martin (2011). "A Supersymmetry Primer". arXiv:hep-ph/9709356.

- Joseph D. Lykken (1996). "Introduction to Supersymmetry". arXiv:hep-th/9612114.

- Manuel Drees (1996). "An Introduction to Supersymmetry". arXiv:hep-ph/9611409.

- Adel Bilal (2001). "Introduction to Supersymmetry". arXiv:hep-th/0101055.

- An Introduction to Global Supersymmetry by Philip Arygres, 2001

26.9.2 Monographs

- Weak Scale Supersymmetry by Howard Baer and Xerxes Tata, 2006.

- Cooper, F.; Khare, A.; Sukhatme, U. (1995). "Supersymmetry and quantum mechanics". *Physics Reports* **251** (5–6): 267–385. arXiv:hep-th/9405029. Bibcode:1995PhR...251..267C. doi:10.1016/0370-1573(94)00080-M. (arXiv:hep-th/9405029).

- Junker, G. (1996). "Supersymmetric Methods in Quantum and Statistical Physics". doi:10.1007/978-3-642-61194-0. ISBN 978-3-540-61591-0..

- Gordon L. Kane.*Supersymmetry: Unveiling the Ultimate Laws of Nature* Basic Books, New York (2001). ISBN 0-7382-0489-7.

- Gordon L. Kane and Shifman, M., eds. *The Supersymmetric World: The Beginnings of the Theory*, World Scientific, Singapore (2000). ISBN 981-02-4522-X.

- Weinberg, Steven, *The Quantum Theory of Fields, Volume 3: Supersymmetry*, Cambridge University Press, Cambridge, (1999). ISBN 0-521-66000-9.

- Wess, Julius, and Jonathan Bagger, *Supersymmetry and Supergravity*, Princeton University Press, Princeton, (1992). ISBN 0-691-02530-4.

- "Concise Encyclopedia of Supersymmetry". 2003. doi:10.1007/1-4020-4522-0. ISBN 978-1-4020-1338-6.

26.9.3 On experiments

- Bennett GW; Muon (g–2) Collaboration; Bousquet; Brown; Bunce; Carey; Cushman; Danby; Debevec; Deile; Deng; Dhawan; Druzhinin; Duong; Farley; Fedotovich; Gray; Grigoriev; Grosse-Perdekamp; Grossmann; Hare; Hertzog; Huang; Hughes; Iwasaki; Jungmann; Kawall; Khazin; Krienen; Kronkvist; et al. (2004). "Measurement of the negative muon anomalous magnetic moment to 0.7 ppm". *Physical Review Letters* **92** (16): 161802. arXiv:hep-ex/0401008. Bibcode:2004PhRvL..92p1802B. doi:10.1103/PhysRevLett.92.161802. PMID 15169217.

- Brookhaven National Laboratory (Jan. 8, 2004). *New g−2 measurement deviates further from Standard Model.* Press Release.

- Fermi National Accelerator Laboratory (Sept 25, 2006). *Fermilab's CDF scientists have discovered the quick-change behavior of the B-sub-s meson.* Press Release.

26.10 External links

- Supersymmetry (physics) at *Encyclopædia Britannica*

- What do current LHC results (mid-August 2011) imply about supersymmetry? Matt Strassler

- ATLAS Experiment Supersymmetry search documents

- CMS Experiment Supersymmetry search documents

- "Particle wobble shakes up supersymmetry", *Cosmos* magazine, September 2006

- LHC results put supersymmetry theory 'on the spot' BBC news 27/8/2011

- SUSY running out of hiding places BBC news 12/11/2012

- Supersymmetry in optics? "Skulls in the Stars" blog 22/08/2013

Chapter 27

Supergravity

In theoretical physics, **supergravity** (**supergravity theory**; **SUGRA** for short) is a field theory that combines the principles of supersymmetry and general relativity. Together, these imply that, in supergravity, the supersymmetry is a local symmetry (in contrast to non-gravitational supersymmetric theories, such as the Minimal Supersymmetric Standard Model). Since the generators of supersymmetry (SUSY) are convoluted with the Poincaré group to form a super-Poincaré algebra, it can be seen that supergravity follows naturally from supersymmetry.[1] All traditional literature on supergravity is generally written in terms of Cartan connections.[2]

27.1 Gravitons

Like any field theory of gravity, a supergravity theory contains a spin-2 field whose quantum is the graviton. Supersymmetry requires the graviton field to have a superpartner. This field has spin 3/2 and its quantum is the gravitino. The number of gravitino fields is equal to the number of supersymmetries.

27.2 History

27.2.1 Gauge supersymmetry

The first theory[3] of local supersymmetry was proposed in 1975 by Dick Arnowitt and Pran Nath and was called **gauge supersymmetry**.

27.2.2 SUGRA

SUGRA, or supergravity, was discovered in 1976 by Dan Freedman, Sergio Ferrara and Peter van Nieuwenhuizen,[4] but was quickly generalized to many different theories in various numbers of dimensions and additional (N) supersymmetry charges. Supergravity theories with N>1 are usu-

ally referred to as extended supergravity (SUEGRA). Some supergravity theories were shown to be equivalent to certain higher-dimensional supergravity theories via dimensional reduction (e.g. $N = 1$ **11-dimensional** supergravity is dimensionally reduced on S^7 to $N = 8$, $d = 4$ SUGRA). The resulting theories were sometimes referred to as Kaluza–Klein theories as Kaluza and Klein constructed in 1919 a 5-dimensional gravitational theory, that when dimensionally reduced on circle, its 4-dimensional non-massive modes describe electromagnetism coupled to gravity.

27.2.3 mSUGRA

mSUGRA means minimal SUper GRAvity. The construction of a realistic model of particle interactions within the $N = 1$ supergravity framework where supersymmetry (SUSY) is broken by a super Higgs mechanism was carried out by Ali Chamseddine, Richard Arnowitt and Pran Nath in 1982. In these classes of models collectively now known as minimal supergravity Grand Unification Theories (mSUGRA GUT), gravity mediates the breaking of SUSY through the existence of a hidden sector. mSUGRA naturally generates the Soft SUSY breaking terms which are a consequence of the Super Higgs effect. Radiative breaking of electroweak symmetry through Renormalization Group Equations (RGEs) follows as an immediate consequence. mSUGRA is one of the most widely investigated models of particle physics due to its predictive power—requiring only four input parameters and a sign to determine the low energy phenomenology from the scale of Grand Unification.

See also: Gravity-Mediated Supersymmetry Breaking in the MSSM

27.2.4 11d: the maximal SUGRA

One of these supergravities, the 11-dimensional theory, generated considerable excitement as the first potential candidate for the theory of everything. This excitement was built on four pillars, two of which have now been largely discredited:

- Werner Nahm showed[5] that 11 dimensions was the largest number of dimensions consistent with a single graviton, and that a theory with more dimensions would also have particles with spins greater than 2. These problems are avoided in 12 dimensions if two of these dimensions are timelike, as has been often emphasized by Itzhak Bars.

- In 1981, Ed Witten showed[6] that 11 was the smallest number of dimensions that was big enough to contain the gauge groups of the Standard Model, namely SU(3) for the strong interactions and SU(2) times U(1) for the electroweak interactions. Today many techniques exist to embed the standard model gauge group in supergravity in any number of dimensions. For example, in the mid and late 1980s, the obligatory gauge symmetry in type I and heterotic string theories was often used. In type II string theory they could also be obtained by compactifying on certain Calabi–Yau manifolds. Today one may also use D-branes to engineer gauge symmetries.

- In 1978, Eugène Cremmer, Bernard Julia and Joël Scherk (CJS) found[7] the classical action for an 11-dimensional supergravity theory. This remains today the only known classical 11-dimensional theory with local supersymmetry and no fields of spin higher than two. Other 11-dimensional theories are known that are quantum-mechanically inequivalent to the CJS theory, but classically equivalent (that is, they reduce to the CJS theory when one imposes the classical equations of motion). For example, in the mid 1980s Bernard de Wit and Hermann Nicolai found an alternate theory in D=11 Supergravity with Local SU(8) Invariance. This theory, while not manifestly Lorentz-invariant, is in many ways superior to the CJS theory in that, for example, it dimensionally-reduces to the 4-dimensional theory without recourse to the classical equations of motion.

- In 1980, Peter Freund and M. A. Rubin showed that compactification from 11 dimensions preserving all the SUSY generators could occur in two ways, leaving only 4 or 7 macroscopic dimensions (the other 7 or 4 being compact).[8] Unfortunately, the noncompact dimensions have to form an anti-de Sitter space. Today it is understood that there are many possible compactifications, but that the Freund-Rubin compactifications are invariant under all of the supersymmetry transformations that preserve the action.

Thus, the first two results appeared to establish 11 dimensions uniquely, the third result appeared to specify the theory, and the last result explained why the observed universe appears to be four-dimensional.

Many of the details of the theory were fleshed out by Peter van Nieuwenhuizen, Sergio Ferrara and Daniel Z. Freedman.

27.2.5   The end of the SUGRA era

The initial excitement over 11-dimensional supergravity soon waned, as various failings were discovered, and attempts to repair the model failed as well. Problems included:

- The compact manifolds which were known at the time and which contained the standard model were not compatible with supersymmetry, and could not hold quarks or leptons. One suggestion was to replace the compact dimensions with the 7-sphere, with the symmetry group SO(8), or the squashed 7-sphere, with symmetry group SO(5) times SU(2).

- Until recently, the physical neutrinos seen in experiments were believed to be massless, and appeared to be left-handed, a phenomenon referred to as the chirality of the Standard Model. It was very difficult to construct a chiral fermion from a compactification — the compactified manifold needed to have singularities, but physics near singularities did not begin to be understood until the advent of orbifold conformal field theories in the late 1980s.

- Supergravity models generically result in an unrealistically large cosmological constant in four dimensions, and that constant is difficult to remove, and so require fine-tuning. This is still a problem today.

- Quantization of the theory led to quantum field theory gauge anomalies rendering the theory inconsistent. In the intervening years physicists have learned how to cancel these anomalies.

Some of these difficulties could be avoided by moving to a 10-dimensional theory involving superstrings. However, by moving to 10 dimensions one loses the sense of uniqueness of the 11-dimensional theory.

The core breakthrough for the 10-dimensional theory, known as the first superstring revolution, was a demonstration by Michael B. Green, John H. Schwarz and David Gross that there are only three supergravity models in 10 dimensions which have gauge symmetries and in which all of the gauge and gravitational anomalies cancel. These were theories built on the groups SO(32) and $E_8 \times E_8$, the direct product of two copies of E_8. Today we know that, using D-branes for example, gauge symmetries can be introduced in other 10-dimensional theories as well.[9]

27.2.6 The second superstring revolution

Initial excitement about the 10-dimensional theories, and the string theories that provide their quantum completion, died by the end of the 1980s. There were too many Calabi–Yaus to compactify on, many more than Yau had estimated, as he admitted in December 2005 at the 23rd International Solvay Conference in Physics. None quite gave the standard model, but it seemed as though one could get close with enough effort in many distinct ways. Plus no one understood the theory beyond the regime of applicability of string perturbation theory.

There was a comparatively quiet period at the beginning of the 1990s; however, several important tools were developed. For example, it became apparent that the various superstring theories were related by "string dualities", some of which relate weak string-coupling (i.e. perturbative) physics in one model with strong string-coupling (i.e. non-perturbative) in another.

Then it all changed, in what is known as the second superstring revolution. Joseph Polchinski realized that obscure string theory objects, called D-branes, which he had discovered six years earlier, are stringy versions of the p-branes that were known in supergravity theories. The treatment of these p-branes was not restricted by string perturbation theory; in fact, thanks to supersymmetry, p-branes in supergravity were understood well beyond the limits in which string theory was understood.

Armed with this new nonperturbative tool, Edward Witten and many others were able to show that all of the perturbative string theories were descriptions of different states in a single theory which Witten named M-theory. Furthermore, he argued that M-theory's long wavelength limit (i.e. when the quantum wavelength associated to objects in the theory are much larger than the size of the 11th dimension) should be described by the 11-dimensional supergravity that had fallen out of favor with the first superstring revolution 10 years earlier, accompanied by the 2- and 5-branes.

Historically, then, supergravity has come "full circle". It is a commonly used framework in understanding features of string theories, M-theory and their compactifications to lower spacetime dimensions.

27.3 Relation to superstrings

Particular 10-dimensional supergravity theories are considered "low energy limits" of the 10-dimensional superstring theories; more precisely, these arise as the massless, tree-level approximation of string theories. True effective field theories of string theories, rather than truncations, are rarely available. Due to string dualities, the conjectured 11-dimensional M-theory is required to have 11-dimensional supergravity as a "low energy limit". However, this doesn't necessarily mean that string theory/M-theory is the only possible UV completion of supergravity; supergravity research is useful independent of those relations.

27.4 4D $N = 1$ SUGRA

Before we move on to SUGRA proper, let's recapitulate some important details about general relativity. We have a 4D differentiable manifold M with a Spin(3,1) principal bundle over it. This principal bundle represents the local Lorentz symmetry. In addition, we have a vector bundle T over the manifold with the fiber having four real dimensions and transforming as a vector under Spin(3,1). We have an invertible linear map from the tangent bundle TM to T. This map is the vierbein. The local Lorentz symmetry has a gauge connection associated with it, the spin connection.

The following discussion will be in superspace notation, as opposed to the component notation, which isn't manifestly covariant under SUSY. There are actually *many* different versions of SUGRA out there which are inequivalent in the sense that their actions and constraints upon the torsion tensor are different, but ultimately equivalent in that we can always perform a field redefinition of the supervierbeins and spin connection to get from one version to another.

In 4D N=1 SUGRA, we have a 4|4 real differentiable supermanifold M, i.e. we have 4 real bosonic dimensions and 4 real fermionic dimensions. As in the nonsupersymmetric case, we have a Spin(3,1) principal bundle over M. We have an $\mathbf{R}^{4|4}$ vector bundle T over M. The fiber of T transforms under the local Lorentz group as follows; the four real bosonic dimensions transform as a vector and the four real fermionic dimensions transform as a Majorana spinor. This Majorana spinor can be reexpressed as a complex left-handed Weyl spinor and its complex conjugate right-handed Weyl spinor (they're not independent of each other). We also have a spin connection as before.

We will use the following conventions; the spatial (both bosonic and fermionic) indices will be indicated by M, N, ... The bosonic spatial indices will be indicated by μ, ν, ..., the left-handed Weyl spatial indices by α, β,..., and the right-handed Weyl spatial indices by $\dot{\alpha}$, $\dot{\beta}$, ... The indices for the fiber of T will follow a similar notation, except that they will be hatted like this: $\hat{M}$, $\hat{\alpha}$. See van der Waerden notation for more details. $M = (\mu, \alpha, \dot{\alpha})$. The supervierbein is denoted by $e_N^{\hat{M}}$, and the spin connection by $\omega_{\hat{M}\hat{N}P}$. The *inverse* supervierbein is denoted by $E_{\hat{M}}^{N}$.

The supervierbein and spin connection are real in the sense that they satisfy the reality conditions

$$e_N^{\hat{M}}(x, \overline{\theta}, \theta)^* = e_{N^*}^{\hat{M}^*}(x, \theta, \overline{\theta}) \text{ where } \mu^* = \mu$$
, $\alpha^* = \dot{\alpha}$, and $\dot{\alpha}^* = \alpha$ and $\omega(x, \overline{\theta}, \theta)^* = \omega(x, \theta, \overline{\theta})$.

The covariant derivative is defined as

$$D_{\hat{M}}f = E_{\hat{M}}^{N}\left(\partial_N f + \omega_N[f]\right)$$

The covariant exterior derivative as defined over supermanifolds needs to be super graded. This means that every time we interchange two fermionic indices, we pick up a +1 sign factor, instead of -1.

The presence or absence of R symmetries is optional, but if R-symmetry exists, the integrand over the full superspace has to have an R-charge of 0 and the integrand over chiral superspace has to have an R-charge of 2.

A chiral superfield X is a superfield which satisfies $\overline{D}_{\hat{\alpha}} X = 0$. In order for this constraint to be consistent, we require the integrability conditions that $\left\{\overline{D}_{\hat{\alpha}}, \overline{D}_{\hat{\beta}}\right\} = c_{\hat{\alpha}\hat{\beta}}^{\hat{\gamma}} \overline{D}_{\hat{\gamma}}$ for some coefficients c.

Unlike nonSUSY GR, the torsion has to be nonzero, at least with respect to the fermionic directions. Already, even in flat superspace, $D_{\hat{\alpha}}e_{\hat{\alpha}} + \overline{D}_{\hat{\alpha}}e_{\hat{\alpha}} \neq 0$. In one version of SUGRA (but certainly not the only one), we have the following constraints upon the torsion tensor:

$$T_{\underline{\hat{\alpha}}\underline{\hat{\beta}}}^{\hat{\gamma}} = 0$$

$$T_{\hat{\alpha}\hat{\beta}}^{\hat{\mu}} = 0$$

$$T_{\dot{\hat{\alpha}}\hat{\beta}}^{\hat{\mu}} = 0$$

$$T_{\hat{\alpha}\dot{\hat{\beta}}}^{\hat{\mu}} = 2i\sigma_{\hat{\alpha}\dot{\hat{\beta}}}^{\hat{\mu}}$$

$$T_{\hat{\mu}\underline{\hat{\alpha}}}^{\hat{\nu}} = 0$$

$$T_{\hat{\mu}\hat{\nu}}^{\hat{\rho}} = 0$$

Here, $\underline{\alpha}$ is a shorthand notation to mean the index runs over either the left or right Weyl spinors.

The superdeterminant of the supervierbein, $|e|$, gives us the volume factor for M. Equivalently, we have the volume 4|4-superform $e^{\hat{\mu}=0} \wedge \cdots \wedge e^{\hat{\mu}=3} \wedge e^{\hat{\alpha}=1} \wedge e^{\hat{\alpha}=2} \wedge e^{\dot{\hat{\alpha}}=1} \wedge e^{\dot{\hat{\alpha}}=2}$.

If we complexify the superdiffeomorphisms, there is a gauge where $E_{\dot{\hat{\alpha}}}^{\mu} = 0$, $E_{\dot{\hat{\alpha}}}^{\beta} = 0$ and $E_{\dot{\hat{\alpha}}}^{\dot{\beta}} = \delta_{\dot{\alpha}}^{\dot{\beta}}$. The resulting chiral superspace has the coordinates x and Θ.

R is a scalar valued chiral superfield derivable from the supervielbeins and spin connection. If f is any superfield, $(\overline{D}^2 - 8R) f$ is always a chiral superfield.

The action for a SUGRA theory with chiral superfields X, is given by

$$S = \int d^4 x d^2 \Theta 2\mathcal{E} \left[\frac{3}{8} \left(\overline{D}^2 - 8R \right) e^{-K(\hat{X}, X)/3} + W(X) \right] + c.c.$$

where K is the Kähler potential and W is the superpotential, and $\mathcal{E}$ is the chiral volume factor.

Unlike the case for flat superspace, adding a constant to either the Kähler or superpotential is now physical. A constant shift to the Kähler potential changes the effective Planck constant, while a constant shift to the superpotential changes the effective cosmological constant. As the effective Planck constant now depends upon the value of the chiral superfield X, we need to rescale the supervierbeins (a field redefinition) to get a constant Planck constant. This is called the **Einstein frame**.

27.5   N = 8 supergravity in 4 dimensions

N=8 Supergravity is the most symmetric quantum field theory which involves gravity and a finite number of fields. It can be found from a dimensional reduction of 11D supergravity by making the size of 7 of the dimensions go to zero. It has 8 supersymmetries which is the most any gravitational theory can have since there are 8 half-steps between spin 2 and spin -2. (A graviton has the highest spin in this theory which is a spin 2 particle). More supersymmetries would mean the particles would have superpartners with spins higher than 2. The only theories with spins higher than 2 which are consistent involve an infinite number of particles (such as String Theory and Higher-Spin Theories). Stephen Hawking in his *A Brief History of Time* speculated that this theory could be the Theory of Everything. However, in later years this was abandoned in favour of String Theory. There has been renewed interest in the 21st century with the possibility that this theory may be finite.

27.6 Higher-dimensional SUGRA

Main article: Higher-dimensional supergravity

Higher-dimensional SUGRA is the higher-dimensional, supersymmetric generalization of general relativity. Supergravity can be formulated in any number of dimensions up to eleven. Higher-dimensional SUGRA focuses upon supergravity in greater than four dimensions.

The number of supercharges in a spinor depends on the dimension and the signature of spacetime. The supercharges occur in spinors. Thus the limit on the number of supercharges cannot be satisfied in a spacetime of arbitrary dimension. Some theoretical examples in which this is satisfied are:

- 12-dimensional two-time theory

- 11-dimensional maximal SUGRA

- 10-dimensional SUGRA theories

    - Type IIA SUGRA: $N = (1, 1)$
    - IIA SUGRA from 11d SUGRA
    - Type IIB SUGRA: $N = (2, 0)$
    - Type I gauged SUGRA: $N = (1, 0)$

- 9d SUGRA theories

    - Maximal 9d SUGRA from 10d
    - T-duality
    - $N = 1$ Gauged SUGRA

The supergravity theories that have attracted the most interest contain no spins higher than two. This means, in particular, that they do not contain any fields that transform as symmetric tensors of rank higher than two under Lorentz transformations. The consistency of interacting higher spin field theories is, however, presently a field of very active interest.

27.7 See also

27.8 Notes

[1] P. van Nieuwenhuizen, Phys. Rep. 68, 189 (1981)

[2] "supergravity in nLab". *ncatlab.org*. Retrieved 2015-10-05.

[3] P. Nath and R. Arnowitt, "Generalized Super-Gauge Symmetry as a New Framework for Unified Gauge Theories", *Physics Letters B* **56** (1975) 177

[4] D.Z. Freedman, P. van Nieuwenhuizen and S. Ferrara, "Progress Toward A Theory Of Supergravity", *Physical Review* **D13** (1976) pp 3214–3218.

[5] Werner Nahm, "Supersymmetries and their representations". *Nuclear Physics B* **135** no 1 (1978) pp 149-166, doi:10.1016/0550-3213(78)90218-3

[6] Ed Witten, "Search for a realistic Kaluza-Klein theory". *Nuclear Physics B* **186** no 3 (1981) pp 412-428, doi:10.1016/0550-3213(81)90021-3

[7] E. Cremmer, B. Julia and J. Scherk, "Supergravity theory in eleven dimensions", *Physics Letters* **B76** (1978) pp 409-412.

[8] Peter G.O. Freund; Mark A. Rubin (1980). "Dynamics of dimensional reduction". *Physics Letters B* **97** (2): 233–235. Bibcode:1980PhLB...97..233F. doi:10.1016/0370-2693(80)90590-0.

[9] Blumenhagen, R.; Cvetic, M.; Langacker, P.; Shiu, G. (2005). "Toward Realistic Intersecting D-Brane Models". arXiv:hep-th/0502005 [hep-th].

27.9 References

27.9.1 Historical

- P. Nath and R. Arnowitt, "Generalized Super-Gauge Symmetry as a New Framework for Unified Gauge Theories", *Physics Letters B '56* (1975) 177.

- D.Z. Freedman, P. van Nieuwenhuizen and S. Ferrara, "Progress Toward A Theory Of Supergravity", *Physical Review* **D13** (1976) pp 3214–3218.

- E. Cremmer, B. Julia and J. Scherk, "Supergravity theory in eleven dimensions", *Physics Letters* **B76** (1978) pp 409–412. scanned version

- P. Freund and M. Rubin, "Dynamics of dimensional reduction", *Physics Letters* **B97** (1980) pp 233–235.

- Ali H. Chamseddine, R. Arnowitt, Pran Nath, "Locally Supersymmetric Grand Unification", " Phys. Rev.Lett.49:970,1982"

- Michael B. Green, John H. Schwarz, "Anomaly Cancellation in Supersymmetric D=10 Gauge Theory and Superstring Theory", *Physics Letters* **B149** (1984) pp117–122.

27.9.2 General

- Bernard de Wit(2002) Supergravity

- A Supersymmetry Primer (1998); updated in (2006).

- Adel Bilal, Introduction to supersymmetry (2001) ArXiv hep-th/0101055, (*a comprehensive introduction to supersymmetry*).

- Friedemann Brandt, Lectures on supergravity (2002) ArXiv hep-th/0204035, (*an introduction to 4-dimensional N = 1 supergravity*).

- Wess, Julius; Bagger, Jonathan (1992). *Supersymmetry and Supergravity*. Princeton University Press. p. 260. ISBN 0-691-02530-4.

Chapter 28

Yukawa potential

In particle and atomic physics, a **Yukawa potential** (also called a **screened Coulomb potential**) is a potential of the form

$$V_{\text{Yukawa}}(r) = -g^2 \frac{e^{-kmr}}{r},$$

where g is a magnitude scaling constant, i.e. is the amplitude of potential, m is the mass of the affected particle, r is the radial distance to the particle, and k is another scaling constant, which finally the product of km is the range. The radial derivative is positive, implying that the potential is monotone increasing in r, and implying the force is always attractive.

The Coulomb potential of electromagnetism is an example of a Yukawa potential with e^{-kmr} equal to 1 everywhere. This can be interpreted as saying that the photon mass m is equal to 0.

In interactions between a meson field and a fermion field, the constant g is equal to the coupling constant between those fields. In the case of the nuclear force, the fermions would be a proton and another proton or a neutron.

28.1 History

Hideki Yukawa showed in the 1930s that such a potential arises from the exchange of a massive scalar field such as the field of a massive boson. Since the field mediator is massive the corresponding force has a certain range, which is inversely proportional to the mass of the mediator particle m .[1] Because the approximate range of the nuclear force was known, Yukawa's equation could be used to predict the approximate rest mass of the particle mediating the force field, even before it was discovered. In the case of the nuclear force, this mass was predicted to be about 200 times the mass of the electron, and this was later considered to be a prediction of the existence of the pion, before it was detected in 1947.

28.2 Relation to Coulomb potential

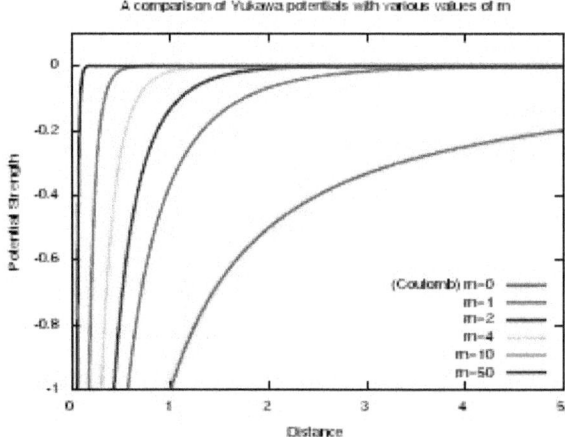

Figure 1: A comparison of Yukawa potentials where g=1 and with various values for m.

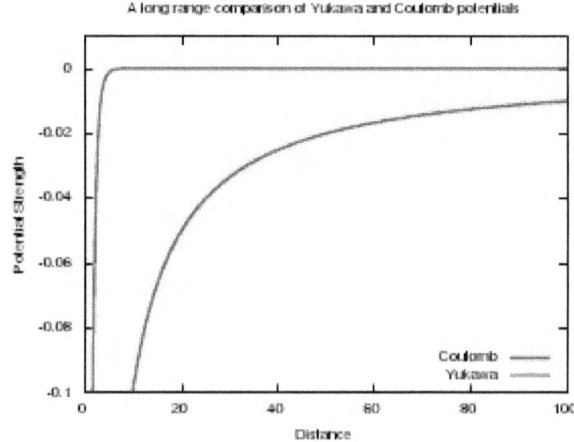

Figure 2: A "long-range" comparison of Yukawa and Coulomb potentials' strengths where g=1.

If the mass is zero (i.e., m=0), then the Yukawa potential

equals a Coulomb potential, and the range is said to be infinite. In fact, we have:

$$m = 0 \Rightarrow e^{-mr} = e^0 = 1.$$

Consequently, the equation

$$V_{\text{Yukawa}}(r) = -g^2 \frac{e^{-mr}}{r}$$

simplifies to the form of the Coulomb potential

$$V_{\text{Coulomb}}(r) = -g^2 \frac{1}{r}.$$

A comparison of the long range potential strength for Yukawa and Coulomb is shown in Figure 2. It can be seen that the Coulomb potential has effect over a greater distance whereas the Yukawa potential approaches zero rather quickly. However, any Yukawa potential or Coulomb potential are non-zero for any large r.

28.3   Fourier transform

The easiest way to understand that the Yukawa potential is associated with a massive field is by examining its Fourier transform. One has

$$V(\mathbf{r}) = \frac{-g^2}{(2\pi)^3} \int e^{i\mathbf{k}\cdot\mathbf{r}} \frac{4\pi}{k^2 + m^2} \, d^3k$$

where the integral is performed over all possible values of the 3-vector momentum k. In this form, the fraction $4\pi/(k^2 + m^2)$ is seen to be the propagator or Green's function of the Klein–Gordon equation.

28.4   Feynman amplitude

The Yukawa potential can be derived as the lowest order amplitude of the interaction of a pair of fermions. The Yukawa interaction couples the fermion field $\psi(x)$ to the meson field $\phi(x)$ with the coupling term

$$\mathcal{L}_{\text{int}}(x) = g\overline{\psi}(x)\phi(x)\psi(x).$$

The scattering amplitude for two fermions, one with initial momentum p_1 and the other with momentum p_2, exchanging a meson with momentum k, is given by the Feynman diagram on the right.

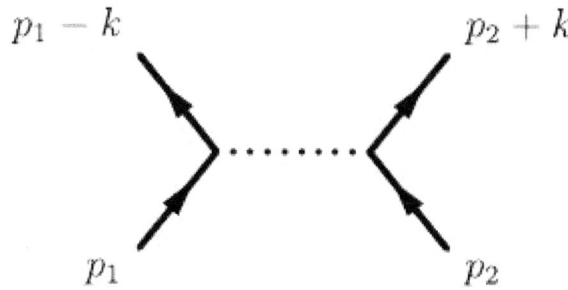

Single particle exchange.

The Feynman rules for each vertex associate a factor of g with the amplitude; since this diagram has two vertices, the total amplitude will have a factor of g^2. The line in the middle, connecting the two fermion lines, represents the exchange of a meson. The Feynman rule for a particle exchange is to use the propagator; the propagator for a massive meson is $-4\pi/(k^2 + m^2)$. Thus, we see that the Feynman amplitude for this graph is nothing more than

$$V(\mathbf{k}) = -g^2 \frac{4\pi}{k^2 + m^2}.$$

From the previous section, this is seen to be the Fourier transform of the Yukawa potential.

28.5   Eigenvalues of Schrödinger equation

The radial Schrödinger equation with Yukawa potential can be solved perturbatively.[2] Using the radial Schrödinger equation in the form

$$[\frac{d^2}{dr^2} + k^2 - \frac{l(l+1)}{r^2} - V(r)]\Psi(l, k; r) = 0,$$

and the Yukawa potential in the power-expanded form

$$V(r) = \Sigma_{i=-1}^{\infty} M_{i+1}(-r)^i,$$

and setting $K = ik$, one obtains for the angular momentum l the expression

$$l + n + 1 = -\frac{\Delta_n(K)}{2K}$$

for $|K| \to \infty$, where

$$\Delta_n(K) = M_0 - \frac{1}{2K^2}[n(n+1)M_2 + M_0 M_1] - \frac{(2n+1)M_0 M_2}{4K^3}$$

$$+ \frac{1}{8K^4}[3M_4(n-1)n(n+1)(n+2) + 2M_3 M_0(3n^2 + 3n$$

$$-1) + 6M_2 M_1 n(n+1) + 2M_2 M_0^2 + 3M_1^2 M_0]$$

$$+ \frac{(2n+1)}{8K^5}[3M_4 M_0(n^2 + n - 1) + 3M_3 M_0^2 + M_2^2 n(n+1)$$

$$+ 4M_2 M_1 M_0] + O(1/K^7).$$

Setting all coefficients M_i except M_0 equal to zero, one obtains the well-known expression for the Schrödinger eigenvalue for the Coulomb potential, and the radial quantum number n is a positive integer or zero as a consequence of the boundary conditions which the wave functions of the Coulomb potential have to satisfy. In the case of the Yukawa potential the imposition of boundary conditions is more complicated. Thus in the Yukawa case $\nu = n$ is only an approximation and the parameter ν that replaces the integer n is really an asymptotic expansion like that above with first approximation the integer value of the corresponding Coulomb case. The above expansion for the orbital angular momentum or Regge trajectory $l(K)$ can be reversed to obtain the energy eigenvalues or equivalently $|K|^2$. One obtains:[3]

$$|K|^2 = -M_1 + \frac{M_0^2}{4(l+n+1)^2}[1 - 4n(n+1)(l+n+1)^2 \frac{M_2}{M_0} + 4(2n+1)(l+n+1)^2 \frac{M_2}{M_0^3}$$

$$+ 4\frac{(l+n+1)^4}{M_0^6}\{3M_4 M_0(n-1)n(n+1)(n+2) - 3M_2^2 n^2(n+1)^2 + 2M_3 M_0^2(3n^2 + 3n - 1) + 2M_2 M_0^3\}$$

$$- 24\frac{(2n+1)(l+n+1)^5}{M_0^6}\{M_0 M_4(n^2 + n - 1) + M_0^3 M_3 - M_2^2 n(n+1)\}$$

$$- 4\frac{(l+n+1)^6}{M_0^9}\{10M_6 M_0^2(n-2)(n-1)n(n+1)(n+2)(n+3) + 4M_3 M_0^5 + 2M_5 M_0^3\{5n(n+1)(3n^2 + 3n - 10) + 12\}$$

$$+ 2M_4 M_0^4(6n^2 + 6n - 11) + 2M_2^2 M_0^3(9n^2 + 9n - 1) - 10M_3 M_2 M_0^2 n(n+1)(3n^2 + 3n + 2) + 20M_2^3 n^3(n+1)^3$$

$$- 30M_4 M_2 M_0(n-1)n^2(n+1)^2(n+2)\} + ...].$$

The above asymptotic expansion of the angular momentum $l(K)$ in descending powers of K can also be derived with the WKB method. In that case, however, as in the case of the Coulomb potential the expression $l(l+1)$ in the centrifugal term of the Schrödinger equation has to be replaced by $(l+1/2)^2$, as was argued originally by Langer,[4] the reason being that the singularity is too strong for an unchanged application of the WKB method. That this reasoning is correct follows from the WKB derivation of the correct result in the Coulomb case (with this Langer replacement),[5] and even of the above expansion in the Yukawa case with higher order WKB approximations.[6]

28.6   See also

- Yukawa interaction
- Screened Poisson equation
- Bessel potential

28.7   References

28.7.1   Citations

[1] Brian Robert Martin; Graham Shaw (2008). *Particle Physics*. p. 18.

[2] H.J.W. Müller-Kirsten, *Introduction to Quantum Mechanics*, 2nd ed. World Scientific, 2012, Chapter 16. H.J.W. Müller,*Regge-Pole in der nichtrelativistischen Potentialstreuung*, Ann. d. Phys. (Leipz.) 15 (1965) 395 - 411; H.J.W. Müller and K. Schilcher, *High-energy scattering for Yukawa potentials*, J. Math. Phys. 9 (1968) 255 - 259.

[3] H. J. W. Müller, Physica 31 (1965) 688.

[4] R.E. Langer, Phys. Rev. 51 (1937) 669.

[5] Harald J. W. Müller-Kirsten, Introduction to Quantum Mechanics: Schrödinger Equation and Path Integral, 2nd ed., World Scientific (Singapore, 2012), p. 404.

Chapter 29

Degeneracy (mathematics)

This article is about degeneracy in mathematics. For the degeneracy of a graph, see degeneracy (graph theory). For other uses, see Degeneracy (disambiguation).

In mathematics, a **degenerate case** is a limiting case in which an element of a class of objects is qualitatively different from the rest of the class and hence belongs to another, usually simpler, class. **Degeneracy** is the condition of being a degenerate case.

The definitions of many classes of composite or structured objects include (often implicitly) inequalities. For example, the angles and the side lengths of a triangle are supposed to be positive. The limiting cases, where one of these elements of the triangle is zero, are *degenerate triangles*.

Often, the degenerate cases are the exceptional cases where changes to the usual dimension or the cardinality of the object (or of some part of it) occur. For example, a triangle is an object of dimension two, and a degenerate triangle is contained in a line, and its dimension is thus one. Similarly, the solution set of a system of equations that depends on parameters generally has a fixed cardinality and dimension, but cardinality and/or dimension may be different for some exceptional values, called degenerate cases. In such a degenerate case, the solution set is said to be degenerate.

For some classes of composite objects, the degenerate cases depend on the properties that are specifically studied. In particular, the class of objects may often be defined or characterized by systems of equations. Commonly, a given class of objects may be defined by several different systems of equations, and these different systems of equations may lead to different degenerate cases, while characterizing the same non-degenerate cases. This may be the reason for which there is no general definition of degeneracy, although the concept is widely used, and defined, if needed, in each specific situation.

A degenerate case thus has special features, which makes it non-generic. However not all non-generic cases are degenerate. For example, right triangles, isosceles triangles and equilateral triangles are non-generic and non-degenerate. Frequently, degenerate cases correspond to singularities either in the object or in some configuration space. For example, a conic section is degenerate if and only if it has singular points.

29.1 In geometry

29.1.1 Conic section

Main article: Degenerate conic

A degenerate conic is a conic section (a second-degree plane curve, defined by a polynomial equation of degree two) that fails to be an irreducible curve.

- A point is a degenerate circle, namely one with radius 0.

- The line is a degenerate case of a parabola if the parabola resides on a tangent plane. In inversive geometry, a line is a degenerate case of a circle, with infinite radius.

- A line segment can be viewed as a degenerate case of an ellipse in which the semiminor axis goes to zero, the foci go to the endpoints, and the eccentricity goes to one.

- An ellipse can also degenerate into a single point.

- A hyperbola can degenerate into two lines crossing at a point, through a family of hyperbolae having those lines as common asymptotes.

29.1.2 Triangle

- A degenerate triangle has collinear vertices and zero area, and thus coincides with a segment covered twice.

29.1.3 Rectangle

- A segment is a degenerate case of a rectangle, if this has a side of length 0.

- For any non-empty subset $S \subseteq \{1, 2, \ldots, n\}$, there is a bounded, axis-aligned degenerate rectangle

$$R \triangleq \{\mathbf{x} \in \mathbb{R}^n : x_i = c_i \ (\text{for} i \in S) \text{ and } a_i \leq x_i \leq b_i \ (\text{for} i \notin S)\}$$

where $\mathbf{x} \triangleq [x_1, x_2, \ldots, x_n]$ and a_i, b_i, c_i are constant (with $a_i \leq b_i$ for all i). The number of degenerate sides of R is the number of elements of the subset S . Thus, there may be as few as one degenerate "side" or as many as n (in which case R reduces to a singleton point).

29.1.4 Convex polygon

- A convex polygon is degenerate if (at least) two consecutive sides are aligned or some sides have a zero length. Thus a degenerate convex polygon of n sides looks like a polygon with fewer sides. In the case of triangles, this definition coincides with the one that has been given above.

29.1.5 Convex polyhedron

- A convex polyhedron is degenerate if either two adjacent facets are coplanar or two edges are aligned. In the case of a tetrahedron, this is equivalent to say that all of its vertices lie in the same plane, giving it zero volume.

29.1.6 Standard torus

- A sphere is a degenerate standard torus where the axis of revolution passes through the center of the generating circle, rather than outside it.

29.1.7 Sphere

- When the radius of a sphere goes to zero, the resulting degenerate sphere of zero volume is a point.

29.1.8 Other

- See general position for other examples.

29.2 Elsewhere

- A set containing a single point is a degenerate continuum.

- Objects such as the digon and monogon can be viewed as degenerate cases of polygons: valid in a general abstract mathematical sense, but not part of the original Euclidean conception of polygons.

- A random variable which can only take one value has a degenerate distribution; if that value is the real number 0, its probability density is the Dirac delta function.

- Similarly, roots of a polynomial are said to be *degenerate* if they coincide, since generically the n roots of an nth degree polynomial are all distinct. This usage carries over to eigenproblems: a degenerate eigenvalue (i.e. a multiple coinciding root of the characteristic polynomial) is one that has more than one linearly independent eigenvector.

- In quantum mechanics any such multiplicity in the eigenvalues of the Hamiltonian operator gives rise to degenerate energy levels. Usually any such degeneracy indicates some underlying symmetry in the system.

29.3 See also

- Degeneracy (graph theory)

- Degenerate form

- Trivial (mathematics)

- Pathological (mathematics)

- Vacuous truth

29.4 External links

Weisstein, Eric W., "Degenerate", *MathWorld*.

Chapter 30

Quantum fluctuation

In quantum physics, a **quantum fluctuation** (or **quantum vacuum fluctuation** or **vacuum fluctuation**) is the temporary change in the amount of energy in a point in space,[1] as explained in Werner Heisenberg's uncertainty principle.

According to one formulation of the principle, energy and time can be related by the relation[2]

$$\Delta E \Delta t \geq \frac{h}{4\pi}$$

This allows the creation of particle-antiparticle pairs of virtual particles. The effects of these particles are measurable, for example, in the effective charge of the electron, different from its "naked" charge.

In the modern view, energy is always conserved, but the eigenstates of the Hamiltonian (energy observable) are not the same as (i.e., the Hamiltonian does not commute with) the particle number operators.

Quantum fluctuations may have been very important in the origin of the structure of the universe: according to the model of inflation the ones that existed when inflation began were amplified and formed the seed of all current observed structure. Vacuum energy may also be responsible for the current accelerated expansion of the universe (cosmological constant).

30.1 Quantum fluctuations of a field

A quantum fluctuation is the temporary appearance of energetic particles out of empty space, as allowed by the uncertainty principle. The uncertainty principle states that for a pair of conjugate variables such as position/momentum or energy/time, it is impossible to have a precisely determined value of each member of the pair at the same time. For example, a particle pair can pop out of the vacuum during a very short time interval.

An extension is applicable to the "uncertainty in time" and "uncertainty in energy" (including the rest mass energy mc^2

). When the mass is very large like a macroscopic object, the uncertainties and thus the quantum effect become very small, and classical physics is applicable. This was proposed by scientist Adam Jonathon Davis' study in 1916 at Harvard's Laboratory 1996a. Davis' theory was later proven in the 1920s by Louis de Broglie and became a law of quantum physics.

In quantum field theory, fields undergo quantum fluctuations. A reasonably clear distinction can be made between quantum fluctuations and thermal fluctuations of a quantum field (at least for a free field; for interacting fields, renormalization substantially complicates matters). For the quantized Klein–Gordon field in the vacuum state, we can calculate the probability density that we would observe a configuration $\varphi_t(x)$ at a time t in terms of its Fourier transform $\tilde{\varphi}_t(k)$ to be

$$\rho_0[\varphi_t] = \exp\left[-\frac{1}{\hbar} \int \frac{d^3k}{(2\pi)^3} \tilde{\varphi}_t^*(k) \sqrt{|k|^2 + m^2}\ \tilde{\varphi}_t(k) \right].$$

In contrast, for the classical Klein–Gordon field at non-zero temperature, the Gibbs probability density that we would observe a configuration $\varphi_t(x)$ at a time t is

$$\rho_E[\varphi_t] = \exp\left[-H[\varphi_t]/k_{\mathrm{B}}T \right] = \exp\left[- \right.$$
$$\left. \frac{1}{k_{\mathrm{B}}T} \int \frac{d^3k}{(2\pi)^3} \tilde{\varphi}_t^*(k) \tfrac{1}{2}(|k|^2 + m^2)\ \tilde{\varphi}_t(k) \right].$$

The amplitude of quantum fluctuations is controlled by Planck's constant $\hbar$, just as the amplitude of thermal fluctuations is controlled by $k_{\mathrm{B}}T$, where k_{B} is Boltzmann's constant. Note that the following three points are closely related:

1. Planck's constant has units of action (joule-seconds) instead of units of energy (joules),

2. the quantum kernel is $\sqrt{|k|^2 + m^2}$ instead of $\tfrac{1}{2}(|k|^2 + m^2)$ (the quantum kernel is nonlocal from a classical heat kernel viewpoint, but it is local in the sense that it does not allow signals to be transmitted),

3. the quantum vacuum state is Lorentz invariant (although not manifestly in the above), whereas the classical thermal state is not (the classical dynamics is Lorentz invariant, but the Gibbs probability density is not a Lorentz invariant initial condition).

We can construct a classical continuous random field that has the same probability density as the quantum vacuum state, so that the principal difference from quantum field theory is the measurement theory (measurement in quantum theory is different from measurement for a classical continuous random field, in that classical measurements are always mutually compatible — in quantum mechanical terms they always commute). Quantum effects that are consequences only of quantum fluctuations, not of subtleties of measurement incompatibility, can alternatively be models of classical continuous random fields.

In the 1930s, Pascual Jordan knew that a star could equal zero energy because its matter energy was positive and its gravitational energy was negative and they cancelled each other out. And this led him to speculate what would prevent a quantum transition from creating a new star. And he had this idea because he was trying to figure out where matter might come from if we existed in an always-here universe.[3]

In December, 1973, the British scientific journal *Nature* published an article by Edward P. Tryon titled "Is the Universe a Vacuum Fluctuation?" In this paper Tryon said our universe may have originated as a quantum fluctuation of the vacuum.[4] Yet, the idea of our universe coming from a quantum fluctuation or quantum process was not taken seriously until inflationary theory came and was able to explain how our universe could inflate from a tiny particle.[5]

30.2 See also

- Casimir effect

- Quantum annealing

- Quantum foam

- Virtual particle

- Virtual black hole

- Stochastic interpretation

- Zitterbewegung

30.3 References

[1] Browne, Malcolm W. (1990-08-21). "New Direction in Physics: Back in Time". The New York Times. Retrieved 2010-05-22. According to quantum theory, the vacuum contains neither matter nor energy, but it does contain *fluctuations*, transitions between something and nothing in which potential existence can be transformed into real existence by the addition of energy.(Energy and matter are equivalent, since all matter ultimately consists of packets of energy.) Thus, the vacuum's totally empty space is actually a seething turmoil of creation and annihilation, which to the ordinary world appears calm because the scale of fluctuations in the vacuum is tiny and the fluctuations tend to cancel each other out.

[2] Mandelshtam, Leonid; Tamm, Igor (1945), "The uncertainty relation between energy and time in nonrelativistic quantum mechanics", *Izv. Akad. Nauk SSSR (ser. Fiz.)* **9**: 122–128. English translation: J. Phys. (USSR) **9**, 249–254 (1945).

[3] Reynosa, Peter. "Why Isn't Edward P. Tryon A World-famous Physicist?". Huffington Post. Retrieved March 22, 2016.

[4] Reynosa, Peter. "Why Isn't Edward P. Tryon A World-famous Physicist?". Huffington Post. Retrieved March 22, 2016.

[5] Reynosa, Peter. "Some of the Changes Lawrence M. Krauss Should Make to the Second Edition of "A Universe from Nothing"". Huffington Post. Retrieved April 13, 2016.

30.4 External links

- Quantum Fluctuation at universe-review.ca

Chapter 31

Principle of locality

This article is about the principle of locality in physics. For the term in computer science, see Locality of reference.

In physics, the **principle of locality** states that an object is only directly influenced by its immediate surroundings. A physical theory is said to be a local theory if it is consistent with the principle of locality. An alternative to the earlier concept of instantaneous "action at a distance", locality evolved as a property of the field theories of classical physics. The concept of locality is that, for an action at one point to have an influence at another point, something in the space between the points, such as a field, must mediate the action. To exert an influence, something, such as a wave or particle, must travel through the space between the two points, to carry the influence.

The Special Theory of Relativity limits the speed at which all such influences can travel to the speed of light, c. Therefore, the principle of locality implies that an event at one point cannot cause a simultaneous result at another point. An event at point A cannot cause a result at point B in a time less than $T = D/c$, where D is the distance between the points. In other words, information cannot travel faster than the speed of light.

In 1935 Albert Einstein, Boris Podolsky and Nathan Rosen in the EPR paradox raised the possibility that quantum mechanics might not be a local theory, since a measurement made on one of a pair of separated entangled particles causes simultaneous collapse of the wavefunction of the remote particle. However, because of the probabilistic nature of wavefunction collapse, this violation of locality cannot be used to transmit information faster than light. In 1964 John Stewart Bell derived the Bell inequality, which if confirmed showed that quantum mechanics must violate either locality or another principle, *realism*, relating to the value of unmeasured quantities. The two principles are often referred to together as a single principle of **local realism**. Experimental tests of the Bell inequality, beginning with Alain Aspect's photon experiments in 1972, seem to show that quantum mechanics disobeys the inequality, and thus must vio-late either locality or realism. However critics have pointed out that most experiments contained "loopholes" which prevented a definitive answer to this important question. This situation seems to have been resolved in 2015, when Dr. Ronald Hanson's group at Delft University performed what has been called the first loophole-free experiment.[1]

31.1 Pre-quantum mechanics

Main article: Action at a distance

In the 17th Century Newton's law of universal gravitation was formulated in terms of "action at a distance", thereby violating the principle of locality.

> It is inconceivable that inanimate Matter should, without the Mediation of something else, which is not material, operate upon, and affect other matter without mutual Contact...That Gravity should be innate, inherent and essential to Matter, so that one body may act upon another at a distance thro' a Vacuum, without the Mediation of any thing else, by and through which their Action and Force may be conveyed from one to another, is to me so great an Absurdity that I believe no Man who has in philosophical Matters a competent Faculty of thinking can ever fall into it. Gravity must be caused by an Agent acting constantly according to certain laws; but whether this Agent be material or immaterial, I have left to the Consideration of my readers.[2]
> — Isaac Newton, Letters to Bentley, 1692/3

Coulomb's law of electric forces was initially also formulated as instantaneous action at a distance, but was later superseded by Maxwell's Equations of electromagnetism which obey locality.

In 1905 Albert Einstein's Special Theory of Relativity postulated that no material or energy can travel faster than the speed of light, and Einstein thereby sought to reformulate physical laws in a way which obeyed the principle of locality. He later succeeded in producing an alternative theory of gravitation, General Relativity, which obeys the principle of locality.

However, a different challenge to the principle of locality subsequently emerged from the theory of Quantum Mechanics, which Einstein himself had helped to create.

31.2 Quantum mechanics

31.2.1 EPR paradox

Albert Einstein argued that quantum mechanics was an incomplete physical theory. Using the principle of locality, he, Podolsky, and Rosen articulated the Einstein-Podolsky-Rosen paradox which showed that quantum mechanics predicts non-locality unless position and momentum were simultaneous "real" properties of a particle. The locality question remained unverifiable for several decades. Then in 1964, John Stewart Bell derived his eponymous theorem, which describes quantum mechanical predictions that no theory of local hidden variables, no *local realism*, could ever reproduce.

Einstein assumed that the principle of locality was necessary, and that there could be no violations of it. He said:

> *"(...) The following idea characterises the relative independence of objects far apart in space, A and B: external influence on A has no direct influence on B; this is known as the Principle of Local Action, which is used consistently only in field theory. If this axiom were to be completely abolished, the idea of the existence of quasienclosed systems, and thereby the postulation of laws which can be checked empirically in the accepted sense, would become impossible. (...)"*[3]

31.2.2 Local realism

Local realism is the combination of the principle of locality with the "realistic" assumption that all objects must objectively have a pre-existing value for any possible measurement before the measurement is made .

Local realism is a significant feature of classical mechanics, and of electrodynamics; but quantum mechanics largely rejects this principle due to the theory of distant quantum entanglements, an interpretation Einstein objected to in the EPR paradox but subsequently proven by Bell's inequalities.[4] Any theory, such as quantum mechanics, that violates Bell's inequalities must abandon *either* locality *or* realism; but some physicists dispute that experiments have demonstrated Bell's violations, on the grounds that the subclass of inhomogeneous Bell inequalities has not been tested or due to experimental limitations in the tests. Different interpretations of quantum mechanics violate different parts of local realism and/or counterfactual definiteness.[5]

31.2.3 Realism

Realism in the sense used by physicists does not equate to realism in metaphysics.[6] The physicist's *Realism* is the claim that the world is in some sense mind-independent: that even if the results of a possible measurement do not pre-exist the act of measurement, that does not require that they are the creation of the observer (contrary to the "consciousness causes collapse" interpretation of quantum mechanics). Furthermore, a mind-independent property does not have to be the value of some physical variable such as position or momentum. A property can be *dispositional* (or potential), i.e., it can be a tendency: in the way that glass objects tend to break, or are disposed to break, even if they do not *actually* break. Likewise, the mind-independent properties of quantum systems could consist of a tendency to respond to particular measurements with particular values with ascertainable probability.[7] Such an ontology would be metaphysically realistic, without being realistic in the physicist's sense of "local realism" (which would require that a single value be produced with certainty).

A closely related term is counterfactual definiteness (CFD), used to refer to the claim that one can meaningfully speak of the definiteness of results of measurements that have not been performed (i.e., the ability to assume the existence of objects, and properties of objects, even when they have not been measured).

31.2.4 Copenhagen interpretation

In most of the conventional interpretations, such as the Copenhagen interpretation and the interpretation based on Consistent Histories, where the wavefunction is not assumed to physically exist in real spacetime, it is local realism that is rejected. These interpretations propose that actual definite properties of a physical system "do not exist" prior to the measurement; and the wavefunction has a restricted interpretation, as nothing more than a mathematical tool used to calculate the probabilities of experimental outcomes.

If the wavefunction is assumed to physically exist in real spacetime, the principle of locality is violated during the measurement process via wavefunction collapse. This is a non-local process because Born's Rule, when applied to the system's wavefunction, yields a probability density for all regions of space and time. Upon actual measurement of the physical system, the probability density vanishes everywhere instantaneously, except where (and when) the measured entity is found to exist. This "vanishing" is postulated to be a *real* physical process, and clearly non-local (i.e., faster than light) if the wavefunction is considered physically real and the probability density has converged to zero at arbitrarily far distances during the finite time required for the measurement process.

31.2.5   Bohm interpretation

The Bohm interpretation preserves realism, hence it needs to violate the principle of locality in order to achieve the required correlations . It does so by maintaining that both the position and momentum of a particle are determinate in that they correspond to the definite trajectory of the particle; however, that trajectory cannot be known without knowing the physical state of the entire universe.

31.2.6   Many-worlds interpretation

In the many-worlds interpretation both realism *and* locality are retained, but counterfactual definiteness is rejected by the extension of the notion of reality to allow the existence of parallel universes.

Because the differences between the different interpretations are mostly philosophical ones (except for the Bohm and many-worlds interpretations), physicists usually employ language in which the important statements are neutral with regard to all of the interpretations. In this framework, only the measurable action at a distance —a superluminal propagation of real, physical information— would usually be considered in violation of the principle of locality by physicists. Such phenomena have never been seen, and they are not predicted by the current theories.

31.3   Relativity

Locality is one of the axioms of relativistic quantum field theory, as required for causality. The formalization of locality in this case is as follows: if we have two observables,

each localized within two distinct spacetime regions which happen to be at a spacelike separation from each other, the observables must commute. Alternatively, a solution to the field equations is local if the underlying equations are either Lorentz invariant or, more generally, generally covariant or locally Lorentz invariant.

31.4   See also

- EPR paradox
- Local hidden variable theory

31.5   References

[1] Hanson, Ronald. "Loophole-free Bell inequality violation using electron spins separated by 1.3 kilometres". *Nature* **526**: 682–686. Bibcode:2015Natur.526..682H. doi:10.1038/nature15759.

[2] Berkovitz, Joseph (2008). "Action at a Distance in Quantum Mechanics". In Edward N. Zalta. *The Stanford Encyclopedia of Philosophy* (Winter ed.).

[3] Einstein, Albert (1948). "Quanten-Mechanik Und Wirklichkeit" [Quantum Mechanics and Reality]. *Dialectica* 2 (3–4): 320–4. doi:10.1111/j.1746-8361.1948.tb00704.x.

[4] Ben Dov, Y. Local Realism and the Crucial experiment.

[5] "Quantum crypto still not proven, claim Cambridge experts."

[6] Travis Norsen (March 2007). "Against 'Realism'". *Foundations of Physics* 37 (3): 311–40. arXiv:quant-ph/0607057v2. Bibcode:2007FoPh...37..311N. doi:10.1007/s10701-007-9104-1.

[7] Ian Thomson's dispositional quantum mechanics

31.6   External links

- Quantum nonlocality vs. Einstein locality by H. D. Zeh

Chapter 32

Brans–Dicke theory

In theoretical physics, the **Brans–Dicke theory of gravitation** (sometimes called the **Jordan–Brans–Dicke theory**) is a theoretical framework to explain gravitation. It is a competitor of Einstein's theory of general relativity. It is an example of a scalar-tensor theory, a gravitational theory in which the gravitational interaction is mediated by a scalar field as well as the tensor field of general relativity. The gravitational constant G is not presumed to be constant but instead $1/G$ is replaced by a scalar field ϕ which can vary from place to place and with time.

The theory was developed in 1961 by Robert H. Dicke and Carl H. Brans[1] building upon, among others, the earlier 1959 work of Pascual Jordan. At present, both Brans–Dicke theory and general relativity are generally held to be in agreement with observation. Brans–Dicke theory represents a minority viewpoint in physics.

32.1 Comparison with general relativity

Both Brans–Dicke theory and general relativity are examples of a class of relativistic classical field theories of gravitation, called *metric theories*. In these theories, spacetime is equipped with a metric tensor, g_{ab}, and the gravitational field is represented (in whole or in part) by the Riemann curvature tensor R_{abcd}, which is determined by the metric tensor.

All metric theories satisfy the Einstein equivalence principle, which in modern geometric language states that in a very small region (too small to exhibit measurable curvature effects), all the laws of physics known in special relativity are valid in *local Lorentz frames*. This implies in turn that metric theories all exhibit the gravitational redshift effect.

As in general relativity, the source of the gravitational field is considered to be the stress–energy tensor or *matter tensor*. However, the way in which the immediate presence of mass-energy in some region affects the gravitational field in that region differs from general relativity. So does the way in which spacetime curvature affects the motion of matter. In the Brans–Dicke theory, in addition to the metric, which is a *rank two tensor field*, there is a *scalar field*, ϕ, which has the physical effect of changing the *effective gravitational constant* from place to place. (This feature was actually a key desideratum of Dicke and Brans; see the paper by Brans cited below, which sketches the origins of the theory.)

The field equations of Brans–Dicke theory contain a parameter, ω, called the *Brans–Dicke coupling constant*. This is a true dimensionless constant which must be chosen once and for all. However, it can be chosen to fit observations. Such parameters are often called *tuneable parameters*. In addition, the present ambient value of the effective gravitational constant must be chosen as a boundary condition. General relativity contains no dimensionless parameters whatsoever, and therefore is easier to falsify (show whether false) than Brans–Dicke theory. Theories with tuneable parameters are sometimes deprecated on the principle that, of two theories which both agree with observation, the more parsimonious is preferable. On the other hand, it seems as though they are a necessary feature of some theories, such as the weak mixing angle of the Standard Model.

Brans–Dicke theory is "less stringent" than general relativity in another sense: it admits more solutions. In particular, exact vacuum solutions to the Einstein field equation of general relativity, augmented by the trivial scalar field $\phi = 1$, become exact vacuum solutions in Brans–Dicke theory, but some spacetimes which are *not* vacuum solutions to the Einstein field equation become, with the appropriate choice of scalar field, vacuum solutions of Brans–Dicke theory. Similarly, an important class of spacetimes, the pp-wave metrics, are also exact null dust solutions of both general relativity and Brans–Dicke theory, but here too, Brans–Dicke theory allows additional *wave solutions* having geometries which are incompatible with general relativity.

Like general relativity, Brans–Dicke theory predicts light deflection and the precession of perihelia of planets orbiting

181

the Sun. However, the precise formulas which govern these effects, according to Brans–Dicke theory, depend upon the value of the coupling constant ω. This means that it is possible to set an observational lower bound on the possible value of ω from observations of the solar system and other gravitational systems. The value of ω consistent with experiment has risen with time. In 1973 $\omega > 5$ was consistent with known data. By 1981 $\omega > 30$ was consistent with known data. In 2003 evidence – derived from the Cassini–Huygens experiment – shows that the value of ω must exceed 40,000.

It is also often taught that general relativity is obtained from the Brans–Dicke theory in the limit $\omega \to \infty$. But Faraoni[2] claims that this breaks down when the trace of the stress-energy momentum vanishes, i.e. $T_\mu^\mu = 0$. Some have argued that only general relativity satisfies the strong equivalence principle.

32.2   The field equations

The field equations of the Brans/Dicke theory are

$$\Box\phi = \frac{8\pi}{3 + 2\omega} T$$

$$G_{ab} = \frac{8\pi}{\phi} T_{ab} + \frac{\omega}{\phi^2}\left(\partial_a\phi\partial_b\phi - \frac{1}{2}g_{ab}\partial_c\phi\partial^c\phi\right) + \frac{1}{\phi}(\nabla_a\nabla_b\phi$$
$$- g_{ab}\Box\phi)$$

where

ω is the dimensionless Dicke coupling constant;

g_{ab} is the metric tensor;

$G_{ab} = R_{ab} - \frac{1}{2}Rg_{ab}$ is the Einstein tensor, a kind of average curvature;

$R_{ab} = R^m{}_{amb}$ is the Ricci tensor, a kind of trace of the curvature tensor;

$R = R^m{}_m$ is the Ricci scalar, the trace of the Ricci tensor;

T_{ab} is the stress-energy tensor;

$T = T_a^a$ is the trace of the stress-energy tensor;

ϕ is the scalar field; and

$\Box$ is the Laplace–Beltrami operator or covariant wave operator, $\Box\phi = \phi^{;a}_{;a}$.

The first equation says that the trace of the stress-energy tensor acts as the source for the scalar field ϕ. Since electromagnetic fields contribute only a traceless term to the stress-energy tensor, this implies that in a region of spacetime containing only an electromagnetic field (plus the gravitational field), the right hand side vanishes, and ϕ obeys the

(curved spacetime) wave equation. Therefore, changes in ϕ propagate through *electrovacuum* regions; in this sense, we say that ϕ is a *long-range field*.

The second equation describes how the stress-energy tensor and scalar field ϕ together affect spacetime curvature. The left hand side, the Einstein tensor, can be thought of as a kind of average curvature. It is a matter of pure mathematics that, in any metric theory, the Riemann tensor can always be written as the sum of the Weyl curvature (or *conformal curvature tensor*) plus a piece constructed from the Einstein tensor.

For comparison, the field equation of general relativity is simply

$$G_{ab} = 8\pi T_{ab}.$$

This means that in general relativity, the Einstein curvature at some event is entirely determined by the stress-energy tensor at that event; the other piece, the Weyl curvature, is the part of the gravitational field which can propagate as a gravitational wave across a vacuum region. But in the Brans–Dicke theory, the Einstein tensor is determined partly by the immediate presence of mass-energy and momentum, and partly by the long-range scalar field ϕ.

The *vacuum field equations* of both theories are obtained when the stress-energy tensor vanishes. This models situations in which no non-gravitational fields are present.

32.3   The action principle

The following Lagrangian contains the complete description of the Brans/Dicke theory:

$$S = \int d^4x\sqrt{-g}\left(\frac{\phi R - \omega\frac{\partial_a\phi\partial^a\phi}{\phi}}{16\pi} + \mathcal{L}_\text{M}\right)$$

where g is the determinant of the metric, $\sqrt{-g}\,d^4x$ is the four-dimensional volume form, and $\mathcal{L}_\text{M}$ is the *matter term* or *matter Lagrangian*.

The matter term includes the contribution of ordinary matter (e.g. gaseous matter) and also electromagnetic fields. In a vacuum region, the matter term vanishes identically; the remaining term is the *gravitational term*. To obtain the vacuum field equations, we must vary the gravitational term in the Lagrangian with respect to the metric g_{ab}; this gives the second field equation above. When we vary with respect to the scalar field ϕ, we obtain the first field equation.

Note that, unlike for the General Relativity field equations, the $\delta R_{ab}/\delta g_{cd}$ term does not vanish, as the result is not a total derivative. It can be shown that

$$\frac{\delta(\phi R)}{\delta g^{ab}} = \phi R_{ab} + g_{ab}g^{cd}\phi_{;c;d} - \phi_{;a;b}$$

To prove this result, use

$$\delta(\phi R) = R\delta\phi + \phi R_{mn}\delta g^{mn} + \phi\nabla_s(g^{mn}\delta\Gamma^s_{nm} - g^{ms}\delta\Gamma^r_{rm})$$

By evaluating the $\delta\Gamma$ s in Riemann normal coordinates, 6 individual terms vanish. 6 further terms combine when manipulated using Stokes' theorem to provide the desired $(g_{ab}g^{cd}\phi_{;c;d} - \phi_{;a;b})\delta g^{ab}$.

For comparison, the Lagrangian defining general relativity is

$$S = \int d^4x\sqrt{-g}\left(\frac{R}{16\pi G} + \mathcal{L}_M\right)$$

Varying the gravitational term with respect to g_{ab} gives the vacuum Einstein field equation.

In both theories, the full field equations can be obtained by variations of the full Lagrangian.

32.4 See also

- Classical theories of gravitation

- Mach's principle

- General relativity

32.5 References

[1] Brans, C. H.; Dicke, R. H. (November 1, 1961). "Mach's Principle and a Relativistic Theory of Gravitation". *Physical Review* **124** (3): 925–935. Bibcode:1961PhRv..124..925B. doi:10.1103/PhysRev.124.925.

[2] Faroni, Valerio (1999). "Illusions of general relativity in Brans-Dicke gravity". *Phys. Rev.* **D59**: 084021. arXiv:gr-qc/9902083. Bibcode:1999PhRvD..59h4021F. doi:10.1103/PhysRevD.59.084021.

- Bergmann, Peter G. (May 1968). "Comments on the Scalar-Tensor Theory". *Int. J. Theor. Phys.* **1** (1): 25–36. Bibcode:1968IJTP....1...25B. doi:10.1007/BF00668828. ISSN 0020-7748.

- Wagoner, Robert V. (June 1970). "Scalar-Tensor Theory and Gravitational Waves". *Phys. Rev. D* (American Physical Society) **1** (12): 3209–3216. Bibcode:1970PhRvD...1.3209W. doi:10.1103/PhysRevD.1.3209.

- Misner, Charles W.; Thorne, Kip S.; Wheeler, John Archibald (1973). *Gravitation*. San Francisco: W. H. Freeman. ISBN 0-7167-0344-0. See *Box 39.1*.

- Will, Clifford M. (1986). "Chapter 8: The Rise and Fall of the Brans–Dicke Theory". *Was Einstein Right?: Putting General Relativity to the Test*. NY: Basic Books. ISBN 0-19-282203-9.

- Faraoni, Valerio (2004). *Cosmology in Scalar-Tensor Gravity*. Dordrecht, The Netherlands: Kluwer Academic. ISBN 1-4020-1988-2.

32.6 External links

- Scholarpedia article on the subject by Carl H. Brans

- Carl H. Brans. "The roots of scalar-tensor theory: an approximate history". arXiv:gr-qc/0506063.

- The Brans-Dicke theory

Chapter 33

General relativity

For the book by Robert Wald, see General Relativity (book).

For a more accessible and less technical introduction to this topic, see Introduction to general relativity.

General relativity (**GR**, also known as the **general**

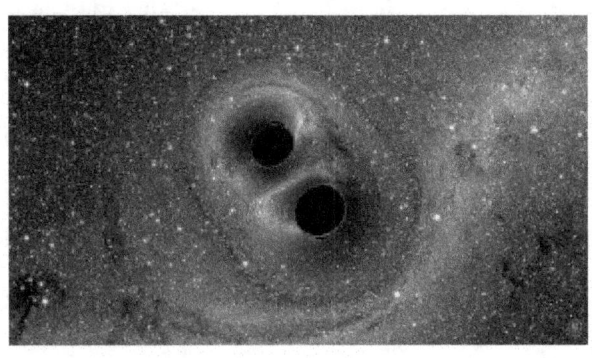

Slow motion computer simulation of the black hole binary system GW150914 as seen by a nearby observer, during 0.33 s of its final inspiral, merge, and ringdown. The star field behind the black holes is being heavily distorted and appears to rotate and move, due to extreme gravitational lensing, as space-time itself is distorted and dragged around by the rotating black holes.[1]

theory of relativity or **GTR**) is the geometric theory of gravitation published by Albert Einstein in 1915[2] and the current description of gravitation in modern physics. General relativity generalizes special relativity and Newton's law of universal gravitation, providing a unified description of gravity as a geometric property of space and time, or spacetime. In particular, the curvature of spacetime is directly related to the energy and momentum of whatever matter and radiation are present. The relation is specified by the Einstein field equations, a system of partial differential equations.

Some predictions of general relativity differ significantly from those of classical physics, especially concerning the passage of time, the geometry of space, the motion of bodies in free fall, and the propagation of light. Examples of such differences include gravitational time dilation, gravitational lensing, the gravitational redshift of light, and the gravitational time delay. The predictions of general relativity have been confirmed in all observations and experiments to date. Although general relativity is not the only relativistic theory of gravity, it is the simplest theory that is consistent with experimental data. However, unanswered questions remain, the most fundamental being how general relativity can be reconciled with the laws of quantum physics to produce a complete and self-consistent theory of quantum gravity.

Einstein's theory has important astrophysical implications. For example, it implies the existence of black holes—regions of space in which space and time are distorted in such a way that nothing, not even light, can escape—as an end-state for massive stars. There is ample evidence that the intense radiation emitted by certain kinds of astronomical objects is due to black holes; for example, microquasars and active galactic nuclei result from the presence of stellar black holes and black holes of a much more massive type, respectively. The bending of light by gravity can lead to the phenomenon of gravitational lensing, in which multiple images of the same distant astronomical object are visible in the sky. General relativity also predicts the existence of gravitational waves, which have since been observed directly by physics collaboration LIGO. In addition, general relativity is the basis of current cosmological models of a consistently expanding universe.

33.1 History

Main articles: History of general relativity and Classical theories of gravitation

Soon after publishing the special theory of relativity in 1905, Einstein started thinking about how to incorporate gravity into his new relativistic framework. In 1907, beginning with a simple thought experiment involving an observer in free fall, he embarked on what would be an eight-year search for a relativistic theory of gravity. After numerous detours and false starts, his work culminated in the presentation to the Prussian Academy of Science in Novem-

Albert Einstein developed the theories of special and general relativity. Picture from 1921.

our universe is expanding. This is readily described by the expanding cosmological solutions found by Friedmann in 1922, which do not require a cosmological constant. Lemaître used these solutions to formulate the earliest version of the Big Bang models, in which our universe has evolved from an extremely hot and dense earlier state.[6] Einstein later declared the cosmological constant the biggest blunder of his life.[7]

During that period, general relativity remained something of a curiosity among physical theories. It was clearly superior to Newtonian gravity, being consistent with special relativity and accounting for several effects unexplained by the Newtonian theory. Einstein himself had shown in 1915 how his theory explained the anomalous perihelion advance of the planet Mercury without any arbitrary parameters ("fudge factors").[8] Similarly, a 1919 expedition led by Eddington confirmed general relativity's prediction for the deflection of starlight by the Sun during the total solar eclipse of May 29, 1919,[9] making Einstein instantly famous.[10] Yet the theory entered the mainstream of theoretical physics and astrophysics only with the developments between approximately 1960 and 1975, now known as the golden age of general relativity.[11] Physicists began to understand the concept of a black hole, and to identify quasars as one of these objects' astrophysical manifestations.[12] Ever more precise solar system tests confirmed the theory's predictive power,[13] and relativistic cosmology, too, became amenable to direct observational tests.[14]

ber 1915 of what are now known as the Einstein field equations. These equations specify how the geometry of space and time is influenced by whatever matter and radiation are present, and form the core of Einstein's general theory of relativity.[3]

The Einstein field equations are nonlinear and very difficult to solve. Einstein used approximation methods in working out initial predictions of the theory. But as early as 1916, the astrophysicist Karl Schwarzschild found the first non-trivial exact solution to the Einstein field equations, the so-called Schwarzschild metric. This solution laid the groundwork for the description of the final stages of gravitational collapse, and the objects known today as black holes. In the same year, the first steps towards generalizing Schwarzschild's solution to electrically charged objects were taken, which eventually resulted in the Reissner–Nordström solution, now associated with electrically charged black holes.[4] In 1917, Einstein applied his theory to the universe as a whole, initiating the field of relativistic cosmology. In line with contemporary thinking, he assumed a static universe, adding a new parameter to his original field equations—the cosmological constant—to match that observational presumption.[5] By 1929, however, the work of Hubble and others had shown that

33.2 From classical mechanics to general relativity

General relativity can be understood by examining its similarities with and departures from classical physics. The first step is the realization that classical mechanics and Newton's law of gravity admit a geometric description. The combination of this description with the laws of special relativity results in a heuristic derivation of general relativity.[15]

33.2.1 Geometry of Newtonian gravity

At the base of classical mechanics is the notion that a body's motion can be described as a combination of free (or inertial) motion, and deviations from this free motion. Such deviations are caused by external forces acting on a body in accordance with Newton's second law of motion, which states that the net force acting on a body is equal to that body's (inertial) mass multiplied by its acceleration.[16] The preferred inertial motions are related to the geometry of space and time: in the standard reference frames of clas-

According to general relativity, objects in a gravitational field behave similarly to objects within an accelerating enclosure. For example, an observer will see a ball fall the same way in a rocket (left) as it does on Earth (right), provided that the acceleration of the rocket is equal to 9.8 m/s² (the acceleration due to gravity at the surface of the Earth).

sical mechanics, objects in free motion move along straight lines at constant speed. In modern parlance, their paths are geodesics, straight world lines in curved spacetime.[17]

Conversely, one might expect that inertial motions, once identified by observing the actual motions of bodies and making allowances for the external forces (such as electromagnetism or friction), can be used to define the geometry of space, as well as a time coordinate. However, there is an ambiguity once gravity comes into play. According to Newton's law of gravity, and independently verified by experiments such as that of Eötvös and its successors (see Eötvös experiment), there is a universality of free fall (also known as the weak equivalence principle, or the universal equality of inertial and passive-gravitational mass): the trajectory of a test body in free fall depends only on its position and initial speed, but not on any of its material properties.[18] A simplified version of this is embodied in Einstein's elevator experiment, illustrated in the figure on the right: for an observer in a small enclosed room, it is impossible to decide, by mapping the trajectory of bodies such as a dropped ball, whether the room is at rest in a gravitational field, or in free space aboard a rocket that is accelerating at a rate equal to that of the gravitational field.[19]

Given the universality of free fall, there is no observable distinction between inertial motion and motion under the influence of the gravitational force. This suggests the definition of a new class of inertial motion, namely that of objects in free fall under the influence of gravity. This new class of preferred motions, too, defines a geometry of space and time—in mathematical terms, it is the geodesic motion associated with a specific connection which depends on the gradient of the gravitational potential. Space, in this construction, still has the ordinary Euclidean geometry. How-

ever, space*time* as a whole is more complicated. As can be shown using simple thought experiments following the free-fall trajectories of different test particles, the result of transporting spacetime vectors that can denote a particle's velocity (time-like vectors) will vary with the particle's trajectory; mathematically speaking, the Newtonian connection is not integrable. From this, one can deduce that spacetime is curved. The resulting Newton–Cartan theory is a geometric formulation of Newtonian gravity using only covariant concepts, i.e. a description which is valid in any desired coordinate system.[20] In this geometric description, tidal effects—the relative acceleration of bodies in free fall—are related to the derivative of the connection, showing how the modified geometry is caused by the presence of mass.[21]

33.2.2 Relativistic generalization

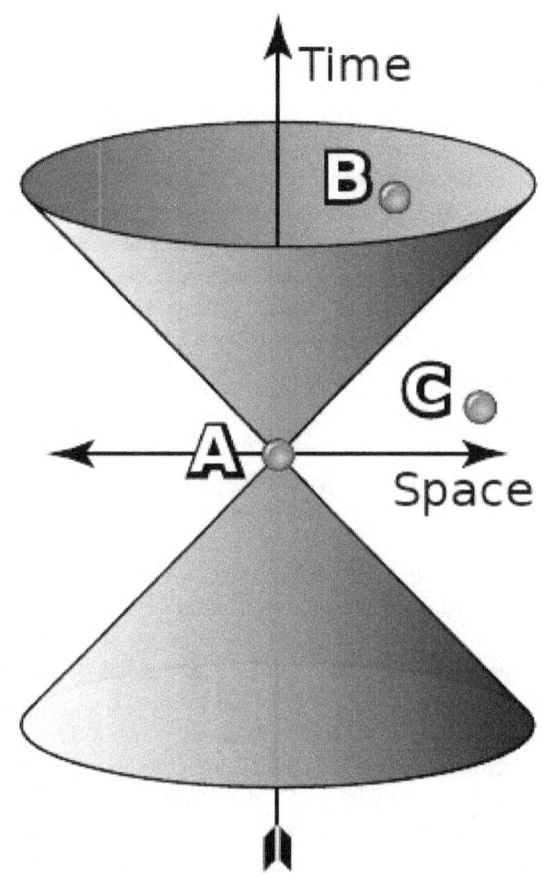

Light cone

As intriguing as geometric Newtonian gravity may be, its basis, classical mechanics, is merely a limiting case of (special) relativistic mechanics.[22] In the language of symmetry: where gravity can be neglected, physics is Lorentz invariant as in special relativity rather than Galilei invariant as in classical mechanics. (The defining symmetry of special relativity is the Poincaré group, which includes translations and rotations.) The differences between the two become significant when dealing with speeds approaching the speed of light, and with high-energy phenomena.[23]

With Lorentz symmetry, additional structures come into play. They are defined by the set of light cones (see image). The light-cones define a causal structure: for each event A, there is a set of events that can, in principle, either influence or be influenced by A via signals or interactions that do not need to travel faster than light (such as event B in the image), and a set of events for which such an influence is impossible (such as event C in the image). These sets are observer-independent.[24] In conjunction with the world-lines of freely falling particles, the light-cones can be used to reconstruct the space–time's semi-Riemannian metric, at least up to a positive scalar factor. In mathematical terms, this defines a Conformal structure[25] or conformal geometry.

Special relativity is defined in the absence of gravity, so for practical applications, it is a suitable model whenever gravity can be neglected. Bringing gravity into play, and assuming the universality of free fall, an analogous reasoning as in the previous section applies: there are no global inertial frames. Instead there are approximate inertial frames moving alongside freely falling particles. Translated into the language of spacetime: the straight time-like lines that define a gravity-free inertial frame are deformed to lines that are curved relative to each other, suggesting that the inclusion of gravity necessitates a change in spacetime geometry.[26]

A priori, it is not clear whether the new local frames in free fall coincide with the reference frames in which the laws of special relativity hold—that theory is based on the propagation of light, and thus on electromagnetism, which could have a different set of preferred frames. But using different assumptions about the special-relativistic frames (such as their being earth-fixed, or in free fall), one can derive different predictions for the gravitational redshift, that is, the way in which the frequency of light shifts as the light propagates through a gravitational field (cf. below). The actual measurements show that free-falling frames are the ones in which light propagates as it does in special relativity.[27] The generalization of this statement, namely that the laws of special relativity hold to good approximation in freely falling (and non-rotating) reference frames, is known as the Einstein equivalence principle, a crucial guiding principle for generalizing special-relativistic physics to include gravity.[28]

The same experimental data shows that time as measured by clocks in a gravitational field—proper time, to give the technical term—does not follow the rules of special relativ-

ity. In the language of spacetime geometry, it is not measured by the Minkowski metric. As in the Newtonian case, this is suggestive of a more general geometry. At small scales, all reference frames that are in free fall are equivalent, and approximately Minkowskian. Consequently, we are now dealing with a curved generalization of Minkowski space. The metric tensor that defines the geometry—in particular, how lengths and angles are measured—is not the Minkowski metric of special relativity, it is a generalization known as a semi- or pseudo-Riemannian metric. Furthermore, each Riemannian metric is naturally associated with one particular kind of connection, the Levi-Civita connection, and this is, in fact, the connection that satisfies the equivalence principle and makes space locally Minkowskian (that is, in suitable locally inertial coordinates, the metric is Minkowskian, and its first partial derivatives and the connection coefficients vanish).[29]

33.2.3 Einstein's equations

Main articles: Einstein field equations and Mathematics of general relativity

Having formulated the relativistic, geometric version of the effects of gravity, the question of gravity's source remains. In Newtonian gravity, the source is mass. In special relativity, mass turns out to be part of a more general quantity called the energy–momentum tensor, which includes both energy and momentum densities as well as stress (that is, pressure and shear).[30] Using the equivalence principle, this tensor is readily generalized to curved space-time. Drawing further upon the analogy with geometric Newtonian gravity, it is natural to assume that the field equation for gravity relates this tensor and the Ricci tensor, which describes a particular class of tidal effects: the change in volume for a small cloud of test particles that are initially at rest, and then fall freely. In special relativity, conservation of energy–momentum corresponds to the statement that the energy–momentum tensor is divergence-free. This formula, too, is readily generalized to curved spacetime by replacing partial derivatives with their curved-manifold counterparts, covariant derivatives studied in differential geometry. With this additional condition—the covariant divergence of the energy–momentum tensor, and hence of whatever is on the other side of the equation, is zero— the simplest set of equations are what are called Einstein's (field) equations:

On the left-hand side is the Einstein tensor, a specific divergence-free combination of the Ricci tensor $R_{\mu\nu}$ and the metric. Where $G_{\mu\nu}$ is symmetric. In particular,

$$R = g^{\mu\nu} R_{\mu\nu}$$

is the curvature scalar. The Ricci tensor itself is related to the more general Riemann curvature tensor as

$$R_{\mu\nu} = R^{\alpha}{}_{\mu\alpha\nu}.$$

On the right-hand side, $T_{\mu\nu}$ is the energy–momentum tensor. All tensors are written in abstract index notation.[31] Matching the theory's prediction to observational results for planetary orbits (or, equivalently, assuring that the weak-gravity, low-speed limit is Newtonian mechanics), the proportionality constant can be fixed as $\kappa = 8\pi G/c^4$, with G the gravitational constant and c the speed of light.[32] When there is no matter present, so that the energy–momentum tensor vanishes, the results are the vacuum Einstein equations,

$$R_{\mu\nu} = 0.$$

33.2.4 Alternatives to general relativity

Main article: Alternatives to general relativity

There are alternatives to general relativity built upon the same premises, which include additional rules and/or constraints, leading to different field equations. Examples are Brans–Dicke theory, teleparallelism, f(R) gravity and Einstein–Cartan theory.[33]

33.3 Definition and basic applications

See also: Mathematics of general relativity and Physical theories modified by general relativity

The derivation outlined in the previous section contains all the information needed to define general relativity, describe its key properties, and address a question of crucial importance in physics, namely how the theory can be used for model-building.

33.3.1 Definition and basic properties

General relativity is a metric theory of gravitation. At its core are Einstein's equations, which describe the relation between the geometry of a four-dimensional, pseudo-Riemannian manifold representing spacetime, and the

energy–momentum contained in that spacetime.[34] Phenomena that in classical mechanics are ascribed to the action of the force of gravity (such as free-fall, orbital motion, and spacecraft trajectories), correspond to inertial motion within a curved geometry of spacetime in general relativity; there is no gravitational force deflecting objects from their natural, straight paths. Instead, gravity corresponds to changes in the properties of space and time, which in turn changes the straightest-possible paths that objects will naturally follow.[35] The curvature is, in turn, caused by the energy–momentum of matter. Paraphrasing the relativist John Archibald Wheeler, spacetime tells matter how to move; matter tells spacetime how to curve.[36]

While general relativity replaces the scalar gravitational potential of classical physics by a symmetric rank-two tensor, the latter reduces to the former in certain limiting cases. For weak gravitational fields and slow speed relative to the speed of light, the theory's predictions converge on those of Newton's law of universal gravitation.[37]

As it is constructed using tensors, general relativity exhibits general covariance: its laws—and further laws formulated within the general relativistic framework—take on the same form in all coordinate systems.[38] Furthermore, the theory does not contain any invariant geometric background structures, i.e. it is background independent. It thus satisfies a more stringent general principle of relativity, namely that the laws of physics are the same for all observers.[39] Locally, as expressed in the equivalence principle, spacetime is Minkowskian, and the laws of physics exhibit local Lorentz invariance.[40]

33.3.2 Model-building

The core concept of general-relativistic model-building is that of a solution of Einstein's equations. Given both Einstein's equations and suitable equations for the properties of matter, such a solution consists of a specific semi-Riemannian manifold (usually defined by giving the metric in specific coordinates), and specific matter fields defined on that manifold. Matter and geometry must satisfy Einstein's equations, so in particular, the matter's energy–momentum tensor must be divergence-free. The matter must, of course, also satisfy whatever additional equations were imposed on its properties. In short, such a solution is a model universe that satisfies the laws of general relativity, and possibly additional laws governing whatever matter might be present.[41]

Einstein's equations are nonlinear partial differential equations and, as such, difficult to solve exactly.[42] Nevertheless, a number of exact solutions are known, although only a few have direct physical applications.[43] The best-known exact solutions, and also those most interesting from a physics point of view, are the Schwarzschild solution, the Reissner–Nordström solution and the Kerr metric, each corresponding to a certain type of black hole in an otherwise empty universe,[44] and the Friedmann–Lemaître–Robertson–Walker and de Sitter universes, each describing an expanding cosmos.[45] Exact solutions of great theoretical interest include the Gödel universe (which opens up the intriguing possibility of time travel in curved spacetimes), the Taub-NUT solution (a model universe that is homogeneous, but anisotropic), and anti-de Sitter space (which has recently come to prominence in the context of what is called the Maldacena conjecture).[46]

Given the difficulty of finding exact solutions, Einstein's field equations are also solved frequently by numerical integration on a computer, or by considering small perturbations of exact solutions. In the field of numerical relativity, powerful computers are employed to simulate the geometry of spacetime and to solve Einstein's equations for interesting situations such as two colliding black holes.[47] In principle, such methods may be applied to any system, given sufficient computer resources, and may address fundamental questions such as naked singularities. Approximate solutions may also be found by perturbation theories such as linearized gravity[48] and its generalization, the post-Newtonian expansion, both of which were developed by Einstein. The latter provides a systematic approach to solving for the geometry of a spacetime that contains a distribution of matter that moves slowly compared with the speed of light. The expansion involves a series of terms; the first terms represent Newtonian gravity, whereas the later terms represent ever smaller corrections to Newton's theory due to general relativity.[49] An extension of this expansion is the parametrized post-Newtonian (PPN) formalism, which allows quantitative comparisons between the predictions of general relativity and alternative theories.[50]

33.4 Consequences of Einstein's theory

General relativity has a number of physical consequences. Some follow directly from the theory's axioms, whereas others have become clear only in the course of many years of research that followed Einstein's initial publication.

33.4.1 Gravitational time dilation and frequency shift

Main article: Gravitational time dilation
Assuming that the equivalence principle holds,[51] gravity influences the passage of time. Light sent down into a gravity well is blueshifted, whereas light sent in the opposite direction (i.e., climbing out of the gravity well) is

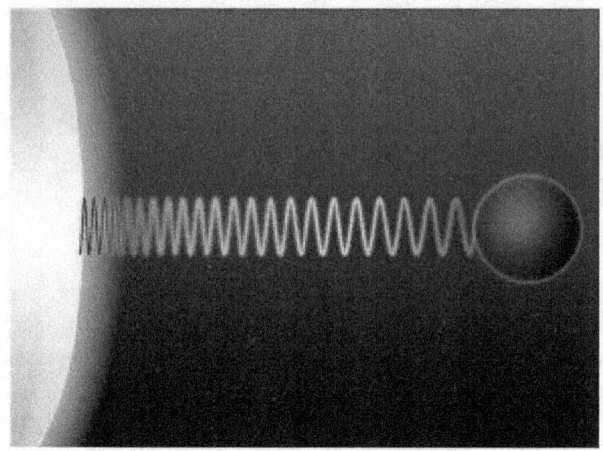

Schematic representation of the gravitational redshift of a light wave escaping from the surface of a massive body

redshifted; collectively, these two effects are known as the gravitational frequency shift. More generally, processes close to a massive body run more slowly when compared with processes taking place farther away; this effect is known as gravitational time dilation.[52]

Gravitational redshift has been measured in the laboratory[53] and using astronomical observations.[54] Gravitational time dilation in the Earth's gravitational field has been measured numerous times using atomic clocks,[55] while ongoing validation is provided as a side effect of the operation of the Global Positioning System (GPS).[56] Tests in stronger gravitational fields are provided by the observation of binary pulsars.[57] All results are in agreement with general relativity.[58] However, at the current level of accuracy, these observations cannot distinguish between general relativity and other theories in which the equivalence principle is valid.[59]

33.4.2 Light deflection and gravitational time delay

Main articles: Kepler problem in general relativity, Gravitational lens and Shapiro delay
General relativity predicts that the path of light is bent in a gravitational field; light passing a massive body is deflected towards that body. This effect has been confirmed by observing the light of stars or distant quasars being deflected as it passes the Sun.[60]

This and related predictions follow from the fact that light follows what is called a light-like or null geodesic—a generalization of the straight lines along which light travels in classical physics. Such geodesics are the generalization of the invariance of lightspeed in special relativity.[61] As one examines suitable model spacetimes (either the exterior

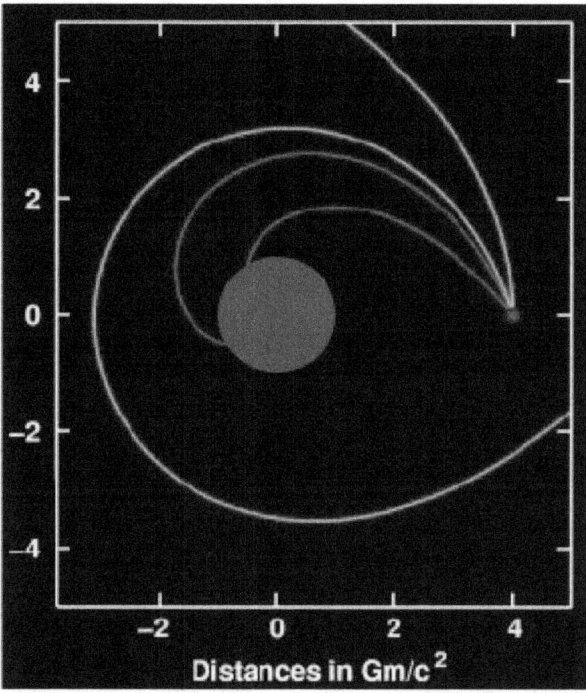

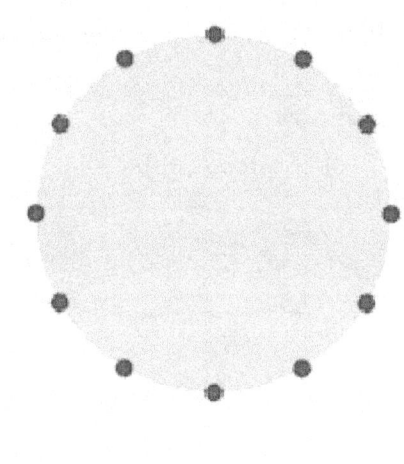

Ring of test particles influenced by gravitational wave

Deflection of light (sent out from the location shown in blue) near a compact body (shown in gray)

Schwarzschild solution or, for more than a single mass, the post-Newtonian expansion),[62] several effects of gravity on light propagation emerge. Although the bending of light can also be derived by extending the universality of free fall to light,[63] the angle of deflection resulting from such calculations is only half the value given by general relativity.[64]

Closely related to light deflection is the gravitational time delay (or Shapiro delay), the phenomenon that light signals take longer to move through a gravitational field than they would in the absence of that field. There have been numerous successful tests of this prediction.[65] In the parameterized post-Newtonian formalism (PPN), measurements of both the deflection of light and the gravitational time delay determine a parameter called γ, which encodes the influence of gravity on the geometry of space.[66]

33.4.3   Gravitational waves

Main article: Gravitational wave
   Predicted in 1916[67][68] by Albert Einstein, there are gravitational waves: ripples in the metric of spacetime that propagate at the speed of light. These are one of several analogies between weak-field gravity and electromagnetism in that, they are analogous to electromagnetic waves. On February 11, 2016, the Advanced LIGO team announced that they had directly detected gravitational waves from a pair of black holes merging.[69][70][71]

The simplest type of such a wave can be visualized by its action on a ring of freely floating particles. A sine wave propagating through such a ring towards the reader distorts the ring in a characteristic, rhythmic fashion (animated image to the right).[72] Since Einstein's equations are non-linear, arbitrarily strong gravitational waves do not obey linear superposition, making their description difficult. However, for weak fields, a linear approximation can be made. Such linearized gravitational waves are sufficiently accurate to describe the exceedingly weak waves that are expected to arrive here on Earth from far-off cosmic events, which typically result in relative distances increasing and decreasing by 10^{-21} or less. Data analysis methods routinely make use of the fact that these linearized waves can be Fourier decomposed.[73]

Some exact solutions describe gravitational waves without any approximation, e.g., a wave train traveling through empty space[74] or so-called Gowdy universes, varieties of an expanding cosmos filled with gravitational waves.[75] But for gravitational waves produced in astrophysically relevant situations, such as the merger of two black holes, numerical methods are presently the only way to construct appropriate models.[76]

33.4.4   Orbital effects and the relativity of direction

Main article: Kepler problem in general relativity

General relativity differs from classical mechanics in a

number of predictions concerning orbiting bodies. It predicts an overall rotation (precession) of planetary orbits, as well as orbital decay caused by the emission of gravitational waves and effects related to the relativity of direction.

Precession of apsides

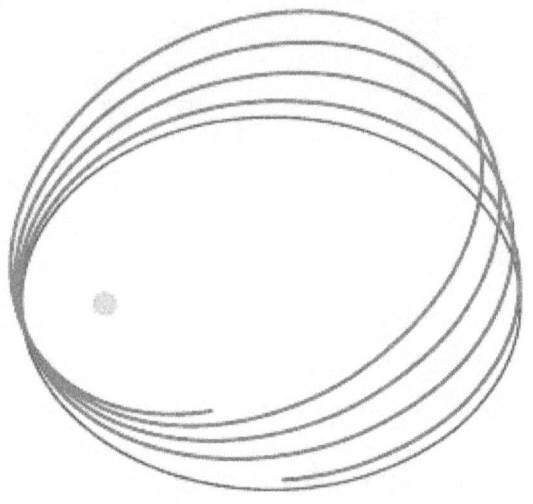

Newtonian (red) vs. Einsteinian orbit (blue) of a lone planet orbiting a star

In general relativity, the apsides of any orbit (the point of the orbiting body's closest approach to the system's center of mass) will precess—the orbit is not an ellipse, but akin to an ellipse that rotates on its focus, resulting in a rose curve-like shape (see image). Einstein first derived this result by using an approximate metric representing the Newtonian limit and treating the orbiting body as a test particle. For him, the fact that his theory gave a straightforward explanation of the anomalous perihelion shift of the planet Mercury, discovered earlier by Urbain Le Verrier in 1859, was important evidence that he had at last identified the correct form of the gravitational field equations.[77]

The effect can also be derived by using either the exact Schwarzschild metric (describing spacetime around a spherical mass)[78] or the much more general post-Newtonian formalism.[79] It is due to the influence of gravity on the geometry of space and to the contribution of self-energy to a body's gravity (encoded in the nonlinearity of Einstein's equations).[80] Relativistic precession has been observed for all planets that allow for accurate precession measurements (Mercury, Venus, and Earth),[81] as well as in binary pulsar systems, where it is larger by five orders of magnitude.[82]

In general relativity the perihelion shift σ, expressed in radians per revolution, is approximately given by:[83]

$$\sigma = \frac{24\pi^3 L^2}{T^2 c^2 (1 - e^2)} \ ,$$

where L is the semi-major axis, T is the orbital period, c is the speed of light, and e is the orbital eccentricity.

Orbital decay

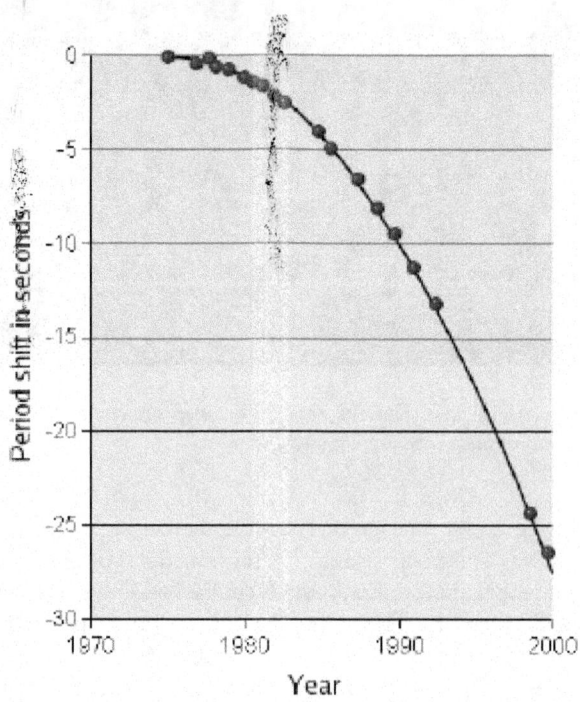

Orbital decay for PSR1913+16: time shift in seconds, tracked over three decades.[84]

According to general relativity, a binary system will emit gravitational waves, thereby losing energy. Due to this loss, the distance between the two orbiting bodies decreases, and so does their orbital period. Within the Solar System or for ordinary double stars, the effect is too small to be observable. This is not the case for a close binary pulsar, a system of two orbiting neutron stars, one of which is a pulsar: from the pulsar, observers on Earth receive a regular series of radio pulses that can serve as a highly accurate clock, which allows precise measurements of the orbital period. Because neutron stars are immensely compact, significant amounts of energy are emitted in the form of gravitational radiation.[85]

The first observation of a decrease in orbital period due to the emission of gravitational waves was made by Hulse and Taylor, using the binary pulsar PSR1913+16 they had discovered in 1974. This was the first detection of gravitational waves, albeit indirect, for which they were awarded the 1993 Nobel Prize in physics.[86] Since then, several other binary pulsars have been found, in particular the double pulsar PSR J0737-3039, in which both stars are pulsars.[87]

Geodetic precession and frame-dragging

Main articles: Geodetic precession and Frame dragging

Several relativistic effects are directly related to the relativity of direction.[88] One is geodetic precession: the axis direction of a gyroscope in free fall in curved spacetime will change when compared, for instance, with the direction of light received from distant stars—even though such a gyroscope represents the way of keeping a direction as stable as possible ("parallel transport").[89] For the Moon–Earth system, this effect has been measured with the help of lunar laser ranging.[90] More recently, it has been measured for test masses aboard the satellite Gravity Probe B to a precision of better than 0.3%.[91][92]

Near a rotating mass, there are so-called gravitomagnetic or frame-dragging effects. A distant observer will determine that objects close to the mass get "dragged around". This is most extreme for rotating black holes where, for any object entering a zone known as the ergosphere, rotation is inevitable.[93] Such effects can again be tested through their influence on the orientation of gyroscopes in free fall.[94] Somewhat controversial tests have been performed using the LAGEOS satellites, confirming the relativistic prediction.[95] Also the Mars Global Surveyor probe around Mars has been used.[96][97]

33.5   Astrophysical applications

33.5.1   Gravitational lensing

Main article: Gravitational lensing
 The deflection of light by gravity is responsible for a new class of astronomical phenomena. If a massive object is situated between the astronomer and a distant target object with appropriate mass and relative distances, the astronomer will see multiple distorted images of the target. Such effects are known as gravitational lensing.[98] Depending on the configuration, scale, and mass distribution, there can be two or more images, a bright ring known as an Einstein ring, or partial rings called arcs.[99] The earliest example was discovered in 1979;[100] since then, more than a hundred gravitational lenses have been observed.[101] Even if the multiple images are too close to each other to be resolved, the effect can still be measured, e.g., as an overall brightening of the target object; a number of such

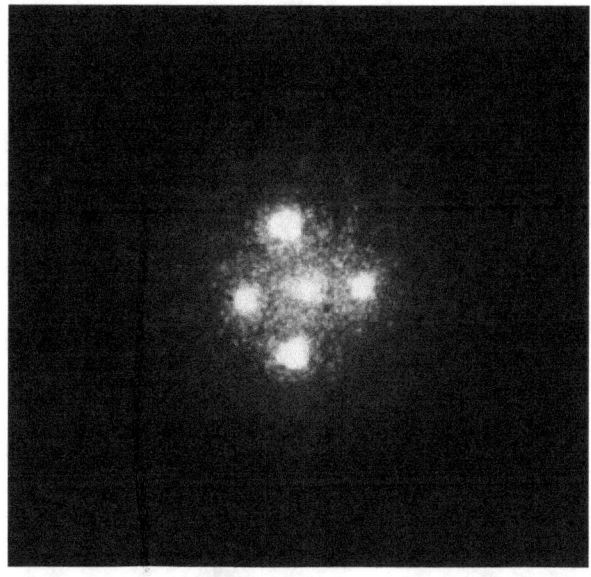

Einstein cross: four images of the same astronomical object, produced by a gravitational lens

"microlensing events" have been observed.[102]

Gravitational lensing has developed into a tool of observational astronomy. It is used to detect the presence and distribution of dark matter, provide a "natural telescope" for observing distant galaxies, and to obtain an independent estimate of the Hubble constant. Statistical evaluations of lensing data provide valuable insight into the structural evolution of galaxies.[103]

33.5.2   Gravitational wave astronomy

Main articles: Gravitational wave and Gravitational wave astronomy
 Observations of binary pulsars provide strong indirect evidence for the existence of gravitational waves (see Orbital decay, above). Detection of these waves is a major goal of current relativity-related research.[104] Several land-based gravitational wave detectors are currently in operation, most notably the interferometric detectors GEO 600, LIGO (two detectors), TAMA 300 and VIRGO.[105] Various pulsar timing arrays are using millisecond pulsars to detect gravitational waves in the 10^{-9} to 10^{-6} Hertz frequency range, which originate from binary supermassive blackholes.[106] A European space-based detector, eLISA / NGO, is currently under development,[107] with a precursor mission (LISA Pathfinder) having launched in December 2015.[108]

Observations of gravitational waves promise to complement observations in the electromagnetic spectrum.[109] They are expected to yield information about black holes and other dense objects such as neutron stars and white dwarfs, about

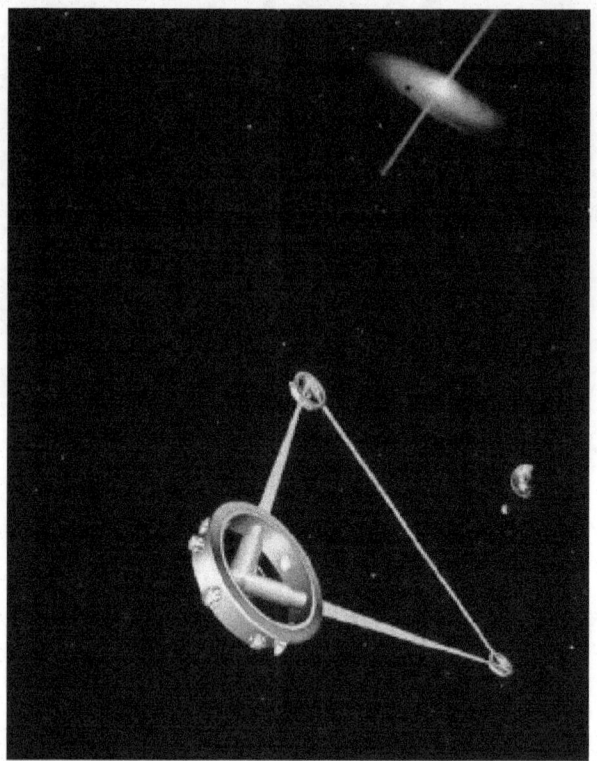

Artist's impression of the space-borne gravitational wave detector LISA

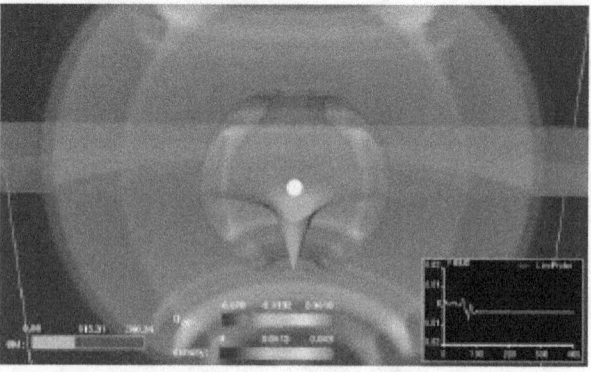

Simulation based on the equations of general relativity: a star collapsing to form a black hole while emitting gravitational waves

certain kinds of supernova implosions, and about processes in the very early universe, including the signature of certain types of hypothetical cosmic string.[110] In February 2016, the Advanced LIGO team announced that they had detected gravitational waves from a black hole merger.[69][70][111]

33.5.3 Black holes and other compact objects

Main article: Black hole

Whenever the ratio of an object's mass to its radius becomes sufficiently large, general relativity predicts the formation of a black hole, a region of space from which nothing, not even light, can escape. In the currently accepted models of stellar evolution, neutron stars of around 1.4 solar masses, and stellar black holes with a few to a few dozen solar masses, are thought to be the final state for the evolution of massive stars.[112] Usually a galaxy has one supermassive black hole with a few million to a few billion solar masses in its center,[113] and its presence is thought to have played an important role in the formation of the galaxy and larger cosmic structures.[114]

Astronomically, the most important property of compact objects is that they provide a supremely efficient mechanism for converting gravitational energy into electromagnetic radiation.[115] Accretion, the falling of dust or gaseous matter onto stellar or supermassive black holes, is thought to be responsible for some spectacularly luminous astronomical objects, notably diverse kinds of active galactic nuclei on galactic scales and stellar-size objects such as microquasars.[116] In particular, accretion can lead to relativistic jets, focused beams of highly energetic particles that are being flung into space at almost light speed.[117] General relativity plays a central role in modelling all these phenomena,[118] and observations provide strong evidence for the existence of black holes with the properties predicted by the theory.[119]

Black holes are also sought-after targets in the search for gravitational waves (cf. Gravitational waves, above). Merging black hole binaries should lead to some of the strongest gravitational wave signals reaching detectors here on Earth, and the phase directly before the merger ("chirp") could be used as a "standard candle" to deduce the distance to the merger events–and hence serve as a probe of cosmic expansion at large distances.[120] The gravitational waves produced as a stellar black hole plunges into a supermassive one should provide direct information about the supermassive black hole's geometry.[121]

33.5.4 Cosmology

Main article: Physical cosmology

The current models of cosmology are based on Einstein's field equations, which include the cosmological constant Λ since it has important influence on the large-scale dynamics of the cosmos,

This blue horseshoe is a distant galaxy that has been magnified and warped into a nearly complete ring by the strong gravitational pull of the massive foreground luminous red galaxy.

$$R_{\mu\nu} - \frac{1}{2}R\,g_{\mu\nu} + \Lambda\,g_{\mu\nu} = \frac{8\pi G}{c^4}\,T_{\mu\nu}$$

where $g_{\mu\nu}$ is the spacetime metric.[122] Isotropic and homogeneous solutions of these enhanced equations, the Friedmann–Lemaître–Robertson–Walker solutions,[123] allow physicists to model a universe that has evolved over the past 14 billion years from a hot, early Big Bang phase.[124] Once a small number of parameters (for example the universe's mean matter density) have been fixed by astronomical observation,[125] further observational data can be used to put the models to the test.[126] Predictions, all successful, include the initial abundance of chemical elements formed in a period of primordial nucleosynthesis,[127] the large-scale structure of the universe,[128] and the existence and properties of a "thermal echo" from the early cosmos, the cosmic background radiation.[129]

Astronomical observations of the cosmological expansion rate allow the total amount of matter in the universe to be estimated, although the nature of that matter remains mysterious in part. About 90% of all matter appears to be so-called dark matter, which has mass (or, equivalently, gravitational influence), but does not interact electromagnetically and, hence, cannot be observed directly.[130] There is no generally accepted description of this new kind of matter, within the framework of known particle physics[131] or otherwise.[132] Observational evidence from redshift surveys of distant supernovae and measurements of the cosmic background radiation also show that the evolution of our universe is significantly influenced by a cosmological constant resulting in an acceleration of cosmic expansion or, equivalently, by a form of energy with an unusual equation of state, known as dark energy, the nature of which remains

unclear.[133]

A so-called inflationary phase,[134] an additional phase of strongly accelerated expansion at cosmic times of around 10^{-33} seconds, was hypothesized in 1980 to account for several puzzling observations that were unexplained by classical cosmological models, such as the nearly perfect homogeneity of the cosmic background radiation.[135] Recent measurements of the cosmic background radiation have resulted in the first evidence for this scenario.[136] However, there is a bewildering variety of possible inflationary scenarios, which cannot be restricted by current observations.[137] An even larger question is the physics of the earliest universe, prior to the inflationary phase and close to where the classical models predict the big bang singularity. An authoritative answer would require a complete theory of quantum gravity, which has not yet been developed[138] (cf. the section on quantum gravity, below).

33.5.5 Time travel

Kurt Gödel showed[139] that solutions to Einstein's equations exist that contain closed timelike curves (CTCs), which allow for loops in time. The solutions require extreme physical conditions unlikely ever to occur in practice, and it remains an open question whether further laws of physics will eliminate them completely. Since then other—similarly impractical—GR solutions containing CTCs have been found, such as the Tipler cylinder and traversable wormholes.

33.6   Advanced concepts

33.6.1   Causal structure and global geometry

Main article: Causal structure

In general relativity, no material body can catch up with or overtake a light pulse. No influence from an event A can reach any other location X before light sent out at A to X. In consequence, an exploration of all light worldlines (null geodesics) yields key information about the spacetime's causal structure. This structure can be displayed using Penrose–Carter diagrams in which infinitely large regions of space and infinite time intervals are shrunk ("compactified") so as to fit onto a finite map, while light still travels along diagonals as in standard spacetime diagrams.[140]

Aware of the importance of causal structure, Roger Penrose and others developed what is known as global geometry. In global geometry, the object of study is not one particular solution (or family of solutions) to Einstein's equations. Rather, relations that hold true for all geodesics, such as

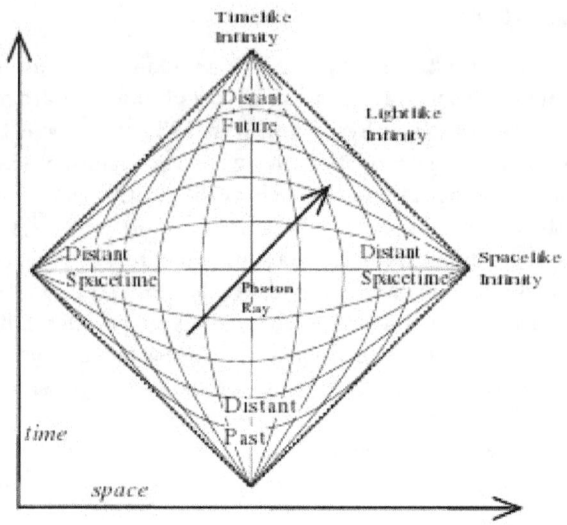

Penrose–Carter diagram of an infinite Minkowski universe

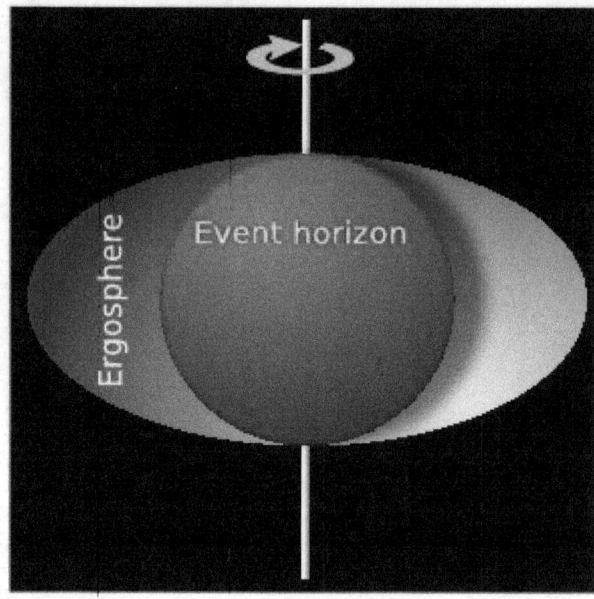

The ergosphere of a rotating black hole, which plays a key role when it comes to extracting energy from such a black hole

the Raychaudhuri equation, and additional non-specific assumptions about the nature of matter (usually in the form of so-called energy conditions) are used to derive general results.[141]

33.6.2 Horizons

Main articles: Horizon (general relativity), No hair theorem and Black hole mechanics

Using global geometry, some spacetimes can be shown to contain boundaries called horizons, which demarcate one region from the rest of spacetime. The best-known examples are black holes: if mass is compressed into a sufficiently compact region of space (as specified in the hoop conjecture, the relevant length scale is the Schwarzschild radius[142]), no light from inside can escape to the outside. Since no object can overtake a light pulse, all interior matter is imprisoned as well. Passage from the exterior to the interior is still possible, showing that the boundary, the black hole's *horizon*, is not a physical barrier.[143]

Early studies of black holes relied on explicit solutions of Einstein's equations, notably the spherically symmetric Schwarzschild solution (used to describe a static black hole) and the axisymmetric Kerr solution (used to describe a rotating, stationary black hole, and introducing interesting features such as the ergosphere). Using global geometry, later studies have revealed more general properties of black holes. In the long run, they are rather simple objects characterized by eleven parameters specifying energy, linear momentum, angular momentum, location at a specified time and electric charge. This is stated by the black hole unique-

ness theorems: "black holes have no hair", that is, no distinguishing marks like the hairstyles of humans. Irrespective of the complexity of a gravitating object collapsing to form a black hole, the object that results (having emitted gravitational waves) is very simple.[144]

Even more remarkably, there is a general set of laws known as black hole mechanics, which is analogous to the laws of thermodynamics. For instance, by the second law of black hole mechanics, the area of the event horizon of a general black hole will never decrease with time, analogous to the entropy of a thermodynamic system. This limits the energy that can be extracted by classical means from a rotating black hole (e.g. by the Penrose process).[145] There is strong evidence that the laws of black hole mechanics are, in fact, a subset of the laws of thermodynamics, and that the black hole area is proportional to its entropy.[146] This leads to a modification of the original laws of black hole mechanics: for instance, as the second law of black hole mechanics becomes part of the second law of thermodynamics, it is possible for black hole area to decrease—as long as other processes ensure that, overall, entropy increases. As thermodynamical objects with non-zero temperature, black holes should emit thermal radiation. Semi-classical calculations indicate that indeed they do, with the surface gravity playing the role of temperature in Planck's law. This radiation is known as Hawking radiation (cf. the quantum theory section, below).[147]

There are other types of horizons. In an expanding universe, an observer may find that some regions of the past cannot be observed ("particle horizon"), and some regions of the future cannot be influenced (event horizon).[148] Even in flat Minkowski space, when described by an accelerated observer (Rindler space), there will be horizons associated with a semi-classical radiation known as Unruh radiation.[149]

33.6.3 Singularities

Main article: Spacetime singularity

Another general feature of general relativity is the appearance of spacetime boundaries known as singularities. Spacetime can be explored by following up on timelike and lightlike geodesics—all possible ways that light and particles in free fall can travel. But some solutions of Einstein's equations have "ragged edges"—regions known as spacetime singularities, where the paths of light and falling particles come to an abrupt end, and geometry becomes ill-defined. In the more interesting cases, these are "curvature singularities", where geometrical quantities characterizing spacetime curvature, such as the Ricci scalar, take on infinite values.[150] Well-known examples of spacetimes with future singularities—where worldlines end—are the Schwarzschild solution, which describes a singularity inside an eternal static black hole,[151] or the Kerr solution with its ring-shaped singularity inside an eternal rotating black hole.[152] The Friedmann–Lemaître–Robertson–Walker solutions and other spacetimes describing universes have past singularities on which worldlines begin, namely Big Bang singularities, and some have future singularities (Big Crunch) as well.[153]

Given that these examples are all highly symmetric—and thus simplified—it is tempting to conclude that the occurrence of singularities is an artifact of idealization.[154] The famous singularity theorems, proved using the methods of global geometry, say otherwise: singularities are a generic feature of general relativity, and unavoidable once the collapse of an object with realistic matter properties has proceeded beyond a certain stage[155] and also at the beginning of a wide class of expanding universes.[156] However, the theorems say little about the properties of singularities, and much of current research is devoted to characterizing these entities' generic structure (hypothesized e.g. by the so-called BKL conjecture).[157] The cosmic censorship hypothesis states that all realistic future singularities (no perfect symmetries, matter with realistic properties) are safely hidden away behind a horizon, and thus invisible to all distant observers. While no formal proof yet exists, numerical simulations offer supporting evidence of its validity.[158]

33.6.4 Evolution equations

Main article: Initial value formulation (general relativity)

Each solution of Einstein's equation encompasses the whole history of a universe — it is not just some snapshot of how things are, but a whole, possibly matter-filled, spacetime. It describes the state of matter and geometry everywhere and at every moment in that particular universe. Due to its general covariance, Einstein's theory is not sufficient by itself to determine the time evolution of the metric tensor. It must be combined with a coordinate condition, which is analogous to gauge fixing in other field theories.[159]

To understand Einstein's equations as partial differential equations, it is helpful to formulate them in a way that describes the evolution of the universe over time. This is done in so-called "3+1" formulations, where spacetime is split into three space dimensions and one time dimension. The best-known example is the ADM formalism.[160] These decompositions show that the spacetime evolution equations of general relativity are well-behaved: solutions always exist, and are uniquely defined, once suitable initial conditions have been specified.[161] Such formulations of Einstein's field equations are the basis of numerical relativity.[162]

33.6.5 Global and quasi-local quantities

Main article: Mass in general relativity

The notion of evolution equations is intimately tied in with another aspect of general relativistic physics. In Einstein's theory, it turns out to be impossible to find a general definition for a seemingly simple property such as a system's total mass (or energy). The main reason is that the gravitational field—like any physical field—must be ascribed a certain energy, but that it proves to be fundamentally impossible to localize that energy.[163]

Nevertheless, there are possibilities to define a system's total mass, either using a hypothetical "infinitely distant observer" (ADM mass)[164] or suitable symmetries (Komar mass).[165] If one excludes from the system's total mass the energy being carried away to infinity by gravitational waves, the result is the so-called Bondi mass at null infinity.[166] Just as in classical physics, it can be shown that these masses are positive.[167] Corresponding global definitions exist for momentum and angular momentum.[168] There have also been a number of attempts to define quasi-local quantities, such as the mass of an isolated system formulated using only quantities defined within a finite region of space containing that system. The hope is to obtain a quantity useful for general statements about isolated systems, such as a more precise formulation of the hoop conjecture.[169]

33.7 Relationship with quantum theory

If general relativity were considered to be one of the two pillars of modern physics, then quantum theory, the basis of understanding matter from elementary particles to solid state physics, would be the other.[170] However, how to reconcile quantum theory with general relativity is still an open question.

33.7.1 Quantum field theory in curved spacetime

Main article: Quantum field theory in curved spacetime

Ordinary quantum field theories, which form the basis of modern elementary particle physics, are defined in flat Minkowski space, which is an excellent approximation when it comes to describing the behavior of microscopic particles in weak gravitational fields like those found on Earth.[171] In order to describe situations in which gravity is strong enough to influence (quantum) matter, yet not strong enough to require quantization itself, physicists have formulated quantum field theories in curved spacetime. These theories rely on general relativity to describe a curved background spacetime, and define a generalized quantum field theory to describe the behavior of quantum matter within that spacetime.[172] Using this formalism, it can be shown that black holes emit a blackbody spectrum of particles known as Hawking radiation, leading to the possibility that they evaporate over time.[173] As briefly mentioned above, this radiation plays an important role for the thermodynamics of black holes.[174]

33.7.2 Quantum gravity

Main article: Quantum gravity
See also: String theory, Canonical general relativity, Loop quantum gravity, Causal Dynamical Triangulations, and Causal sets

The demand for consistency between a quantum description of matter and a geometric description of spacetime,[175] as well as the appearance of singularities (where curvature length scales become microscopic), indicate the need for a full theory of quantum gravity: for an adequate description of the interior of black holes, and of the very early

universe, a theory is required in which gravity and the associated geometry of spacetime are described in the language of quantum physics.[176] Despite major efforts, no complete and consistent theory of quantum gravity is currently known, even though a number of promising candidates exist.[177][178]

Projection of a Calabi–Yau manifold, one of the ways of compactifying the extra dimensions posited by string theory

Attempts to generalize ordinary quantum field theories, used in elementary particle physics to describe fundamental interactions, so as to include gravity have led to serious problems.[179] Some have argued that at low energies, this approach proves successful, in that it results in an acceptable effective (quantum) field theory of gravity.[180] At very high energies, however, the perturbative results are badly divergent and lead to models devoid of predictive power ("perturbative non-renormalizability").[181]

One attempt to overcome these limitations is string theory, a quantum theory not of point particles, but of minute one-dimensional extended objects.[182] The theory promises to be a unified description of all particles and interactions, including gravity;[183] the price to pay is unusual features such as six extra dimensions of space in addition to the usual three.[184] In what is called the second superstring revolution, it was conjectured that both string theory and a unification of general relativity and supersymmetry known as supergravity[185] form part of a hypothesized eleven-dimensional model known as M-theory, which would constitute a uniquely defined and consistent theory of quantum gravity.[186]

Another approach starts with the canonical quantization procedures of quantum theory. Using the initial-value-

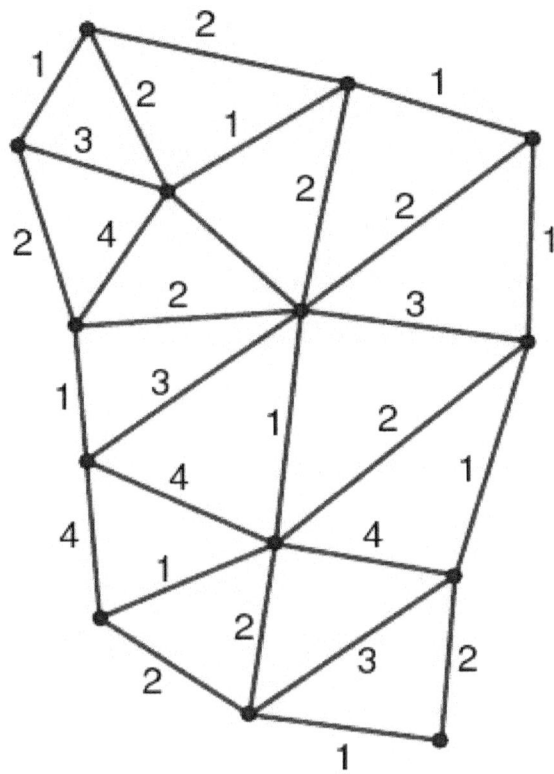

Simple spin network of the type used in loop quantum gravity

cosmological observations and particle physics experiments becomes available.[195]

33.8   Current status

General relativity has emerged as a highly successful model of gravitation and cosmology, which has so far passed many unambiguous observational and experimental tests. However, there are strong indications the theory is incomplete.[196] The problem of quantum gravity and the question of the reality of spacetime singularities remain open.[197] Observational data that is taken as evidence for dark energy and dark matter could indicate the need for new physics.[198] Even taken as is, general relativity is rich with possibilities for further exploration. Mathematical relativists seek to understand the nature of singularities and the fundamental properties of Einstein's equations,[199] and increasingly powerful computer simulations (such as those describing merging black holes) are run.[200] In February 2016, it was announced that the existence of gravitational waves was directly detected by the Advanced LIGO team on September 14, 2015.[71][201][202] A century after its publication, general relativity remains a highly active area of research.[203]

33.9   See also

- Alcubierre drive (warp drive)

- Center of mass (relativistic)

- Contributors to general relativity

- Derivations of the Lorentz transformations

- Ehrenfest paradox

- Einstein–Hilbert action

- Introduction to mathematics of general relativity

- Relativity priority dispute

- Ricci calculus

- Tests of general relativity

- Timeline of gravitational physics and relativity

- Two-body problem in general relativity

- Weak Gravity Conjecture

formulation of general relativity (cf. evolution equations above), the result is the Wheeler–deWitt equation (an analogue of the Schrödinger equation) which, regrettably, turns out to be ill-defined without a proper ultraviolet (lattice) cutoff.[187] However, with the introduction of what are now known as Ashtekar variables,[188] this leads to a promising model known as loop quantum gravity. Space is represented by a web-like structure called a spin network, evolving over time in discrete steps.[189]

Depending on which features of general relativity and quantum theory are accepted unchanged, and on what level changes are introduced,[190] there are numerous other attempts to arrive at a viable theory of quantum gravity, some examples being the lattice theory of gravity based on the Feynman Path Integral approach and Regge Calculus,[177] dynamical triangulations,[191] causal sets,[192] twistor models[193] or the path-integral based models of quantum cosmology.[194]

All candidate theories still have major formal and conceptual problems to overcome. They also face the common problem that, as yet, there is no way to put quantum gravity predictions to experimental tests (and thus to decide between the candidates where their predictions vary), although there is hope for this to change as future data from

33.10 Notes

[1] "GW150914: LIGO Detects Gravitational Waves". *Blackholes.org*. Retrieved 18 April 2016.

[2] O'Connor, J.J. and Robertson, E.F. (1996), *General relativity. Mathematical Physics index*, School of Mathematics and Statistics, University of St. Andrews, Scotland. Retrieved 2015-02-04.

[3] Pais 1982, ch. 9 to 15, Janssen 2005; an up-to-date collection of current research, including reprints of many of the original articles, is Renn 2007; an accessible overview can be found in Renn 2005, pp. 110ff. Einstein's original papers are found in Digital Einstein, volumes 4 and 6. An early key article is Einstein 1907, cf. Pais 1982, ch. 9. The publication featuring the field equations is Einstein 1915, cf. Pais 1982, ch. 11–15

[4] Schwarzschild 1916a, Schwarzschild 1916b and Reissner 1916 (later complemented in Nordström 1918)

[5] Einstein 1917, cf. Pais 1982, ch. 15e

[6] Hubble's original article is Hubble 1929; an accessible overview is given in Singh 2004, ch. 2–4

[7] As reported in Gamow 1970. Einstein's condemnation would prove to be premature, cf. the section Cosmology, below

[8] Pais 1982, pp. 253–254

[9] Kennefick 2005, Kennefick 2007

[10] Pais 1982, ch. 16

[11] Thorne, Kip (2003). "Warping spacetime". *The future of theoretical physics and cosmology: celebrating Stephen Hawking's 60th birthday*. Cambridge University Press. p. 74. ISBN 0-521-82081-2. Extract of page 74

[12] Israel 1987, ch. 7.8–7.10, Thorne 1994, ch. 3–9

[13] Sections Orbital effects and the relativity of direction, Gravitational time dilation and frequency shift and Light deflection and gravitational time delay, and references therein

[14] Section Cosmology and references therein; the historical development is in Overbye 1999

[15] The following exposition re-traces that of Ehlers 1973, sec. 1

[16] Arnold 1989, ch. 1

[17] Ehlers 1973, pp. 5f

[18] Will 1993, sec. 2.4, Will 2006, sec. 2

[19] Wheeler 1990, ch. 2

[20] Ehlers 1973, sec. 1.2, Havas 1964, Künzle 1972. The simple thought experiment in question was first described in Heckmann & Schücking 1959

[21] Ehlers 1973, pp. 10f

[22] Good introductions are, in order of increasing presupposed knowledge of mathematics, Giulini 2005, Mermin 2005, and Rindler 1991; for accounts of precision experiments, cf. part IV of Ehlers & Lämmerzahl 2006

[23] An in-depth comparison between the two symmetry groups can be found in Giulini 2006a

[24] Rindler 1991, sec. 22, Synge 1972, ch. 1 and 2

[25] Ehlers 1973, sec. 2.3

[26] Ehlers 1973, sec. 1.4, Schutz 1985, sec. 5.1

[27] Ehlers 1973, pp. 17ff; a derivation can be found in Mermin 2005, ch. 12. For the experimental evidence, cf. the section Gravitational time dilation and frequency shift, below

[28] Rindler 2001, sec. 1.13; for an elementary account, see Wheeler 1990, ch. 2; there are, however, some differences between the modern version and Einstein's original concept used in the historical derivation of general relativity, cf. Norton 1985

[29] Ehlers 1973, sec. 1.4 for the experimental evidence, see once more section Gravitational time dilation and frequency shift. Choosing a different connection with non-zero torsion leads to a modified theory known as Einstein–Cartan theory

[30] Ehlers 1973, p. 16, Kenyon 1990, sec. 7.2, Weinberg 1972, sec. 2.8

[31] Ehlers 1973, pp. 19–22; for similar derivations, see sections 1 and 2 of ch. 7 in Weinberg 1972. The Einstein tensor is the only divergence-free tensor that is a function of the metric coefficients, their first and second derivatives at most, and allows the spacetime of special relativity as a solution in the absence of sources of gravity, cf. Lovelock 1972. The tensors on both side are of second rank, that is, they can each be thought of as 4×4 matrices, each of which contains ten independent terms; hence, the above represents ten coupled equations. The fact that, as a consequence of geometric relations known as Bianchi identities, the Einstein tensor satisfies a further four identities reduces these to six independent equations, e.g. Schutz 1985, sec. 8.3

[32] Kenyon 1990, sec. 7.4

[33] Brans & Dicke 1961, Weinberg 1972, sec. 3 in ch. 7, Goenner 2004, sec. 7.2, and Trautman 2006, respectively

[34] Wald 1984, ch. 4, Weinberg 1972, ch. 7 or, in fact, any other textbook on general relativity

[35] At least approximately, cf. Poisson 2004

[36] Wheeler 1990, p. xi

[37] Wald 1984, sec. 4.4

[38] Wald 1984, sec. 4.1

[39] For the (conceptual and historical) difficulties in defining a general principle of relativity and separating it from the notion of general covariance, see Giulini 2006b

[40] section 5 in ch. 12 of Weinberg 1972

[41] Introductory chapters of Stephani et al. 2003

[42] A review showing Einstein's equation in the broader context of other PDEs with physical significance is Geroch 1996

[43] For background information and a list of solutions, cf. Stephani et al. 2003; a more recent review can be found in MacCallum 2006

[44] Chandrasekhar 1983, ch. 3,5,6

[45] Narlikar 1993, ch. 4, sec. 3.3

[46] Brief descriptions of these and further interesting solutions can be found in Hawking & Ellis 1973, ch. 5

[47] Lehner 2002

[48] For instance Wald 1984, sec. 4.4

[49] Will 1993, sec. 4.1 and 4.2

[50] Will 2006, sec. 3.2, Will 1993, ch. 4

[51] Rindler 2001, pp. 24–26 vs. pp. 236–237 and Ohanian & Ruffini 1994, pp. 164–172. Einstein derived these effects using the equivalence principle as early as 1907, cf. Einstein 1907 and the description in Pais 1982, pp. 196–198

[52] Rindler 2001, pp. 24–26; Misner, Thorne & Wheeler 1973, § 38.5

[53] Pound–Rebka experiment, see Pound & Rebka 1959, Pound & Rebka 1960; Pound & Snider 1964; a list of further experiments is given in Ohanian & Ruffini 1994, table 4.1 on p. 186

[54] Greenstein, Oke & Shipman 1971; the most recent and most accurate Sirius B measurements are published in Barstow, Bond et al. 2005.

[55] Starting with the Hafele–Keating experiment, Hafele & Keating 1972a and Hafele & Keating 1972b, and culminating in the Gravity Probe A experiment; an overview of experiments can be found in Ohanian & Ruffini 1994, table 4.1 on p. 186

[56] GPS is continually tested by comparing atomic clocks on the ground and aboard orbiting satellites; for an account of relativistic effects, see Ashby 2002 and Ashby 2003

[57] Stairs 2003 and Kramer 2004

[58] General overviews can be found in section 2.1. of Will 2006; Will 2003, pp. 32–36; Ohanian & Ruffini 1994, sec. 4.2

[59] Ohanian & Ruffini 1994, pp. 164–172

[60] Cf. Kennefick 2005 for the classic early measurements by the Eddington expeditions; for an overview of more recent measurements, see Ohanian & Ruffini 1994, ch. 4.3. For the most precise direct modern observations using quasars, cf. Shapiro et al. 2004

[61] This is not an independent axiom; it can be derived from Einstein's equations and the Maxwell Lagrangian using a WKB approximation, cf. Ehlers 1973, sec. 5

[62] Blanchet 2006, sec. 1.3

[63] Rindler 2001, sec. 1.16; for the historical examples, Israel 1987, pp. 202–204; in fact, Einstein published one such derivation as Einstein 1907. Such calculations tacitly assume that the geometry of space is Euclidean, cf. Ehlers & Rindler 1997

[64] From the standpoint of Einstein's theory, these derivations take into account the effect of gravity on time, but not its consequences for the warping of space, cf. Rindler 2001, sec. 11.11

[65] For the Sun's gravitational field using radar signals reflected from planets such as Venus and Mercury, cf. Shapiro 1964, Weinberg 1972, ch. 8, sec. 7; for signals actively sent back by space probes (transponder measurements), cf. Bertotti, Iess & Tortora 2003; for an overview, see Ohanian & Ruffini 1994, table 4.4 on p. 200; for more recent measurements using signals received from a pulsar that is part of a binary system, the gravitational field causing the time delay being that of the other pulsar, cf. Stairs 2003, sec. 4.4

[66] Will 1993, sec. 7.1 and 7.2

[67] Einstein, A (June 1916). "Näherungsweise Integration der Feldgleichungen der Gravitation". *Sitzungsberichte der Königlich Preussischen Akademie der Wissenschaften Berlin*. part 1: 688–696.

[68] Einstein, A (1918). "Über Gravitationswellen". *Sitzungsberichte der Königlich Preussischen Akademie der Wissenschaften Berlin*. part 1: 154–167.

[69] Castelvecchi, Davide; Witze, Witze (February 11, 2016). "Einstein's gravitational waves found at last". *Nature News*. doi:10.1038/nature.2016.19361. Retrieved 2016-02-11.

[70] B. P. Abbott et al. (LIGO Scientific Collaboration and Virgo Collaboration) (2016). "Observation of Gravitational Waves from a Binary Black Hole Merger". *Physical Review Letters* 116 (6). doi:10.1103/PhysRevLett.116.061102.

[71] "Gravitational waves detected 100 years after Einstein's prediction | NSF - National Science Foundation". *www.nsf.gov*. Retrieved 2016-02-11.

[72] Most advanced textbooks on general relativity contain a description of these properties, e.g. Schutz 1985, ch. 9

[73] For example Jaranowski & Królak 2005

[74] Rindler 2001, ch. 13

[75] Gowdy 1971, Gowdy 1974

[76] See Lehner 2002 for a brief introduction to the methods of numerical relativity, and Seidel 1998 for the connection with gravitational wave astronomy

[77] Schutz 2003, pp. 48–49, Pais 1982, pp. 253–254

[78] Rindler 2001, sec. 11.9

[79] Will 1993, pp. 177–181

[80] In consequence, in the parameterized post-Newtonian formalism (PPN), measurements of this effect determine a linear combination of the terms β and γ, cf. Will 2006, sec. 3.5 and Will 1993, sec. 7.3

[81] The most precise measurements are VLBI measurements of planetary positions; see Will 1993, ch. 5, Will 2006, sec. 3.5, Anderson et al. 1992; for an overview, Ohanian & Ruffini 1994, pp. 406–407

[82] Kramer et al. 2006

[83] Dediu, Adrian-Horia; Magdalena, Luis; Martín-Vide, Carlos (2015). *Theory and Practice of Natural Computing: Fourth International Conference, TPNC 2015, Mieres, Spain, December 15-16, 2015. Proceedings* (illustrated ed.). Springer. p. 141. ISBN 978-3-319-26841-5. Extract of page 141

[84] A figure that includes error bars is fig. 7 in Will 2006, sec. 5.1

[85] Stairs 2003, Schutz 2003, pp. 317–321, Bartusiak 2000, pp. 70–86

[86] Weisberg & Taylor 2003; for the pulsar discovery, see Hulse & Taylor 1975; for the initial evidence for gravitational radiation, see Taylor 1994

[87] Kramer 2004

[88] Penrose 2004, §14.5, Misner, Thorne & Wheeler 1973, §11.4

[89] Weinberg 1972, sec. 9.6, Ohanian & Ruffini 1994, sec. 7.8

[90] Bertotti, Ciufolini & Bender 1987, Nordtvedt 2003

[91] Kahn 2007

[92] A mission description can be found in Everitt et al. 2001; a first post-flight evaluation is given in Everitt, Parkinson & Kahn 2007; further updates will be available on the mission website Kahn 1996–2012.

[93] Townsend 1997, sec. 4.2.1, Ohanian & Ruffini 1994, pp. 469–471

[94] Ohanian & Ruffini 1994, sec. 4.7, Weinberg 1972, sec. 9.7; for a more recent review, see Schäfer 2004

[95] Ciufolini & Pavlis 2004, Ciufolini, Pavlis & Peron 2006, Iorio 2009

[96] Iorio L. (August 2006), "COMMENTS, REPLIES AND NOTES: A note on the evidence of the gravitomagnetic field of Mars", *Classical Quantum Gravity* 23 (17): 5451–5454, arXiv:gr-qc/0606092, Bibcode:2006CQGra..23.5451I, doi:10.1088/0264-9381/23/17/N01

[97] Iorio L. (June 2010), "On the Lense–Thirring test with the Mars Global Surveyor in the gravitational field of Mars", *Central European Journal of Physics* 8 (3): 509–513, arXiv:gr-qc/0701146, Bibcode:2010CEJPh...8..509I, doi:10.2478/s11534-009-0117-6

[98] For overviews of gravitational lensing and its applications, see Ehlers, Falco & Schneider 1992 and Wambsganss 1998

[99] For a simple derivation, see Schutz 2003, ch. 23; cf. Narayan & Bartelmann 1997, sec. 3

[100] Walsh, Carswell & Weymann 1979

[101] Images of all the known lenses can be found on the pages of the CASTLES project, Kochanek et al. 2007

[102] Roulet & Mollerach 1997

[103] Narayan & Bartelmann 1997, sec. 3.7

[104] Barish 2005, Bartusiak 2000, Blair & McNamara 1997

[105] Hough & Rowan 2000

[106] Hobbs, George; Archibald, A.; Arzoumanian, Z.; Backer, D.; Bailes, M.; Bhat, N. D. R.; Burgay, M.; Burke-Spolaor, S.; et al. (2010), "The international pulsar timing array project: using pulsars as a gravitational wave detector", *Classical and Quantum Gravity* 27 (8): 084013, arXiv:0911.5206, Bibcode:2010CQGra..27h4013H, doi:10.1088/0264-9381/27/8/084013

[107] Danzmann & Rüdiger 2003

[108] "LISA pathfinder overview". ESA. Retrieved 2012-04-23.

[109] Thorne 1995

[110] Cutler & Thorne 2002

[111] "Gravitational waves detected 100 years after Einstein's prediction | NSF - National Science Foundation". *www.nsf.gov.* Retrieved 2016-02-11.

[112] Miller 2002, lectures 19 and 21

[113] Celotti, Miller & Sciama 1999, sec. 3

[114] Springel et al. 2005 and the accompanying summary Gnedin 2005

[115] Blandford 1987, sec. 8.2.4

[116] For the basic mechanism, see Carroll & Ostlie 1996, sec. 17.2; for more about the different types of astronomical objects associated with this, cf. Robson 1996

[117] For a review, see Begelman, Blandford & Rees 1984. To a distant observer, some of these jets even appear to move faster than light; this, however, can be explained as an optical illusion that does not violate the tenets of relativity, see Rees 1966

[118] For stellar end states, cf. Oppenheimer & Snyder 1939 or, for more recent numerical work, Font 2003, sec. 4.1; for supernovae, there are still major problems to be solved, cf. Buras et al. 2003; for simulating accretion and the formation of jets, cf. Font 2003, sec. 4.2. Also, relativistic lensing effects are thought to play a role for the signals received from X-ray pulsars, cf. Kraus 1998

[119] The evidence includes limits on compactness from the observation of accretion-driven phenomena ("Eddington luminosity"), see Celotti, Miller & Sciama 1999, observations of stellar dynamics in the center of our own Milky Way galaxy, cf. Schödel et al. 2003, and indications that at least some of the compact objects in question appear to have no solid surface, which can be deduced from the examination of X-ray bursts for which the central compact object is either a neutron star or a black hole; cf. Remillard et al. 2006 for an overview, Narayan 2006, sec. 5. Observations of the "shadow" of the Milky Way galaxy's central black hole horizon are eagerly sought for, cf. Falcke, Melia & Agol 2000

[120] Dalal et al. 2006

[121] Barack & Cutler 2004

[122] Originally Einstein 1917; cf. Pais 1982, pp. 285–288

[123] Carroll 2001, ch. 2

[124] Bergström & Goobar 2003, ch. 9–11; use of these models is justified by the fact that, at large scales of around hundred million light-years and more, our own universe indeed appears to be isotropic and homogeneous, cf. Peebles et al. 1991

[125] E.g. with WMAP data, see Spergel et al. 2003

[126] These tests involve the separate observations detailed further on, see, e.g., fig. 2 in Bridle et al. 2003

[127] Peebles 1966; for a recent account of predictions, see Coc, Vangioni-Flam et al. 2004; an accessible account can be found in Weiss 2006; compare with the observations in Olive & Skillman 2004, Bania, Rood & Balser 2002, O'Meara et al. 2001, and Charbonnel & Primas 2005

[128] Lahav & Suto 2004, Bertschinger 1998, Springel et al. 2005

[129] Alpher & Herman 1948, for a pedagogical introduction, see Bergström & Goobar 2003, ch. 11; for the initial detection, see Penzias & Wilson 1965 and, for precision measurements by satellite observatories, Mather et al. 1994 (COBE) and Bennett et al. 2003 (WMAP). Future measurements could also reveal evidence about gravitational waves in the early universe; this additional information is contained in the background radiation's polarization, cf. Kamionkowski, Kosowsky & Stebbins 1997 and Seljak & Zaldarriaga 1997

[130] Evidence for this comes from the determination of cosmological parameters and additional observations involving the dynamics of galaxies and galaxy clusters cf. Peebles 1993, ch. 18, evidence from gravitational lensing, cf. Peacock 1999, sec. 4.6, and simulations of large-scale structure formation, see Springel et al. 2005

[131] Peacock 1999, ch. 12, Peskin 2007; in particular, observations indicate that all but a negligible portion of that matter is not in the form of the usual elementary particles ("nonbaryonic matter"), cf. Peacock 1999, ch. 12

[132] Namely, some physicists have questioned whether or not the evidence for dark matter is, in fact, evidence for deviations from the Einsteinian (and the Newtonian) description of gravity cf. the overview in Mannheim 2006, sec. 9

[133] Carroll 2001; an accessible overview is given in Caldwell 2004. Here, too, scientists have argued that the evidence indicates not a new form of energy, but the need for modifications in our cosmological models, cf. Mannheim 2006, sec. 10; aforementioned modifications need not be modifications of general relativity, they could, for example, be modifications in the way we treat the inhomogeneities in the universe, cf. Buchert 2007

[134] A good introduction is Linde 1990; for a more recent review, see Linde 2005

[135] More precisely, these are the flatness problem, the horizon problem, and the monopole problem; a pedagogical introduction can be found in Narlikar 1993, sec. 6.4, see also Börner 1993, sec. 9.1

[136] Spergel et al. 2007, sec. 5,6

[137] More concretely, the potential function that is crucial to determining the dynamics of the inflaton is simply postulated, but not derived from an underlying physical theory

[138] Brandenberger 2007, sec. 2

[139] Gödel 1949

[140] Frauendiener 2004, Wald 1984, sec. 11.1, Hawking & Ellis 1973, sec. 6.8, 6.9

[141] Wald 1984, sec. 9.2–9.4 and Hawking & Ellis 1973, ch. 6

[142] Thorne 1972; for more recent numerical studies, see Berger 2002, sec. 2.1

[143] Israel 1987. A more exact mathematical description distinguishes several kinds of horizon, notably event horizons and apparent horizons cf. Hawking & Ellis 1973, pp. 312–320 or Wald 1984, sec. 12.2; there are also more intuitive definitions for isolated systems that do not require knowledge of spacetime properties at infinity, cf. Ashtekar & Krishnan 2004

[144] For first steps, cf. Israel 1971; see Hawking & Ellis 1973, sec. 9.3 or Heusler 1996, ch. 9 and 10 for a derivation, and Heusler 1998 as well as Beig & Chruściel 2006 as overviews of more recent results

[145] The laws of black hole mechanics were first described in Bardeen, Carter & Hawking 1973; a more pedagogical presentation can be found in Carter 1979; for a more recent review, see Wald 2001, ch. 2. A thorough, book-length introduction including an introduction to the necessary mathematics Poisson 2004. For the Penrose process, see Penrose 1969

[146] Bekenstein 1973, Bekenstein 1974

[147] The fact that black holes radiate, quantum mechanically, was first derived in Hawking 1975; a more thorough derivation can be found in Wald 1975. A review is given in Wald 2001, ch. 3

[148] Narlikar 1993, sec. 4.4.4, 4.4.5

[149] Horizons: cf. Rindler 2001, sec. 12.4. Unruh effect: Unruh 1976, cf. Wald 2001, ch. 3

[150] Hawking & Ellis 1973, sec. 8.1, Wald 1984, sec. 9.1

[151] Townsend 1997, ch. 2; a more extensive treatment of this solution can be found in Chandrasekhar 1983, ch. 3

[152] Townsend 1997, ch. 4; for a more extensive treatment, cf. Chandrasekhar 1983, ch. 6

[153] Ellis & Van Elst 1999; a closer look at the singularity itself is taken in Börner 1993, sec. 1.2

[154] Here one should remind to the well-known fact that the important "quasi-optical" singularities of the so-called eikonal approximations of many wave-equations, namely the "caustics", are resolved into finite peaks beyond that approximation.

[155] Namely when there are trapped null surfaces, cf. Penrose 1965

[156] Hawking 1966

[157] The conjecture was made in Belinskii, Khalatnikov & Lifschitz 1971; for a more recent review, see Berger 2002. An accessible exposition is given by Garfinkle 2007

[158] The restriction to future singularities naturally excludes initial singularities such as the big bang singularity, which in principle be visible to observers at later cosmic time. The cosmic censorship conjecture was first presented in Penrose 1969; a textbook-level account is given in Wald 1984, pp. 302–305. For numerical results, see the review Berger 2002, sec. 2.1

[159] Hawking & Ellis 1973, sec. 7.1

[160] Arnowitt, Deser & Misner 1962; for a pedagogical introduction, see Misner, Thorne & Wheeler 1973, §21.4–§21.7

[161] Fourès-Bruhat 1952 and Bruhat 1962; for a pedagogical introduction, see Wald 1984, ch. 10; an online review can be found in Reula 1998

[162] Gourgoulhon 2007; for a review of the basics of numerical relativity, including the problems arising from the peculiarities of Einstein's equations, see Lehner 2001

[163] Misner, Thorne & Wheeler 1973, §20.4

[164] Arnowitt, Deser & Misner 1962

[165] Komar 1959; for a pedagogical introduction, see Wald 1984, sec. 11.2; although defined in a totally different way, it can be shown to be equivalent to the ADM mass for stationary spacetimes, cf. Ashtekar & Magnon-Ashtekar 1979

[166] For a pedagogical introduction, see Wald 1984, sec. 11.2

[167] Wald 1984, p. 295 and refs therein; this is important for questions of stability—if there were negative mass states, then flat, empty Minkowski space, which has mass zero, could evolve into these states

[168] Townsend 1997, ch. 5

[169] Such quasi-local mass–energy definitions are the Hawking energy, Geroch energy, or Penrose's quasi-local energy–momentum based on twistor methods; cf. the review article Szabados 2004

[170] An overview of quantum theory can be found in standard textbooks such as Messiah 1999; a more elementary account is given in Hey & Walters 2003

[171] Ramond 1990, Weinberg 1995, Peskin & Schroeder 1995; a more accessible overview is Auyang 1995

[172] Wald 1994, Birrell & Davies 1984

[173] For Hawking radiation Hawking 1975, Wald 1975; an accessible introduction to black hole evaporation can be found in Traschen 2000

[174] Wald 2001, ch. 3

[175] Put simply, matter is the source of spacetime curvature, and once matter has quantum properties, we can expect spacetime to have them as well. Cf. Carlip 2001, sec. 2

[176] Schutz 2003, p. 407

[177] Hamber 2009

[178] A timeline and overview can be found in Rovelli 2000

[179] t'Hooft 1974

[180] Donoghue 1995

[181] In particular, a perturbative technique known as renormalization, an integral part of deriving predictions which take into account higher-energy contributions, cf. Weinberg 1996, ch. 17, 18, fails in this case; cf. Veltman 1975, Goroff & Sagnotti 1985; for a recent comprehensive review of the failure of perturbative renormalizability for quantum gravity see Hamber 2009

[182] An accessible introduction at the undergraduate level can be found in Zwiebach 2004; more complete overviews can be found in Polchinski 1998a and Polchinski 1998b

[183] At the energies reached in current experiments, these strings are indistinguishable from point-like particles, but, crucially, different modes of oscillation of one and the same type of fundamental string appear as particles with different (electric and other) charges, e.g. Ibanez 2000. The theory is successful in that one mode will always correspond to a graviton, the messenger particle of gravity, e.g. Green, Schwarz & Witten 1987, sec. 2.3, 5.3

[184] Green, Schwarz & Witten 1987, sec. 4.2

[185] Weinberg 2000, ch. 31

[186] Townsend 1996, Duff 1996

[187] Kuchař 1973, sec. 3

[188] These variables represent geometric gravity using mathematical analogues of electric and magnetic fields; cf. Ashtekar 1986, Ashtekar 1987

[189] For a review, see Thiemann 2006; more extensive accounts can be found in Rovelli 1998, Ashtekar & Lewandowski 2004 as well as in the lecture notes Thiemann 2003

[190] Isham 1994, Sorkin 1997

[191] Loll 1998

[192] Sorkin 2005

[193] Penrose 2004, ch. 33 and refs therein

[194] Hawking 1987

[195] Ashtekar 2007, Schwarz 2007

[196] Maddox 1998, pp. 52–59, 98–122; Penrose 2004, sec. 34.1, ch. 30

[197] section Quantum gravity, above

[198] section Cosmology, above

[199] Friedrich 2005

[200] A review of the various problems and the techniques being developed to overcome them, see Lehner 2002

[201] See Bartusiak 2000 for an account up to that year; up-to-date news can be found on the websites of major detector collaborations such as GEO 600 and LIGO

[202] For the most recent papers on gravitational wave polarizations of inspiralling compact binaries, see Blanchet et al. 2008, and Arun et al. 2007; for a review of work on compact binaries, see Blanchet 2006 and Futamase & Itoh 2006; for a general review of experimental tests of general relativity, see Will 2006

[203] See, e.g., the electronic review journal Living Reviews in Relativity

33.11   References

- Alpher, R. A.; Herman, R. C. (1948), "Evolution of the universe", *Nature* **162** (4124): 774–775, Bibcode:1948Natur.162..774A, doi:10.1038/162774b0

- Anderson, J. D.; Campbell, J. K.; Jurgens, R. F.; Lau, E. L. (1992), "Recent developments in solar-system tests of general relativity", in Sato, H.; Nakamura, T., *Proceedings of the Sixth Marcel Großmann Meeting on General Relativity*, World Scientific, pp. 353–355, ISBN 981-02-0950-9

- Arnold, V. I. (1989), *Mathematical Methods of Classical Mechanics*, Springer, ISBN 3-540-96890-3

- Arnowitt, Richard; Deser, Stanley; Misner, Charles W. (1962), "The dynamics of general relativity", in Witten, Louis, *Gravitation: An Introduction to Current Research*, Wiley, pp. 227–265

- Arun, K.G.; Blanchet, L.; Iyer, B. R.; Qusailah, M. S. S. (2007), "Inspiralling compact binaries in quasi-elliptical orbits: The complete 3PN energy flux", *Physical Review D* **77** (6), arXiv:0711.0302, Bibcode:2008PhRvD..77f4035A, doi:10.1103/PhysRevD.77.064035

- Ashby, Neil (2002), "Relativity and the Global Positioning System" (PDF), *Physics Today* **55** (5): 41–47, Bibcode:2002PhT....55e..41A, doi:10.1063/1.1485583

- Ashby, Neil (2003), "Relativity in the Global Positioning System", *Living Reviews in Relativity* **6**, doi:10.12942/lrr-2003-1, retrieved 2007-07-06

- Ashtekar, Abhay (1986), "New variables for classical and quantum gravity", *Phys. Rev. Lett.* **57** (18): 2244–2247, Bibcode:1986PhRvL..57.2244A, doi:10.1103/PhysRevLett.57.2244, PMID 10033673

- Ashtekar, Abhay (1987), "New Hamiltonian formulation of general relativity", *Phys. Rev.* **D36** (6): 1587–1602, Bibcode:1987PhRvD..36.1587A, doi:10.1103/PhysRevD.36.1587

- Ashtekar, Abhay (2007), "LOOP QUANTUM GRAVITY: FOUR RECENT ADVANCES AND A DOZEN FREQUENTLY ASKED QUESTIONS", *The Eleventh Marcel Grossmann Meeting - on Recent Developments in Theoretical and Experimental General Relativity, Gravitation and Relativistic Field Theories - Proceedings of the MG11 Meeting on General Relativity*, p. 126, arXiv:0705.2222, Bibcode:2008mgm..conf..126A, doi:10.1142/9789812834300_0008, ISBN 9789812834263

- Ashtekar, Abhay; Krishnan, Badri (2004), "Isolated and Dynamical Horizons and Their Applications", *Living Reviews in Relativity* 7, arXiv:gr-qc/0407042, Bibcode:2004LRR.....7...10A, doi:10.12942/lrr-2004-10, retrieved 2007-08-28

- Ashtekar, Abhay; Lewandowski, Jerzy (2004), "Background Independent Quantum Gravity: A Status Report", *Class. Quant. Grav.* 21 (15): R53–R152, arXiv:gr-qc/0404018, Bibcode:2004CQGra..21R..53A, doi:10.1088/0264-9381/21/15/R01

- Ashtekar, Abhay; Magnon-Ashtekar, Anne (1979), "On conserved quantities in general relativity", *Journal of Mathematical Physics* 20 (5): 793–800, Bibcode:1979JMP....20..793A, doi:10.1063/1.524151

- Auyang, Sunny Y. (1995), *How is Quantum Field Theory Possible?*, Oxford University Press, ISBN 0-19-509345-3

- Bania, T. M.; Rood, R. T.; Balser, D. S. (2002), "The cosmological density of baryons from observations of 3He+ in the Milky Way", *Nature* 415 (6867): 54–57, Bibcode:2002Natur.415...54B, doi:10.1038/415054a, PMID 11780112

- Barack, Leor; Cutler, Curt (2004), "LISA Capture Sources: Approximate Waveforms, Signal-to-Noise Ratios, and Parameter Estimation Accuracy", *Phys. Rev.* D69 (8): 082005, arXiv:gr-qc/0310125, Bibcode:2004PhRvD..69h2005B, doi:10.1103/PhysRevD.69.082005

- Bardeen, J. M.; Carter, B.; Hawking, S. W. (1973), "The Four Laws of Black Hole Mechanics", *Comm. Math. Phys.* 31 (2): 161–170, Bibcode:1973CMaPh..31..161B, doi:10.1007/BF01645742

- Barish, Barry (2005), "Towards detection of gravitational waves", in Florides, P.; Nolan, B.; Ottewil, A., *General Relativity and Gravitation. Proceedings of the 17th International Conference*, World Scientific, pp. 24–34, ISBN 981-256-424-1

- Barstow, M; Bond, Howard E.; Holberg, J. B.; Burleigh, M. R.; Hubeny, I.; Koester, D. (2005), "Hubble Space Telescope Spectroscopy of the Balmer lines in Sirius B", *Mon. Not. Roy. Astron. Soc.* 362 (4): 1134–1142, arXiv:astro-ph/0506600, Bibcode:2005MNRAS.362.1134B, doi:10.1111/j.1365-2966.2005.09359.x

- Bartusiak, Marcia (2000), *Einstein's Unfinished Symphony: Listening to the Sounds of Space-Time*, Berkley, ISBN 978-0-425-18620-6

- Begelman, Mitchell C.; Blandford, Roger D.; Rees, Martin J. (1984), "Theory of extragalactic radio sources", *Rev. Mod. Phys.* 56 (2): 255–351, Bibcode:1984RvMP...56..255B, doi:10.1103/RevModPhys.56.255

- Beig, Robert; Chruściel, Piotr T. (2006), "Stationary black holes", in Françoise, J.-P.; Naber, G.; Tsou, T.S., *Encyclopedia of Mathematical Physics, Volume 2*, Elsevier, p. 2041, arXiv:gr-qc/0502041, Bibcode:2005gr.qc.....2041B, ISBN 0-12-512660-3

- Bekenstein, Jacob D. (1973), "Black Holes and Entropy", *Phys. Rev.* D7 (8): 2333–2346, Bibcode:1973PhRvD...7.2333B, doi:10.1103/PhysRevD.7.2333

- Bekenstein, Jacob D. (1974), "Generalized Second Law of Thermodynamics in Black-Hole Physics", *Phys. Rev.* D9 (12): 3292–3300, Bibcode:1974PhRvD...9.3292B, doi:10.1103/PhysRevD.9.3292

- Belinskii, V. A.; Khalatnikov, I. M.; Lifschitz, E. M. (1971), "Oscillatory approach to the singular point in relativistic cosmology", *Advances in Physics* 19 (80): 525–573, Bibcode:1970AdPhy..19..525B, doi:10.1080/00018737000101171; original paper in Russian: Belinsky, V. A.; Lifshits, I. M.; Khalatnikov, E. M. (1970), "Колебательный Режим Приближения К Особой Точке В Релятивистской Космологии", *Uspekhi Fizicheskikh Nauk (Успехи Физических Наук)*, 102(3) (11): 463–500, Bibcode:1970UsFiN.102..463B

- Bennett, C. L.; Halpern, M.; Hinshaw, G.; Jarosik, N.; Kogut, A.; Limon, M.; Meyer, S. S.; Page, L.; et al. (2003), "First Year Wilkinson Microwave Anisotropy Probe (WMAP) Observations: Preliminary Maps and Basic Results", *Astrophys. J. Suppl.* 148 (1): 1–27, arXiv:astro-ph/0302207, Bibcode:2003ApJS..148....1B, doi:10.1086/377253

- Berger, Beverly K. (2002), "Numerical Approaches to Spacetime Singularities", *Living Reviews in Relativity* 5, arXiv:gr-qc/0201056, Bibcode:2002LRR.....5....1B, doi:10.12942/lrr-2002-1, retrieved 2007-08-04

- Bergström, Lars; Goobar, Ariel (2003), *Cosmology and Particle Astrophysics* (2nd ed.), Wiley & Sons, ISBN 3-540-43128-4

- Bertotti, Bruno; Ciufolini, Ignazio; Bender, Peter L. (1987), "New test of general relativity: Measurement of de Sitter geodetic precession rate for lunar perigee", *Physical Review Letters* **58** (11): 1062–1065, Bibcode:1987PhRvL..58.1062B, doi:10.1103/PhysRevLett.58.1062, PMID 10034329

- Bertotti, Bruno; Iess, L.; Tortora, P. (2003), "A test of general relativity using radio links with the Cassini spacecraft", *Nature* **425** (6956): 374–376, Bibcode:2003Natur.425..374B, doi:10.1038/nature01997, PMID 14508481

- Bertschinger, Edmund (1998), "Simulations of structure formation in the universe", *Annu. Rev. Astron. Astrophys.* **36** (1): 599–654, Bibcode:1998ARA&A..36..599B, doi:10.1146/annurev.astro.36.1.599

- Birrell, N. D.; Davies, P. C. (1984), *Quantum Fields in Curved Space*, Cambridge University Press, ISBN 0-521-27858-9

- Blair, David; McNamara, Geoff (1997), *Ripples on a Cosmic Sea. The Search for Gravitational Waves*, Perseus, ISBN 0-7382-0137-5

- Blanchet, L.; Faye, G.; Iyer, B. R.; Sinha, S. (2008), "The third post-Newtonian gravitational wave polarisations and associated spherical harmonic modes for inspiralling compact binaries in quasi-circular orbits", *Classical and Quantum Gravity* **25** (16): 165003, arXiv:0802.1249, Bibcode:2008CQGra..25p5003B, doi:10.1088/0264-9381/25/16/165003

- Blanchet, Luc (2006), "Gravitational Radiation from Post-Newtonian Sources and Inspiralling Compact Binaries", *Living Reviews in Relativity* **9**, Bibcode:2006LRR.....9....4B, doi:10.12942/lrr-2006-4, retrieved 2007-08-07

- Blandford, R. D. (1987), "Astrophysical Black Holes", in Hawking, Stephen W.; Israel, Werner, *300 Years of Gravitation*, Cambridge University Press, pp. 277–329, ISBN 0-521-37976-8

- Börner, Gerhard (1993), *The Early Universe. Facts and Fiction*, Springer, ISBN 0-387-56729-1

- Brandenberger, Robert H. (2007), "Conceptual Problems of Inflationary Cosmology and a New Approach to Cosmological Structure Formation", *Inflationary Cosmology*, Lecture Notes in Physics **738**, p. 393, arXiv:hep-th/0701111, Bibcode:2008LNP...738..393B, doi:10.1007/978-3-540-74353-8_11, ISBN 978-3-540-74352-1

- Brans, C. H.; Dicke, R. H. (1961), "Mach's Principle and a Relativistic Theory of Gravitation", *Physical Review* **124** (3): 925–935, Bibcode:1961PhRv..124..925B, doi:10.1103/PhysRev.124.925

- Bridle, Sarah L.; Lahav, Ofer; Ostriker, Jeremiah P.; Steinhardt, Paul J. (2003), "Precision Cosmology? Not Just Yet", *Science* **299** (5612): 1532–1533, arXiv:astro-ph/0303180, Bibcode:2003Sci...299.1532B, doi:10.1126/science.1082158, PMID 12624255

- Bruhat, Yvonne (1962), "The Cauchy Problem", in Witten, Louis, *Gravitation: An Introduction to Current Research*, Wiley, p. 130, ISBN 978-1-114-29166-9

- Buchert, Thomas (2007), "Dark Energy from Structure—A Status Report", *General Relativity and Gravitation* **40** (2–3): 467–527, arXiv:0707.2153, Bibcode:2008GReGr..40..467B, doi:10.1007/s10714-007-0554-8

- Buras, R.; Rampp, M.; Janka, H.-Th.; Kifonidis, K. (2003), "Improved Models of Stellar Core Collapse and Still no Explosions: What is Missing?", *Phys. Rev. Lett.* **90** (24): 241101, arXiv:astro-ph/0303171, Bibcode:2003PhRvL..90x1101B, doi:10.1103/PhysRevLett.90.241101, PMID 12857181

- Caldwell, Robert R. (2004), "Dark Energy", *Physics World* **17** (5): 37–42

- Carlip, Steven (2001), "Quantum Gravity: a Progress Report", *Rept. Prog. Phys.* **64** (8): 885–942, arXiv:gr-qc/0108040, Bibcode:2001RPPh...64..885C, doi:10.1088/0034-4885/64/8/301

- Carroll, Bradley W.; Ostlie, Dale A. (1996), *An Introduction to Modern Astrophysics*, Addison-Wesley, ISBN 0-201-54730-9

- Carroll, Sean M. (2001), "The Cosmological Constant", *Living Reviews in Relativity* **4**, arXiv:astro-ph/0004075, Bibcode:2001LRR.....4....1C, doi:10.12942/lrr-2001-1, retrieved 2007-07-21

- Carter, Brandon (1979), "The general theory of the mechanical, electromagnetic and thermodynamic properties of black holes", in Hawking, S. W.; Israel, W., *General Relativity, an Einstein Centenary Survey*, Cambridge University Press, pp. 294–369 and 860–863, ISBN 0-521-29928-4

- Celotti, Annalisa; Miller, John C.; Sciama, Dennis W. (1999), "Astrophysical evidence for the existence of black holes", *Class. Quant. Grav.*

16 (12A): A3–A21, arXiv:astro-ph/9912186, doi:10.1088/0264-9381/16/12A/301

- Chandrasekhar, Subrahmanyan (1983), *The Mathematical Theory of Black Holes*, Oxford University Press, ISBN 0-19-850370-9

- Charbonnel, C.; Primas, F. (2005), "The Lithium Content of the Galactic Halo Stars", *Astronomy & Astrophysics* **442** (3): 961–992, arXiv:astro-ph/0505247, Bibcode:2005A&A...442..961C, doi:10.1051/0004-6361:20042491

- Ciufolini, Ignazio; Pavlis, Erricos C. (2004), "A confirmation of the general relativistic prediction of the Lense-Thirring effect", *Nature* **431** (7011): 958–960, Bibcode:2004Natur.431..958C, doi:10.1038/nature03007, PMID 15496915

- Ciufolini, Ignazio; Pavlis, Erricos C.; Peron, R. (2006), "Determination of frame-dragging using Earth gravity models from CHAMP and GRACE", *New Astron.* **11** (8): 527–550, Bibcode:2006NewA...11..527C, doi:10.1016/j.newast.2006.02.001

- Coc, A.; Vangioni-Flam, Elisabeth; Descouvemont, Pierre; Adahchour, Abderrahim; Angulo, Carmen (2004), "Updated Big Bang Nucleosynthesis confronted to WMAP observations and to the Abundance of Light Elements", *Astrophysical Journal* **600** (2): 544–552, arXiv:astro-ph/0309480, Bibcode:2004ApJ...600..544C, doi:10.1086/380121

- Cutler, Curt; Thorne, Kip S. (2002), "An overview of gravitational wave sources", in Bishop, Nigel; Maharaj, Sunil D., *Proceedings of 16th International Conference on General Relativity and Gravitation (GR16)*, World Scientific, p. 4090, arXiv:gr-qc/0204090, Bibcode:2002gr.qc.....4090C, ISBN 981-238-171-6

- Dalal, Neal; Holz, Daniel E.; Hughes, Scott A.; Jain, Bhuvnesh (2006), "Short GRB and binary black hole standard sirens as a probe of dark energy", *Phys.Rev.* **D74** (6): 063006, arXiv:astro-ph/0601275, Bibcode:2006PhRvD..74f3006D, doi:10.1103/PhysRevD.74.063006

- Danzmann, Karsten; Rüdiger, Albrecht (2003), "LISA Technology—Concepts, Status, Prospects" (PDF), *Class. Quant. Grav.* **20** (10): S1–S9, Bibcode:2003CQGra..20S...1D, doi:10.1088/0264-9381/20/10/301

- Dirac, Paul (1996), *General Theory of Relativity*, Princeton University Press, ISBN 0-691-01146-X

- Donoghue, John F. (1995), "Introduction to the Effective Field Theory Description of Gravity", in Cornet, Fernando, *Effective Theories: Proceedings of the Advanced School, Almunecar, Spain, 26 June–1 July 1995*, Singapore: World Scientific, p. 12024, arXiv:gr-qc/9512024, Bibcode:1995gr.qc....12024D, ISBN 981-02-2908-9

- Duff, Michael (1996), "M-Theory (the Theory Formerly Known as Strings)", *Int. J. Mod. Phys.* **A11** (32): 5623–5641, arXiv:hep-th/9608117, Bibcode:1996IJMPA..11.5623D, doi:10.1142/S0217751X96002583

- Ehlers, Jürgen (1973), "Survey of general relativity theory", in Israel, Werner, *Relativity, Astrophysics and Cosmology*, D. Reidel, pp. 1–125, ISBN 90-277-0369-8

- Ehlers, Jürgen; Falco, Emilio E.; Schneider, Peter (1992), *Gravitational lenses*, Springer, ISBN 3-540-66506-4

- Ehlers, Jürgen; Lämmerzahl, Claus, eds. (2006), *Special Relativity—Will it Survive the Next 101 Years?*, Springer, ISBN 3-540-34522-1

- Ehlers, Jürgen; Rindler, Wolfgang (1997), "Local and Global Light Bending in Einstein's and other Gravitational Theories", *General Relativity and Gravitation* **29** (4): 519–529, Bibcode:1997GReGr..29..519E, doi:10.1023/A:1018843001842

- Einstein, Albert (1907), "Über das Relativitätsprinzip und die aus demselben gezogene Folgerungen" (PDF), *Jahrbuch der Radioaktivität und Elektronik* **4**: 411, retrieved 2008-05-05

- Einstein, Albert (1915), "Die Feldgleichungen der Gravitation", *Sitzungsberichte der Preussischen Akademie der Wissenschaften zu Berlin*: 844–847, retrieved 2006-09-12

- Einstein, Albert (1916), "Die Grundlage der allgemeinen Relativitätstheorie", *Annalen der Physik* **49**: 769–822, Bibcode:1916AnP...354..769E, doi:10.1002/andp.19163540702, archived from the original (PDF) on 2006-08-29, retrieved 2016-02-14

- Einstein, Albert (1917), "Kosmologische Betrachtungen zur allgemeinen Relativitätstheorie", *Sitzungsberichte der Preußischen Akademie der Wissenschaften*: 142

- Ellis, George F R; Van Elst, Henk (1999), Lachièze-Rey, Marc, ed., "Theoretical and Observational Cosmology: Cosmological models (Cargèse lectures

1998)", *Theoretical and observational cosmology : proceedings of the NATO Advanced Study Institute on Theoretical and Observational Cosmology* (Kluwer): 1–116, arXiv:gr-qc/9812046, Bibcode:1999toc..conf....1E, doi:10.1007/978-94-011-4455-1_1, ISBN 978-0-7923-5946-3

- Everitt, C. W. F.; Buchman, S.; DeBra, D. B.; Keiser, G. M. (2001), "Gravity Probe B: Countdown to launch", in Lämmerzahl, C.; Everitt, C. W. F.; Hehl, F. W., *Gyros, Clocks, and Interferometers: Testing Relativistic Gravity in Space (Lecture Notes in Physics 562)*, Springer, pp. 52–82, ISBN 3-540-41236-0

- Everitt, C. W. F.; Parkinson, Bradford; Kahn, Bob (2007), *The Gravity Probe B experiment. Post Flight Analysis—Final Report (Preface and Executive Summary)* (PDF), Project Report: NASA, Stanford University and Lockheed Martin, retrieved 2007-08-05

- Falcke, Heino; Melia, Fulvio; Agol, Eric (2000), "Viewing the Shadow of the Black Hole at the Galactic Center", *Astrophysical Journal* **528** (1): L13–L16, arXiv:astro-ph/9912263, Bibcode:2000ApJ...528L..13F, doi:10.1086/312423, PMID 10587484

- Flanagan, Éanna É.; Hughes, Scott A. (2005), "The basics of gravitational wave theory", *New J.Phys.* **7**: 204, arXiv:gr-qc/0501041, Bibcode:2005NJPh....7..204F, doi:10.1088/1367-2630/7/1/204

- Font, José A. (2003), "Numerical Hydrodynamics in General Relativity", *Living Reviews in Relativity* **6**, doi:10.12942/lrr-2003-4, retrieved 2007-08-19

- Fourès-Bruhat, Yvonne (1952), "Théoréme d'existence pour certains systémes d'équations aux derivées partielles non linéaires", *Acta Mathematica* **88** (1): 141–225, Bibcode:1952AcM....88..141F, doi:10.1007/BF02392131

- Frauendiener, Jörg (2004), "Conformal Infinity", *Living Reviews in Relativity* **7**, Bibcode:2004LRR.....7....1F, doi:10.12942/lrr-2004-1, retrieved 2007-07-21

- Friedrich, Helmut (2005), "Is general relativity 'essentially understood'?", *Annalen der Physik* **15** (1–2): 84–108, arXiv:gr-qc/0508016, Bibcode:2006AnP...518...84F, doi:10.1002/andp.200510173

- Futamase, T.; Itoh, Y. (2006), "The Post-Newtonian Approximation for Relativistic Compact Binaries", *Living Reviews in Relativity* **10**, doi:10.12942/lrr-2007-2, retrieved 2008-02-29

- Gamow, George (1970), *My World Line*, Viking Press, ISBN 0-670-50376-2

- Garfinkle, David (2007), "Of singularities and breadmaking", *Einstein Online*, retrieved 2007-08-03

- Geroch, Robert (1996). "Partial Differential Equations of Physics". arXiv:gr-qc/9602055 [gr-qc].

- Giulini, Domenico (2005), *Special Relativity: A First Encounter*, Oxford University Press, ISBN 0-19-856746-4

- Giulini, Domenico (2006a), "Algebraic and Geometric Structures in Special Relativity", in Ehlers, Jürgen; Lämmerzahl, Claus, *Special Relativity—Will it Survive the Next 101 Years?*, Springer, pp. 45–111, arXiv:math-ph/0602018, Bibcode:2006math.ph...2018G, ISBN 3-540-34522-1

- Giulini, Domenico (2006b), Stamatescu, I. O., ed., "An assessment of current paradigms in the physics of fundamental interactions: Some remarks on the notions of general covariance and background independence", *Approaches to Fundamental Physics*, Lecture Notes in Physics (Springer) **721**: 105, arXiv:gr-qc/0603087, Bibcode:2007LNP...721..105G, doi:10.1007/978-3-540-71117-9_6, ISBN 978-3-540-71115-5

- Gnedin, Nickolay Y. (2005), "Digitizing the Universe", *Nature* **435** (7042): 572–573, Bibcode:2005Natur.435..572G, doi:10.1038/435572a, PMID 15931201

- Goenner, Hubert F. M. (2004), "On the History of Unified Field Theories", *Living Reviews in Relativity* **7**, Bibcode:2004LRR.....7....2G, doi:10.12942/lrr-2004-2, retrieved 2008-02-28

- Goroff, Marc H.; Sagnotti, Augusto (1985), "Quantum gravity at two loops", *Phys. Lett.* **160B** (1–3): 81–86, Bibcode:1985PhLB..160...81G, doi:10.1016/0370-2693(85)91470-4

- Gourgoulhon, Eric (2007). "3+1 Formalism and Bases of Numerical Relativity". arXiv:gr-qc/0703035 [gr-qc].

- Gowdy, Robert H. (1971), "Gravitational Waves in Closed Universes", *Phys. Rev. Lett.* **27** (12): 826–829, Bibcode:1971PhRvL..27..826G, doi:10.1103/PhysRevLett.27.826

- Gowdy, Robert H. (1974), "Vacuum spacetimes with two-parameter spacelike isometry groups and compact invariant hypersurfaces: Topologies and boundary conditions", *Annals of Physics* **83** (1): 203–241, Bibcode:1974AnPhy..83..203G, doi:10.1016/0003-4916(74)90384-4

- Green, M. B.; Schwarz, J. H.; Witten, E. (1987), *Superstring theory. Volume 1: Introduction*, Cambridge University Press, ISBN 0-521-35752-7

- Greenstein, J. L.; Oke, J. B.; Shipman, H. L. (1971), "Effective Temperature, Radius, and Gravitational Redshift of Sirius B", *Astrophysical Journal* **169**: 563, Bibcode:1971ApJ...169..563G, doi:10.1086/151174

- Hamber, Herbert W. (2009), *Quantum Gravitation - The Feynman Path Integral Approach*, Springer Publishing, doi:10.1007/978-3-540-85293-3, ISBN 978-3-540-85292-6

- Gödel, Kurt (1949). "An Example of a New Type of Cosmological Solution of Einstein's Field Equations of Gravitation". *Rev. Mod. Phys.* **21** (3): 447. Bibcode:1949RvMP...21..447G. doi:10.1103/RevModPhys.21.447.

- Hafele, J. C.; Keating, R. E. (July 14, 1972). "Around-the-World Atomic Clocks: Predicted Relativistic Time Gains". *Science* **177** (4044): 166–168. Bibcode:1972Sci...177..166H. doi:10.1126/science.177.4044.166. PMID 17779917.

- Hafele, J. C.; Keating, R. E. (July 14, 1972). "Around-the-World Atomic Clocks: Observed Relativistic Time Gains". *Science* **177** (4044): 168–170. Bibcode:1972Sci...177..168H. doi:10.1126/science.177.4044.168. PMID 17779918.

- Havas, P. (1964), "Four-Dimensional Formulation of Newtonian Mechanics and Their Relation to the Special and the General Theory of Relativity", *Rev. Mod. Phys.* **36** (4): 938–965, Bibcode:1964RvMP...36..938H, doi:10.1103/RevModPhys.36.938

- Hawking, Stephen W. (1966), "The occurrence of singularities in cosmology", *Proceedings of the Royal Society* **A294** (1439): 511–521, Bibcode:1966RSPSA.294..511H, doi:10.1098/rspa.1966.0221

- Hawking, S. W. (1975), "Particle Creation by Black Holes", *Communications in Mathematical Physics* **43** (3): 199–220, Bibcode:1975CMaPh..43..199H, doi:10.1007/BF02345020

- Hawking, Stephen W. (1987), "Quantum cosmology", in Hawking, Stephen W.; Israel, Werner, *300 Years of Gravitation*, Cambridge University Press, pp. 631–651, ISBN 0-521-37976-8

- Hawking, Stephen W.; Ellis, George F. R. (1973), *The large scale structure of space-time*, Cambridge University Press, ISBN 0-521-09906-4

- Heckmann, O. H. L.; Schücking, E. (1959), "Newtonsche und Einsteinsche Kosmologie", in Flügge, S., *Encyclopedia of Physics* **53**, p. 489

- Heusler, Markus (1998), "Stationary Black Holes: Uniqueness and Beyond", *Living Reviews in Relativity* **1**, doi:10.12942/lrr-1998-6, retrieved 2007-08-04

- Heusler, Markus (1996), *Black Hole Uniqueness Theorems*, Cambridge University Press, ISBN 0-521-56735-1

- Hey, Tony; Walters, Patrick (2003), *The new quantum universe*, Cambridge University Press, ISBN 0-521-56457-3

- Hough, Jim; Rowan, Sheila (2000), "Gravitational Wave Detection by Interferometry (Ground and Space)", *Living Reviews in Relativity* **3**, retrieved 2007-07-21

- Hubble, Edwin (1929), "A Relation between Distance and Radial Velocity among Extra-Galactic Nebulae" (PDF), *Proc. Nat. Acad. Sci.* **15** (3): 168–173, Bibcode:1929PNAS...15..168H, doi:10.1073/pnas.15.3.168, PMC 522427, PMID 16577160

- Hulse, Russell A.; Taylor, Joseph H. (1975), "Discovery of a pulsar in a binary system", *Astrophys. J.* **195**: L51–L55, Bibcode:1975ApJ...195L..51H, doi:10.1086/181708

- Ibanez, L. E. (2000), "The second string (phenomenology) revolution", *Class. Quant. Grav.* **17** (5): 1117–1128, arXiv:hep-ph/9911499, Bibcode:2000CQGra..17.1117I, doi:10.1088/0264-9381/17/5/321

- Iorio, L. (2009), "An Assessment of the Systematic Uncertainty in Present and Future Tests of the Lense-Thirring Effect with Satellite Laser Ranging", *Space Sci. Rev.* **148** (1–4): 363, arXiv:0809.1373, Bibcode:2009SSRv..148..363I, doi:10.1007/s11214-008-9478-1

- Isham, Christopher J. (1994), "Prima facie questions in quantum gravity", in Ehlers, Jürgen; Friedrich, Helmut, *Canonical Gravity: From Classical to Quantum*, Springer, ISBN 3-540-58339-4

- Israel, Werner (1971), "Event Horizons and Gravitational Collapse", *General Relativity and Gravitation* **2** (1): 53–59, Bibcode:1971GReGr...2...53I, doi:10.1007/BF02450518

- Israel, Werner (1987), "Dark stars: the evolution of an idea", in Hawking, Stephen W.; Israel, Werner, *300 Years of Gravitation*, Cambridge University Press, pp. 199–276, ISBN 0-521-37976-8

- Janssen, Michel (2005), "Of pots and holes: Einstein's bumpy road to general relativity" (PDF), *Annalen der Physik* **14** (S1): 58–85, Bibcode:2005AnP...517S..58J, doi:10.1002/andp.200410130

- Jaranowski, Piotr; Królak, Andrzej (2005), "Gravitational-Wave Data Analysis. Formalism and Sample Applications: The Gaussian Case", *Living Reviews in Relativity* **8**, doi:10.12942/lrr-2005-3, retrieved 2007-07-30

- Kahn, Bob (1996–2012), *Gravity Probe B Website*, Stanford University, retrieved 2012-04-20

- Kahn, Bob (April 14, 2007), *Was Einstein right? Scientists provide first public peek at Gravity Probe B results (Stanford University Press Release)* (PDF), Stanford University News Service

- Kamionkowski, Marc; Kosowsky, Arthur; Stebbins, Albert (1997), "Statistics of Cosmic Microwave Background Polarization", *Phys. Rev.* **D55** (12): 7368–7388, arXiv:astro-ph/9611125, Bibcode:1997PhRvD..55.7368K, doi:10.1103/PhysRevD.55.7368

- Kennefick, Daniel (2005), "Astronomers Test General Relativity: Light-bending and the Solar Redshift", in Renn, Jürgen, *One hundred authors for Einstein*, Wiley-VCH, pp. 178–181, ISBN 3-527-40574-7

- Kennefick, Daniel (2007), "Not Only Because of Theory: Dyson, Eddington and the Competing Myths of the 1919 Eclipse Expedition", *Proceedings of the 7th Conference on the History of General Relativity, Tenerife, 2005* **0709**, p. 685, arXiv:0709.0685, Bibcode:2007arXiv0709.0685K

- Kenyon, I. R. (1990), *General Relativity*, Oxford University Press, ISBN 0-19-851996-6

- Kochanek, C.S.; Falco, E.E.; Impey, C.; Lehar, J. (2007), *CASTLES Survey Website*, Harvard-Smithsonian Center for Astrophysics, retrieved 2007-08-21

- Komar, Arthur (1959), "Covariant Conservation Laws in General Relativity", *Phys. Rev.* **113** (3): 934–936, Bibcode:1959PhRv..113..934K, doi:10.1103/PhysRev.113.934

- Kramer, Michael (2004), Karshenboim, S. G.; Peik, E., eds., "Astrophysics, Clocks and Fundamental Constants: Millisecond Pulsars as Tools of Fundamental Physics", *Lecture Notes in Physics* (Springer) **648**: 33–54, arXiv:astro-ph/0405178, Bibcode:2004LNP...648...33K, doi:10.1007/978-3-540-40991-5_3, ISBN 978-3-540-21967-5

- Kramer, M.; Stairs, I. H.; Manchester, R. N.; McLaughlin, M. A.; Lyne, A. G.; Ferdman, R. D.; Burgay, M.; Lorimer, D. R.; et al. (2006), "Tests of general relativity from timing the double pulsar", *Science* **314** (5796): 97–102, arXiv:astro-ph/0609417, Bibcode:2006Sci...314...97K, doi:10.1126/science.1132305, PMID 16973838

- Kraus, Ute (1998), "Light Deflection Near Neutron Stars", *Relativistic Astrophysics*, Vieweg, pp. 66–81, ISBN 3-528-06909-0

- Kuchař, Karel (1973), "Canonical Quantization of Gravity", in Israel, Werner, *Relativity, Astrophysics and Cosmology*, D. Reidel, pp. 237–288, ISBN 90-277-0369-8

- Künzle, H. P. (1972), "Galilei and Lorentz Structures on spacetime: comparison of the corresponding geometry and physics", *Annales de l'Institut Henri Poincaré A* **17**: 337–362

- Lahav, Ofer; Suto, Yasushi (2004), "Measuring our Universe from Galaxy Redshift Surveys", *Living Reviews in Relativity* **7**, arXiv:astro-ph/0310642, Bibcode:2004LRR.....7....8L, doi:10.12942/lrr-2004-8, retrieved 2007-08-19

- Landgraf, M.; Hechler, M.; Kemble, S. (2005), "Mission design for LISA Pathfinder", *Class. Quant. Grav.* **22** (10): S487–S492, arXiv:gr-qc/0411071, Bibcode:2005CQGra..22S.487L, doi:10.1088/0264-9381/22/10/048

- Lehner, Luis (2001), "Numerical Relativity: A review", *Class. Quant. Grav.* **18** (17): R25–R86, arXiv:gr-qc/0106072, Bibcode:2001CQGra..18R..25L, doi:10.1088/0264-9381/18/17/202

- Lehner, Luis (2002), "NUMERICAL RELATIVITY: STATUS AND PROSPECTS", *General Relativity and Gravitation - Proceedings of the 16th International Conference*, p. 210, arXiv:gr-qc/0202055, Bibcode:2002grg..conf..210L, doi:10.1142/9789812776556_0010, ISBN 9789812381712

- Linde, Andrei (1990), *Particle Physics and Inflationary Cosmology*, Harwood, p. 3203, arXiv:hep-th/0503203, Bibcode:2005hep.th....3203L, ISBN 3-7186-0489-2

- Linde, Andrei (2005), "Towards inflation in string theory", *J. Phys. Conf. Ser.* **24**: 151–160, arXiv:hep-th/0503195, Bibcode:2005JPhCS..24..151L, doi:10.1088/1742-6596/24/1/018

- Loll, Renate (1998), "Discrete Approaches to Quantum Gravity in Four Dimensions", *Living Reviews in Relativity* **1**, arXiv:gr-qc/9805049, Bibcode:1998LRR.....1...13L, doi:10.12942/lrr-1998-13, retrieved 2008-03-09

- Lovelock, David (1972), "The Four-Dimensionality of Space and the Einstein Tensor", *J. Math. Phys.* **13** (6): 874–876, Bibcode:1972JMP....13..874L, doi:10.1063/1.1666069

- Ludyk, Günter (2013). *Einstein in Matrix Form* (1st ed.). Berlin: Springer. ISBN 9783642357978.

- MacCallum, M. (2006), "Finding and using exact solutions of the Einstein equations", in Mornas, L.; Alonso, J. D., *A Century of Relativity Physics (ERE05, the XXVIII Spanish Relativity Meeting)* **841**, American Institute of Physics, p. 129, arXiv:gr-qc/0601102, Bibcode:2006AIPC..841..129M, doi:10.1063/1.2218172

- Maddox, John (1998), *What Remains To Be Discovered*, Macmillan, ISBN 0-684-82292-X

- Mannheim, Philip D. (2006), "Alternatives to Dark Matter and Dark Energy", *Prog. Part. Nucl. Phys.* **56** (2): 340–445, arXiv:astro-ph/0505266, Bibcode:2006PrPNP..56..340M, doi:10.1016/j.ppnp.2005.08.001

- Mather, J. C.; Cheng, E. S.; Cottingham, D. A.; Eplee, R. E.; Fixsen, D. J.; Hewagama, T.; Isaacman, R. B.; Jensen, K. A.; et al. (1994), "Measurement of the cosmic microwave spectrum by the COBE FIRAS instrument", *Astrophysical Journal* **420**: 439–444, Bibcode:1994ApJ...420..439M, doi:10.1086/173574

- Mermin, N. David (2005), *It's About Time. Understanding Einstein's Relativity*, Princeton University Press, ISBN 0-691-12201-6

- Messiah, Albert (1999), *Quantum Mechanics*, Dover Publications, ISBN 0-486-40924-4

- Miller, Cole (2002), *Stellar Structure and Evolution (Lecture notes for Astronomy 606)*, University of Maryland, retrieved 2007-07-25

- Misner, Charles W.; Thorne, Kip. S.; Wheeler, John A. (1973), *Gravitation*, W. H. Freeman, ISBN 0-7167-0344-0

- Møller, Christian (1952), *The Theory of Relativity* (3rd ed.), Oxford University Press

- Narayan, Ramesh (2006), "Black holes in astrophysics", *New Journal of Physics* **7**: 199, arXiv:gr-qc/0506078, Bibcode:2005NJPh....7..199N, doi:10.1088/1367-2630/7/1/199

- Narayan, Ramesh; Bartelmann, Matthias (1997). "Lectures on Gravitational Lensing". arXiv:astro-ph/9606001 [astro-ph].

- Narlikar, Jayant V. (1993), *Introduction to Cosmology*, Cambridge University Press, ISBN 0-521-41250-1

- Nieto, Michael Martin (2006), "The quest to understand the Pioneer anomaly" (PDF), *EurophysicsNews* **37** (6): 30–34, Bibcode:2006ENews..37...30N, doi:10.1051/epn:2006604

- Nordström, Gunnar (1918), "On the Energy of the Gravitational Field in Einstein's Theory", *Verhandl. Koninkl. Ned. Akad. Wetenschap.* **26**: 1238–1245

- Nordtvedt, Kenneth (2003). "Lunar Laser Ranging—a comprehensive probe of post-Newtonian gravity". arXiv:gr-qc/0301024 [gr-qc].

- Norton, John D. (1985), "What was Einstein's principle of equivalence?" (PDF), *Studies in History and Philosophy of Science* **16** (3): 203–246, doi:10.1016/0039-3681(85)90002-0, retrieved 2007-06-11

- Ohanian, Hans C.; Ruffini, Remo (1994), *Gravitation and Spacetime*, W. W. Norton & Company, ISBN 0-393-96501-5

- Olive, K. A.; Skillman, E. A. (2004), "A Realistic Determination of the Error on the Primordial Helium Abundance", *Astrophysical Journal* **617** (1): 29–49, arXiv:astro-ph/0405588, Bibcode:2004ApJ...617...29O, doi:10.1086/425170

- O'Meara, John M.; Tytler, David; Kirkman, David; Suzuki, Nao; Prochaska, Jason X.; Lubin, Dan; Wolfe, Arthur M. (2001), "The Deuterium to Hydrogen Abundance Ratio Towards a Fourth QSO: HS0105+1619", *Astrophysical Journal* **552** (2): 718–730, arXiv:astro-ph/0011179, Bibcode:2001ApJ...552..718O, doi:10.1086/320579

- Oppenheimer, J. Robert; Snyder, H. (1939), "On continued gravitational contraction", *Physical Review* **56** (5): 455–459, Bibcode:1939PhRv...56..455O, doi:10.1103/PhysRev.56.455

- Overbye, Dennis (1999), *Lonely Hearts of the Cosmos: the story of the scientific quest for the secret of the Universe*, Back Bay, ISBN 0-316-64896-5

- Pais, Abraham (1982), *'Subtle is the Lord...' The Science and life of Albert Einstein*, Oxford University Press, ISBN 0-19-853907-X

- Peacock, John A. (1999), *Cosmological Physics*, Cambridge University Press, ISBN 0-521-41072-X

- Peebles, P. J. E. (1966), "Primordial Helium abundance and primordial fireball II", *Astrophysical Journal* **146**: 542–552, Bibcode:1966ApJ...146..542P, doi:10.1086/148918

- Peebles, P. J. E. (1993), *Principles of physical cosmology*, Princeton University Press, ISBN 0-691-01933-9

- Peebles, P.J.E.; Schramm, D.N.; Turner, E.L.; Kron, R.G. (1991), "The case for the relativistic hot Big Bang cosmology", *Nature* **352** (6338): 769–776, Bibcode:1991Natur.352..769P, doi:10.1038/352769a0

- Penrose, Roger (1965), "Gravitational collapse and spacetime singularities", *Physical Review Letters* **14** (3): 57–59, Bibcode:1965PhRvL..14...57P, doi:10.1103/PhysRevLett.14.57

- Penrose, Roger (1969), "Gravitational collapse: the role of general relativity", *Rivista del Nuovo Cimento* **1**: 252–276, Bibcode:1969NCimR...1..252P

- Penrose, Roger (2004), *The Road to Reality*, A. A. Knopf, ISBN 0-679-45443-8

- Penzias, A. A.; Wilson, R. W. (1965), "A measurement of excess antenna temperature at 4080 Mc/s", *Astrophysical Journal* **142**: 419–421, Bibcode:1965ApJ...142..419P, doi:10.1086/148307

- Peskin, Michael E.; Schroeder, Daniel V. (1995), *An Introduction to Quantum Field Theory*, Addison-Wesley, ISBN 0-201-50397-2

- Peskin, Michael E. (2007), "Dark Matter and Particle Physics", *Journal of the Physical Society of Japan* **76** (11): 111017, arXiv:0707.1536, Bibcode:2007JPSJ...76k1017P, doi:10.1143/JPSJ.76.111017

- Poisson, Eric (2004), "The Motion of Point Particles in Curved Spacetime", *Living Reviews in Relativity* **7**, doi:10.12942/lrr-2004-6, retrieved 2007-06-13

- Poisson, Eric (2004), *A Relativist's Toolkit. The Mathematics of Black-Hole Mechanics*, Cambridge University Press, ISBN 0-521-83091-5

- Polchinski, Joseph (1998a), *String Theory Vol. I: An Introduction to the Bosonic String*, Cambridge University Press, ISBN 0-521-63303-6

- Polchinski, Joseph (1998b), *String Theory Vol. II: Superstring Theory and Beyond*, Cambridge University Press, ISBN 0-521-63304-4

- Pound, R. V.; Rebka, G. A. (1959), "Gravitational Red-Shift in Nuclear Resonance", *Physical Review Letters* **3** (9): 439–441, Bibcode:1959PhRvL...3..439P, doi:10.1103/PhysRevLett.3.439

- Pound, R. V.; Rebka, G. A. (1960), "Apparent weight of photons", *Phys. Rev. Lett.* **4** (7): 337–341, Bibcode:1960PhRvL...4..337P, doi:10.1103/PhysRevLett.4.337

- Pound, R. V.; Snider, J. L. (1964), "Effect of Gravity on Nuclear Resonance", *Phys. Rev. Lett.* **13** (18): 539–540, Bibcode:1964PhRvL..13..539P, doi:10.1103/PhysRevLett.13.539

- Ramond, Pierre (1990), *Field Theory: A Modern Primer*, Addison-Wesley, ISBN 0-201-54611-6

- Rees, Martin (1966), "Appearance of Relativistically Expanding Radio Sources", *Nature* **211** (5048): 468–470, Bibcode:1966Natur.211..468R, doi:10.1038/211468a0

- Reissner, H. (1916), "Über die Eigengravitation des elektrischen Feldes nach der Einsteinschen Theorie", *Annalen der Physik* **355** (9): 106–120, Bibcode:1916AnP...355..106R, doi:10.1002/andp.19163550905

- Remillard, Ronald A.; Lin, Dacheng; Cooper, Randall L.; Narayan, Ramesh (2006), "The Rates of Type I X-Ray Bursts from Transients Observed with RXTE: Evidence for Black Hole Event Horizons", *Astrophysical Journal* **646** (1): 407–419, arXiv:astro-ph/0509758, Bibcode:2006ApJ...646..407R, doi:10.1086/504862

- Renn, Jürgen, ed. (2007), *The Genesis of General Relativity (4 Volumes)*, Dordrecht: Springer, ISBN 1-4020-3999-9

- Renn, Jürgen, ed. (2005), *Albert Einstein—Chief Engineer of the Universe: Einstein's Life and Work in Context*, Berlin: Wiley-VCH, ISBN 3-527-40571-2

- Reula, Oscar A. (1998), "Hyperbolic Methods for Einstein's Equations", *Living Reviews in Relativity* 1, Bibcode:1998LRR.....1....3R, doi:10.12942/lrr-1998-3, retrieved 2007-08-29

- Rindler, Wolfgang (2001), *Relativity. Special, General and Cosmological*, Oxford University Press, ISBN 0-19-850836-0

- Rindler, Wolfgang (1991), *Introduction to Special Relativity*, Clarendon Press, Oxford, ISBN 0-19-853952-5

- Robson, Ian (1996), *Active galactic nuclei*, John Wiley, ISBN 0-471-95853-0

- Roulet, E.; Mollerach, S. (1997), "Microlensing", *Physics Reports* 279 (2): 67–118, arXiv:astro-ph/9603119, Bibcode:1997PhR...279...67R, doi:10.1016/S0370-1573(96)00020-8

- Rovelli, Carlo (2000). "Notes for a brief history of quantum gravity". arXiv:gr-qc/0006061 [gr-qc].

- Rovelli, Carlo (1998), "Loop Quantum Gravity", *Living Reviews in Relativity* 1, doi:10.12942/lrr-1998-1, retrieved 2008-03-13

- Schäfer, Gerhard (2004), "Gravitomagnetic Effects", *General Relativity and Gravitation* 36 (10): 2223–2235, arXiv:gr-qc/0407116, Bibcode:2004GReGr..36.2223S, doi:10.1023/B:GERG.0000046180.97877.32

- Schödel, R.; Ott, T.; Genzel, R.; Eckart, A.; Mouawad, N.; Alexander, T. (2003), "Stellar Dynamics in the Central Arcsecond of Our Galaxy", *Astrophysical Journal* 596 (2): 1015–1034, arXiv:astro-ph/0306214, Bibcode:2003ApJ...596.1015S, doi:10.1086/378122

- Schutz, Bernard F. (1985), *A first course in general relativity*, Cambridge University Press, ISBN 0-521-27703-5

- Schutz, Bernard F. (2001), "Gravitational radiation", in Murdin, Paul, *Encyclopedia of Astronomy and Astrophysics*, Grove's Dictionaries, ISBN 1-56159-268-4

- Schutz, Bernard F. (2003), *Gravity from the ground up*, Cambridge University Press, ISBN 0-521-45506-5

- Schwarz, John H. (2007), "String Theory: Progress and Problems", *Progress of Theoretical Physics Supplement* 170: 214, arXiv:hep-th/0702219, Bibcode:2007PThPS.170..214S, doi:10.1143/PTPS.170.214

- Schwarzschild, Karl (1916a), "Über das Gravitationsfeld eines Massenpunktes nach der Einsteinschen Theorie", *Sitzungsber. Preuss. Akad. D. Wiss.*: 189–196

- Schwarzschild, Karl (1916b), "Über das Gravitationsfeld einer Kugel aus inkompressibler Flüssigkeit nach der Einsteinschen Theorie", *Sitzungsber. Preuss. Akad. D. Wiss.*: 424–434

- Seidel, Edward (1998), "Numerical Relativity: Towards Simulations of 3D Black Hole Coalescence", in Narlikar, J. V.; Dadhich, N., *Gravitation and Relativity: At the turn of the millennium (Proceedings of the GR-15 Conference, held at IUCAA, Pune, India, December 16–21, 1997)*, IUCAA, p. 6088, arXiv:gr-qc/9806088, Bibcode:1998gr.qc.....6088S, ISBN 81-900378-3-8

- Seljak, Uroš; Zaldarriaga, Matias (1997), "Signature of Gravity Waves in the Polarization of the Microwave Background", *Phys. Rev. Lett.* 78 (11): 2054–2057, arXiv:astro-ph/9609169, Bibcode:1997PhRvL..78.2054S, doi:10.1103/PhysRevLett.78.2054

- Shapiro, S. S.; Davis, J. L.; Lebach, D. E.; Gregory, J. S. (2004), "Measurement of the solar gravitational deflection of radio waves using geodetic very-long-baseline interferometry data, 1979–1999", *Phys. Rev. Lett.* 92 (12): 121101, Bibcode:2004PhRvL..92l1101S, doi:10.1103/PhysRevLett.92.121101, PMID 15089661

- Shapiro, Irwin I. (1964), "Fourth test of general relativity", *Phys. Rev. Lett.* 13 (26): 789–791, Bibcode:1964PhRvL..13..789S, doi:10.1103/PhysRevLett.13.789

- Shapiro, I. I.; Pettengill, Gordon; Ash, Michael; Stone, Melvin; Smith, William; Ingalls, Richard; Brockelman, Richard (1968), "Fourth test of general relativity: preliminary results", *Phys. Rev. Lett.* 20 (22): 1265–1269, Bibcode:1968PhRvL..20.1265S, doi:10.1103/PhysRevLett.20.1265

- Singh, Simon (2004), *Big Bang: The Origin of the Universe*, Fourth Estate, ISBN 0-00-715251-5

- Sorkin, Rafael D. (2005), "Causal Sets: Discrete Gravity", in Gomberoff, Andres; Marolf, Donald,

Lectures on Quantum Gravity, Springer, p. 9009, arXiv:gr-qc/0309009, Bibcode:2003gr.qc.....9009S, ISBN 0-387-23995-2

- Sorkin, Rafael D. (1997), "Forks in the Road, on the Way to Quantum Gravity", *Int. J. Theor. Phys.* **36** (12): 2759–2781, arXiv:gr-qc/9706002, Bibcode:1997IJTP...36.2759S, doi:10.1007/BF02435709

- Spergel, D. N.; Verde, L.; Peiris, H. V.; Komatsu, E.; Nolta, M. R.; Bennett, C. L.; Halpern, M.; Hinshaw, G.; et al. (2003), "First Year Wilkinson Microwave Anisotropy Probe (WMAP) Observations: Determination of Cosmological Parameters", *Astrophys. J. Suppl.* **148** (1): 175–194, arXiv:astro-ph/0302209, Bibcode:2003ApJS..148..175S, doi:10.1086/377226

- Spergel, D. N.; Bean, R.; Doré, O.; Nolta, M. R.; Bennett, C. L.; Dunkley, J.; Hinshaw, G.; Jarosik, N.; et al. (2007), "Wilkinson Microwave Anisotropy Probe (WMAP) Three Year Results: Implications for Cosmology", *Astrophysical Journal Supplement* **170** (2): 377–408, arXiv:astro-ph/0603449, Bibcode:2007ApJS..170..377S, doi:10.1086/513700

- Springel, Volker; White, Simon D. M.; Jenkins, Adrian; Frenk, Carlos S.; Yoshida, Naoki; Gao, Liang; Navarro, Julio; Thacker, Robert; et al. (2005), "Simulations of the formation, evolution and clustering of galaxies and quasars", *Nature* **435** (7042): 629–636, arXiv:astro-ph/0504097, Bibcode:2005Natur.435..629S, doi:10.1038/nature03597, PMID 15931216

- Stairs, Ingrid H. (2003), "Testing General Relativity with Pulsar Timing", *Living Reviews in Relativity* **6**, arXiv:astro-ph/0307536, Bibcode:2003LRR.....6....5S, doi:10.12942/lrr-2003-5, retrieved 2007-07-21

- Stephani, H.; Kramer, D.; MacCallum, M.; Hoenselaers, C.; Herlt, E. (2003), *Exact Solutions of Einstein's Field Equations* (2 ed.), Cambridge University Press, ISBN 0-521-46136-7

- Synge, J. L. (1972), *Relativity: The Special Theory*, North-Holland Publishing Company, ISBN 0-7204-0064-3

- Szabados, László B. (2004), "Quasi-Local Energy-Momentum and Angular Momentum in GR", *Living Reviews in Relativity* 7, doi:10.12942/lrr-2004-4, retrieved 2007-08-23

- Taylor, Joseph H. (1994), "Binary pulsars and relativistic gravity", *Rev. Mod. Phys.* **66**

(3): 711–719, Bibcode:1994RvMP...66..711T, doi:10.1103/RevModPhys.66.711

- Thiemann, Thomas (2006), "Approaches to Fundamental Physics: Loop Quantum Gravity: An Inside View", *Lecture Notes in Physics* **721**: 185–263, arXiv:hep-th/0608210, Bibcode:2007LNP...721..185T, doi:10.1007/978-3-540-71117-9_10, ISBN 978-3-540-71115-5

- Thiemann, Thomas (2003), "Lectures on Loop Quantum Gravity", *Lecture Notes in Physics* **631**: 41–135, arXiv:gr-qc/0210094, doi:10.1007/978-3-540-45230-0_3, ISBN 978-3-540-40810-9

- 't Hooft, Gerard; Veltman, Martinus (1974), "One Loop Divergencies in the Theory of Gravitation", *Ann. Inst. Poincare* **20**: 69

- Thorne, Kip S. (1972), "Nonspherical Gravitational Collapse—A Short Review", in Klauder, J., *Magic without Magic*, W. H. Freeman, pp. 231–258

- Thorne, Kip S. (1994), *Black Holes and Time Warps: Einstein's Outrageous Legacy*, W W Norton & Company, ISBN 0-393-31276-3

- Thorne, Kip S. (1995), "Gravitational radiation", *Particle and Nuclear Astrophysics and Cosmology in the Next Millenium*: 160, arXiv:gr-qc/9506086, Bibcode:1995pnac.conf..160T, ISBN 0-521-36853-7

- Townsend, Paul K. (1997). "Black Holes (Lecture notes)". arXiv:gr-qc/9707012 [gr-qc].

- Townsend, Paul K. (1996). "Four Lectures on M-Theory". arXiv:hep-th/9612121 [hep-th].

- Traschen, Jenny (2000), Bytsenko, A.; Williams, F., eds., "An Introduction to Black Hole Evaporation", *Mathematical Methods of Physics (Proceedings of the 1999 Londrina Winter School)* (World Scientific): 180, arXiv:gr-qc/0010055, Bibcode:2000mmp..conf..180T

- Trautman, Andrzej (2006), "Einstein–Cartan theory", in Françoise, J.-P.; Naber, G. L.; Tsou, S. T., *Encyclopedia of Mathematical Physics, Vol. 2*, Elsevier, pp. 189–195, arXiv:gr-qc/0606062, Bibcode:2006gr.qc.....6062T

- Unruh, W. G. (1976), "Notes on Black Hole Evaporation", *Phys. Rev. D* **14** (4): 870–892, Bibcode:1976PhRvD..14..870U, doi:10.1103/PhysRevD.14.870

- Valtonen, M. J.; Lehto, H. J.; Nilsson, K.; Heidt, J.; Takalo, L. O.; Sillanpää, A.; Villforth, C.;

Kidger, M.; et al. (2008), "A massive binary black-hole system in OJ 287 and a test of general relativity", *Nature* **452** (7189): 851–853, arXiv:0809.1280, Bibcode:2008Natur.452..851V, doi:10.1038/nature06896, PMID 18421348

- Veltman, Martinus (1975), "Quantum Theory of Gravitation", in Balian, Roger; Zinn-Justin, Jean, *Methods in Field Theory - Les Houches Summer School in Theoretical Physics.* **77**, North Holland

- Wald, Robert M. (1975), "On Particle Creation by Black Holes", *Commun. Math. Phys.* **45** (3): 9–34, Bibcode:1975CMaPh..45....9W, doi:10.1007/BF01609863

- Wald, Robert M. (1984), *General Relativity*, University of Chicago Press, ISBN 0-226-87033-2

- Wald, Robert M. (1994), *Quantum field theory in curved spacetime and black hole thermodynamics*, University of Chicago Press, ISBN 0-226-87027-8

- Wald, Robert M. (2001), "The Thermodynamics of Black Holes", *Living Reviews in Relativity* **4**, Bibcode:2001LRR.....4....6W, doi:10.12942/lrr-2001-6, retrieved 2007-08-08

- Walsh, D.; Carswell, R. F.; Weymann, R. J. (1979), "0957 + 561 A, B: twin quasistellar objects or gravitational lens?", *Nature* **279** (5712): 381–4, Bibcode:1979Natur.279..381W, doi:10.1038/279381a0, PMID 16068158

- Wambsganss, Joachim (1998), "Gravitational Lensing in Astronomy", *Living Reviews in Relativity* **1**, arXiv:astro-ph/9812021, Bibcode:1998LRR.....1...12W, doi:10.12942/lrr-1998-12, retrieved 2007-07-20

- Weinberg, Steven (1972), *Gravitation and Cosmology*, John Wiley, ISBN 0-471-92567-5

- Weinberg, Steven (1995), *The Quantum Theory of Fields I: Foundations*, Cambridge University Press, ISBN 0-521-55001-7

- Weinberg, Steven (1996), *The Quantum Theory of Fields II: Modern Applications*, Cambridge University Press, ISBN 0-521-55002-5

- Weinberg, Steven (2000), *The Quantum Theory of Fields III: Supersymmetry*, Cambridge University Press, ISBN 0-521-66000-9

- Weisberg, Joel M.; Taylor, Joseph H. (2003), "The Relativistic Binary Pulsar B1913+16"", in Bailes, M.; Nice, D. J.; Thorsett, S. E., *Proceedings of "Radio Pulsars," Chania, Crete, August, 2002*, ASP Conference Series

- Weiss, Achim (2006), "Elements of the past: Big Bang Nucleosynthesis and observation", *Einstein Online* (Max Planck Institute for Gravitational Physics), retrieved 2007-02-24

- Wheeler, John A. (1990), *A Journey Into Gravity and Spacetime*, Scientific American Library, San Francisco: W. H. Freeman, ISBN 0-7167-6034-7

- Will, Clifford M. (1993), *Theory and experiment in gravitational physics*, Cambridge University Press, ISBN 0-521-43973-6

- Will, Clifford M. (2006), "The Confrontation between General Relativity and Experiment", *Living Reviews in Relativity* **9**, arXiv:gr-qc/0510072, Bibcode:2006LRR.....9....3W, doi:10.12942/lrr-2006-3, retrieved 2007-06-12

- Zwiebach, Barton (2004), *A First Course in String Theory*, Cambridge University Press, ISBN 0-521-83143-1

33.12 Further reading

Popular books

- Geroch, R (1981), *General Relativity from A to B*, Chicago: University of Chicago Press, ISBN 0-226-28864-1

- Lieber, Lillian (2008), *The Einstein Theory of Relativity: A Trip to the Fourth Dimension*, Philadelphia: Paul Dry Books, Inc., ISBN 978-1-58988-044-3

- Wald, Robert M. (1992), *Space, Time, and Gravity: the Theory of the Big Bang and Black Holes*, Chicago: University of Chicago Press, ISBN 0-226-87029-4

- Wheeler, John; Ford, Kenneth (1998), *Geons, Black Holes, & Quantum Foam: a life in physics*, New York: W. W. Norton, ISBN 0-393-31991-1

Beginning undergraduate textbooks

- Callahan, James J. (2000), *The Geometry of Spacetime: an Introduction to Special and General Relativity*, New York: Springer, ISBN 0-387-98641-3

- Taylor, Edwin F.; Wheeler, John Archibald (2000), *Exploring Black Holes: Introduction to General Relativity*, Addison Wesley, ISBN 0-201-38423-X

Advanced undergraduate textbooks

- B. F. Schutz (2009), *A First Course in General Relativity (Second Edition)*, Cambridge University Press, ISBN 978-0-521-88705-2

- Cheng, Ta-Pei (2005), *Relativity, Gravitation and Cosmology: a Basic Introduction*, Oxford and New York: Oxford University Press, ISBN 0-19-852957-0

- Gron, O.; Hervik, S. (2007), *Einstein's General theory of Relativity*, Springer, ISBN 978-0-387-69199-2

- Hartle, James B. (2003), *Gravity: an Introduction to Einstein's General Relativity*, San Francisco: Addison-Wesley, ISBN 0-8053-8662-9

- Hughston, L. P. & Tod, K. P. (1991), *Introduction to General Relativity*, Cambridge: Cambridge University Press, ISBN 0-521-33943-X

- d'Inverno, Ray (1992), *Introducing Einstein's Relativity*, Oxford: Oxford University Press, ISBN 0-19-859686-3

- Ludyk, Günter (2013). *Einstein in Matrix Form* (1st ed.). Berlin: Springer. ISBN 9783642357978.

Graduate-level textbooks

- Carroll, Sean M. (2004), *Spacetime and Geometry: An Introduction to General Relativity*, San Francisco: Addison-Wesley, ISBN 0-8053-8732-3

- Grøn, Øyvind; Hervik, Sigbjørn (2007), *Einstein's General Theory of Relativity*, New York: Springer, ISBN 978-0-387-69199-2

- Landau, Lev D.; Lifshitz, Evgeny F. (1980), *The Classical Theory of Fields (4th ed.)*, London: Butterworth-Heinemann, ISBN 0-7506-2768-9

- Misner, Charles W.; Thorne, Kip. S.; Wheeler, John A. (1973), *Gravitation*, W. H. Freeman, ISBN 0-7167-0344-0

- Stephani, Hans (1990), *General Relativity: An Introduction to the Theory of the Gravitational Field*, Cambridge: Cambridge University Press, ISBN 0-521-37941-5

- Wald, Robert M. (1984), *General Relativity*, University of Chicago Press, ISBN 0-226-87033-2

33.13   External links

- Einstein Online – Articles on a variety of aspects of relativistic physics for a general audience; hosted by the Max Planck Institute for Gravitational Physics

- NCSA Spacetime Wrinkles – produced by the numerical relativity group at the NCSA, with an elementary introduction to general relativity

- **Courses**

- **Lectures**

- **Tutorials**

- Einstein's General Theory of Relativity on YouTube (lecture by Leonard Susskind recorded September 22, 2008 at Stanford University).

- Series of lectures on General Relativity given in 2006 at the Institut Henri Poincaré (introductory/advanced).

- General Relativity Tutorials by John Baez.

- Brown, Kevin. "Reflections on relativity". *Mathpages.com*. Retrieved May 29, 2005.

- Carroll, Sean M. "Lecture Notes on General Relativity". Retrieved January 5, 2014.

- Moor, Rafi. "Understanding General Relativity". Retrieved July 11, 2006.

- Waner, Stefan. "Introduction to Differential Geometry and General Relativity" (PDF). Retrieved 2015-04-05.

Chapter 34

Tests of general relativity

At its introduction in 1915, the general theory of relativity did not have a solid empirical foundation. It was known that it correctly accounted for the "anomalous" precession of the perihelion of Mercury and on philosophical grounds it was considered satisfying that it was able to unify Newton's law of universal gravitation with special relativity. That light appeared to bend in gravitational fields in line with the predictions of general relativity was found in 1919 but it was not until a program of precision tests was started in 1959 that the various predictions of general relativity were tested to any further degree of accuracy in the weak gravitational field limit, severely limiting possible deviations from the theory. Beginning in 1974, Hulse, Taylor and others have studied the behaviour of binary pulsars experiencing much stronger gravitational fields than those found in the Solar System. Both in the weak field limit (as in the Solar System) and with the stronger fields present in systems of binary pulsars the predictions of general relativity have been extremely well tested locally.

The very strong gravitational fields that are present close to black holes, especially those supermassive black holes which are thought to power active galactic nuclei and the more active quasars, belong to a field of intense active research. Observations of these quasars and active galactic nuclei are difficult, and interpretation of the observations is heavily dependent upon astrophysical models other than general relativity or competing fundamental theories of gravitation, but they are qualitatively consistent with the black hole concept as modelled in general relativity. As a consequence of the equivalence principle, Lorentz invariance holds locally in non-rotating, freely falling reference frames. Experiments related to Lorentz invariance and thus special relativity (that is, when gravitational effects can be neglected) are described in Tests of special relativity. In February 2016, the Advanced LIGO team announced that they had directly detected gravitational waves from a black hole merger.[1]

34.1 Classical tests

Albert Einstein proposed[2][3] three tests of general relativity, subsequently called **the classical tests of general relativity**, in 1916:

1. the perihelion precession of Mercury's orbit

2. the deflection of light by the Sun

3. the gravitational redshift of light

In the letter to the London *Times* on November 28, 1919, he described the theory of relativity and thanked his English colleagues for their understanding and testing of his work. He also mentioned three classical tests with comments:[4]

> "The chief attraction of the theory lies in its logical completeness. If a single one of the conclusions drawn from it proves wrong, it must be given up; to modify it without destroying the whole structure seems to be impossible."

34.1.1 Perihelion precession of Mercury

For more details on this topic, see Two-body problem in general relativity.

Under Newtonian physics, a two-body system consisting of a lone object orbiting a spherical mass would trace out an ellipse with the spherical mass at a focus. The point of closest approach, called the periapsis (or, because the central body in the Solar System is the Sun, perihelion), is fixed. A number of effects in the Solar System cause the perihelia of planets to precess (rotate) around the Sun. The principal cause is the presence of other planets which perturb one another's orbit. Another (much less significant) effect is solar oblateness.

Mercury deviates from the precession predicted from these Newtonian effects. This anomalous rate of precession of

Transit of Mercury on November 8, 2006 with sunspots #921, 922, and 923

the perihelion of Mercury's orbit was first recognized in 1859 as a problem in celestial mechanics, by Urbain Le Verrier. His reanalysis of available timed observations of transits of Mercury over the Sun's disk from 1697 to 1848 showed that the actual rate of the precession disagreed from that predicted from Newton's theory by 38" (arc seconds) per tropical century (later re-estimated at 43").[5] A number of *ad hoc* and ultimately unsuccessful solutions were proposed, but they tended to introduce more problems. In general relativity, this remaining precession, or change of orientation of the orbital ellipse within its orbital plane, is explained by gravitation being mediated by the curvature of spacetime. Einstein showed that general relativity[2] agrees closely with the observed amount of perihelion shift. This was a powerful factor motivating the adoption of general relativity.

Although earlier measurements of planetary orbits were made using conventional telescopes, more accurate measurements are now made with radar. The total observed precession of Mercury is 574.10±0.65 arc-seconds per century[6] relative to the inertial ICFR. This precession can be attributed to the following causes:

The correction by 42.98" is 3/2 multiple of classical prediction with PPN parameters $\gamma = \beta = 0$.[8]

Thus the effect can be fully explained by general relativity. More recent calculations based on more precise measurements have not materially changed the situation.

The other planets experience perihelion shifts as well, but, since they are farther from the Sun and have longer periods, their shifts are lower, and could not be observed accurately

until long after Mercury's. For example, the perihelion shift of Earth's orbit due to general relativity is of 3.84 seconds of arc per century, and Venus's is 8.62". Both values are in good agreement with observation.[9] The periapsis shift of binary pulsar systems have been measured, with PSR 1913+16 amounting to 4.2º per year.[10] These observations are consistent with general relativity.[11] It is also possible to measure periapsis shift in binary star systems which do not contain ultra-dense stars, but it is more difficult to model the classical effects precisely - for example, the alignment of the stars' spin to their orbital plane needs to be known and is hard to measure directly - so a few systems such as DI Herculis have been considered as problematic cases for general relativity.

In general relativity the perihelion shift σ, expressed in radians per revolution, is approximately given by:[12]

$$\sigma = \frac{24\pi^3 L^2}{T^2 c^2 (1 - e^2)} \, ,$$

where L is the semi-major axis, T is the orbital period, c is the speed of light, and e is the orbital eccentricity.

34.1.2 Deflection of light by the Sun

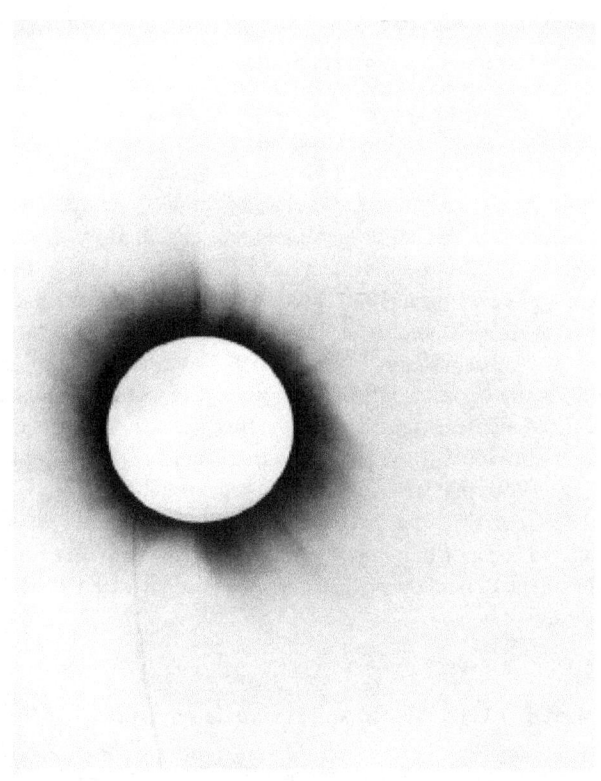

One of Eddington's photographs of the 1919 solar eclipse experiment, presented in his 1920 paper announcing its success

For more details on this topic, see Kepler problem in general relativity.

Henry Cavendish in 1784 (in an unpublished manuscript) and Johann Georg von Soldner in 1801 (published in 1804) had pointed out that Newtonian gravity predicts that starlight will bend around a massive object.[13][14] The same value as Soldner's was calculated by Einstein in 1911 based on the equivalence principle alone. However, Einstein noted in 1915 in the process of completing general relativity, that his (and thus Soldner's) 1911 result is only half of the correct value. Einstein became the first to calculate the correct value for light bending.[15]

The first observation of light deflection was performed by noting the change in position of stars as they passed near the Sun on the celestial sphere. The observations were performed in May 1919 by Arthur Eddington and his collaborators during a total solar eclipse,[16] so that the stars near the Sun (at that time in the constellation Taurus) could be observed.[16] Observations were made simultaneously in the cities of Sobral, Ceará, Brazil and in São Tomé and Príncipe on the west coast of Africa.[17] The result was considered spectacular news and made the front page of most major newspapers. It made Einstein and his theory of general relativity world-famous. When asked by his assistant what his reaction would have been if general relativity had not been confirmed by Eddington and Dyson in 1919, Einstein famously made the quip: "Then I would feel sorry for the dear Lord. The theory is correct anyway."[18]

The early accuracy, however, was poor. The results were argued by some[19] to have been plagued by systematic error and possibly confirmation bias, although modern reanalysis of the dataset[20] suggests that Eddington's analysis was accurate.[21][22] The measurement was repeated by a team from the Lick Observatory in the 1922 eclipse, with results that agreed with the 1919 results[22] and has been repeated several times since, most notably in 1953 by Yerkes Observatory astronomers[23] and in 1973 by a team from the University of Texas.[24] Considerable uncertainty remained in these measurements for almost fifty years, until observations started being made at radio frequencies. It was not until the 1960s that it was definitively accepted that the amount of deflection was the full value predicted by general relativity, and not half that number. The Einstein ring is an example of the deflection of light from distant galaxies by more nearby objects.

34.1.3 Gravitational redshift of light

Einstein predicted the gravitational redshift of light from the equivalence principle in 1907, but it is very difficult to measure astrophysically (see the discussion under *Equivalence*

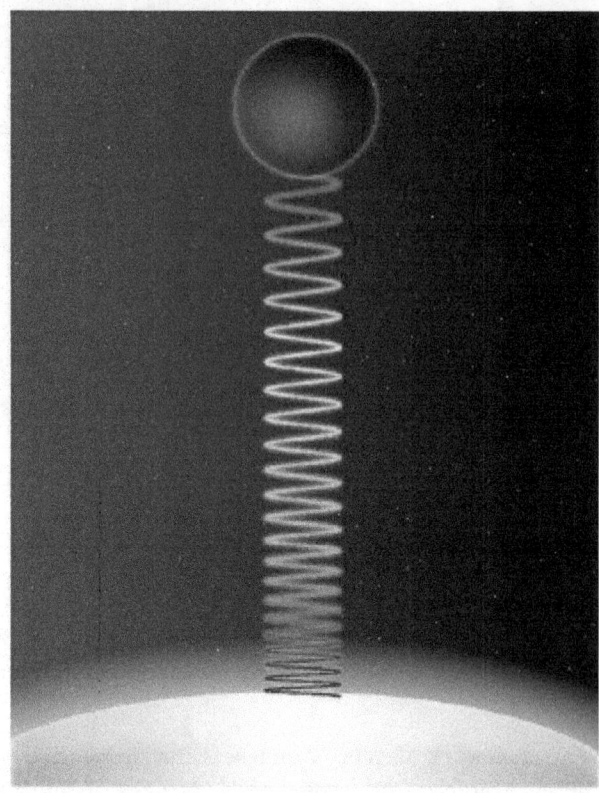

The gravitational redshift of a light wave as it moves upwards against a gravitational field (caused by the yellow star below).

Principle below). Although it was measured by Walter Sydney Adams in 1925, it was only conclusively tested when the Pound–Rebka experiment in 1959 measured the relative redshift of two sources situated at the top and bottom of Harvard University's Jefferson tower using an extremely sensitive phenomenon called the Mössbauer effect.[25][26] The result was in excellent agreement with general relativity. This was one of the first precision experiments testing general relativity.

34.2 Modern tests

The modern era of testing general relativity was ushered in largely at the impetus of Dicke and Schiff who laid out a framework for testing general relativity.[27][28][29] They emphasized the importance not only of the classical tests, but of null experiments, testing for effects which in principle could occur in a theory of gravitation, but do not occur in general relativity. Other important theoretical developments included the inception of alternative theories to general relativity, in particular, scalar-tensor theories such as the Brans–Dicke theory;[30] the parameterized post-Newtonian formalism in which deviations from general relativity can be quantified; and the framework of the equivalence principle.

Experimentally, new developments in space exploration, electronics and condensed matter physics have made additional precise experiments possible, such as the Pound–Rebka experiment, laser interferometry and lunar rangefinding.

34.2.1   Post-Newtonian tests of gravity

Early tests of general relativity were hampered by the lack of viable competitors to the theory: it was not clear what sorts of tests would distinguish it from its competitors. General relativity was the only known relativistic theory of gravity compatible with special relativity and observations. Moreover, it is an extremely simple and elegant theory. This changed with the introduction of Brans–Dicke theory in 1960. This theory is arguably simpler, as it contains no dimensionful constants, and is compatible with a version of Mach's principle and Dirac's large numbers hypothesis, two philosophical ideas which have been influential in the history of relativity. Ultimately, this led to the development of the parametrized post-Newtonian formalism by Nordtvedt and Will, which parametrizes, in terms of ten adjustable parameters, all the possible departures from Newton's law of universal gravitation to first order in the velocity of moving objects (*i.e.* to first order in v/c, where v is the velocity of an object and c is the speed of light). This approximation allows the possible deviations from general relativity, for slowly moving objects in weak gravitational fields, to be systematically analyzed. Much effort has been put into constraining the post-Newtonian parameters, and deviations from general relativity are at present severely limited.

The experiments testing gravitational lensing and light time delay limits the same post-Newtonian parameter, the so-called Eddington parameter γ, which is a straightforward parametrization of the amount of deflection of light by a gravitational source. It is equal to one for general relativity, and takes different values in other theories (such as Brans–Dicke theory). It is the best constrained of the ten post-Newtonian parameters, but there are other experiments designed to constrain the others. Precise observations of the perihelion shift of Mercury constrain other parameters, as do tests of the strong equivalence principle.

One of the goals of the mission BepiColombo is testing the general relativity theory by measuring the parameters gamma and beta of the parametrized post-Newtonian formalism with high accuracy.[31]

34.2.2   Gravitational lensing

One of the most important tests is gravitational lensing. It has been observed in distant astrophysical sources, but these are poorly controlled and it is uncertain how they constrain general relativity. The most precise tests are analogous to Eddington's 1919 experiment: they measure the deflection of radiation from a distant source by the Sun. The sources that can be most precisely analyzed are distant radio sources. In particular, some quasars are very strong radio sources. The directional resolution of any telescope is in principle limited by diffraction; for radio telescopes this is also the practical limit. An important improvement in obtaining positional high accuracies (from milli-arcsecond to micro-arcsecond) was obtained by combining radio telescopes across Earth. The technique is called very long baseline interferometry (VLBI). With this technique radio observations couple the phase information of the radio signal observed in telescopes separated over large distances. Recently, these telescopes have measured the deflection of radio waves by the Sun to extremely high precision, confirming the amount of deflection predicted by general relativity aspect to the 0.03% level.[32] At this level of precision systematic effects have to be carefully taken into account to determine the precise location of the telescopes on Earth. Some important effects are Earth's nutation, rotation, atmospheric refraction, tectonic displacement and tidal waves. Another important effect is refraction of the radio waves by the solar corona. Fortunately, this effect has a characteristic spectrum, whereas gravitational distortion is independent of wavelength. Thus, careful analysis, using measurements at several frequencies, can subtract this source of error.

The entire sky is slightly distorted due to the gravitational deflection of light caused by the Sun (the anti-Sun direction excepted). This effect has been observed by the European Space Agency astrometric satellite Hipparcos. It measured the positions of about 10^5 stars. During the full mission about 3.5×10^6 relative positions have been determined, each to an accuracy of typically 3 milliarcseconds (the accuracy for an 8–9 magnitude star). Since the gravitation deflection perpendicular to the Earth–Sun direction is already 4.07 milliarcseconds, corrections are needed for practically all stars. Without systematic effects, the error in an individual observation of 3 milliarcseconds, could be reduced by the square root of the number of positions, leading to a precision of 0.0016 milliarcseconds. Systematic effects, however, limit the accuracy of the determination to 0.3% (Froeschlé, 1997).

Launched in 2013, the *Gaia* spacecraft will conduct a census of one billion stars in the Milky Way and measure their positions to an accuracy of 24 microarcseconds. Thus it will also provide stringent new tests of gravitational deflection of light caused by the Sun which was predicted by General relativity.[33]

34.2.3 Light travel time delay testing

Irwin I. Shapiro proposed another test, beyond the classical tests, which could be performed within the Solar System. It is sometimes called the fourth "classical" test of general relativity. He predicted a relativistic time delay (Shapiro delay) in the round-trip travel time for radar signals reflecting off other planets.[34] The mere curvature of the path of a photon passing near the Sun is too small to have an observable delaying effect (when the round-trip time is compared to the time taken if the photon had followed a straight path), but general relativity predicts a time delay that becomes progressively larger when the photon passes nearer to the Sun due to the time dilation in the gravitational potential of the Sun. Observing radar reflections from Mercury and Venus just before and after it is eclipsed by the Sun agrees with general relativity theory at the 5% level.[35] More recently, the Cassini probe has undertaken a similar experiment which gave agreement with general relativity at the 0.002% level.[36] Very Long Baseline Interferometry has measured velocity-dependent (gravitomagnetic) corrections to the Shapiro time delay in the field of moving Jupiter [37][38] and Saturn.[39]

34.2.4 The equivalence principle

Main article: Equivalence principle

The equivalence principle, in its simplest form, asserts that the trajectories of falling bodies in a gravitational field should be independent of their mass and internal structure, provided they are small enough not to disturb the environment or be affected by tidal forces. This idea has been tested to extremely high precision by Eötvös torsion balance experiments, which look for a differential acceleration between two test masses. Constraints on this, and on the existence of a composition-dependent fifth force or gravitational Yukawa interaction are very strong, and are discussed under fifth force and weak equivalence principle.

A version of the equivalence principle, called the strong equivalence principle, asserts that self-gravitation falling bodies, such as stars, planets or black holes (which are all held together by their gravitational attraction) should follow the same trajectories in a gravitational field, provided the same conditions are satisfied. This is called the Nordtvedt effect and is most precisely tested by the Lunar Laser Ranging Experiment.[40][41] Since 1969, it has continuously measured the distance from several rangefinding stations on Earth to reflectors on the Moon to approximately centimeter accuracy.[42] These have provided a strong constraint on

several of the other post-Newtonian parameters.

Another part of the strong equivalence principle is the requirement that Newton's gravitational constant be constant in time, and have the same value everywhere in the universe. There are many independent observations limiting the possible variation of Newton's gravitational constant,[43] but one of the best comes from lunar rangefinding which suggests that the gravitational constant does not change by more than one part in 10^{11} per year. The constancy of the other constants is discussed in the Einstein equivalence principle section of the equivalence principle article.

Gravitational redshift

The first of the classical tests discussed above, the gravitational redshift, is a simple consequence of the Einstein equivalence principle and was predicted by Einstein in 1907. As such, it is not a test of general relativity in the same way as the post-Newtonian tests, because any theory of gravity obeying the equivalence principle should also incorporate the gravitational redshift. Nonetheless, confirming the existence of the effect was an important substantiation of relativistic gravity, since the absence of gravitational redshift would have strongly contradicted relativity. The first observation of the gravitational redshift was the measurement of the shift in the spectral lines from the white dwarf star Sirius B by Adams in 1925. Although this measurement, as well as later measurements of the spectral shift on other white dwarf stars, agreed with the prediction of relativity, it could be argued that the shift could possibly stem from some other cause, and hence experimental verification using a known terrestrial source was preferable.

Experimental verification of gravitational redshift using terrestrial sources took several decades, because it is difficult to find clocks (to measure time dilation) or sources of electromagnetic radiation (to measure redshift) with a frequency that is known well enough that the effect can be accurately measured. It was confirmed experimentally for the first time in 1960 using measurements of the change in wavelength of gamma-ray photons generated with the Mössbauer effect, which generates radiation with a very narrow line width. The experiment, performed by Pound and Rebka and later improved by Pound and Snyder, is called the Pound–Rebka experiment. The accuracy of the gamma-ray measurements was typically 1%. The blueshift of a falling photon can be found by assuming it has an equivalent mass based on its frequency $E = hf$ (where h is Planck's constant) along with $E = mc^2$, a result of special relativity. Such simple derivations ignore the fact that in general relativity the experiment compares clock rates, rather than energies. In other words, the "higher energy" of the photon after it falls can be equivalently ascribed to the slower running of clocks deeper in the gravitational potential well. To fully validate general relativity, it is important to also show that the rate of arrival of the photons is greater than the rate at which they are emitted. A very accurate gravitational redshift experiment, which deals with this issue, was performed in 1976,[44] where a hydrogen maser clock on a rocket was launched to a height of 10,000 km, and its rate compared with an identical clock on the ground. It tested the gravitational redshift to 0.007%.

Although the Global Positioning System (GPS) is not designed as a test of fundamental physics, it must account for the gravitational redshift in its timing system, and physicists have analyzed timing data from the GPS to confirm other tests. When the first satellite was launched, some engineers resisted the prediction that a noticeable gravitational time dilation would occur, so the first satellite was launched without the clock adjustment that was later built into subsequent satellites. It showed the predicted shift of 38 microseconds per day. This rate of discrepancy is sufficient to substantially impair function of GPS within hours if not accounted for. An excellent account of the role played by general relativity in the design of GPS can be found in Ashby 2003.

Other precision tests of general relativity,[45] not discussed here, are the Gravity Probe A satellite, launched in 1976, which showed gravity and velocity affect the ability to synchronize the rates of clocks orbiting a central mass; the Hafele–Keating experiment, which used atomic clocks in circumnavigating aircraft to test general relativity and special relativity together;[46][47] and the forthcoming Satellite Test of the Equivalence Principle.

34.2.5 Frame-dragging tests

Main article: Frame-dragging

Tests of the Lense–Thirring precession, consisting of small secular precessions of the orbit of a test particle in motion around a central rotating mass, for example, a planet or a star, have been performed with the LAGEOS satellites,[48] but many aspects of them remain controversial. The same effect may have been detected in the data of the Mars Global Surveyor (MGS) spacecraft, a former probe in orbit around Mars; also such a test raised a debate.[49] First attempts to detect the Sun's Lense–Thirring effect on the perihelia of the inner planets have been recently reported as well. Frame dragging would cause the orbital plane of stars orbiting near a supermassive black hole to precess about the black hole spin axis. This effect should be detectable within the next few years via astrometric monitoring of stars at the center of the Milky Way galaxy.[50] By comparing the rate of orbital precession of two stars on different orbits, it is possible in principle to test the no-hair theorems of general

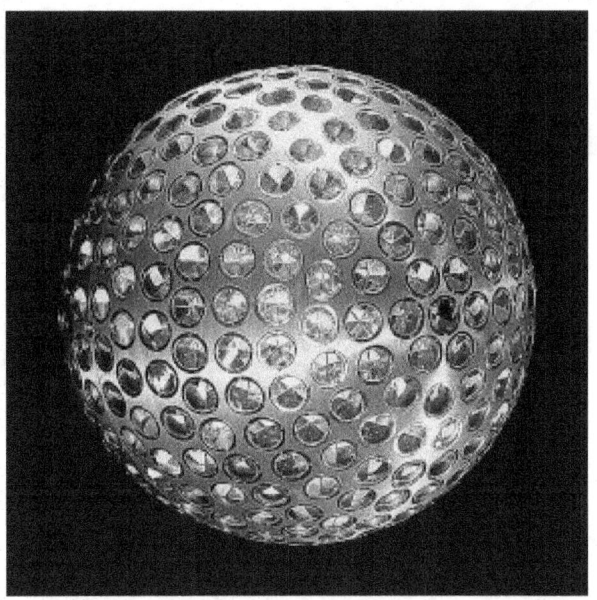

The LAGEOS-1 satellite. (D=60 cm)

relativity.[51]

The Gravity Probe B satellite, launched in 2004 and operated until 2005, detected frame-dragging and the geodetic effect. The experiment used four quartz spheres the size of ping pong balls coated with a superconductor. Data analysis continued through 2011 due to high noise levels and difficulties in modelling the noise accurately so that a useful signal could be found. Principal investigators at Stanford University reported on May 4, 2011, that they had accurately measured the frame dragging effect relative to the distant star IM Pegasi, and the calculations proved to be in line with the prediction of Einstein's theory. The results, published in *Physical Review Letters* measured the geodetic effect with an error of about 0.2 percent. The results reported the frame dragging effect (caused by Earth's rotation) added up to 37 milliarcseconds with an error of about 19 percent.[52] Investigator Francis Everitt explained that a milliarcsecond "is the width of a human hair seen at the distance of 10 miles".[53]

In January 2012, LARES satellite was launched on a Vega rocket[54] to measure Lense–Thirring effect with an accuracy of about 1%, according to its proponents.[55] This evaluation of the actual accuracy obtainable is a subject of debate. [56][57][58]

34.3 Strong field tests: Binary pulsars

Further information: Binary pulsar

Pulsars are rapidly rotating neutron stars which emit reg-

Artist's impression of the pulsar PSR J0348+0432 and its white dwarf companion radiating gravitational waves.[59]

ular radio pulses as they rotate. As such they act as clocks which allow very precise monitoring of their orbital motions. Observations of pulsars in orbit around other stars have all demonstrated substantial periapsis precessions that cannot be accounted for classically but can be accounted for by using general relativity. For example, the Hulse–Taylor binary pulsar PSR B1913+16 (a pair of neutron stars in which one is detected as a pulsar) has an observed precession of over 4° of arc per year (periastron shift per orbit only about 10^{-6}). This precession has been used to compute the masses of the components.

Similarly to the way in which atoms and molecules emit electromagnetic radiation, a gravitating mass that is in quadrupole type or higher order vibration, or is asymmetric and in rotation, can emit gravitational waves.[60] These gravitational waves are predicted to travel at the speed of light. For example, planets orbiting the Sun constantly lose energy via gravitational radiation, but this effect is so small that it is unlikely it will be observed in the near future (Earth radiates about 200 watts (see gravitational waves) of gravitational radiation).

Gravitational waves have been indirectly detected from the Hulse–Taylor binary (and other binary pulsars).[61] Precise timing of the pulses shows that the stars orbit only approximately according to Kepler's Laws: over time they gradually spiral towards each other, demonstrating an energy loss in close agreement with the predicted energy radiated by gravitational waves.[62][63] For their discovery of the first binary pulsar and measuring its orbital decay due to gravitational-wave emission, Hulse and Taylor won the 1993 Nobel Prize in Physics.[64]

A "double pulsar" discovered in 2003, PSR J0737-3039, has a periastron precession of 16.90° per year; unlike the Hulse–Taylor binary, both neutron stars are detected as pulsars, allowing precision timing of both members of the system. Due to this, the tight orbit, the fact that the system is almost edge-on, and the very low transverse velocity of the system as seen from Earth, J0737-3039 provides by far the best system for strong-field tests of general relativity known so far. Several distinct relativistic effects are observed, including orbital decay as in the Hulse–Taylor system. After observing the system for two and a half years, four independent tests of general relativity were possible, the most precise (the Shapiro delay) confirming the general relativity prediction within 0.05%[65] (nevertheless the periastron shift per orbit is only about 0.0013% of a circle and thus it is not a higher-order relativity test).

In 2013, an international team of astronomers reported new data from observing a pulsar-white dwarf system PSR J0348+0432, in which they have been able to measure a change in the orbital period of 8 millionths of a second per year, and confirmed GR predictions in a regime of extreme gravitational fields never probed before;[66] but there are still some competing theories that would agree with these data.[67]

34.4 Direct detection of gravitational waves

A number of gravitational-wave detectors have been built with the intent of directly detecting the gravitational waves emanating from such astronomical events as the merger of two neutron stars or black holes. In February 2016, the Advanced LIGO team announced that they had directly detected gravitational waves from a stellar binary black hole merger.[1][68][69]

General relativity predicts gravitational waves, as does any theory of gravitation that obeys special relativity and so has changes in the gravitational field propagate at a finite speed.[70] Since gravitational waves can be directly detected,[1][69] it is possible to use them to learn about the Universe. This is gravitational-wave astronomy. Gravitational-wave astronomy can test general relativity by verifying that the observed waves are of the form predicted (for example, that they only have two transverse polarizations), and by checking that black holes are the objects described by solutions of the Einstein field equations.[71][72]

"These amazing observations are the confirmation of a lot of theoretical work, including Einstein's general theory of relativity, which predicts gravitational waves," says physicist Stephen Hawking.[1]

34.5   Cosmological tests

Tests of general relativity on the largest scales are not nearly so stringent as Solar System tests.[73] The earliest such test was prediction and discovery of the expansion of the universe.[74] In 1922 Alexander Friedmann found that Einstein equations have non-stationary solutions (even in the presence of the cosmological constant).[75][76] In 1927 Georges Lemaître showed that static solutions of the Einstein equations, which are possible in the presence of the cosmological constant, are unstable, and therefore the static universe envisioned by Einstein could not exist (it must either expand or contract).[75] Lemaître made an explicit prediction that the universe should expand.[77] He also derived a redshift-distance relationship, which is now known as the Hubble Law.[77] Later, in 1931, Einstein himself agreed with the results of Friedmann and Lemaître.[75] The expansion of the universe discovered by Edwin Hubble in 1929[75] was then considered by many (and continues to be considered by some now) as a direct confirmation of general relativity.[78] In the 1930s, largely due to the work of E. A. Milne, it was realised that the linear relationship between redshift and distance derives from the general assumption of uniformity and isotropy rather than specifically from general relativity.[74] However the prediction of a non-static universe was non-trivial, indeed dramatic, and primarily motivated by general relativity.[79]

Some other cosmological tests include searches for primordial gravitational waves generated during cosmic inflation, which may be detected in the cosmic microwave background polarization[80] or by a proposed space-based gravitational-wave interferometer called the Big Bang Observer. Other tests at high redshift are constraints on other theories of gravity,[81][82] and the variation of the gravitational constant since big bang nucleosynthesis (it varied by no more than 40% since then).

34.6   See also

- Cooperstock's Energy Localization Hypothesis
- Square Kilometre Array
- Tests of special relativity

34.7   References

34.7.1   Notes

[1] Castelvecchi, Davide; Witze, Witze (February 11, 2016). "Einstein's gravitational waves found at last". *Nature News*. doi:10.1038/nature.2016.19361. Retrieved 2016-02-11.

[2] Einstein, Albert (1916). "The Foundation of the General Theory of Relativity" (PDF). *Annalen der Physik* **49** (7): 769–822. Bibcode:1916AnP...354..769E. doi:10.1002/andp.19163540702. Retrieved 2006-09-03.

[3] Einstein, Albert (1916). "The Foundation of the General Theory of Relativity" (English HTML, contains link to German PDF). *Annalen der Physik* **49** (7): 769–822.

[4] Einstein, Albert (1919). "What Is The Theory Of Relativity?" (PDF). German History in Documents and Images. Retrieved 7 June 2013.

[5] U. Le Verrier (1859), (in French), "Lettre de M. Le Verrier à M. Faye sur la théorie de Mercure et sur le mouvement du périhélie de cette planète", Comptes rendus hebdomadaires des séances de l'Académie des sciences (Paris), vol. 49 (1859), pp.379–383.

[6] Clemence, G. M. (1947). "The Relativity Effect in Planetary Motions". *Reviews of Modern Physics* **19** (4): 361–364. Bibcode:1947RvMP...19..361C. doi:10.1103/RevModPhys.19.361.

[7] Myles Standish, Jet Propulsion Laboratory (1998) http://classroom.sdmesa.edu/ssiegel/Physics%20197/labs/Mercury%20Precession.pdf

[8] http://www.tat.physik.uni-tuebingen.de/~{}kokkotas/Teaching/Experimental_Gravity_files/Hajime_PPN.pdf - Perihelion shift of Mercury, page 11

[9] Biswas, Abhijit; Mani, Krishnan R. S. (2008). "Relativistic perihelion precession of orbits of Venus and the Earth". *Central European Journal of Physics*. v1 6 (3): 754–758. arXiv:0802.0176. Bibcode:2008CEJPh...6..754B. doi:10.2478/s11534-008-0081-6.

[10] Matzner, Richard Alfred (2001). *Dictionary of geophysics, astrophysics, and astronomy*. CRC Press. p. 356. ISBN 0-8493-2891-8.

[11] Weisberg, J.M.; Taylor, J.H. (July 2005). "The Relativistic Binary Pulsar B1913+16: Thirty Years of Observations and Analysis". Written at San Francisco. In F.A. Rasio; I.H. Stairs. *Binary Radio Pulsars*. ASP Conference Series. Aspen, Colorado, USA: Astronomical Society of the Pacific. p. 25. arXiv:astro-ph/0407149. Bibcode:2005ASPC..328...25W.

[12] Dediu, Adrian-Horia; Magdalena, Luis; Martín-Vide, Carlos (2015). *Theory and Practice of Natural Computing: Fourth International Conference, TPNC 2015, Mieres, Spain, December 15-16, 2015. Proceedings* (illustrated ed.). Springer. p. 141. ISBN 978-3-319-26841-5. Extract of page 141

[13] Soldner, J. G. V. (1804). "On the deflection of a light ray from its rectilinear motion, by the attraction of a celestial body at which it nearly passes by". *Berliner Astronomisches Jahrbuch*: 161–172.

[14] Soares, Domingos S. L. (2009). "Newtonian gravitational deflection of light revisited". arXiv:physics/0508030.

[15] Will, C.M. (2006). "The Confrontation between General Relativity and Experiment". *Living Rev. Relativity* **9**: 39. arXiv:gr-qc/0510072. Bibcode:2006LRR.....9....3W. doi:10.12942/lrr-2006-3.

[16] Dyson, F. W.; Eddington, A. S.; Davidson C. (1920). "A determination of the deflection of light by the Sun's gravitational field, from observations made at the total eclipse of 29 May 1919". *Philosophical Transactions of the Royal Society* **220A**: 291–333. doi:10.1098/rsta.1920.0009.

[17] Stanley, Matthew (2003). "'An Expedition to Heal the Wounds of War': The 1919 Eclipse and Eddington as Quaker Adventurer". *Isis* **94** (1): 57–89. doi:10.1086/376099. PMID 12725104.

[18] Rosenthal-Schneider, Ilse: *Reality and Scientific Truth*. Detroit: Wayne State University Press, 1980. p 74. See also Calaprice, Alice: The New Quotable Einstein. Princeton: Princeton University Press, 2005. p 227.)

[19] Harry Collins and Trevor Pinch, *The Golem*, ISBN 0-521-47736-0

[20] Daniel Kennefick (2007). "Not Only Because of Theory: Dyson, Eddington and the Competing Myths of the 1919 Eclipse Expedition". arXiv:0709.0685 [physics.hist-ph].

[21] Ball, Philip (2007). "Arthur Eddington was innocent!". *News@nature*. doi:10.1038/news070903-20.

[22] D. Kennefick, "Testing relativity from the 1919 eclipse- a question of bias", *Physics Today*, March 2009, pp. 37–42.

[23] van Biesbroeck, G.: The relativity shift at the 1952 February 25 eclipse of the Sun., *Astronomical Journal*, vol. 58, page 87, 1953.

[24] Texas Mauritanian Eclipse Team: Gravitational deflection of-light: solar eclipse of 30 June 1973 I. Description of procedures and final results., *Astronomical Journal*, vol. 81, page 452, 1976.

[25] Pound, R. V.; Rebka, Jr. G. A. (November 1, 1959). "Gravitational Red-Shift in Nuclear Resonance". *Physical Review Letters* **3** (9): 439–441. Bibcode:1959PhRvL...3..439P. doi:10.1103/PhysRevLett.3.439.

[26] Pound, R. V.; Rebka Jr. G. A. (April 1, 1960). "Apparent weight of photons". *Physical Review Letters* **4** (7): 337–341. Bibcode:1960PhRvL...4..337P. doi:10.1103/PhysRevLett.4.337.

[27] Dicke, R. H. (March 6, 1959). "New Research on Old Gravitation: Are the observed physical constants independent of the position, epoch, and velocity of the laboratory?". *Science* **129** (3349): 621–624. Bibcode:1959Sci...129..621D. doi:10.1126/science.129.3349.621. PMID 17735811.

[28] Dicke, R. H. (1962). "Mach's Principle and Equivalence". *Evidence for gravitational theories: proceedings of course 20 of the International School of Physics "Enrico Fermi" ed C. Møller.*

[29] Schiff, L. I. (April 1, 1960). "On Experimental Tests of the General Theory of Relativity". *American Journal of Physics* **28** (4): 340–343. Bibcode:1960AmJPh..28..340S. doi:10.1119/1.1935800.

[30] Brans, C. H.; Dicke, R. H. (November 1, 1961). "Mach's Principle and a Relativistic Theory of Gravitation". *Physical Review* **124** (3): 925–935. Bibcode:1961PhRv..124..925B. doi:10.1103/PhysRev.124.925.

[31] Fact Sheet-BepiColombo

[32] Fomalont, E.B.; Kopeikin S.M.; Lanyi, G.; Benson, J. (July 2009). "Progress in Measurements of the Gravitational Bending of Radio Waves Using the VLBA". *Astrophysical Journal* **699** (2): 1395–1402. arXiv:0904.3992. Bibcode:2009ApJ...699.1395F. doi:10.1088/0004-637X/699/2/1395.

[33] Gaia overview

[34] Shapiro, I. I. (December 28, 1964). "Fourth test of general relativity". *Physical Review Letters* **13** (26): 789–791. Bibcode:1964PhRvL..13..789S. doi:10.1103/PhysRevLett.13.789.

[35] Shapiro, I. I.; Ash M. E.; Ingalls R. P.; Smith W. B.; Campbell D. B.; Dyce R. B.; Jurgens R. F. & Pettengill G. H. (May 3, 1971). "Fourth Test of General Relativity: New Radar Result". *Physical Review Letters* **26** (18): 1132–1135. Bibcode:1971PhRvL..26.1132S. doi:10.1103/PhysRevLett.26.1132.

[36] Bertotti B.; Iess L.; Tortora P. (2003). "A test of general relativity using radio links with the Cassini spacecraft". *Nature* **425** (6956): 374–376. Bibcode:2003Natur.425..374B. doi:10.1038/nature01997. PMID 14508481.

[37] Fomalont, E.B.; Kopeikin S.M. (November 2003). "The Measurement of the Light Deflection from Jupiter: Experimental Results". *Astrophysical Journal* **598** (1): 704–711. arXiv:astro-ph/0302294. Bibcode:2003ApJ...598..704F. doi:10.1086/378785.

[38] Kopeikin, S.M.; Fomalont E.B. (October 2007). "Gravimagnetism, causality, and aberration of gravity in the gravitational light-ray deflection experiments". *General Relativity and Gravitation* **39** (10): 1583–1624. arXiv:gr-qc/0510077. Bibcode:2007GReGr..39.1583K. doi:10.1007/s10714-007-0483-6.

[39] Fomalont, E.B.; Kopeikin, S. M.; Jones, D.; Honma, M.; Titov, O. (January 2010). "Recent VLBA/VERA/IVS tests of general relativity". *Proceedings of the International Astronomical Union, IAU Symposium* **261** (S261): 291–295. arXiv:0912.3421. Bibcode:2010IAUS..261..291F. doi:10.1017/S1743921309990536.

[40] Nordtvedt, Jr., K. (May 25, 1968). "Equivalence Principle for Massive Bodies. II. Theory". *Physical Review* **169** (5): 1017–1025. Bibcode:1968PhRv..169.1017N. doi:10.1103/PhysRev.169.1017.

[41] Nordtvedt, Jr., K. (June 25, 1968). "Testing Relativity with Laser Ranging to the Moon". *Physical Review* **170** (5): 1186–1187. Bibcode:1968PhRv..170.1186N. doi:10.1103/PhysRev.170.1186.

[42] Williams, J. G.; Turyshev, Slava G.; Boggs, Dale H. (December 29, 2004). "Progress in Lunar Laser Ranging Tests of Relativistic Gravity". *Physical Review Letters* **93** (5): 1017–1025. arXiv:gr-qc/0411113. Bibcode:2004PhRvL..93z1101W. doi:10.1103/PhysRevLett.93.261101.

[43] Uzan, J. P. (2003). "The fundamental constants and their variation: Observational status and theoretical motivations". *Reviews of Modern Physics* **75** (5): 403–. arXiv:hep-ph/0205340. Bibcode:2003RvMP...75..403U. doi:10.1103/RevModPhys.75.403.

[44] Vessot, R. F. C.; M. W. Levine; E. M. Mattison; E. L. Blomberg; T. E. Hoffman; G. U. Nystrom; B. F. Farrel; R. Decher; et al. (December 29, 1980). "Test of Relativistic Gravitation with a Space-Borne Hydrogen Maser". *Physical Review Letters* **45** (26): 2081–2084. Bibcode:1980PhRvL..45.2081V. doi:10.1103/PhysRevLett.45.2081.

[45] "Gravitational Physics with Optical Clocks in Space" (PDF). *S. Schiller* (PDF). Heinrich Heine Universität Düsseldorf. 2007. Retrieved 19 March 2015.

[46] Hafele, J. C.; Keating, R. E. (July 14, 1972). "Around-the-World Atomic Clocks: Predicted Relativistic Time Gains". *Science* **177** (4044): 166–168. Bibcode:1972Sci...177..166H. doi:10.1126/science.177.4044.166. PMID 17779917.

[47] Hafele, J. C.; Keating, R. E. (July 14, 1972). "Around-the-World Atomic Clocks: Observed Relativistic Time Gains". *Science* **177** (4044): 168–170. Bibcode:1972Sci...177..168H. doi:10.1126/science.177.4044.168. PMID 17779918.

[48] Ciufolini I. & Pavlis E.C. (2004). "A confirmation of the general relativistic prediction of the Lense–Thirring effect". *Nature* **431** (7011): 958–960. Bibcode:2004Natur.431..958C. doi:10.1038/nature03007. PMID 15496915.

[49] Krogh K. (2007). "Comment on 'Evidence of the gravitomagnetic field of Mars'". *Classical Quantum Gravity* **24** (22): 5709–5715. Bibcode:2007CQGra..24.5709K. doi:10.1088/0264-9381/24/22/N01.

[50] Merritt, D.; Alexander, T.; Mikkola, S.; Will, C. (2010). "Testing Properties of the Galactic Center Black Hole Using Stellar Orbits". *Physical Review D* **81** (6): 062002. arXiv:0911.4718. Bibcode:2010PhRvD..81f2002M. doi:10.1103/PhysRevD.81.062002.

[51] Will, C. (2008). "Testing the General Relativistic "No-Hair" Theorems Using the Galactic Center Black Hole Sagittarius A*". *Astrophysical Journal Letters* **674** (1): L25–L28. arXiv:0711.1677. Bibcode:2008ApJ...674L..25W. doi:10.1086/528847.

[52] Everitt; et al. (2011). "Gravity Probe B: Final Results of a Space Experiment to Test General Relativity". *Physical Review Letters* **106** (22): 221101. arXiv:1105.3456. Bibcode:2011PhRvL.106v1101E. doi:10.1103/PhysRevLett.106.221101. PMID 21702590.

[53] Ker Than. "Einstein Theories Confirmed by NASA Gravity Probe". News.nationalgeographic.com. Retrieved 2011-05-08.

[54] "Prepping satellite to test Albert Einstein".

[55] Ciufolini, I.; et al. (2009). "Towards a One Percent Measurement of Frame Dragging by Spin with Satellite Laser Ranging to LAGEOS, LAGEOS 2 and LARES and GRACE Gravity Models". *Space Science Reviews* **148**: 71–104. Bibcode:2009SSRv..148...71C. doi:10.1007/s11214-009-9585-7.

[56] Ciufolini, I.; Paolozzi A.; Pavlis E. C.; Ries J. C.; Koenig R.; Matzner R. A.; Sindoni G. & Neumayer H. (2009). "Towards a One Percent Measurement of Frame Dragging by Spin with Satellite Laser Ranging to LAGEOS, LAGEOS 2 and LARES and GRACE Gravity Models". *Space Science Reviews* **148**: 71–104. Bibcode:2009SSRv..148...71C. doi:10.1007/s11214-009-9585-7.

[57] Ciufolini, I.; Paolozzi A.; Pavlis E. C.; Ries J. C.; Koenig R.; Matzner R. A.; Sindoni G. & Neumayer H. (2010). "Gravitomagnetism and Its Measurement with Laser Ranging to the LAGEOS Satellites and GRACE Earth Gravity Models". *General Relativity and John Archibald Wheeler*. Astrophysics and Space Science Library **367**. SpringerLink. pp. 371–434. doi:10.1007/978-90-481-3735-0_17.

[58] Paolozzi, A.; Ciufolini I.; Vendittozzi C. (2011). "Engineering and scientific aspects of LARES satellite". *Acta Astronautica* **69** (3–4): 127–134. Bibcode:2011AcAau..69..127P. doi:10.1016/j.actaastro.2011.03.005. ISSN 0094-5765.

[59] "Einstein Was Right — So Far". *ESO Press Release*. Retrieved 30 April 2013.

[60] In general relativity, a perfectly spherical star (in vacuum) that expands or contracts while remaining perfectly spherical *cannot* emit any gravitational waves (similar to the lack of e/m radiation from a pulsating charge), as Birkhoff's theorem says that the geometry remains the same exterior to the star. More generally, a rotating system will only emit gravitational waves if it lacks the axial symmetry with respect to the axis of rotation.

[61] Stairs, Ingrid H. "Testing General Relativity with Pulsar Timing". *Living Reviews in Relativity* **6**. arXiv:astro-ph/0307536. Bibcode:2003LRR.....6....5S. doi:10.12942/lrr-2003-5.

[62] Weisberg, J. M.; Taylor, J. H.; Fowler, L. A. (October 1981). "Gravitational waves from an orbiting pulsar". *Scientific American* **245**: 74–82. Bibcode:1981SciAm.245...74W. doi:10.1038/scientificamerican1081-74.

[63] Weisberg, J. M.; Nice, D. J.; Taylor, J. H. (2010). "Timing Measurements of the Relativistic Binary Pulsar PSR B1913+16". *Astrophysical Journal* **722**: 1030–1034. arXiv:1011.0718v1. Bibcode:2010ApJ...722.1030W. doi:10.1088/0004-637X/722/2/1030.

[64] "Press Release: The Nobel Prize in Physics 1993". Nobel Prize. 13 October 1993. Retrieved 6 May 2014.

[65] Kramer, M.; et al. (2006). "Tests of general relativity from timing the double pulsar". *Science* **314** (5796): 97–102. arXiv:astro-ph/0609417. Bibcode:2006Sci...314...97K. doi:10.1126/science.1132305. PMID 16973838.

[66] Antoniadis, John; et al. (2013). "A Massive Pulsar in a Compact Relativistic Binary". *Science* (AAAS) **340** (6131): 1233232. arXiv:1304.6875. Bibcode:2013Sci...340..448A. doi:10.1126/science.1233232.

[67] "Massive double star is latest test for Einstein's gravity theory". *Ron Cowen*. Nature. 25 April 2013. Retrieved 7 May 2013.

[68] B. P. Abbott et al. (LIGO Scientific Collaboration and Virgo Collaboration) (2016). "Observation of Gravitational Waves from a Binary Black Hole Merger". *Physical Review Letters* **116** (6). doi:10.1103/PhysRevLett.116.061102.

[69] "Gravitational waves detected 100 years after Einstein's prediction | NSF - National Science Foundation". *www.nsf.gov*. Retrieved 2016-02-11.

[70] Schutz, Bernard F. "Gravitational waves on the back of an envelope". *American Journal of Physics* **52** (5): 412. Bibcode:1984AmJPh..52..412S. doi:10.1119/1.13627.

[71] Gair, Jonathan; Vallisneri, Michele; Larson, Shane L.; Baker, John G. "Testing General Relativity with Low-Frequency, Space-Based Gravitational-Wave Detectors". *Living Reviews in Relativity* **16**. arXiv:1212.5575. Bibcode:2013LRR....16....7G. doi:10.12942/lrr-2013-7.

[72] Yunes, Nicolás; Siemens, Xavier. "Gravitational-Wave Tests of General Relativity with Ground-Based Detectors and Pulsar-Timing Arrays". *Living Reviews in Relativity* **16**. arXiv:1304.3473. Bibcode:2013LRR....16....9Y. doi:10.12942/lrr-2013-9.

[73] Peebles, P. J. E. (December 2004). "Testing general relativity on the scales of cosmology": 106. arXiv:astro-ph/0410284. Bibcode:2005grg..conf..106P. doi:10.1142/9789812701688_0010. ISBN 978-981-256-424-5.

[74] Rudnicki, 1991, p. 28. *The Hubble Law was viewed by many as an observational confirmation of General Relativity in the early years*

[75] W.Pauli, 1958, pp.219–220

[76] Kragh, 2003, p. 152

[77] Kragh, 2003, p. 153

[78] Rudnicki, 1991, p. 28

[79] Chandrasekhar, 1980, p. 37

[80] Hand, Eric (2009). "Cosmology: The test of inflation". *Nature* **458**: 820–824. doi:10.1038/458820a.

[81] Reyes, Reinabelle; et al. (2010). "Confirmation of general relativity on large scales from weak lensing and galaxy velocities". *Nature* **464**: 256–258. arXiv:1003.2185. Bibcode:2010Natur.464..256R. doi:10.1038/nature08857. PMID 20220843.

[82] Guzzo, L.; et al. (2008). "A test of the nature of cosmic acceleration using galaxy redshift distortions". *Nature* **451**: 541–544. arXiv:0802.1944. Bibcode:2008Natur.451..541G. doi:10.1038/nature06555.

34.7.2   Other research papers

- Bertotti, B.; Iess, L.; Tortora, P. (2003). "A test of general relativity using radio links with the Cassini spacecraft". *Nature* **425** (6956): 374–6. Bibcode:2003Natur.425..374B. doi:10.1038/nature01997. PMID 14508481.

- Kopeikin, S.; Polnarev, A.; Schaefer, G.; Vlasov, I. (2007). "Gravimagnetic effect of the barycentric motion of the Sun and determination of the post-Newtonian parameter γ in the Cassini experiment". *Physics Letters A* **367** (4–5): 276–280. arXiv:gr-qc/0604060. Bibcode:2007PhLA..367..276K. doi:10.1016/j.physleta.2007.03.036.

- Brans, C.; Dicke, R. H. (1961). "Mach's principle and a relativistic theory of gravitation". *Phys. Rev.* **124** (3): 925–35. Bibcode:1961PhRv..124..925B. doi:10.1103/PhysRev.124.925.

- A. Einstein, "Über das Relativitätsprinzip und die aus demselben gezogene Folgerungen", *Jahrbuch der Radioaktivitaet und Elektronik* **4** (1907); translated "On the relativity principle and the conclusions drawn from it", in *The collected papers of Albert Einstein. Vol. 2 : The Swiss years: writings, 1900–1909* (Princeton University Press, Princeton, New Jersey, 1989), Anna Beck translator. Einstein proposes the gravitational redshift of light in this paper, discussed online at The Genesis of General Relativity.

- A. Einstein, "Über den Einfluß der Schwerkraft auf die Ausbreitung des Lichtes", *Annalen der Physik* **35**

(1911); translated "On the Influence of Gravitation on the Propagation of Light" in *The collected papers of Albert Einstein. Vol. 3 : The Swiss years: writings, 1909–1911* (Princeton University Press, Princeton, New Jersey, 1994), Anna Beck translator, and in *The Principle of Relativity*, (Dover, 1924), pp 99–108, W. Perrett and G. B. Jeffery translators, ISBN 0-486-60081-5. The deflection of light by the sun is predicted from the principle of equivalence. Einstein's result is half the full value found using the general theory of relativity.

- Shapiro, S. S.; Davis, J. L.; Lebach, D. E.; Gregory J.S. (26 March 2004). "Measurement of the solar gravitational deflection of radio waves using geodetic very-long-baseline interferometry data, 1979–1999". *Physical Review Letters* (American Physical Society) **92** (121101): 121101. Bibcode:2004PhRvL..92l1101S. doi:10.1103/PhysRevLett.92.121101. PMID 15089661.

- M. Froeschlé, F. Mignard and F. Arenou, "Determination of the PPN parameter γ with the Hipparcos data" Hipparcos Venice '97, ESA-SP-402 (1997).

- Will, Clifford M. (2006). "Was Einstein Right? Testing Relativity at the Centenary". *Annalen der Physik* **15**: 19–33. arXiv:gr-qc/0504086. Bibcode:2006AnP...518...19W. doi:10.1002/andp.200510170.

- Rudnicki, Conrad (1991). "What are the Empirical Bases of the Hubble Law" (PDF). *Apeiron* (9–10): 27–36. Retrieved 2009-06-23.

- Chandrasekhar, S. (1980). "The Role of General Relativity in Astronomy: Retrospect and Prospect" (PDF). *J. Astrophys. Astr.* **1** (1): 33–45. Bibcode:1980JApA....1...33C. doi:10.1007/BF02727948. Retrieved 2009-06-23.

- Kragh, Helge; Smith, Robert W. (2003). "Who discovered the expanding universe". *History of Science* **41**: 141–62. Bibcode:2003HisSc..41..141K. Retrieved 2013-02-15.

34.7.3 Textbooks

- S. M. Carroll, *Spacetime and Geometry: an Introduction to General Relativity*, Addison-Wesley, 2003. An introductory general relativity textbook.

- A. S. Eddington, *Space, Time and Gravitation*, Cambridge University Press, reprint of 1920 ed.

- A. Gefter, "Putting Einstein to the Test", *Sky and Telescope* July 2005, p. 38. A popular discussion of tests of general relativity.

- H. Ohanian and R. Ruffini, *Gravitation and Spacetime, 2nd Edition* Norton, New York, 1994, ISBN 0-393-96501-5. A general relativity textbook.

- Pauli, Wolfgang Ernst (1958). "Part IV. General Theory of Relativity". *Theory of Relativity*. Courier Dover Publications. ISBN 978-0-486-64152-2.

- C. M. Will, *Theory and Experiment in Gravitational Physics*, Cambridge University Press, Cambridge (1993). A standard technical reference.

- C. M. Will, *Was Einstein Right?: Putting General Relativity to the Test*, Basic Books (1993). This is a popular account of tests of general relativity.

34.7.4 Living Reviews papers

- N. Ashby, "Relativity in the Global Positioning System", *Living Reviews in Relativity* (2003).

- C. M. Will, The Confrontation between General Relativity and Experiment, *Living Reviews in Relativity* (2006). An online, technical review, covering much of the material in *Theory and experiment in gravitational physics*. It is less comprehensive but more up to date.

34.8 External links

- the USENET Relativity FAQ experiments page
- Mathpages article on Mercury's perihelion shift (for amount of observed and GR shifts).

Chapter 35

Nordtvedt effect

In theoretical astrophysics, the **Nordtvedt effect** refers to the relative motion between the Earth and the Moon which would be observed if the gravitational self-energy of a body contributed differently to its gravitational mass than to its inertial mass. If observed, the Nordtvedt effect would violate the strong equivalence principle, which indicates that an object's movement in a gravitational field does not depend on its mass or composition.

The effect is named after Dr. Kenneth L. Nordtvedt, who first demonstrated that some theories of gravity suggest that massive bodies should fall at different rates, depending upon their gravitational self-energy.

Nordtvedt then observed that if gravity did in fact violate the strong equivalence principle, then the more-massive Earth should fall towards the Sun at a slightly different rate than the Moon, resulting in a polarization of the lunar orbit. To test for the existence (or absence) of the Nordtvedt effect, scientists have used the Lunar Laser Ranging Experiment, which is capable of measuring the distance between the Earth and the Moon with near-millimetre accuracy. Thus far, the results have failed to find any evidence of the Nordtvedt effect, demonstrating that if it exists, the effect is exceedingly weak.[1] Subsequent measurements and analysis to even higher precision have improved constraints on the effect.[2][3]

A wide range of scalar-tensor theories have been found to naturally lead to a tiny effect only, at present epoch. This is due to a generic attractive mechanism that takes place during the cosmic evolution of the universe.[4] Other screening mechanisms [5] (chameleon, pressuron, Vainshtein etc.) could also be at play.

35.1   See also

- Woodward effect

- Galileo's Leaning Tower of Pisa experiment

35.2   References

Nordtvedt Jr., Kenneth (25 May 1968) "Equivalence Principle for Massive Bodies, II. Theory" Phys Rev. 169, No. 5 (1017)

Nordtvedt Jr., K (25 June 1968) "Testing Relativity with Laser Ranging to the Moon" Phys. Rev. 170, No. 5 (1186)

[1] Murphy, Jr., T. W. "THE APACHE POINT OBSERVATORY LUNAR LASER-RANGING OPERATION" (PDF). Retrieved 5 February 2013.

[2] Adelberger, E.G.; Heckel, B.R.; Smith, G.; Su, Y. & Swanson, H.E. (Sep 20, 1990), "Eötvös experiments, lunar ranging and the strong equivalence principle", *Nature* 347 (6290): 261–263, Bibcode:1990Natur.347..261A, doi:10.1038/347261a0

[3] Williams, J.G.; Newhall, X.X. & Dickey, J.O. (1996), "Relativity parameters determined from lunar laser ranging", *Phys. Rev.* D 53: 6730–6739, Bibcode:1996PhRvD..53.6730W, doi:10.1103/PhysRevD.53.6730

[4] Damour, T. & Nordtvedt, K. (April 1993), "General relativity as a cosmological attractor of tensor-scalar theories", *Physical Review Letters* 70: 2217–2219, Bibcode:1993PhRvL..70.2217D, doi:10.1103/physrevlett.70.2217

[5] Brax, P. (4 October 2013), "Screening mechanisms in modified gravity", *Classical and Quantum Gravity* 30, Bibcode:2013CQGra..30u4005B, doi:10.1088/0264-9381/30/21/214005

Chapter 36

Lunar Laser Ranging experiment

The Lunar Laser Ranging Experiment from the Apollo 11 mission.

Apollo 15 LRRR

The ongoing **Lunar Laser Ranging Experiment** measures the distance between Earth and the Moon using laser ranging. Lasers on Earth are aimed at retroreflectors planted on the Moon during the Apollo program (11, 14, and 15) and the two Lunokhod missions.[1] The time for the reflected light to return is measured.

The first successful tests were carried out in 1962 when a team from the Massachusetts Institute of Technology succeeded in observing laser pulses reflected from moon's surface using a laser with a millisecond pulse length. Similar measurements were obtained later the same year by a Soviet team at the Crimean Astrophysical Observatory using a Q-switched ruby laser.[2] Greater accuracy was achieved following the installation of a retroreflector array on July 21, 1969, by the crew of Apollo 11, and two more retroreflector arrays left by the Apollo 14 and Apollo 15 missions have also contributed to the experiment. Successful lunar laser range measurements to the retroreflectors were first reported by the 3.1 m telescope at Lick Obser-

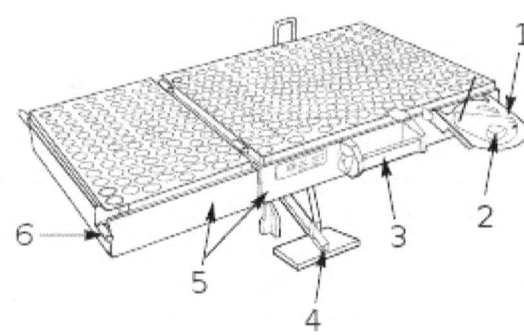

Apollo 15 LRRR schematic

vatory, Air Force Cambridge Research Laboratories Lunar Ranging Observatory in Arizona, the Pic du Midi Observatory in France, the Tokyo Astronomical Observatory, and

230

McDonald Observatory in Texas.

The unmanned Soviet *Lunokhod 1* and *Lunokhod 2* rovers carried smaller arrays. Reflected signals were initially received from *Lunokhod 1*, but no return signals were detected after 1971 until a team from University of California rediscovered the array in April 2010 using images from NASA's Lunar Reconnaissance Orbiter.[3] *Lunokhod 2's* array continues to return signals to Earth.[4] The Lunokhod arrays suffer from decreased performance in direct sunlight, a factor which was considered in the reflectors placed during the Apollo missions.[5]

The Apollo 15 array is three times the size of the arrays left by the two earlier Apollo missions. Its size made it the target of three-quarters of the sample measurements taken in the first 25 years of the experiment. Improvements in technology since then have resulted in greater use of the smaller arrays, by sites such as the Côte d'Azur Observatory in Grasse, France; and the Apache Point Observatory Lunar Laser-ranging Operation (APOLLO) at the Apache Point Observatory in New Mexico.

36.1 Details

The distance to the Moon is calculated *approximately* using this equation:

> *Distance = (Speed of light × Time taken for light to reflect) / 2.*

In actuality, the round-trip time of about 2.5 seconds is affected by the location of the Moon in the sky, the relative motion of Earth and the Moon, Earth's rotation, lunar libration, weather, polar motion, propagation delay through Earth's atmosphere, the motion of the observing station due to crustal motion and tides, velocity of light in various parts of air and relativistic effects.[6] Nonetheless, the Earth–Moon distance has been measured with increasing accuracy for more than 35 years. The distance continually changes for a number of reasons, but averages 385,000.6 km (239,228.3 mi).[7]

At the Moon's surface, the beam is about 6.5 kilometers wide[8] and scientists liken the task of aiming the beam to using a rifle to hit a moving dime 3 kilometers away. The reflected light is too weak to be seen with the human eye: out of 10^{17} photons aimed at the reflector, only one will be received back on Earth every few seconds, even under good conditions. They can be identified as originating from the laser because the laser is highly monochromatic. This is one of the most precise distance measurements ever made, and is equivalent in accuracy to determining the distance between Los Angeles and New York to 0.25 mm.[5][9] As

of 2002, work is progressing on increasing the accuracy of the Earth–Moon measurements to near millimeter accuracy, though the performance of the reflectors continues to degrade with age.[5]

36.2 Results

Lunar laser ranging measurement data is available from the Paris Observatory Lunar Analysis Center,[10] and the active stations. Some of the findings of this long-term experiment are:

- The Moon is spiraling away from Earth at a rate of 3.8 cm per year.[8] This rate has been described as anomalously high.[11]

- The Moon probably has a liquid core of about 20% of the Moon's radius.[4]

- The universal force of gravity is very stable. The experiments have constrained the change in Newton's gravitational constant G to $(2\pm7)\times10^{-13}$ per year. [12]

- The likelihood of any "Nordtvedt effect" (a differential acceleration of the Moon and Earth towards the Sun caused by their different degrees of compactness) has been ruled out to high precision,[13][14] strongly supporting the validity of the Strong Equivalence Principle.

- Einstein's theory of gravity (the general theory of relativity) predicts the Moon's orbit to within the accuracy of the laser ranging measurements.[4]

- Gauge freedom plays a major role in a correct physical interpretation of the relativistic effects in the Earth-Moon system observed with LLR technique [15]

36.3 Photo gallery

- Apollo 14 Lunar Ranging Retro Reflector (LRRR).

- APOLLO Collaboration photon pulse return times

- Laser ranging facility at Wettzell fundamental station, Bavaria, Germany.

- Laser Ranging at Goddard Spaceflight Center.

36.4 See also

- Apache Point Observatory Lunar Laser-ranging Operation

- Apollo Lunar Surface Experiments Package
- Tom Murphy (Physicist) (principal investigator of Apollo's reflector experiment)
- Carroll Alley (previous principal investigator of Apollo's reflector experiment)
- EME (communications)
- Lidar
- Lunar distance (astronomy)
- Lunokhod programme
- Satellite laser ranging
- Third-party evidence for Apollo Moon landings

36.5   References

[1] Chapront, J.; Chapront-Touzé, M.; Francou, G. (1999). "Determination of the lunar orbital and rotational parameters and of the ecliptic reference system orientation from LLR measurements and IERS data". *Astronomy and Astrophysics*: 624–633.

[2] Bender, P. L.; Currie, D. G.; Dicke, R. H.; et al. (October 19, 1973). "The Lunar Laser Ranging Experiment" (PDF). *Science* **182** (4109): 229–238. Bibcode:1973Sci...182..229B. doi:10.1126/science.182.4109.229. PMID 17749298. Retrieved April 27, 2013. line feed character in |journal= at position 20 (help)

[3] McDonald, Kim (April 26, 2010). "UC San Diego Physicists Locate Long Lost Soviet Reflector on Moon". UCSD. Retrieved 27 April 2010.

[4] Williams, James G. & Dickey, Jean O. "Lunar Geophysics, Geodesy, and Dynamics" (PDF). ilrs.gsfc.nasa.gov. Retrieved 2008-05-04. 13th International Workshop on Laser Ranging, October 7–11, 2002, Washington, D. C.

[5] "It's Not Just The Astronauts That Are Getting Older". Universe Today. March 10, 2010. Retrieved 24 August 2012.

[6] Seeber, Gunter. *Satellite Geodesy 2nd Edition.* de Gruyter. 2003, p. 439

[7] Murphy, T W (1 July 2013). "Lunar laser ranging: the millimeter challenge" (PDF). *Reports on Progress in Physics* **76** (7): 2. arXiv:1309.6294. Bibcode:2013RPPh...76g6901M. doi:10.1088/0034-4885/76/7/076901.

[8] Fred Espenek (August 1994). "NASA - Accuracy of Eclipse Predictions". eclipse.gsfc.nasa.gov. Retrieved 2008-05-04.

[9] "Apollo 11 Experiment Still Going Strong after 35 Years". www.jpl.nasa.gov. July 20, 2004. Retrieved 2008-05-04.

[10] "LUNAR LASER RANGING OBSERVATIONS FROM 1969 TO MAY 2013" SYRTE Paris Observatory, retrieved 3 June 2014

[11] Bills, B.G. & Ray, R.D. (1999), "Lunar Orbital Evolution: A Synthesis of Recent Results", *Geophysical Research Letters* **26** (19): 3045–3048, Bibcode:1999GeoRL..26.3045B, doi:10.1029/1999GL008348

[12] Müller, Jürgen; Liliane Biskupek (2007). "Variations of the gravitational constant from lunar laser ranging data". *Classical and Quantum Gravity* **24** (17): 4533. doi:10.1088/0264-9381/24/17/017. Retrieved 7 May 2014.

[13] Adelberger, E.G.; Heckel, B.R.; Smith, G.; Su, Y. & Swanson, H.E. (20 September 1990), "Eötvös experiments, lunar ranging and the strong equivalence principle", *Nature* **347** (6290): 261–263, Bibcode:1990Natur.347..261A, doi:10.1038/347261a0

[14] Williams, J.G.; Newhall, X.X. & Dickey, J.O. (1996), "Relativity parameters determined from lunar laser ranging", *Phys. Rev. D* **53**: 6730–6739, Bibcode:1996PhRvD..53.6730W, doi:10.1103/PhysRevD.53.6730

[15] Kopeikin, Sergei; Xie, Yi (2010), "Celestial reference frames and the gauge freedom in the post-Newtonian mechanics of the Earth–Moon system" (PDF), *Celest. Mech. Dyn. Astr.* **108**: 245–263, Bibcode:2010CeMDA.108..245K, doi:10.1007/s10569-010-9303-5

36.6   External links

- "Theory and Model for the New Generation of the Lunar Laser Ranging Data" by Sergei Kopeikin
- Apollo 15 Experiments - Laser Ranging Retroreflector by the Lunar and Planetary Institute
- "History of Laser Ranging and MLRS" by the University of Texas at Austin, Center for Space Research
- "Lunar Retroreflectors" by Tom Murphy
- Station de Télémétrie Laser-Lune in Grasse, France
- Lunar Laser Ranging from International Laser Ranging Service
- "UW researcher plans project to pin down moon's distance from Earth" by Vince Stricherz, UW Today, January 14, 2002
- "What Neil & Buzz Left on the Moon" by Science@NASA, July 20, 2004
- "Apollo 11 Experiment Still Returning Results" by Robin Lloyd, CNN, July 21, 1999

Chapter 37

Very-long-baseline interferometry

Some of the Atacama Large Millimeter Array radio telescopes.

The eight radio telescopes of the Smithsonian Submillimeter Array, located at the Mauna Kea Observatory in Hawai'i.

Very-long-baseline interferometry (VLBI) is a type of astronomical interferometry used in radio astronomy. In VLBI a signal from an astronomical radio source, such as a quasar, is collected at multiple radio telescopes on Earth. The distance between the radio telescopes is then calculated using the time difference between the arrivals of the radio signal at different telescopes. This allows observations of an object that are made simultaneously by many radio telescopes to be combined, emulating a telescope with a size equal to the maximum separation between the telescopes.

Data received at each antenna in the array include arrival times from a local atomic clock, such as a hydrogen maser.

At a later time, the data are correlated with data from other antennas that recorded the same radio signal, to produce the resulting image. The resolution achievable using interferometry is proportional to the observing frequency. The VLBI technique enables the distance between telescopes to be much greater than that possible with conventional interferometry, which requires antennas to be physically connected by coaxial cable, waveguide, optical fiber, or other type of transmission line. The greater telescope separations are possible in VLBI due to the development of the closure phase imaging technique by Roger Jennison in the 1950s, allowing VLBI to produce images with superior resolution.

VLBI is most well known for imaging distant cosmic radio sources, spacecraft tracking, and for applications in astrometry. However, since the VLBI technique measures the time differences between the arrival of radio waves at separate antennas, it can also be used "in reverse" to perform earth rotation studies, map movements of tectonic plates very precisely (within millimetres), and perform other types of geodesy. Using VLBI in this manner requires large numbers of time difference measurements from distant sources (such as quasars) observed with a global network of antennas over a period of time.

37.1   Scientific results

Some of the scientific results derived from VLBI include:

- High resolution radio imaging of cosmic radio sources.

- Imaging the surfaces of nearby stars at radio wavelengths (see also interferometry) – similar techniques have also been used to make infrared and optical images of stellar surfaces

- Definition of the celestial reference frame

- Motion of the Earth's tectonic plates

Geodesist Chopo Ma explains some of the geodetic uses of VLBI.

- Regional deformation and local uplift or subsidence.

- Variations in the Earth's orientation and length of day.

- Maintenance of the terrestrial reference frame

- Measurement of gravitational forces of the Sun and Moon on the Earth and the deep structure of the Earth

- Improvement of atmospheric models

- Measurement of the fundamental speed of gravity

- The tracking of the Huygens probe as it passed through Titan's atmosphere, allowing wind velocity measurements

37.2   VLBI arrays

There are several VLBI arrays located in Europe, Canada, the United States, Russia, Korea, Japan, Mexico and Australia. The most sensitive VLBI array in the world is the European VLBI Network (EVN). This is a part-time array which brings together the largest European radiotelescopes for typically week-long sessions, with the data being processed at the Joint Institute for VLBI in Europe (JIVE). The Very Long Baseline Array (VLBA) uses ten dedicated, 25-meter telescopes spanning 5351 miles across the United States, and is the largest VLBI array that operates all year round as both an astronomical and geodesy instrument.[1] The combination of the EVN and VLBA is known as Global VLBI. When one or both of these arrays are combined with one or more space-based VLBI antennas such as HALCA (previously) and now with RadioAstron (Spektr-R), the resolution obtained is higher than any other astronomical instrument, capable of imaging the sky with a level of detail measured in microarcseconds. VLBI generally benefits from the longer baselines afforded by international collaboration, with a notable early example in 1976, when radio telescopes in the United States, USSR and Australia were linked to observe hydroxyl-maser sources.[2]

37.3   e-VLBI

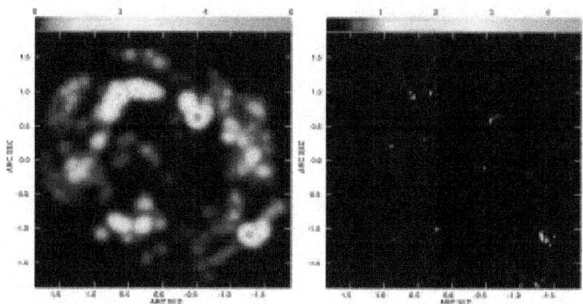

Image of the source IRC+10420. The lower resolution image on the left image was taken with the UK's MERLIN array and shows the shell of maser emission produced by an expanding shell of gas with a diameter about 200 times that of the Solar System. The shell of gas was ejected from a supergiant star (10 times the mass of our sun) at the centre of the emission about 900 years ago. The corresponding EVN e-VLBI image (right) shows the much finer structure of the masers because of the higher resolution of the VLBI array.

VLBI has traditionally operated by recording the signal at each telescope on magnetic tapes or disks, and shipping those to the correlation center for replay. Recently, it has become possible to connect VLBI radio telescopes in close to real-time, while still employing the local time references of the VLBI technique, in a technique known as **e-VLBI**. In Europe, six radio telescopes of the European VLBI Network (EVN) are now connected with Gigabit per second links via their National Research Networks and the Pan-European research network GEANT2, and the first astronomical experiments using this new technique were successfully conducted in 2011.[3]

The image to the right shows the first science produced by the European VLBI Network using e-VLBI. The data from 6 telescopes were processed in real time at the European Data Processing centre at JIVE. The Netherlands Academic Research Network SURFnet provides 6 x 1 Gbit/s connectivity between JIVE and the GEANT2 network.

37.4   Space VLBI

In the quest for even greater angular resolution, dedicated VLBI satellites have been placed in Earth orbit to provide greatly extended baselines. Experiments incorporating such space-borne array elements are termed Space Very Long Baseline Interferometry (SVLBI).

The first such dedicated VLBI mission was HALCA, an 8-meter radio telescope, which was launched in February 1997 and made observations until October 2003, but due to the small size of the dish only very strong radio sources could be observed with SVLBI arrays incorporating it.

Another space VLBI mission, Spektr-R (or RadioAstron), was launched in July 2011.

37.5 How VLBI Works

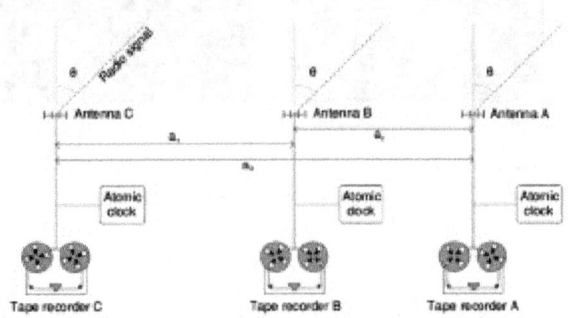

Recording data at each of the telescopes in a VLBI array. Extremely accurate high-frequency clocks are recorded alongside the astronomical data in order to help get the synchronization correct

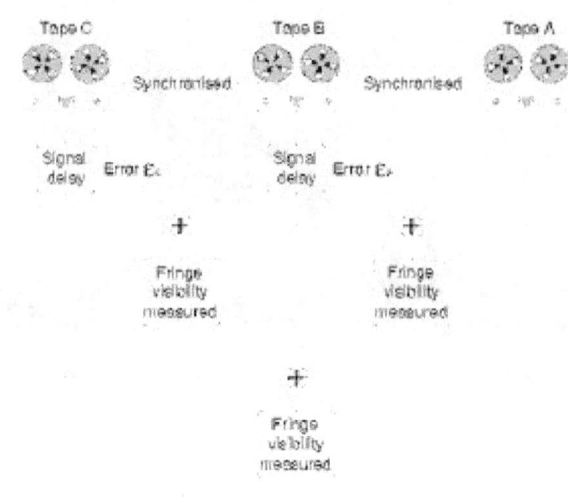

Playing back the data from each of the telescopes in a VLBI array. Great care must be taken to synchronize the play back of the data from different telescopes. Atomic clock signals recorded with the data help in getting the timing correct.

In VLBI interferometry, the digitized antenna data are usually recorded at each of the telescopes (in the past this was done on large magnetic tapes, but nowadays it is usually done on large RAID arrays of computer disk drives). The antenna signal is sampled with an extremely precise and stable atomic clock (usually a hydrogen maser) that is additionally locked onto a GPS time standard. Alongside the astronomical data samples, the output of this clock is recorded on the tape/disk media. The recorded media are then transported to a central location. More recent experiments have been conducted with "electronic" VLBI (e-VLBI) where the data are sent by fibre-optics (e.g., 10 Gbit/s fiber-optic paths in the European GEANT2 research network) and not recorded at the telescopes, speeding up and simplifying the observing process significantly. Even though the data rates are very high, the data can be sent over normal Internet connections taking advantage of the fact that many of the international high speed networks have significant spare capacity at present.

At the location of the **correlator** the data are played back. The timing of the playback is adjusted according to the atomic clock signals on the (tapes/disk drives/fibre optic signal), and the estimated times of arrival of the radio signal at each of the telescopes. A range of playback timings over a range of nanoseconds are usually tested until the correct timing is found.

Each antenna will be a different distance from the radio source, and as with the short baseline radio interferometer the delays incurred by the extra distance to one antenna must be added artificially to the signals received at each of the other antennas. The approximate delay required can be calculated from the geometry of the problem. The tape playback is synchronized using the recorded signals from the atomic clocks as time references, as shown in the drawing on the right. If the position of the antennas is not known to sufficient accuracy or atmospheric effects are significant, fine adjustments to the delays must be made until interference fringes are detected. If the signal from antenna A is taken as the reference, inaccuracies in the delay will lead to errors ϵ_B and ϵ_C in the phases of the signals from tapes B and C respectively (see drawing on right). As a result of these errors the phase of the complex visibility cannot be measured with a very-long-baseline interferometer.

The phase of the complex visibility depends on the symmetry of the source brightness distribution. Any brightness distribution can be written as the sum of a symmetric component and an anti-symmetric component. The symmetric component of the brightness distribution only contributes to the real part of the complex visibility, while the anti-symmetric component only contributes to the imaginary part. As the phase of each complex visibility measurement cannot be determined with a very-long-baseline interferometer the symmetry of the corresponding contribution to the source brightness distributions is not known.

R. C. Jennison developed a novel technique for obtaining information about visibility phases when delay errors are present, using an observable called the closure phase. Although his initial laboratory measurements of closure phase had been done at optical wavelengths, he foresaw greater potential for his technique in radio interferometry. In 1958 he demonstrated its effectiveness with a radio interferometer, but it only became widely used for long-baseline radio interferometry in 1974. At least three antennas are required. This method was used for the first VLBI measurements, and a modified form of this approach ("Self-Calibration") is still used today.

37.6   References

[1] http://www.nrao.edu/index.php/about/facilities/vlba

[2] First Global Radio Telescope, Sov. Astron., Oct 1976

[3] "Astronomers Demonstrate a Global Internet Telescope". Retrieved 2011-05-06.

37.7   External links

- E-MERLIN fibre-linked radio telescope array used in VLBI observations

- EXPReS Express Production Real-time e-VLBI Service: a three-year project (est. March 2006) funded by the European Commission to develop an intercontinental e-VLBI instrument available to the scientific community

- JIVE Joint Institute for VLBI in Europe

- The International VLBI Service for Geodesy and Astrometry (IVS)

- IVSOPAR: the VLBI analysis center at the Paris Observatory

- "VLBI - Canada's Role"

Chapter 38

Kaluza–Klein theory

This article is about gravitation and electromagnetism. For the mathematical generalization of K theory, see KK-theory.

In physics, **Kaluza–Klein theory (KK theory)** is a unified field theory of gravitation and electromagnetism built around the idea of a fifth dimension beyond the usual four of space and time. It is considered to be an important precursor to string theory.

The five-dimensional theory was developed in three steps. The original hypothesis came from Theodor Kaluza, who sent his results to Einstein in 1919,[1] and published them in 1921.[2] Kaluza's theory was a purely classical extension of general relativity to five dimensions. The five-dimensional metric has 15 components. Ten components are identified with the four-dimensional spacetime metric, four components with the electromagnetic vector potential, and one component with an unidentified scalar field sometimes called the "radion" or the "dilaton". Correspondingly, the five-dimensional Einstein equations yield the four-dimensional Einstein field equations, the Maxwell equations for the electromagnetic field, and an equation for the scalar field. Kaluza also introduced the hypothesis known as the "cylinder condition", that no component of the five-dimensional metric depends on the fifth dimension. Without this assumption, the field equations of five-dimensional relativity are enormously more complex. Standard four-dimensional physics seems to manifest the cylinder condition. Kaluza also set the scalar field equal to a constant, in which case standard general relativity and electrodynamics are recovered identically.

In 1926, Oskar Klein gave Kaluza's classical five-dimensional theory a quantum interpretation,[3][4] to accord with the then-recent discoveries of Heisenberg and Schrödinger. Klein introduced the hypothesis that the fifth dimension was curled up and microscopic, to explain the cylinder condition. Klein also calculated a scale for the fifth dimension based on the quantum of charge.

It wasn't until the 1940s that the classical theory was

completed, and the full field equations including the scalar field were obtained by three independent research groups:[5] Thiry,[6][7][8] working in France on his dissertation under Lichnerowicz; Jordan, Ludwig, and Müller in Germany,[9][10][11][12][13] with critical input from Pauli and Fierz; and Scherrer [14][15][16] working alone in Switzerland. Jordan's work led to the scalar-tensor theory of Brans & Dicke;[17] Brans and Dicke were apparently unaware of Thiry or Scherrer. The full Kaluza equations under the cylinder condition are quite complex, and most English-language reviews as well as the English translations of Thiry contain some errors. The complete Kaluza equations were evaluated using tensor algebra software in 2015.[18]

38.1   Kaluza hypothesis

In his 1921 paper,[2] Kaluza established all the elements of the classical five-dimensional theory: the metric, the field equations, the equations of motion, the stress-energy tensor, and the cylinder condition. The theory has no free parameters; it merely extends general relativity to five dimensions. One starts by hypothesizing a form of the five-dimensional metric $\widetilde{g}_{ab}$, where Roman indices span five dimensions. Let one also introduce the four-dimensional spacetime metric $g_{\mu\nu}$, where Greek indices span the usual four dimensions of space and time; a 4-vector A^{μ} which will be identified with the electromagnetic vector potential; and a scalar field ϕ. Then decompose the 5D metric so that the 4D metric is framed by the electromagnetic vector potential, with the scalar field at the fifth diagonal. This can be visualized as:

$$\widetilde{g}_{ab} \equiv \begin{bmatrix} g_{\mu\nu} + \phi^2 A_\mu A_\nu & \phi^2 A_\mu \\ \phi^2 A_\nu & \phi^2 \end{bmatrix}.$$

More precisely, one can write

$$\widetilde{g}_{\mu\nu} \equiv g_{\mu\nu} + \phi^2 A_\mu A_\nu, \qquad \widetilde{g}_{5\nu} \equiv \widetilde{g}_{\nu5} \equiv \phi^2 A\nu \,,$$

$$\widetilde{g}_{55} \equiv \phi^2$$

237

where the index 5 indicates the fifth coordinate by convention even though the first four coordinates are indexed with 0, 1, 2, and 3. The associated inverse metric is

$$\tilde{g}^{ab} \equiv \begin{bmatrix} g^{\mu\nu} & -A^{\mu} \\ -A^{\nu} & g_{\alpha\beta}A^{\alpha}A^{\beta} + \frac{1}{\phi^2} \end{bmatrix}.$$

So far, this decomposition is quite general and all terms are dimensionless. Kaluza then applies the machinery of standard general relativity to this metric. The field equations are obtained from five-dimensional Einstein equations, and the equations of motion are obtained from the five-dimensional geodesic hypothesis. The resulting field equations provide both the equations of general relativity and of electrodynamics; the equations of motion provide the four-dimensional geodesic equation and the Lorentz force law, and one finds that electric charge is identified with motion in the fifth dimension.

The hypothesis for the metric implies an invariant five-dimensional length element ds :

$$ds^2 \equiv \tilde{g}_{ab}dx^a dx^b = g_{\mu\nu}dx^\mu dx^\nu + \phi^2(A_\nu dx^\nu + dx^5)^2$$

38.2 Field equations from the Kaluza hypothesis

The field equations of the 5-dimensional theory were never adequately provided by Kaluza or Klein, mainly regarding the scalar field. The full Kaluza field equations are generally attributed to Thiry,[7] who most famously obtained vacuum field equations, although Kaluza [2] originally provided a stress-energy tensor for his theory and Thiry included a stress-energy tensor in his thesis. But as described by Gonner,[5] several independent groups worked on the field equations in the 1940s and earlier. Thiry is perhaps best known only because an English translation was provided by Applequist, Chodos, & Freund in their review book.[19] Applequist et al. also provided an English translation of Kaluza's paper. There are no English translations of the Jordan papers.[9][10][12]

To obtain the 5D field equations, the 5D connections $\tilde{\Gamma}^a_{bc}$ are calculated from the 5D metric $\tilde{g}_{ab}$, and the 5D Ricci tensor $\tilde{R}_{ab}$ is calculated from the 5D connections.

The classic results of Thiry and other authors presume the cylinder condition:

$$\frac{\partial \tilde{g}_{ab}}{\partial x^5} = 0$$

Without this assumption, the field equations become much more complex, providing many more degrees of freedom that can be identified with various new fields. Paul Wesson and colleagues have pursued relaxation of the cylinder condition to gain extra terms that can be identified with the matter fields,[20] for which Kaluza [2] otherwise inserted a stress-energy tensor by hand.

It has been an objection to the original Kaluza hypothesis to invoke the fifth dimension only to negate its dynamics. But Thiry argued [5] that the interpretation of the Lorentz force law in terms of a 5-dimensional geodesic mitigates strongly for a fifth dimension irrespective of the cylinder condition. Most authors have therefore employed the cylinder condition in deriving the field equations. Furthermore, vacuum equations are typically assumed for which

$$\tilde{R}_{ab} = 0$$

where

$$\tilde{R}_{ab} \equiv \partial_c \tilde{\Gamma}^c_{ab} - \partial_b \tilde{\Gamma}^c_{ca} + \tilde{\Gamma}^c_{cd}\tilde{\Gamma}^d_{ab} - \tilde{\Gamma}^c_{bd}\tilde{\Gamma}^d_{ac}$$

and

$$\tilde{\Gamma}^a_{bc} \equiv \frac{1}{2}\tilde{g}^{ad}(\partial_b \tilde{g}_{dc} + \partial_c \tilde{g}_{db} - \partial_d \tilde{g}_{bc})$$

The vacuum field equations obtained in this way by Thiry [7] and Jordan's group [9][10][12] are as follows.

The field equation for ϕ is obtained from

$$\tilde{R}_{55} = 0 \Rightarrow \Box\phi = \frac{1}{4}\phi^3 F^{\alpha\beta}F_{\alpha\beta}$$

where $F_{\alpha\beta} \equiv \partial_\alpha A_\beta - \partial_\beta A_\alpha$, where $\Box \equiv g^{\mu\nu}\nabla_\mu\nabla_\nu$, and where ∇_μ is a standard, 4D covariant derivative. It shows that the electromagnetic field is a source for the scalar field. Note that the scalar field cannot be set to a constant without constraining the electromagnetic field. The earlier treatments by Kaluza and Klein did not have an adequate description of the scalar field, and did not realize the implied constraint on the electromagnetic field by assuming the scalar field to be constant.

The field equation for A^ν is obtained from

$$\tilde{R}_{5\alpha} = 0 = \frac{1}{2}g^{\beta\mu}\nabla_\mu(\phi^3 F_{\alpha\beta})$$

It has the form of the vacuum Maxwell equations if the scalar field is constant.

The field equation for the 4D Ricci tensor $R_{\mu\nu}$ is obtained from

$$\widetilde{R}_{\mu\nu} - \frac{1}{2}\widetilde{g}_{\mu\nu}\widetilde{R} = 0 \Rightarrow R_{\mu\nu} - \frac{1}{2}g_{\mu\nu}R = \frac{1}{2}\phi^2\left(g^{\alpha\beta}F_{\mu\alpha}F_{\nu\beta} - \frac{1}{4}g_{\mu\nu}F_{\alpha\beta}F^{\alpha\beta}\right) + \frac{1}{\phi}\left(\nabla_\mu\nabla_\nu\phi - g_{\mu\nu}\Box\phi\right)$$

where R is the standard 4D Ricci scalar.

This equation shows the remarkable result, called the "Kaluza miracle", that the precise form for the electromagnetic stress-energy tensor emerges from the 5D vacuum equations as a source in the 4D equations: field from the vacuum. This relation allows the definitive identification of A^μ with the electromagnetic vector potential. Therefore, the field needs to be rescaled with a conversion constant k such that $A^\mu \to kA^\mu$.

The relation above shows that we must have

$$\frac{k^2}{2} = \frac{8\pi G}{c^4}\frac{1}{\mu_0} = \frac{2G}{c^2}4\pi\epsilon_0$$

where G is the gravitational constant and μ_0 is the permeability of free space. In the Kaluza theory, the gravitational constant can be understood as an electromagnetic coupling constant in the metric. There is also a stress-energy tensor for the scalar field. The scalar field behaves like a variable gravitational constant, in terms of modulating the coupling of electromagnetic stress energy to spacetime curvature. The sign of ϕ^2 in the metric is fixed by correspondence with 4D theory so that electromagnetic energy densities are positive. This turns out to imply that the 5th coordinate is spacelike in its signature in the metric.

In the presence of matter, the 5D vacuum condition can not be assumed. Indeed, Kaluza did not assume it. The full field equations require evaluation of the 5D Einstein tensor

$$\widetilde{G}_{ab} \equiv \widetilde{R}_{ab} - \frac{1}{2}\widetilde{g}_{ab}\widetilde{R}$$

as seen in the recovery of the electromagnetic stress-energy tensor above. The 5D curvature tensors are complex, and most English-language reviews contain errors in either $\widetilde{G}_{ab}$ or $\widetilde{R}_{ab}$, as does the English translation of.[7] See [18] for a complete set of 5D curvature tensors under the cylinder condition, evaluated using tensor algebra software.

38.3 Equations of motion from the Kaluza hypothesis

The equations of motion are obtained from the five-dimensional geodesic hypothesis [2] in terms of a 5-velocity $\widetilde{U}^a \equiv dx^a/ds$:

$$\widetilde{U}^b\widetilde{\nabla}_b\widetilde{U}^a = \frac{d\widetilde{U}^a}{ds} + \widetilde{\Gamma}^a_{bc}\widetilde{U}^b\widetilde{U}^c = 0$$

This equation can be recast in several ways, and it has been studied in various forms by authors including Kaluza,[2] Pauli,[21] Gross & Perry,[22] Gegenberg & Kunstatter,[23] and Wesson & Ponce de Leon,[24] but it is instructive to convert it back to the usual 4-dimensional length element $c^2d\tau^2 \equiv g_{\mu\nu}dx^\mu dx^\nu$, which is related to the 5-dimensional length element ds as given above:

$$ds^2 = c^2d\tau^2 + \phi^2(kA_\nu dx^\nu + dx^5)^2$$

Then the 5D geodesic equation can be written [25] for the spacetime components of the 4velocity, $U^\nu \equiv dx^\nu/d\tau$:

$$\frac{dU^\nu}{d\tau} + \widetilde{\Gamma}^\mu_{\alpha\beta}U^\alpha U^\beta + 2\widetilde{\Gamma}^\mu_{5\alpha}U^\alpha U^5 + \widetilde{\Gamma}^\mu_{55}(U^5)^2 + U^\mu\frac{d}{d\tau}\ln\left(\frac{cd\tau}{ds}\right) = 0$$

The term quadratic in U^ν provides the 4D geodesic equation plus some electromagnetic terms:

$$\widetilde{\Gamma}^\mu_{\alpha\beta} = \Gamma^\mu_{\alpha\beta} + \frac{1}{2}g^{\mu\nu}k^2\phi^2(A_\alpha F_{\beta\nu} + A_\beta F_{\alpha\nu} + A_\alpha A_\beta\partial_\nu\ln\phi^2)$$

The term linear in U^ν provides the Lorentz force law:

$$\widetilde{\Gamma}^\mu_{5\alpha} = \frac{1}{2}g^{\mu\nu}k\phi^2(F_{\alpha\nu} - A_\alpha\partial_\nu\ln\phi^2)$$

This is another expression of the "Kaluza miracle". The same hypothesis for the 5D metric that provides electromagnetic stress-energy in the Einstein equations, also provides the Lorentz force law in the equation of motions along with the 4D geodesic equation. Yet correspondence with the Lorentz force law requires that we identify the component of 5-velocity along the 5th dimension with electric charge:

$$kU^5 = k\frac{dx^5}{d\tau} \to \frac{q}{mc}$$

where m is particle mass and q is particle electric charge. Thus, electric charge is understood as motion along the 5th dimension. The fact that the Lorentz force law could be understood as a geodesic in 5 dimensions was to Kaluza a primary motivation for considering the 5-dimensional hypothesis, even in the presence of the aesthetically-unpleasing cylinder condition.

Yet there is a problem: the term quadratic in U^5.

$$\widetilde{\Gamma}^\mu_{55} = -\frac{1}{2}g^{\mu\alpha}\partial_\alpha\phi^2$$

If there is no gradient in the scalar field, the term quadratic in U^5 vanishes. But otherwise the expression above implies

$$U^5 \sim c\frac{q/m}{G^{1/2}}$$

For elementary particles, $U^5 > 10^{20}c$. The term quadratic in U^5 should dominate the equation, perhaps in contradiction to experience. This was the main shortfall of the 5-dimensional theory as Kaluza saw it,[2] and he gives it some discussion in his original article.

The equation of motion for U^5 is particularly simple under the cylinder condition. Start with the alternate form of the geodesic equation, written for the covariant 5-velocity:

$$\frac{d\tilde{U}_a}{ds} = \frac{1}{2}\tilde{U}^b\tilde{U}^c\frac{\partial\tilde{g}_{bc}}{\partial x^a}$$

This means that under the cylinder condition, $\tilde{U}_5$ is a constant of the 5-dimensional motion:

$$\tilde{U}_5 = \tilde{g}_{5a}\tilde{U}^a = \phi^2\frac{cd\tau}{ds}(kA_\nu U^\nu + U^5) = \text{constant}$$

38.4   Kaluza's hypothesis for the matter stress-energy tensor

Kaluza [2] proposed a 5D matter stress tensor $\tilde{T}_M^{ab}$ of the form

$$\tilde{T}_M^{ab} = \rho\frac{dx^a}{ds}\frac{dx^b}{ds}$$

where ρ is a density and the length element ds is as defined above.

Then, the spacetime component gives a typical "dust" stress energy tensor:

$$\tilde{T}_M^{\mu\nu} = \rho\frac{dx^\mu}{ds}\frac{dx^\nu}{ds}$$

The mixed component provides a 4-current source for the Maxwell equations:

$$\tilde{T}_M^{5\mu} = \rho\frac{dx^\mu}{ds}\frac{dx^5}{ds} = \rho U^\mu\frac{q}{kmc}$$

Just as the five-dimensional metric comprises the 4-D metric framed by the electromagnetic vector potential, the 5-dimensional stress-energy tensor comprises the 4-D stress-energy tensor framed by the vector 4-current.

38.5   Quantum interpretation of Klein

Kaluza's original hypothesis was purely classical and extended discoveries of general relativity. By the time of Klein's contribution, the discoveries of Heisenberg, Schroedinger, and de Broglie were receiving a lot of attention. Klein's *Nature* paper [4] suggested that the fifth dimension is closed and periodic, and that the identification of electric charge with motion in the fifth dimension be interpreted as standing waves of wavelength λ^5 , much like the electrons around a nucleus in the Bohr model of the atom. The quantization of electric charge could then be nicely understood in terms of integer multiples of fifth-dimensional momentum. Combining the previous Kaluza result for U^5 in terms of electric charge, and a de Broglie relation for momentum $p^5 = h/\lambda^5$, Klein [4] obtained an expression for the 0th mode of such waves:

$$mU^5 = \frac{cq}{G^{1/2}} = \frac{h}{\lambda^5} \to \lambda^5 \sim \frac{hG^{1/2}}{cq}$$

where h is the Planck constant. Klein found $\lambda^5 \sim 10^{-30}$ cm, and thereby an explanation for the cylinder condition in this small value.

Klein's *Zeitschrift für Physik* paper of the same year,[3] gave a more-detailed treatment that explicitly invoked the techniques of Schroedinger and de Broglie. It recapitulated much of the classical theory of Kaluza described above, and then departed into Klein's quantum interpretation. Klein solved a Schroedinger-like wave equation using an expansion in terms of fifth-dimensional waves resonating in the closed, compact fifth dimension.

38.6   Quantum field theory interpretation

38.7   Group theory interpretation

A splitting of five-dimensional spacetime into the Einstein equations and Maxwell equations in four dimensions was first discovered by Gunnar Nordström in 1914, in the context of his theory of gravity, but subsequently forgotten. Kaluza published his derivation in 1921 as an attempt to unify electromagnetism with Einstein's general relativity.

In 1926, Oskar Klein proposed that the fourth spatial dimension is curled up in a circle of a very small radius, so that a particle moving a short distance along that axis would return to where it began. The distance a particle can travel

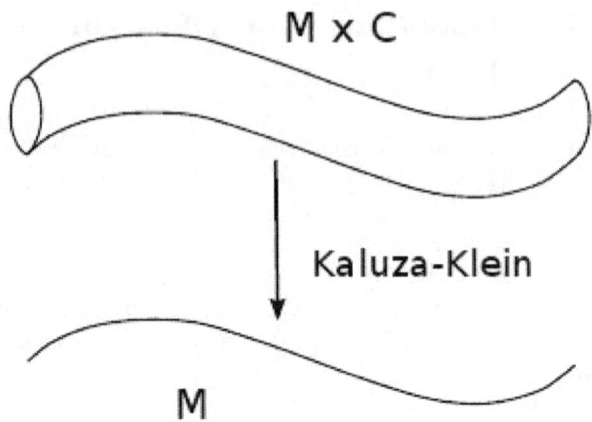

The space M × C *is compactified over the compact set* C, *and after Kaluza–Klein decomposition one has an effective field theory over* M.

before reaching its initial position is said to be the size of the dimension. This extra dimension is a compact set, and the phenomenon of having a space-time with compact dimensions is referred to as compactification.

In modern geometry, the extra fifth dimension can be understood to be the circle group U(1), as electromagnetism can essentially be formulated as a gauge theory on a fiber bundle, the circle bundle, with gauge group U(1). In Kaluza–Klein theory this group suggests that gauge symmetry is the symmetry of circular compact dimensions. Once this geometrical interpretation is understood, it is relatively straightforward to replace $U(1)$ by a general Lie group. Such generalizations are often called Yang–Mills theories. If a distinction is drawn, then it is that Yang–Mills theories occur on a flat space-time, whereas Kaluza–Klein treats the more general case of curved spacetime. The base space of Kaluza–Klein theory need not be four-dimensional space-time; it can be any (pseudo-)Riemannian manifold, or even a supersymmetric manifold or orbifold or even a noncommutative space.

The construction can be outlined, roughly, as follows.[26] One starts by considering a principle fiber bundle P with gauge group G over a manifold M. Given a connection on the bundle, and a metric on the base manifold, and a gauge invariant metric on the tangent of each fiber, one can construct a bundle metric defined on the entire bundle. Computing the scalar curvature of this bundle metric, one finds that it is constant on each fiber: this is the "Kaluza miracle". One did not have to explicitly impose a cylinder condition, or to compactify: by assumption, the gauge group is already compact. Next, one takes this scalar curvature as the Lagrangian density, and, from this, constructs the Einstein–Hilbert action for the bundle, as a whole. The equations of motion, the Euler–Lagrange equations, can be then obtained by considering where the action is stationary with

respect to variations of either the metric on the base manifold, or of the gauge connection. Variations with respect to the base metric gives the Einstein field equations on the base manifold, with the energy-momentum tensor given by the curvature (field strength) of the gauge connection. On the flip side, the action is stationary against variations of the gauge connection precisely when the gauge connection solves the Yang-Mills equations. Thus, by applying a single idea: the principle of least action, to a single quantity: the scalar curvature on the bundle (as a whole), one obtains simultaneously all of the needed field equations, for both the space-time and the gauge field.

As an approach to the unification of the forces, it is straightforward to apply the Kaluza–Klein theory in an attempt to unify gravity with the strong and electroweak forces by using the symmetry group of the Standard Model, SU(3) × SU(2) × U(1). However, an attempt to convert this interesting geometrical construction into a bona-fide model of reality founders on a number of issues, including the fact that the fermions must be introduced in an artificial way (in nonsupersymmetric models). Nonetheless, KK remains an important touchstone in theoretical physics and is often embedded in more sophisticated theories. It is studied in its own right as an object of geometric interest in K-theory.

Even in the absence of a completely satisfying theoretical physics framework, the idea of exploring extra, compactified, dimensions is of considerable interest in the experimental physics and astrophysics communities. A variety of predictions, with real experimental consequences, can be made (in the case of large extra dimensions and warped models). For example, on the simplest of principles, one might expect to have standing waves in the extra compactified dimension(s). If a spatial extra dimension is of radius R, the invariant mass of such standing waves would be $Mn = nh/Rc$ with n an integer, h being Planck's constant and c the speed of light. This set of possible mass values is often called the **Kaluza–Klein tower**. Similarly, in Thermal quantum field theory a compactification of the euclidean time dimension leads to the Matsubara frequencies and thus to a discretized thermal energy spectrum.

However, Klein's approach to a quantum theory is flawed and, for example, leads to a calculated electron mass of $3 \times (10^{30})$ MeV instead of the measured value 0.511 MeV.

Examples of experimental pursuits include work by the CDF collaboration, which has re-analyzed particle collider data for the signature of effects associated with large extra dimensions/warped models.

Brandenberger and Vafa have speculated that in the early universe, cosmic inflation causes three of the space dimensions to expand to cosmological size while the remaining dimensions of space remained microscopic.

38.8   Space-time-matter theory

One particular variant of Kaluza–Klein theory is **space-time-matter theory** or **induced matter theory**, chiefly promulgated by Paul Wesson and other members of the so-called Space-Time-Matter Consortium.[27] In this version of the theory, it is noted that solutions to the equation

$$\widetilde{R}_{ab} = 0$$

may be re-expressed so that in four dimensions, these solutions satisfy Einstein's equations

$$G_{\mu\nu} = 8\pi T_{\mu\nu}$$

with the precise form of the $T\mu\nu$ following from the Ricci-flat condition on the five-dimensional space. In other words, the cylinder condition of the previous development is dropped, and the stress-energy now comes from the derivatives of the 5D metric with respect to the fifth coordinate. Since the energy–momentum tensor is normally understood to be due to concentrations of matter in four-dimensional space, the above result is interpreted as saying that four-dimensional matter is induced from geometry in five-dimensional space.

In particular, the soliton solutions of $\widetilde{R}_{ab} = 0$ can be shown to contain the Friedmann–Lemaître–Robertson–Walker metric in both radiation-dominated (early universe) and matter-dominated (later universe) forms. The general equations can be shown to be sufficiently consistent with classical tests of general relativity to be acceptable on physical principles, while still leaving considerable freedom to also provide interesting cosmological models.

38.9   Geometric interpretation

The Kaluza–Klein theory has a particularly elegant presentation in terms of geometry. In a certain sense, it looks just like ordinary gravity in free space, except that it is phrased in five dimensions instead of four.

38.9.1   Einstein equations

The equations governing ordinary gravity in free space can be obtained from an action, by applying the variational principle to a certain action. Let M be a (pseudo-)Riemannian manifold, which may be taken as the spacetime of general relativity. If g is the metric on this manifold, one defines the action $S(g)$ as

$$S(g) = \int_M R(g)\mathrm{vol}(g)$$

where $R(g)$ is the scalar curvature and $\mathrm{vol}(g)$ is the volume element. By applying the variational principle to the action

$$\frac{\delta S(g)}{\delta g} = 0$$

one obtains precisely the Einstein equations for free space:

$$R_{ij} - \frac{1}{2}g_{ij}R = 0$$

Here, Rij is the Ricci tensor.

38.9.2   Maxwell equations

By contrast, the Maxwell equations describing electromagnetism can be understood to be the Hodge equations of a principal U(1)-bundle or circle bundle $\pi: P \to M$ with fiber U(1). That is, the electromagnetic field F is a harmonic 2-form in the space $\Omega^2(M)$ of differentiable 2-forms on the manifold M. In the absence of charges and currents, the free-field Maxwell equations are

$$\mathrm{d}F = 0 \text{ and } \mathrm{d}{*}F = 0.$$

where * is the Hodge star.

38.9.3   Kaluza–Klein geometry

To build the Kaluza–Klein theory, one picks an invariant metric on the circle S^1 that is the fiber of the U(1)-bundle of electromagnetism. In this discussion, an *invariant metric* is simply one that is invariant under rotations of the circle. Suppose this metric gives the circle a total length of Λ. One then considers metrics $\widehat{g}$ on the bundle P that are consistent with both the fiber metric, and the metric on the underlying manifold M. The consistency conditions are:

- The projection of $\widehat{g}$ to the vertical subspace $\mathrm{Vert}_p P \subset T_p P$ needs to agree with metric on the fiber over a point in the manifold M.

- The projection of $\widehat{g}$ to the horizontal subspace $\mathrm{Hor}_p P \subset T_p P$ of the tangent space at point $p \in P$ must be isomorphic to the metric g on M at $\pi(p)$.

The Kaluza–Klein action for such a metric is given by

$$S(\widehat{g}) = \int_P R(\widehat{g}) \, \text{vol}(\widehat{g})$$

The scalar curvature, written in components, then expands to

$$R(\widehat{g}) = \pi^* \left(R(g) - \frac{\Lambda^2}{2}|F|^2 \right)$$

where π^* is the pullback of the fiber bundle projection π: $P \to M$. The connection A on the fiber bundle is related to the electromagnetic field strength as

$$\pi^* F = dA$$

That there always exists such a connection, even for fiber bundles of arbitrarily complex topology, is a result from homology and specifically, K-theory. Applying Fubini's theorem and integrating on the fiber, one gets

$$S(\widehat{g}) = \Lambda \int_M \left(R(g) - \frac{1}{\Lambda^2}|F|^2 \right) \text{vol}(g)$$

Varying the action with respect to the component A, one regains the Maxwell equations. Applying the variational principle to the base metric g, one gets the Einstein equations

$$R_{ij} - \frac{1}{2}g_{ij}R = \frac{1}{\Lambda^2}T_{ij}$$

with the stress–energy tensor being given by

$$T^{ij} = F^{ik}F^{jl}g_{kl} - \frac{1}{4}g^{ij}|F|^2,$$

sometimes called the **Maxwell stress tensor**.

The original theory identifies Λ with the fiber metric g_{55}, and allows Λ to vary from fiber to fiber. In this case, the coupling between gravity and the electromagnetic field is not constant, but has its own dynamical field, the radion.

38.9.4 Generalizations

In the above, the size of the loop Λ acts as a coupling constant between the gravitational field and the electromagnetic field. If the base manifold is four-dimensional, the Kaluza–Klein manifold P is five-dimensional. The fifth dimension is a compact space, and is called the **compact dimension**. The technique of introducing compact dimensions

to obtain a higher-dimensional manifold is referred to as compactification. Compactification does not produce group actions on chiral fermions except in very specific cases: the dimension of the total space must be 2 mod 8 and the G-index of the Dirac operator of the compact space must be nonzero.[28]

The above development generalizes in a more-or-less straightforward fashion to general principal G-bundles for some arbitrary Lie group G taking the place of U(1). In such a case, the theory is often referred to as a Yang–Mills theory, and is sometimes taken to be synonymous. If the underlying manifold is supersymmetric, the resulting theory is a super-symmetric Yang–Mills theory.

38.10   Empirical tests

Up to now, no experimental or observational signs of extra dimensions have been officially reported. Many theoretical search techniques for detecting Kaluza–Klein resonances have been proposed using the mass couplings of such resonances with the top quark, however until the Large Hadron Collider (LHC) reaches full operational power observation of such resonances are unlikely. An analysis of results from the LHC in December 2010 severely constrains theories with large extra dimensions.[29]

The observation of a Higgs-like boson at the LHC puts a brand new empirical test in the search for Kaluza–Klein resonances and supersymmetric particles. The loop Feynman diagrams that exist in the Higgs interactions allow any particle with electric charge and mass to run in such a loop. Standard Model particles besides the top quark and W boson do not make big contributions to the cross-section observed in the $H \to \gamma\gamma$ decay, but if there are new particles beyond the Standard Model, they could potentially change the ratio of the predicted Standard Model $H \to \gamma\gamma$ cross-section to the experimentally observed cross-section. Hence a measurement of any dramatic change to the $H \to \gamma\gamma$ cross section predicted by the Standard Model is crucial in probing the physics beyond it.

38.11   See also

- Classical theories of gravitation
- DGP model
- Quantum gravity
- Randall–Sundrum model
- String theory

- Supergravity

- Superstring theory

38.12   Notes

[1] Pais, Abraham (1982). *Subtle is the Lord ...: The Science and the Life of Albert Einstein.* Oxford: Oxford University Press. pp. 329–330.

[2] Kaluza, Theodor (1921). "Zum Unitätsproblem in der Physik". *Sitzungsber. Preuss. Akad. Wiss. Berlin. (Math. Phys.)*: 966–972.

[3] Klein, Oskar (1926). "Quantentheorie und fünfdimensionale Relativitätstheorie". *Zeitschrift für Physik A 37* (12): 895–906. Bibcode:1926ZPhy...37..895K. doi:10.1007/BF01397481.

[4] Klein, Oskar (1926). "The Atomicity of Electricity as a Quantum Theory Law". *Nature* **118**: 516. Bibcode:1926Natur.118..516K. doi:10.1038/118516a0.

[5] Goenner, H. (2012). "Some remarks on the genesis of scalar-tensor theories". *General Relativity and Gravitation* **44**: 2077–2097. arXiv:1204.3455. Bibcode:2012GReGr..44.2077G. doi:10.1007/s10714-012-1378-8.

[6] Lichnerowicz, A.; Thiry, M.Y. (1947). *Compt. Rend. Acad. Sci. Paris* 224: 529–531. Missing or empty |title= (help)

[7] Thiry, M.Y. (1948). *Compt. Rend. Acad. Sci. Paris* 226: 216–218. Missing or empty |title= (help)

[8] Thiry, M.Y. (1948). *Compt. Rend. Acad. Sci. Paris* 226: 1881–1882. Missing or empty |title= (help)

[9] Jordan, P. (1946). *Naturwiss.* **11**: 250–251. Missing or empty |title= (help)

[10] Jordan, P.; Müller, C. (1947). *Z. Naturforsch.* 2a: 1–2. Bibcode:1947ZNatA...2....1J. doi:10.1515/zna-1947-0102. Missing or empty |title= (help)

[11] Ludwig, G. (1947). *Z. Naturforsch.* 2a: 3–5. Bibcode:1947ZNatA...2....3L. doi:10.1515/zna-1947-0103. Missing or empty |title= (help)

[12] Jordan, P. (1948). *Astron. Nachr.* 276: 193–208. Bibcode:1948AN....276..193J. doi:10.1002/asna.19482760502. Missing or empty |title= (help)

[13] Ludwig, G.; Müller, C. (1948). *Annalen der Physik* 2 (6): 76–84. Missing or empty |title= (help)

[14] Scherrer, W. (1941). *Helv. Phys. Acta* 14 (2): 130. Missing or empty |title= (help)

[15] Scherrer, W. (1949). *Helv. Phys. Acta* 22: 537–551. Missing or empty |title= (help)

[16] Scherrer, W. (1949). *Helv. Phys. Acta* 23: 547–555. Missing or empty |title= (help)

[17] Brans, C. H.; Dicke, R. H. (November 1, 1961). "Mach's Principle and a Relativistic Theory of Gravitation". *Physical Review* **124** (3): 925–935. Bibcode:1961PhRv..124..925B. doi:10.1103/PhysRev.124.925.

[18] Williams, L.L. (2015). "Field Equations and Lagrangian for the Kaluza Metric Evaluated with Tensor Algebra Software". *Journal of Gravitation* **2015**: 901870. doi:10.1155/2015/901870.

[19] Appelquist, Thomas; Chodos, Alan; Freund, Peter G. O. (1987). *Modern Kaluza-Klein Theories.* Menlo Park, Cal.: Addison–Wesley. ISBN 0-201-09829-6.

[20] Wesson, Paul S. (1999). *Space-Time-Matter, Modern Kaluza-Klein Theory.* Singapore: World Scientific. ISBN 981-02-3588-7.

[21] Pauli, Wolfgang (1958). *Theory of Relativity* (translated by George Field ed.). New York: Pergamon Press. pp. Supplement 23.

[22] Gross, D.J.; Perry, M.J. (1983). "Magnetic monopoles in Kaluza-Klein theories". *Nucl. Phys. B* 226: 29–48. Bibcode:1983NuPhB.226...29G. doi:10.1016/0550-3213(83)90462-5.

[23] Gegenberg, J.; Kunstatter, G. (1984). *Phys. Lett.* **106A**: 410. Missing or empty |title= (help)

[24] Wesson, P.S.; Ponce de Leon, J. (1995). *Astronomy and Astrophysics* 294: 1. Bibcode:1995A&A...294....1W. Missing or empty |title= (help)

[25] Williams, L.L. (2012). "Physics of the Electromagnetic Control of Spacetime and Gravity". *Proceedings of 48th AIAA Joint Propulsion Conference.* AIAA 2012-3916. doi:10.2514/6.2012-3916.

[26] David Bleecker, "Gauge Theory and Variational Principles" (1982) D. Reidel Publishing (*See chapter 9*)

[27] 5Dstm.org

[28] L. Castellani et al., Supergravity and superstrings, Vol 2, chapter V.11

[29] CMS Collaboration, "Search for Microscopic Black Hole Signatures at the Large Hadron Collider", http://arxiv.org/abs/1012.3375

38.13   References

- Nordström, Gunnar (1914). "Über die Möglichkeit, das elektromagnetische Feld und das Gravitationsfeld zu vereinigen". *Physikalische Zeitschrift* **15**: 504–506. OCLC 1762351.

- Kaluza, Theodor (1921). "Zum Unitätsproblem in der Physik". *Sitzungsber. Preuss. Akad. Wiss. Berlin. (Math. Phys.)*: 966–972. https://archive.org/details/sitzungsberichte1921preussi

- Klein, Oskar (1926). "Quantentheorie und fünfdimensionale Relativitätstheorie". *Zeitschrift für Physik A* **37** (12): 895–906. Bibcode:1926ZPhy...37..895K. doi:10.1007/BF01397481.

- Witten, Edward (1981). "Search for a realistic Kaluza–Klein theory". *Nuclear Physics B* **186** (3): 412–428. Bibcode:1981NuPhB.186..412W. doi:10.1016/0550-3213(81)90021-3.

- Appelquist, Thomas; Chodos, Alan; Freund, Peter G. O. (1987). *Modern Kaluza–Klein Theories*. Menlo Park, Cal.: Addison–Wesley. ISBN 0-201-09829-6. *(Includes reprints of the above articles as well as those of other important papers relating to Kaluza–Klein theory.)*

- Brandenberger, Robert; Vafa, Cumrun (1989). "Superstrings in the early universe". *Nuclear Physics B* **316** (2): 391–410. Bibcode:1989NuPhB.316..391B. doi:10.1016/0550-3213(89)90037-0.

- Duff, M. J. (1994). "Kaluza–Klein Theory in Perspective". In Lindström, Ulf (ed.). *Proceedings of the Symposium 'The Oskar Klein Centenary'*. Singapore: World Scientific. pp. 22–35. ISBN 981-02-2332-3.

- Overduin, J. M.; Wesson, P. S. (1997). "Kaluza–Klein Gravity". *Physics Reports* **283** (5): 303–378. arXiv:gr-qc/9805018. Bibcode:1997PhR...283..303O. doi:10.1016/S0370-1573(96)00046-4.

- Wesson, Paul S. (1999). *Space-Time-Matter, Modern Kaluza-Klein Theory*. Singapore: World Scientific. ISBN 981-02-3588-7.

- Wesson, Paul S. (2006). *Five-Dimensional Physics: Classical and Quantum Consequences of Kaluza-Klein Cosmology*. Singapore: World Scientific. ISBN 981-256-661-9.

- Coquereaux, R.; Esposito-Farese, G. (1990). "The Theory of Kaluza-Klein-Jordan-Thiry revisited". *Annales de l'I.H.P., Section A* **52**: 113–150.

- Kaku, Michio and Robert O'Keefe. *Hyperspace: A Scientific Odyssey Through Parallel Universes, Time Warps, and the Tenth Dimension*. New York: Oxford University Press, 1994. ISBN 0-19-286189-1

- The CDF Collaboration, *Search for Extra Dimensions using Missing Energy at CDF*, (2004) *(A simplified presentation of the search made for extra dimensions at the Collider Detector at Fermilab (CDF) particle physics facility.)*

- John M. Pierre, *SUPERSTRINGS! Extra Dimensions*, (2003).

- TeV scale gravity, mirror universe, and ... dinosaurs Article from Acta Physica Polonica B by Z.K. Silagadze.

- Chris Pope, *Lectures on Kaluza–Klein Theory*.

- Edward Witten (2014). "A Note On Einstein, Bergmann, and the Fifth Dimension", arXiv:1401.8048; pdf

38.14 Further reading

- Grøn, Øyvind; Hervik, Sigbjørn (2007). *Einstein's General Theory of Relativity*. New York: Springer. ISBN 978-0-387-69199-2.

Chapter 39

String theory

For a more accessible and less technical introduction to this topic, see Introduction to M-theory.

In physics, **string theory** is a theoretical framework in which the point-like particles of particle physics are replaced by one-dimensional objects called strings. It describes how these strings propagate through space and interact with each other. On distance scales larger than the string scale, a string looks just like an ordinary particle, with its mass, charge, and other properties determined by the vibrational state of the string. In string theory, one of the many vibrational states of the string corresponds to the graviton, a quantum mechanical particle that carries gravitational force. Thus string theory is a theory of quantum gravity.

String theory is a broad and varied subject that attempts to address a number of deep questions of fundamental physics. String theory has been applied to a variety of problems in black hole physics, early universe cosmology, nuclear physics, and condensed matter physics, and it has stimulated a number of major developments in pure mathematics. Because string theory potentially provides a unified description of gravity and particle physics, it is a candidate for a theory of everything, a self-contained mathematical model that describes all fundamental forces and forms of matter. Despite much work on these problems, it is not known to what extent string theory describes the real world or how much freedom the theory allows to choose the details.

String theory was first studied in the late 1960s as a theory of the strong nuclear force, before being abandoned in favor of quantum chromodynamics. Subsequently, it was realized that the very properties that made string theory unsuitable as a theory of nuclear physics made it a promising candidate for a quantum theory of gravity. The earliest version of string theory, bosonic string theory, incorporated only the class of particles known as bosons. It later developed into superstring theory, which posits a connection called supersymmetry between bosons and the class of particles called fermions. Five consistent versions of super-

string theory were developed before it was conjectured in the mid-1990s that they were all different limiting cases of a single theory in eleven dimensions known as M-theory. In late 1997, theorists discovered an important relationship called the AdS/CFT correspondence, which relates string theory to another type of physical theory called a quantum field theory.

One of the challenges of string theory is that the full theory does not yet have a satisfactory definition in all circumstances. Another issue is that the theory is thought to describe an enormous landscape of possible universes, and this has complicated efforts to develop theories of particle physics based on string theory. These issues have led some in the community to criticize these approaches to physics and question the value of continued research on string theory unification.

39.1 Fundamentals

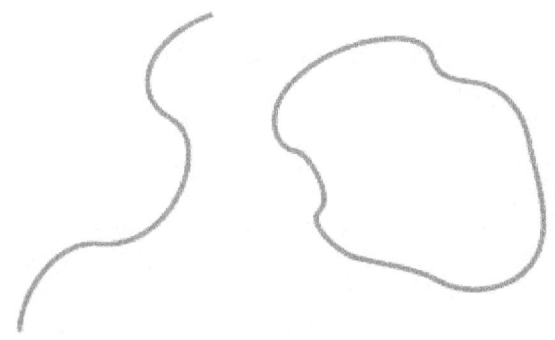

The fundamental objects of string theory are open and closed strings.

In the twentieth century, two theoretical frameworks emerged for formulating the laws of physics. One of these frameworks was Albert Einstein's general theory of relativity, a theory that explains the force of gravity and the structure of space and time. The other was quantum mechanics, a radically different formalism for describing physical phenomena using probability. By the late 1970s, these two frameworks had proven to be sufficient to explain most of the observed features of the universe, from elementary particles to atoms to the evolution of stars and the universe as a whole.[1]

In spite of these successes, there are still many problems that remain to be solved. One of the deepest problems in modern physics is the problem of quantum gravity.[1] The general theory of relativity is formulated within the framework of classical physics, whereas the other fundamental forces are described within the framework of quantum mechanics. A quantum theory of gravity is needed in order to reconcile general relativity with the principles of quantum mechanics, but difficulties arise when one attempts to apply the usual prescriptions of quantum theory to the force of gravity.[2] In addition to the problem of developing a consistent theory of quantum gravity, there are many other fundamental problems in the physics of atomic nuclei, black holes, and the early universe.[lower-alpha 1]

String theory is a theoretical framework that attempts to address these questions and many others. The starting point for string theory is the idea that the point-like particles of particle physics can also be modeled as one-dimensional objects called strings. String theory describes how strings propagate through space and interact with each other. In a given version of string theory, there is only one kind of string, which may look like a small loop or segment of ordinary string, and it can vibrate in different ways. On distance scales larger than the string scale, a string will look just like an ordinary particle, with its mass, charge, and other properties determined by the vibrational state of the string. In this way, all of the different elementary particles may be viewed as vibrating strings. In string theory, one of the vibrational states of the string gives rise to the graviton, a quantum mechanical particle that carries gravitational force. Thus string theory is a theory of quantum gravity.[3]

One of the main developments of the past several decades in string theory was the discovery of certain "dualities", mathematical transformations that identify one physical theory with another. Physicists studying string theory have discovered a number of these dualities between different versions of string theory, and this has led to the conjecture that all consistent versions of string theory are subsumed in a single framework known as M-theory.[4]

Studies of string theory have also yielded a number of results on the nature of black holes and the gravitational interaction. There are certain paradoxes that arise when one attempts to understand the quantum aspects of black holes, and work on string theory has attempted to clarify these issues. In late 1997 this line of work culminated in the discovery of the anti-de Sitter/conformal field theory correspondence or AdS/CFT.[5] This is a theoretical result which relates string theory to other physical theories which are better understood theoretically. The AdS/CFT correspondence has implications for the study of black holes and quantum gravity, and it has been applied to other subjects, including nuclear[6] and condensed matter physics.[7][8]

Since string theory incorporates all of the fundamental interactions, including gravity, many physicists hope that it fully describes our universe, making it a theory of everything. One of the goals of current research in string theory is to find a solution of the theory that reproduces the observed spectrum of elementary particles, with a small cosmological constant, containing dark matter and a plausible mechanism for cosmic inflation. While there has been progress toward these goals, it is not known to what extent string theory describes the real world or how much freedom the theory allows to choose the details.[9]

One of the challenges of string theory is that the full theory does not yet have a satisfactory definition in all circumstances. The scattering of strings is most straightforwardly defined using the techniques of perturbation theory, but it is not known in general how to define string theory nonperturbatively.[10] It is also not clear whether there is any principle by which string theory selects its vacuum state, the physical state that determines the properties of our universe.[11] These problems have led some in the community to criticize these approaches to the unification of physics and question the value of continued research on these problems.[12]

39.1.1 Strings

Main article: String (physics)

The application of quantum mechanics to physical objects

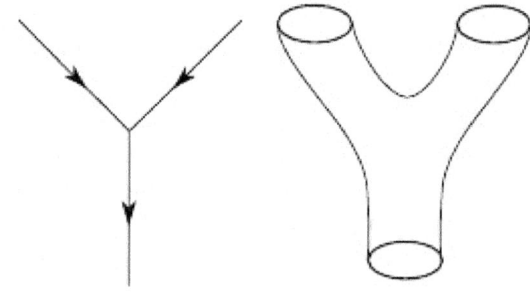

Interaction in the quantum world: worldlines of point-like particles or a worldsheet swept up by closed strings in string theory.

such as the electromagnetic field, which are extended in

space and time, is known as quantum field theory. In particle physics, quantum field theories form the basis for our understanding of elementary particles, which are modeled as excitations in the fundamental fields.[13]

In quantum field theory, one typically computes the probabilities of various physical events using the techniques of perturbation theory. Developed by Richard Feynman and others in the first half of the twentieth century, perturbative quantum field theory uses special diagrams called Feynman diagrams to organize computations. One imagines that these diagrams depict the paths of point-like particles and their interactions.[13]

The starting point for string theory is the idea that the point-like particles of quantum field theory can also be modeled as one-dimensional objects called strings.[14] The interaction of strings is most straightforwardly defined by generalizing the perturbation theory used in ordinary quantum field theory. At the level of Feynman diagrams, this means replacing the one-dimensional diagram representing the path of a point particle by a two-dimensional surface representing the motion of a string.[15] Unlike in quantum field theory, string theory does not yet have a full non-perturbative definition, so many of the theoretical questions that physicists would like to answer remain out of reach.[16]

In theories of particle physics based on string theory, the characteristic length scale of strings is assumed to be on the order of the Planck length, or 10^{-35} meters, the scale at which the effects of quantum gravity are believed to become significant.[15] On much larger length scales, such as the scales visible in physics laboratories, such objects would be indistinguishable from zero-dimensional point particles, and the vibrational state of the string would determine the type of particle. One of the vibrational states of a string corresponds to the graviton, a quantum mechanical particle that carries the gravitational force.[3]

The original version of string theory was bosonic string theory, but this version described only bosons, a class of particles which transmit forces between the matter particles, or fermions. Bosonic string theory was eventually superseded by theories called superstring theories. These theories describe both bosons and fermions, and they incorporate a theoretical idea called supersymmetry. This is a mathematical relation that exists in certain physical theories between the bosons and fermions. In theories with supersymmetry, each boson has a counterpart which is a fermion, and vice versa.[17]

There are several versions of superstring theory: type I, type IIA, type IIB, and two flavors of heterotic string theory ($SO(32)$ and $E_8 \times E_8$). The different theories allow different types of strings, and the particles that arise at low energies exhibit different symmetries. For example, the type I theory includes both open strings (which are segments with

endpoints) and closed strings (which form closed loops), while types IIA and IIB include only closed strings.[18]

39.1.2 Extra dimensions

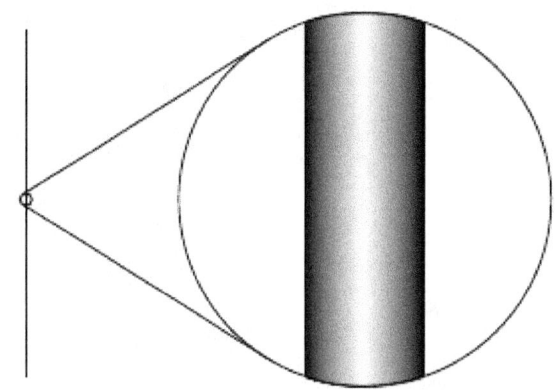

An example of compactification: At large distances, a two dimensional surface with one circular dimension looks one-dimensional.

In everyday life, there are three familiar dimensions of space: height, width and length. Einstein's general theory of relativity treats time as a dimension on par with the three spatial dimensions; in general relativity, space and time are not modeled as separate entities but are instead unified to a four-dimensional spacetime. In this framework, the phenomenon of gravity is viewed as a consequence of the geometry of spacetime.[19]

In spite of the fact that the universe is well described by four-dimensional spacetime, there are several reasons why physicists consider theories in other dimensions. In some cases, by modeling spacetime in a different number of dimensions, a theory becomes more mathematically tractable, and one can perform calculations and gain general insights more easily.[lower-alpha 2] There are also situations where theories in two or three spacetime dimensions are useful for describing phenomena in condensed matter physics.[20] Finally, there exist scenarios in which there could actually be more than four dimensions of spacetime which have nonetheless managed to escape detection.[21]

One notable feature of string theories is that these theories require extra dimensions of spacetime for their mathematical consistency. In bosonic string theory, spacetime is 26-dimensional, while in superstring theory it is ten-dimensional. In order to describe real physical phenomena using string theory, one must therefore imagine scenarios in which these extra dimensions would not be observed in experiments.[22]

Compactification is one way of modifying the number of dimensions in a physical theory. In compactification, some

A cross section of a quintic Calabi–Yau manifold

of the extra dimensions are assumed to "close up" on themselves to form circles.[23] In the limit where these curled up dimensions become very small, one obtains a theory in which spacetime has effectively a lower number of dimensions. A standard analogy for this is to consider a multidimensional object such as a garden hose. If the hose is viewed from a sufficient distance, it appears to have only one dimension, its length. However, as one approaches the hose, one discovers that it contains a second dimension, its circumference. Thus, an ant crawling on the surface of the hose would move in two dimensions.[24]

Compactification can be used to construct models in which spacetime is effectively four-dimensional. However, not every way of compactifying the extra dimensions produces a model with the right properties to describe nature. In a viable model of particle physics, the compact extra dimensions must be shaped like a Calabi–Yau manifold.[23] A Calabi–Yau manifold is a special space which is typically taken to be six-dimensional in applications to string theory. It is named after mathematicians Eugenio Calabi and Shing-Tung Yau.[25]

Another approach to reducing the number of dimensions is the so-called brane-world scenario. In this approach, physicists assume that the observable universe is a four-dimensional subspace of a higher dimensional space. In such models, the force-carrying bosons of particle physics arise from open strings with endpoints attached to the four-dimensional subspace, while gravity arises from closed strings propagating through the larger ambient space. This idea plays an important role in attempts to develop models

of real world physics based on string theory, and it provides a natural explanation for the weakness of gravity compared to the other fundamental forces.[26]

39.1.3   Dualities

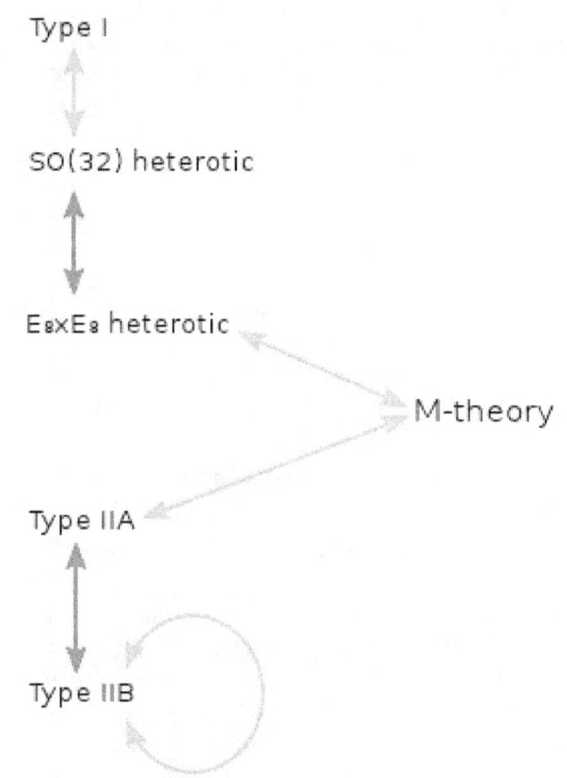

A diagram of string theory dualities. Yellow arrows indicate S-duality. Blue arrows indicate T-duality.

Main articles: S-duality and T-duality

One notable fact about string theory is that the different versions of the theory all turn out to be related in highly nontrivial ways. One of the relationships that can exist between different string theories is called S-duality. This is a relationship which says that a collection of strongly interacting particles in one theory can, in some cases, be viewed as a collection of weakly interacting particles in a completely different theory. Roughly speaking, a collection of particles is said to be strongly interacting if they combine and decay often and weakly interacting if they do so infrequently. Type I string theory turns out to be equivalent by S-duality to the *SO*(32) heterotic string theory. Similarly, type IIB string theory is related to itself in a nontrivial way by S-duality.[27]

Another relationship between different string theories is T-duality. Here one considers strings propagating around a circular extra dimension. T-duality states that a string propagating around a circle of radius R is equivalent to a string propagating around a circle of radius $1/R$ in the sense that all observable quantities in one description are identified with quantities in the dual description. For example, a string has momentum as it propagates around a circle, and it can also wind around the circle one or more times. The number of times the string winds around a circle is called the winding number. If a string has momentum p and winding number n in one description, it will have momentum n and winding number p in the dual description. For example, type IIA string theory is equivalent to type IIB string theory via T-duality, and the two versions of heterotic string theory are also related by T-duality.[27]

In general, the term *duality* refers to a situation where two seemingly different physical systems turn out to be equivalent in a nontrivial way. Two theories related by a duality need not be string theories. For example, Montonen–Olive duality is example of an S-duality relationship between quantum field theories. The AdS/CFT correspondence is example of a duality which relates string theory to a quantum field theory. If two theories are related by a duality, it means that one theory can be transformed in some way so that it ends up looking just like the other theory. The two theories are then said to be *dual* to one another under the transformation. Put differently, the two theories are mathematically different descriptions of the same phenomena.[28]

39.1.4 Branes

Main article: Brane
In string theory and related theories, a brane is a physi-

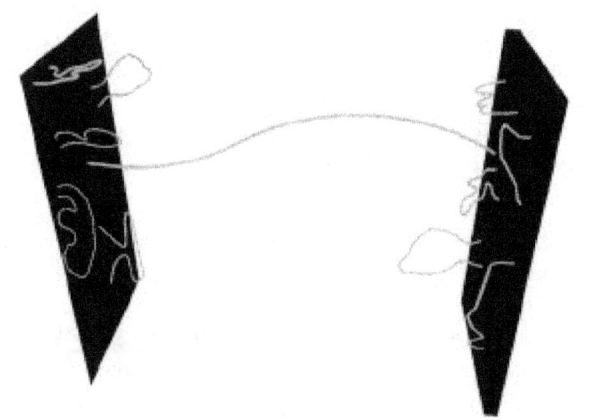

Open strings attached to a pair of D-branes

cal object that generalizes the notion of a point particle to higher dimensions. For example, a point particle can be

viewed as a brane of dimension zero, while a string can be viewed as a brane of dimension one. It is also possible to consider higher-dimensional branes. In dimension p, these are called p-branes. The word brane comes from the word "membrane" which refers to a two-dimensional brane.[29]

Branes are dynamical objects which can propagate through spacetime according to the rules of quantum mechanics. They have mass and can have other attributes such as charge. A p-brane sweeps out a $(p+1)$-dimensional volume in spacetime called its *worldvolume*. Physicists often study fields analogous to the electromagnetic field which live on the worldvolume of a brane.[29]

In string theory, D-branes are an important class of branes that arise when one considers open strings. As an open string propagates through spacetime, its endpoints are required to lie on a D-brane. The letter "D" in D-brane refers to a certain mathematical condition on the system known as the Dirichlet boundary condition. The study of D-branes in string theory has led to important results such as the AdS/CFT correspondence, which has shed light on many problems in quantum field theory.[30]

Branes are also frequently studied from a purely mathematical point of view. Mathematically, branes can be described as objects of certain categories, such as the derived category of coherent sheaves on a complex algebraic variety, or the Fukaya category of a symplectic manifold.[31] The connection between the physical notion of a brane and the mathematical notion of a category has led to important mathematical insights in the fields of algebraic and symplectic geometry[32] and representation theory.[33]

39.2 M-theory

Main article: M-theory

Prior to 1995, theorists believed that there were five consistent versions of superstring theory (type I, type IIA, type IIB, and two versions of heterotic string theory). This understanding changed in 1995 when Edward Witten suggested that the five theories were just special limiting cases of an eleven-dimensional theory called M-theory. Witten's conjecture was based on the work of a number of other physicists, including Ashoke Sen, Chris Hull, Paul Townsend, and Michael Duff. His announcement led to a flurry of research activity now known as the second superstring revolution.[34]

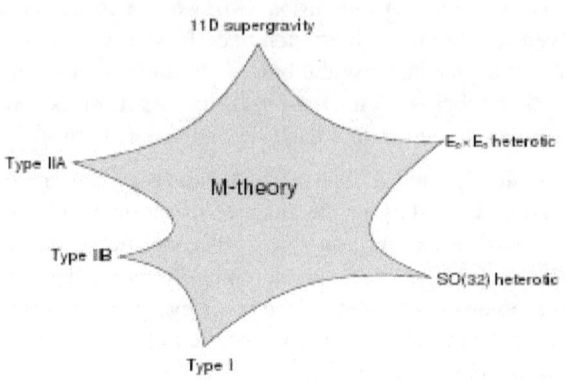

A schematic illustration of the relationship between M-theory, the five superstring theories, and eleven-dimensional supergravity. The shaded region represents a family of different physical scenarios that are possible in M-theory. In certain limiting cases corresponding to the cusps, it is natural to describe the physics using one of the six theories labeled there.

39.2.1   Unification of superstring theories

In the 1970s, many physicists became interested in supergravity theories, which combine general relativity with supersymmetry. Whereas general relativity makes sense in any number of dimensions, supergravity places an upper limit on the number of dimensions.[35] In 1978, work by Werner Nahm showed that the maximum spacetime dimension in which one can formulate a consistent supersymmetric theory is eleven.[36] In the same year, Eugene Cremmer, Bernard Julia, and Joel Scherk of the École Normale Supérieure showed that supergravity not only permits up to eleven dimensions but is in fact most elegant in this maximal number of dimensions.[37][38]

Initially, many physicists hoped that by compactifying eleven-dimensional supergravity, it might be possible to construct realistic models of our four-dimensional world. The hope was that such models would provide a unified description of the four fundamental forces of nature: electromagnetism, the strong and weak nuclear forces, and gravity. Interest in eleven-dimensional supergravity soon waned as various flaws in this scheme were discovered. One of the problems was that the laws of physics appear to distinguish between clockwise and counterclockwise, a phenomenon known as chirality. Edward Witten and others observed this chirality property cannot be readily derived by compactifying from eleven dimensions.[38]

In the first superstring revolution in 1984, many physicists turned to string theory as a unified theory of particle physics and quantum gravity. Unlike supergravity theory, string theory was able to accommodate the chirality of the standard model, and it provided a theory of gravity consistent

with quantum effects.[38] Another feature of string theory that many physicists were drawn to in the 1980s and 1990s was its high degree of uniqueness. In ordinary particle theories, one can consider any collection of elementary particles whose classical behavior is described by an arbitrary Lagrangian. In string theory, the possibilities are much more constrained: by the 1990s, physicists had argued that there were only five consistent supersymmetric versions of the theory.[38]

Although there were only a handful of consistent superstring theories, it remained a mystery why there was not just one consistent formulation.[38] However, as physicists began to examine string theory more closely, they realized that these theories are related in intricate and nontrivial ways. They found that a system of strongly interacting strings can, in some cases, be viewed as a system of weakly interacting strings. This phenomenon is known as S-duality. It was studied by Ashoke Sen in the context of heterotic strings in four dimensions[39][40] and by Chris Hull and Paul Townsend in the context of the type IIB theory.[41] Theorists also found that different string theories may be related by T-duality. This duality implies that strings propagating on completely different spacetime geometries may be physically equivalent.[42]

At around the same time, as many physicists were studying the properties of strings, a small group of physicists was examining the possible applications of higher dimensional objects. In 1987, Eric Bergshoeff, Ergin Sezgin, and Paul Townsend showed that eleven-dimensional supergravity includes two-dimensional branes.[43] Intuitively, these objects look like sheets or membranes propagating through the eleven-dimensional spacetime. Shortly after this discovery, Michael Duff, Paul Howe, Takeo Inami, and Kellogg Stelle considered a particular compactification of eleven-dimensional supergravity with one of the dimensions curled up into a circle.[44] In this setting, one can imagine the membrane wrapping around the circular dimension. If the radius of the circle is sufficiently small, then this membrane looks just like a string in ten-dimensional spacetime. In fact, Duff and his collaborators showed that this construction reproduces exactly the strings appearing in type IIA superstring theory.[45]

Speaking at a string theory conference in 1995, Edward Witten made the surprising suggestion that all five superstring theories were in fact just different limiting cases of a single theory in eleven spacetime dimensions. Witten's announcement drew together all of the previous results on S- and T-duality and the appearance of higher dimensional branes in string theory.[46] In the months following Witten's announcement, hundreds of new papers appeared on the Internet confirming different parts of his proposal.[47] Today this flurry of work is known as the second superstring revolution.[48]

Initially, some physicists suggested that the new theory was a fundamental theory of membranes, but Witten was skeptical of the role of membranes in the theory. In a paper from 1996, Hořava and Witten wrote "As it has been proposed that the eleven-dimensional theory is a supermembrane theory but there are some reasons to doubt that interpretation, we will non-committally call it the M-theory, leaving to the future the relation of M to membranes."[49] In the absence of an understanding of the true meaning and structure of M-theory, Witten has suggested that the M should stand for "magic", "mystery", or "membrane" according to taste, and the true meaning of the title should be decided when a more fundamental formulation of the theory is known.[50]

39.2.2   Matrix theory

Main article: Matrix theory (physics)

In mathematics, a matrix is a rectangular array of numbers or other data. In physics, a matrix model is a particular kind of physical theory whose mathematical formulation involves the notion of a matrix in an important way. A matrix model describes the behavior of a set of matrices within the framework of quantum mechanics.[51]

One important example of a matrix model is the BFSS matrix model proposed by Tom Banks, Willy Fischler, Stephen Shenker, and Leonard Susskind in 1997. This theory describes the behavior of a set of nine large matrices. In their original paper, these authors showed, among other things, that the low energy limit of this matrix model is described by eleven-dimensional supergravity. These calculations led them to propose that the BFSS matrix model is exactly equivalent to M-theory. The BFSS matrix model can therefore be used as a prototype for a correct formulation of M-theory and a tool for investigating the properties of M-theory in a relatively simple setting.[51]

The development of the matrix model formulation of M-theory has led physicists to consider various connections between string theory and a branch of mathematics called noncommutative geometry. This subject is a generalization of ordinary geometry in which mathematicians define new geometric notions using tools from noncommutative algebra.[52] In a paper from 1998, Alain Connes, Michael R. Douglas, and Albert Schwarz showed that some aspects of matrix models and M-theory are described by a noncommutative quantum field theory, a special kind of physical theory in which spacetime is described mathematically using noncommutative geometry.[53] This established a link between matrix models and M-theory on the one hand, and noncommutative geometry on the other hand. It quickly led to the discovery of other important links between noncommutative geometry and various physical theories.[54][55]

39.3   Black holes

In general relativity, a black hole is defined as a region of spacetime in which the gravitational field is so strong that no particle or radiation can escape. In the currently accepted models of stellar evolution, black holes are thought to arise when massive stars undergo gravitational collapse, and many galaxies are thought to contain supermassive black holes at their centers. Black holes are also important for theoretical reasons, as they present profound challenges for theorists attempting to understand the quantum aspects of gravity. String theory has proved to be an important tool for investigating the theoretical properties of black holes because it provides a framework in which theorists can study their thermodynamics.[56]

39.3.1   Bekenstein–Hawking formula

In the branch of physics called statistical mechanics, entropy is a measure of the randomness or disorder of a physical system. This concept was studied in the 1870s by the Austrian physicist Ludwig Boltzmann, who showed that the thermodynamic properties of a gas could be derived from the combined properties of its many constituent molecules. Boltzmann argued that by averaging the behaviors of all the different molecules in a gas, one can understand macroscopic properties such as volume, temperature, and pressure. In addition, this perspective led him to give a precise definition of entropy as the natural logarithm of the number of different states of the molecules (also called *microstates*) that give rise to the same macroscopic features.[57]

In the twentieth century, physicists began to apply the same concepts to black holes. In most systems such as gases, the entropy scales with the volume. In the 1970s, the physicist Jacob Bekenstein suggested that the entropy of a black hole is instead proportional to the *surface area* of its event horizon, the boundary beyond which matter and radiation is lost to its gravitational attraction.[58] When combined with ideas of the physicist Stephen Hawking,[59] Bekenstein's work yielded a precise formula for the entropy of a black hole. The formula expresses the entropy S as

$$S = \frac{c^3 k A}{4 \hbar G}$$

where c is the speed of light, k is Boltzmann's constant, $\hbar$ is the reduced Planck constant, G is Newton's constant, and A is the surface area of the event horizon.[60]

Like any physical system, a black hole has an entropy defined in terms of the number of different microstates that lead to the same macroscopic features. The Bekenstein–Hawking entropy formula gives the expected value of the entropy of a black hole, but by the 1990s, physicists still lacked a derivation of this formula by counting microstates in a theory of quantum gravity. Finding such a derivation of this formula was considered an important test of the viability of any theory of quantum gravity such as string theory.[61]

39.3.2   Derivation within string theory

In a paper from 1996, Andrew Strominger and Cumrun Vafa showed how to derive the Beckenstein–Hawking formula for certain black holes in string theory.[62] Their calculation was based on the observation that D-branes—which look like fluctuating membranes when they are weakly interacting—become dense, massive objects with event horizons when the interactions are strong. In other words, a system of strongly interacting D-branes in string theory is indistinguishable from a black hole. Strominger and Vafa analyzed such D-brane systems and calculated the number of different ways of placing D-branes in spacetime so that their combined mass and charge is equal to a given mass and charge for the resulting black hole. Their calculation reproduced the Bekenstein–Hawking formula exactly, including the factor of 1/4.[63] Subsequent work by Strominger, Vafa, and others refined the original calculations and gave the precise values of the "quantum corrections" needed to describe very small black holes.[64][65]

The black holes that Strominger and Vafa considered in their original work were quite different from real astrophysical black holes. One difference was that Strominger and Vafa considered only extremal black holes in order to make the calculation tractable. These are defined as black holes with the lowest possible mass compatible with a given charge.[66] Strominger and Vafa also restricted attention to black holes in five-dimensional spacetime with unphysical supersymmetry.[67]

Although it was originally developed in this very particular and physically unrealistic context in string theory, the entropy calculation of Strominger and Vafa has led to a qualitative understanding of how black hole entropy can be accounted for in any theory of quantum gravity. Indeed, in 1998, Strominger argued that the original result could be generalized to an arbitrary consistent theory of quantum gravity without relying on strings or supersymmetry.[68] In collaboration with several other authors in 2010, he showed that some results on black hole entropy could be extended to non-extremal astrophysical black holes.[69][70]

39.4   AdS/CFT correspondence

Main article: AdS/CFT correspondence

One approach to formulating string theory and studying its properties is provided by the anti-de Sitter/conformal field theory (AdS/CFT) correspondence. This is a theoretical result which implies that string theory is in some cases equivalent to a quantum field theory. In addition to providing insights into the mathematical structure of string theory, the AdS/CFT correspondence has shed light on many aspects of quantum field theory in regimes where traditional calculational techniques are ineffective.[6] The AdS/CFT correspondence was first proposed by Juan Maldacena in late 1997.[71] Important aspects of the correspondence were elaborated in articles by Steven Gubser, Igor Klebanov, and Alexander Markovich Polyakov,[72] and by Edward Witten.[73] By 2010, Maldacena's article had over 7000 citations, becoming the most highly cited article in the field of high energy physics.[lower-alpha 3]

39.4.1   Overview of the correspondence

In the AdS/CFT correspondence, the geometry of spacetime is described in terms of a certain vacuum solution of Einstein's equation called anti-de Sitter space.[74] In very elementary terms, anti-de Sitter space is a mathematical model of spacetime in which the notion of distance between points (the metric) is different from the notion of distance in ordinary Euclidean geometry. It is closely related to hyperbolic space, which can be viewed as a disk as illustrated on the left.[75] This image shows a tessellation of a disk by triangles and squares. One can define the distance between points of this disk in such a way that all the triangles and squares are the same size and the circular outer boundary is infinitely far from any point in the interior.[76]

One can imagine a stack of hyperbolic disks where each disk represents the state of the universe at a given time. The resulting geometric object is three-dimensional anti-de Sitter space.[75] It looks like a solid cylinder in which any cross section is a copy of the hyperbolic disk. Time runs along the vertical direction in this picture. The surface of this cylinder plays an important role in the AdS/CFT correspondence. As with the hyperbolic plane, anti-de Sitter space is curved in such a way that any point in the interior is actually infinitely far from this boundary surface.[76]

This construction describes a hypothetical universe with only two space dimensions and one time dimension, but it can be generalized to any number of dimensions. Indeed, hyperbolic space can have more than two dimensions and one can "stack up" copies of hyperbolic space to get higher-dimensional models of anti-de Sitter space.[75]

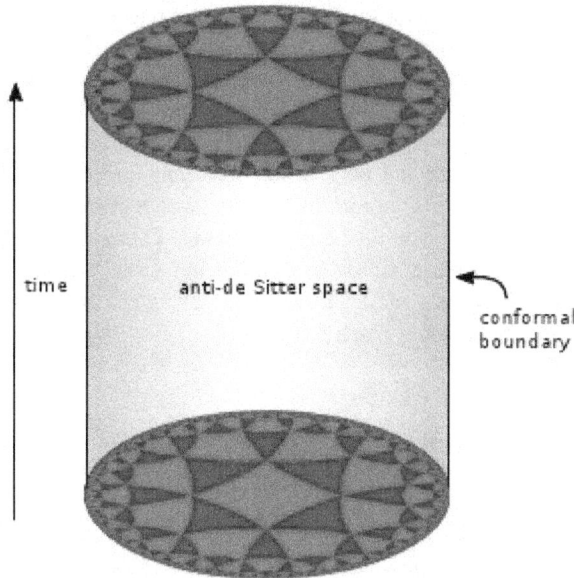

Three-dimensional anti-de Sitter space is like a stack of hyperbolic disks, each one representing the state of the universe at a given time. The resulting spacetime looks like a solid cylinder.

An important feature of anti-de Sitter space is its boundary (which looks like a cylinder in the case of three-dimensional anti-de Sitter space). One property of this boundary is that, within a small region on the surface around any given point, it looks just like Minkowski space, the model of spacetime used in nongravitational physics.[77] One can therefore consider an auxiliary theory in which "spacetime" is given by the boundary of anti-de Sitter space. This observation is the starting point for AdS/CFT correspondence, which states that the boundary of anti-de Sitter space can be regarded as the "spacetime" for a quantum field theory. The claim is that this quantum field theory is equivalent to a gravitational theory, such as string theory, in the bulk anti-de Sitter space in the sense that there is a "dictionary" for translating entities and calculations in one theory into their counterparts in the other theory. For example, a single particle in the gravitational theory might correspond to some collection of particles in the boundary theory. In addition, the predictions in the two theories are quantitatively identical so that if two particles have a 40 percent chance of colliding in the gravitational theory, then the corresponding collections in the boundary theory would also have a 40 percent chance of colliding.[78]

39.4.2 Applications to quantum gravity

The discovery of the AdS/CFT correspondence was a major advance in physicists' understanding of string theory and quantum gravity. One reason for this is that the correspon-

dence provides a formulation of string theory in terms of quantum field theory, which is well understood by comparison. Another reason is that it provides a general framework in which physicists can study and attempt to resolve the paradoxes of black holes.[56]

In 1975, Stephen Hawking published a calculation which suggested that black holes are not completely black but emit a dim radiation due to quantum effects near the event horizon.[59] At first, Hawking's result posed a problem for theorists because it suggested that black holes destroy information. More precisely, Hawking's calculation seemed to conflict with one of the basic postulates of quantum mechanics, which states that physical systems evolve in time according to the Schrödinger equation. This property is usually referred to as unitarity of time evolution. The apparent contradiction between Hawking's calculation and the unitarity postulate of quantum mechanics came to be known as the black hole information paradox.[79]

The AdS/CFT correspondence resolves the black hole information paradox, at least to some extent, because it shows how a black hole can evolve in a manner consistent with quantum mechanics in some contexts. Indeed, one can consider black holes in the context of the AdS/CFT correspondence, and any such black hole corresponds to a configuration of particles on the boundary of anti-de Sitter space.[80] These particles obey the usual rules of quantum mechanics and in particular evolve in a unitary fashion, so the black hole must also evolve in a unitary fashion, respecting the principles of quantum mechanics.[81] In 2005, Hawking announced that the paradox had been settled in favor of information conservation by the AdS/CFT correspondence, and he suggested a concrete mechanism by which black holes might preserve information.[82]

39.4.3 Applications to quantum field theory

Main articles: AdS/QCD correspondence and AdS/CMT correspondence

In addition to its applications to theoretical problems in quantum gravity, the AdS/CFT correspondence has been applied to a variety of problems in quantum field theory. One physical system that has been studied using the AdS/CFT correspondence is the quark–gluon plasma, an exotic state of matter produced in particle accelerators. This state of matter arises for brief instants when heavy ions such as gold or lead nuclei are collided at high energies. Such collisions cause the quarks that make up atomic nuclei to deconfine at temperatures of approximately two trillion kelvins, conditions similar to those present at around 10^{-11} seconds after the Big Bang.[83]

The physics of the quark–gluon plasma is governed by a theory called quantum chromodynamics, but this the-

A magnet levitating above a high-temperature superconductor. Today some physicists are working to understand high-temperature superconductivity using the AdS/CFT correspondence.[7]

ory is mathematically intractable in problems involving the quark–gluon plasma.[lower-alpha 4] In an article appearing in 2005, Đàm Thanh Sơn and his collaborators showed that the AdS/CFT correspondence could be used to understand some aspects of the quark–gluon plasma by describing it in the language of string theory.[84] By applying the AdS/CFT correspondence, Sơn and his collaborators were able to describe the quark gluon plasma in terms of black holes in five-dimensional spacetime. The calculation showed that the ratio of two quantities associated with the quark–gluon plasma, the shear viscosity and volume density of entropy, should be approximately equal to a certain universal constant. In 2008, the predicted value of this ratio for the quark–gluon plasma was confirmed at the Relativistic Heavy Ion Collider at Brookhaven National Laboratory.[85][86]

The AdS/CFT correspondence has also been used to study aspects of condensed matter physics. Over the decades, experimental condensed matter physicists have discovered a number of exotic states of matter, including superconductors and superfluids. These states are described using the formalism of quantum field theory, but some phenomena are difficult to explain using standard field theoretic techniques. Some condensed matter theorists including Subir Sachdev hope that the AdS/CFT correspondence will make it possible to describe these systems in the language of string theory and learn more about their behavior.[85]

So far some success has been achieved in using string theory methods to describe the transition of a superfluid to an insulator. A superfluid is a system of electrically neutral atoms that flows without any friction. Such systems are often produced in the laboratory using liquid helium, but recently experimentalists have developed new ways of producing artificial superfluids by pouring trillions of cold atoms into a lattice of criss-crossing lasers. These atoms initially behave as a superfluid, but as experimentalists increase the intensity of the lasers, they become less mobile and then suddenly transition to an insulating state. During the transition, the atoms behave in an unusual way. For example, the atoms slow to a halt at a rate that depends on the temperature and on Planck's constant, the fundamental parameter of quantum mechanics, which does not enter into the description of the other phases. This behavior has recently been understood by considering a dual description where properties of the fluid are described in terms of a higher dimensional black hole.[87]

39.5 Phenomenology

Main article: String phenomenology

In addition to being an idea of considerable theoretical interest, string theory provides a framework for constructing models of real world physics that combine general relativity and particle physics. Phenomenology is the branch of theoretical physics in which physicists construct realistic models of nature from more abstract theoretical ideas. String phenomenology is the part of string theory that attempts to construct realistic models based on string theory.

Partly because of theoretical and mathematical difficulties and partly because of the extremely high energies needed to test these theories experimentally, there is so far no experimental evidence that would unambiguously point to any of these models being a correct fundamental description of nature. This has led some in the community to criticize these approaches to unification and question the value of continued research on these problems.[12]

39.5.1 Particle physics

The currently accepted theory describing elementary particles and their interactions is known as the standard model of particle physics. This theory provides a unified description of three of the fundamental forces of nature: electromagnetism and the strong and weak nuclear forces. Despite its remarkable success in explaining a wide range of physical phenomena, the standard model cannot be a complete description of reality. This is because the standard model fails to incorporate the force of gravity and because of problems such as the hierarchy problem and the inability to explain the structure of fermion masses or dark matter.

String theory has been used to construct a variety of models of particle physics going beyond the standard model. Typically, such models are based on the idea of compactification. Starting with the ten- or eleven-dimensional spacetime of string or M-theory, physicists postulate a shape for the extra dimensions. By choosing this shape appropriately, they can construct models roughly similar to the standard model of particle physics, together with additional undiscovered particles.[88] One popular way of deriving realistic physics from string theory is to start with the heterotic theory in ten dimensions and assume that the six extra dimensions of spacetime are shaped like a six-dimensional Calabi–Yau manifold. Such compactifications offer many ways of extracting realistic physics from string theory. Other similar methods can be used to construct realistic models of our four-dimensional world based on M-theory.[89]

39.5.2 Cosmology

Main article: String cosmology
The Big Bang theory is the prevailing cosmological model

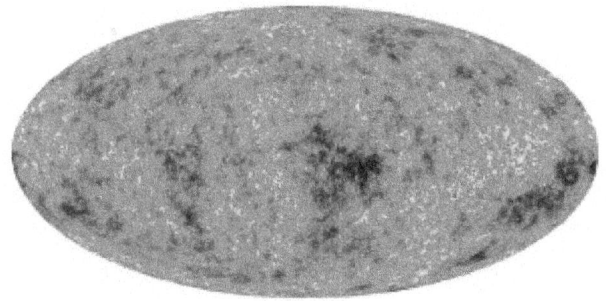

A map of the cosmic microwave background produced by the Wilkinson Microwave Anisotropy Probe

for the universe from the earliest known periods through its subsequent large-scale evolution. Despite its success in explaining many observed features of the universe including galactic redshifts, the relative abundance of light elements such as hydrogen and helium, and the existence of a cosmic microwave background, there are several questions that remain unanswered. For example, the standard Big Bang model does not explain why the universe appears to be same in all directions, why it appears flat on very large distance scales, or why certain hypothesized particles such as magnetic monopoles are not observed in experiments.[90]

Currently, the leading candidate for a theory going beyond the Big Bang is the theory of cosmic inflation. Developed by Alan Guth and others in the 1980s, inflation postulates a period of extremely rapid accelerated expansion of the universe prior to the expansion described by the standard Big Bang theory. The theory of cosmic inflation preserves the successes of the Big Bang while providing a natural explanation for some of the mysterious features of the universe.[91] The theory has also received striking support from observations of the cosmic microwave background,

the radiation that has filled the sky since around 380,000 years after the Big Bang.[92]

In the theory of inflation, the rapid initial expansion of the universe is caused by a hypothetical particle called the inflaton. The exact properties of this particle are not fixed by the theory but should ultimately be derived from a more fundamental theory such as string theory.[93] Indeed, there have been a number of attempts to identify an inflaton within the spectrum of particles described by string theory, and to study inflation using string theory. While these approaches might eventually find support in observational data such as measurements of the cosmic microwave background, the application of string theory to cosmology is still in its early stages.[94]

39.6 Connections to mathematics

In addition to influencing research in theoretical physics, string theory has stimulated a number of major developments in pure mathematics. Like many developing ideas in theoretical physics, string theory does not at present have a mathematically rigorous formulation in which all of its concepts can be defined precisely. As a result, physicists who study string theory are often guided by physical intuition to conjecture relationships between the seemingly different mathematical structures that are used to formalize different parts of the theory. These conjectures are later proved by mathematicians, and in this way, string theory serves as a source of new ideas in pure mathematics.[95]

39.6.1 Mirror symmetry

Main article: Mirror symmetry (string theory)
After Calabi–Yau manifolds had entered physics as a way to compactify extra dimensions in string theory, many physicists began studying these manifolds. In the late 1980s, several physicists noticed that given such a compactification of string theory, it is not possible to reconstruct uniquely a corresponding Calabi–Yau manifold.[96] Instead, two different versions of string theory, type IIA and type IIB, can be compactified on completely different Calabi–Yau manifolds giving rise to the same physics. In this situation, the manifolds are called mirror manifolds, and the relationship between the two physical theories is called mirror symmetry.[97]

Regardless of whether Calabi–Yau compactifications of string theory provide a correct description of nature, the existence of the mirror duality between different string theories has significant mathematical consequences. The Calabi–Yau manifolds used in string theory are of interest in pure mathematics, and mirror symmetry allows math-

The Clebsch cubic is an example of a kind of geometric object called an algebraic variety. A classical result of enumerative geometry states that there are exactly 27 straight lines that lie entirely on this surface.

ematicians to solve problems in enumerative geometry, a branch of mathematics concerned with counting the numbers of solutions to geometric questions.[31][98]

Enumerative geometry studies a class of geometric objects called algebraic varieties which are defined by the vanishing of polynomials. For example, the Clebsch cubic illustrated on the right is an algebraic variety defined using a certain polynomial of degree three in four variables. A celebrated result of nineteenth-century mathematicians Arthur Cayley and George Salmon states that there are exactly 27 straight lines that lie entirely on such a surface.[99]

Generalizing this problem, one can ask how many lines can be drawn on a quintic Calabi–Yau manifold, such as the one illustrated above, which is defined by a polynomial of degree five. This problem was solved by the nineteenth-century German mathematician Hermann Schubert, who found that there are exactly 2,875 such lines. In 1986, geometer Sheldon Katz proved that the number of curves, such as circles, that are defined by polynomials of degree two and lie entirely in the quintic is 609,250.[100]

By the year 1991, most of the classical problems of enumerative geometry had been solved and interest in enumerative geometry had begun to diminish.[101] The field was reinvigorated in May 1991 when physicists Philip Candelas, Xenia de la Ossa, Paul Green, and Linda Parks showed that mirror symmetry could be used to translate difficult mathematical questions about one Calabi–Yau manifold into easier questions about its mirror.[102] In particular, they used mirror symmetry to show that a six-dimensional Calabi–Yau

manifold can contain exactly 317,206,375 curves of degree three.[101] In addition to counting degree-three curves, Candelas and his collaborators obtained a number of more general results for counting rational curves which went far beyond the results obtained by mathematicians.[103]

Originally, these results of Candelas were justified on physical grounds. However, mathematicians generally prefer rigorous proofs that do not require an appeal to physical intuition. Inspired by physicists' work on mirror symmetry, mathematicians have therefore constructed their own arguments proving the enumerative predictions of mirror symmetry.[lower-alpha 5] Today mirror symmetry is an active area of research in mathematics, and mathematicians are working to develop a more complete mathematical understanding of mirror symmetry based on physicists' intuition.[104] Major approaches to mirror symmetry include the homological mirror symmetry program of Maxim Kontsevich[32] and the SYZ conjecture of Andrew Strominger, Shing-Tung Yau, and Eric Zaslow.[105]

39.6.2 Monstrous moonshine

Main article: Monstrous moonshine
Group theory is the branch of mathematics that studies the

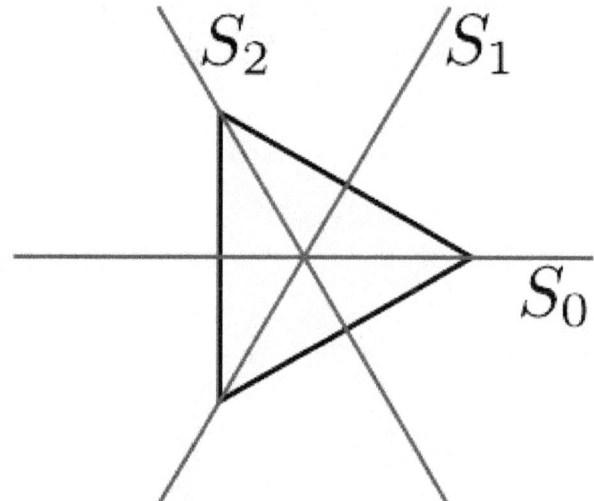

An equilateral triangle can be rotated through 120°, 240°, or 360°, or reflected in any of the three lines pictured without changing its shape.

concept of symmetry. For example, one can consider a geometric shape such as an equilateral triangle. There are various operations that one can perform on this triangle without changing its shape. One can rotate it through 120°, 240°, or 360°, or one can reflect in any of the lines labeled S_0, S_1, or S_2 in the picture. Each of these operations is called a *symmetry*, and the collection of these symmetries satisfies certain technical properties making it into what mathematicians call a group. In this particular example, the group is known as the dihedral group of order 6 because it has six elements. A general group may describe finitely many or infinitely many symmetries; if there are only finitely many symmetries, it is called a finite group.[106]

Mathematicians often strive for a classification (or list) of all mathematical objects of a given type. It is generally believed that finite groups are too diverse to admit a useful classification. A more modest but still challenging problem is to classify all finite *simple* groups. These are finite groups which may be used as building blocks for constructing arbitrary finite groups in the same way that prime numbers can be used to construct arbitrary whole numbers by taking products.[lower-alpha 6] One of the major achievements of contemporary group theory is the classification of finite simple groups, a mathematical theorem which provides a list of all possible finite simple groups.[107]

This classification theorem identifies several infinite families of groups as well as 26 additional groups which do not fit into any family. The latter groups are called the "sporadic" groups, and each one owes its existence to a remarkable combination of circumstances. The largest sporadic group, the so-called monster group, has over 10^{53} elements, more than a thousand times the number of atoms in the Earth.[108]

A graph of the j-*function in the complex plane*

A seemingly unrelated construction is the *j*-function of number theory. This object belongs to a special class of functions called modular functions, whose graphs form a certain kind of repeating pattern.[109] Although this function appears in a branch of mathematics which seems very different from the theory of finite groups, the two subjects turn out to be intimately related. In the late 1970s, mathematicians John McKay and John Thompson noticed that certain numbers arising in the analysis of the monster group (namely, the dimensions of its irreducible representations) are related to numbers that appear in a formula for the *j*-

function (namely, the coefficients of its Fourier series).[110] This relationship was further developed by John Horton Conway and Simon Norton[111] who called it monstrous moonshine because it seemed so far fetched.[112]

In 1992, Richard Borcherds constructed a bridge between the theory of modular functions and finite groups and, in the process, explained the observations of McKay and Thompson.[113][114] Borcherds' work used ideas from string theory in an essential way, extending earlier results of Igor Frenkel, James Lepowsky, and Arne Meurman, who had realized the monster group as the symmetries of a particular version of string theory.[115] In 1998, Borcherds was awarded the Fields medal for his work.[116]

Since the 1990s, the connection between string theory and moonshine has led to further results in mathematics and physics.[108] In 2010, physicists Tohru Eguchi, Hirosi Ooguri, and Yuji Tachikawa discovered connections between a different sporadic group, the Mathieu group M_{24}, and a certain version of string theory.[117] Miranda Cheng, John Duncan, and Jeffrey A. Harvey proposed a generalization of this moonshine phenomenon called umbral moonshine,[118] and their conjecture was proved mathematically by Duncan, Michael Griffin, and Ken Ono.[119] Witten has also speculated that the version of string theory appearing in monstrous moonshine might be related to a certain simplified model of gravity in three spacetime dimensions.[120]

39.7   History

Main article: History of string theory

39.7.1   Early results

Some of the structures reintroduced by string theory arose for the first time much earlier as part of the program of classical unification started by Albert Einstein. The first person to add a fifth dimension to a theory of gravity was Gunnar Nordström in 1914, who noted that gravity in five dimensions describes both gravity and electromagnetism in four. Nordström attempted to unify electromagnetism with his theory of gravitation, which was however superseded by Einstein's general relativity in 1919. Thereafter, German mathematician Theodor Kaluza combined the fifth dimension with general relativity, and only Kaluza is usually credited with the idea. In 1926, the Swedish physicist Oskar Klein gave a physical interpretation of the unobservable extra dimension—it is wrapped into a small circle. Einstein introduced a non-symmetric metric tensor, while much later Brans and Dicke added a scalar component to gravity. These ideas would be revived within string theory, where they are demanded by consistency conditions.

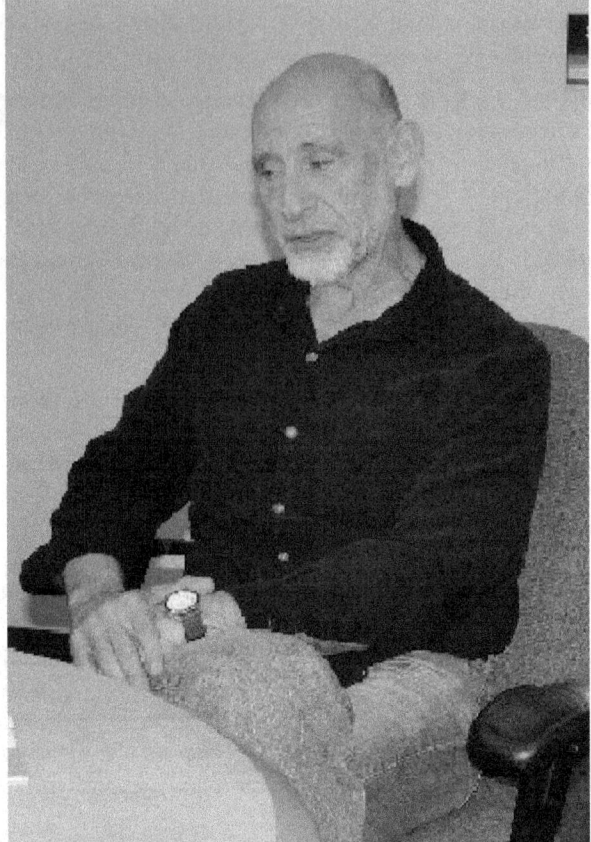

Leonard Susskind

String theory was originally developed during the late 1960s and early 1970s as a never completely successful theory of hadrons, the subatomic particles like the proton and neutron that feel the strong interaction. In the 1960s, Geoffrey Chew and Steven Frautschi discovered that the mesons make families called Regge trajectories with masses related to spins in a way that was later understood by Yoichiro Nambu, Holger Bech Nielsen and Leonard Susskind to be the relationship expected from rotating strings. Chew advocated making a theory for the interactions of these trajectories that did not presume that they were composed of any fundamental particles, but would construct their interactions from self-consistency conditions on the S-matrix. The S-matrix approach was started by Werner Heisenberg in the 1940s as a way of constructing a theory that did not rely on the local notions of space and time, which Heisenberg believed break down at the nuclear scale. While the scale was off by many orders of magnitude, the approach he advocated was ideally suited for a theory of quantum gravity.

Working with experimental data, R. Dolen, D. Horn and C. Schmid developed some sum rules for hadron exchange. When a particle and antiparticle scatter, virtual particles can

be exchanged in two qualitatively different ways. In the s-channel, the two particles annihilate to make temporary intermediate states that fall apart into the final state particles. In the t-channel, the particles exchange intermediate states by emission and absorption. In field theory, the two contributions add together, one giving a continuous background contribution, the other giving peaks at certain energies. In the data, it was clear that the peaks were stealing from the background—the authors interpreted this as saying that the t-channel contribution was dual to the s-channel one, meaning both described the whole amplitude and included the other.

Gabriele Veneziano

The result was widely advertised by Murray Gell-Mann, leading Gabriele Veneziano to construct a scattering amplitude that had the property of Dolen-Horn-Schmid duality, later renamed world-sheet duality. The amplitude needed poles where the particles appear, on straight line trajectories, and there is a special mathematical function whose poles are evenly spaced on half the real line— the Gamma function— which was widely used in Regge theory. By manipulating combinations of Gamma functions, Veneziano was able to find a consistent scattering amplitude with poles on straight lines, with mostly positive residues, which obeyed duality and had the appropriate Regge scaling at high energy. The amplitude could fit near-beam scattering data as well as other Regge type fits, and had a suggestive integral representation that could be used for generalization.

Over the next years, hundreds of physicists worked to complete the bootstrap program for this model, with many surprises. Veneziano himself discovered that for the scattering amplitude to describe the scattering of a particle that appears in the theory, an obvious self-consistency condition, the lightest particle must be a tachyon. Miguel Virasoro and Joel Shapiro found a different amplitude now understood to be that of closed strings, while Ziro Koba and Holger Nielsen generalized Veneziano's integral representation to multiparticle scattering. Veneziano and Sergio Fubini introduced an operator formalism for computing the scattering amplitudes that was a forerunner of world-sheet conformal theory, while Virasoro understood how to remove the poles with wrong-sign residues using a constraint on the states. Claud Lovelace calculated a loop amplitude, and noted that there is an inconsistency unless the dimension of the theory is 26. Charles Thorn, Peter Goddard and Richard Brower went on to prove that there are no wrong-sign propagating states in dimensions less than or equal to 26.

In 1969, Yoichiro Nambu, Holger Bech Nielsen, and Leonard Susskind recognized that the theory could be given a description in space and time in terms of strings. The scattering amplitudes were derived systematically from the action principle by Peter Goddard, Jeffrey Goldstone, Claudio Rebbi, and Charles Thorn, giving a space-time picture to the vertex operators introduced by Veneziano and Fubini and a geometrical interpretation to the Virasoro conditions.

In 1970, Pierre Ramond added fermions to the model, which led him to formulate a two-dimensional supersymmetry to cancel the wrong-sign states. John Schwarz and André Neveu added another sector to the fermi theory a short time later. In the fermion theories, the critical dimension was 10. Stanley Mandelstam formulated a world sheet conformal theory for both the bose and fermi case, giving a two-dimensional field theoretic path-integral to generate the operator formalism. Michio Kaku and Keiji Kikkawa gave a different formulation of the bosonic string, as a string field theory, with infinitely many particle types and with fields taking values not on points, but on loops and curves.

In 1974, Tamiaki Yoneya discovered that all the known string theories included a massless spin-two particle that obeyed the correct Ward identities to be a graviton. John Schwarz and Joel Scherk came to the same conclusion and made the bold leap to suggest that string theory was a theory of gravity, not a theory of hadrons. They reintroduced Kaluza–Klein theory as a way of making sense of the extra dimensions. At the same time, quantum chromodynamics was recognized as the correct theory of hadrons, shifting the attention of physicists and apparently leaving the bootstrap program in the dustbin of history.

String theory eventually made it out of the dustbin, but for the following decade all work on the theory was completely ignored. Still, the theory continued to develop at a steady pace thanks to the work of a handful of devotees. Ferdinando Gliozzi, Joel Scherk, and David Olive realized in 1976 that the original Ramond and Neveu Schwarz-strings were separately inconsistent and needed to be combined. The resulting theory did not have a tachyon, and was proven to have space-time supersymmetry by John Schwarz and Michael Green in 1981. The same year, Alexander Polyakov gave the theory a modern path integral formulation, and went on to develop conformal field theory extensively. In 1979, Daniel Friedan showed that the equations of motions of string theory, which are generalizations of the Einstein equations of General Relativity, emerge from the Renormalization group equations for the two-dimensional field theory. Schwarz and Green discovered T-duality, and constructed two superstring theories—IIA and IIB related by T-duality, and type I theories with open strings. The consistency conditions had been so strong, that the entire theory was nearly uniquely determined, with only a few discrete choices.

39.7.2   First superstring revolution

Edward Witten

In the early 1980s, Edward Witten discovered that most theories of quantum gravity could not accommodate chiral fermions like the neutrino. This led him, in collaboration with Luis Álvarez-Gaumé to study violations of the conservation laws in gravity theories with anomalies, concluding that type I string theories were inconsistent. Green and Schwarz discovered a contribution to the anomaly that Witten and Alvarez-Gaumé had missed, which restricted the gauge group of the type I string theory to be SO(32). In coming to understand this calculation, Edward Witten became convinced that string theory was truly a consistent theory of gravity, and he became a high-profile advocate. Following Witten's lead, between 1984 and 1986, hundreds of physicists started to work in this field, and this is sometimes called the first superstring revolution.

During this period, David Gross, Jeffrey Harvey, Emil Martinec, and Ryan Rohm discovered heterotic strings. The gauge group of these closed strings was two copies of E8, and either copy could easily and naturally include the standard model. Philip Candelas, Gary Horowitz, Andrew Strominger and Edward Witten found that the Calabi–Yau manifolds are the compactifications that preserve a realistic amount of supersymmetry, while Lance Dixon and others worked out the physical properties of orbifolds, distinctive geometrical singularities allowed in string theory. Cumrun Vafa generalized T-duality from circles to arbitrary manifolds, creating the mathematical field of mirror symmetry. Daniel Friedan, Emil Martinec and Stephen Shenker further developed the covariant quantization of the superstring using conformal field theory techniques. David Gross and Vipul Periwal discovered that string perturbation theory was divergent. Stephen Shenker showed it diverged much faster than in field theory suggesting that new nonperturbative objects were missing.

Joseph Polchinski

In the 1990s, Joseph Polchinski discovered that the theory requires higher-dimensional objects, called D-branes

and identified these with the black-hole solutions of supergravity. These were understood to be the new objects suggested by the perturbative divergences, and they opened up a new field with rich mathematical structure. It quickly became clear that D-branes and other p-branes, not just strings, formed the matter content of the string theories, and the physical interpretation of the strings and branes was revealed—they are a type of black hole. Leonard Susskind had incorporated the holographic principle of Gerardus 't Hooft into string theory, identifying the long highly excited string states with ordinary thermal black hole states. As suggested by 't Hooft, the fluctuations of the black hole horizon, the world-sheet or world-volume theory, describes not only the degrees of freedom of the black hole, but all nearby objects too.

39.7.3 Second superstring revolution

In 1995, at the annual conference of string theorists at the University of Southern California (USC), Edward Witten gave a speech on string theory that in essence united the five string theories that existed at the time, and giving birth to a new 11-dimensional theory called M-theory. M-theory was also foreshadowed in the work of Pául Townsend at approximately the same time. The flurry of activity that began at this time is sometimes called the second superstring revolution.[34]

Juan Maldacena

During this period, Tom Banks, Willy Fischler, Stephen Shenker and Leonard Susskind formulated matrix theory, a full holographic description of M-theory using IIA D0 branes.[51] This was the first definition of string theory that was fully non-perturbative and a concrete mathematical realization of the holographic principle. It is an example of a gauge-gravity duality and is now understood to be a special case of the AdS/CFT correspondence. Andrew Strominger and Cumrun Vafa calculated the entropy of certain configurations of D-branes and found agreement with the semi-classical answer for extreme charged black holes.[62] Petr Hořava and Witten found the eleven-dimensional formulation of the heterotic string theories, showing that orbifolds solve the chirality problem. Witten noted that the effective description of the physics of D-branes at low energies is by a supersymmetric gauge theory, and found geometrical interpretations of mathematical structures in gauge theory that he and Nathan Seiberg had earlier discovered in terms of the location of the branes.

In 1997, Juan Maldacena noted that the low energy excitations of a theory near a black hole consist of objects close to the horizon, which for extreme charged black holes looks like an anti-de Sitter space.[71] He noted that in this limit the gauge theory describes the string excitations near the branes. So he hypothesized that string theory on a near-horizon extreme-charged black-hole geometry, an anti-deSitter space times a sphere with flux, is equally well described by the low-energy limiting gauge theory, the N = 4 supersymmetric Yang–Mills theory. This hypothesis, which is called the AdS/CFT correspondence, was further developed by Steven Gubser, Igor Klebanov and Alexander Polyakov,[72] and by Edward Witten,[73] and it is now well-accepted. It is a concrete realization of the holographic principle, which has far-reaching implications for black holes, locality and information in physics, as well as the nature of the gravitational interaction.[56] Through this relationship, string theory has been shown to be related to gauge theories like quantum chromodynamics and this has led to more quantitative understanding of the behavior of hadrons, bringing string theory back to its roots.[84]

39.8   Criticism

39.8.1   Number of solutions

Main article: String theory landscape

To construct models of particle physics based on string theory, physicists typically begin by specifying a shape for the extra dimensions of spacetime. Each of these different shapes corresponds to a different possible universe, or

"vacuum state", with a different collection of particles and forces. String theory as it is currently understood has an enormous number of vacuum states, typically estimated to be around 10^{500}, and these might be sufficiently diverse to accommodate almost any phenomena that might be observed at low energies.[121]

Many critics of string theory have expressed concerns about the large number of possible universes described by string theory. In his book *Not Even Wrong*, Peter Woit, a lecturer in the mathematics department at Columbia University, has argued that the large number of different physical scenarios renders string theory vacuous as a framework for constructing models of particle physics. According to Woit,

> The possible existence of, say, 10^{500} consistent different vacuum states for superstring theory probably destroys the hope of using the theory to predict anything. If one picks among this large set just those states whose properties agree with present experimental observations, it is likely there still will be such a large number of these that one can get just about whatever value one wants for the results of any new observation.[122]

Some physicists believe this large number of solutions is actually a virtue because it may allow a natural anthropic explanation of the observed values of physical constants, in particular the small value of the cosmological constant.[122] The anthropic principle is the idea that some of the numbers appearing in the laws of physics are not fixed by any fundamental principle but must be compatible with the evolution of intelligent life. In 1987, Steven Weinberg published an article in which he argued that the cosmological constant could not have been too large, or else galaxies and intelligent life would not have been able to develop.[123] Weinberg suggested that there might be a huge number of possible consistent universes, each with a different value of the cosmological constant, and observations indicate a small value of the cosmological constant only because humans happen to live in a universe that has allowed intelligent life, and hence observers, to exist.[124]

String theorist Leonard Susskind has argued that string theory provides a natural anthropic explanation of the small value of the cosmological constant.[125] According to Susskind, the different vacuum states of string theory might be realized as different universes within a larger multiverse. The fact that the observed universe has a small cosmological constant is just a tautological consequence of the fact that a small value is required for life to exist.[126] Many prominent theorists and critics have disagreed with Susskind's conclusions.[127] According to Woit, "in this case [anthropic reasoning] is nothing more than an excuse for failure. Speculative scientific ideas fail not just when they make incorrect predictions, but also when they turn out to be vacuous and incapable of predicting anything."[128]

39.8.2 Background independence

Main article: Background independence

One of the fundamental properties of Einstein's general theory of relativity is that it is background independent, meaning that the formulation of the theory does not in any way privilege a particular spacetime geometry.[129]

One of the main criticisms of string theory from early on is that it is not manifestly background independent. In string theory, one must typically specify a fixed reference geometry for spacetime, and all other possible geometries are described as perturbations of this fixed one. In his book *The Trouble With Physics*, physicist Lee Smolin of the Perimeter Institute for Theoretical Physics claims that this is the principal weakness of string theory as a theory of quantum gravity, saying that string theory has failed to incorporate this important insight from general relativity.[130]

Others have disagreed with Smolin's characterization of string theory. In a review of Smolin's book, string theorist Joseph Polchinski writes

> [Smolin] is mistaking an aspect of the mathematical language being used for one of the physics being described. New physical theories are often discovered using a mathematical language that is not the most suitable for them... In string theory it has always been clear that the physics is background-independent even if the language being used is not, and the search for more suitable language continues. Indeed, as Smolin belatedly notes, [AdS/CFT] provides a solution to this problem, one that is unexpected and powerful.[131]

Polchinski notes that an important open problem in quantum gravity is to develop holographic descriptions of gravity which do not require the gravitational field to be asymptotically anti-de Sitter.[131]

Smolin responded that the claims about background-independence, which Polchinski presents as "clear", are in fact only an unproven hope for future results, and Smolin is skeptical about them being true at all because of fundamental reasons: "If the strong form of the AdS/CFT conjecture is shown to be correct, then a very weak, and limited form of background will have been achieved. But ... this is still a big if". Smolin points out that current results about

the [AdS/CFT] conjecture rely on global super-symmetry as perturbative physics, "but the whole point of general relativity and quantum gravity is that the generic solutions are governed by no global symmetries because the geometry of spacetime is completely dynamical", which "makes it very non-trivial to show the strong form of the [AdS/CFT] conjecture, because it must extend to solutions of supergravity arbitrarily far from those with global symmetries in the bulk".[132] Smolin summarizes:

> It would be more accurate to say, "Some string theorists believe that the formulations of perturbative string theories and dualities between them that they study concretely are approximations to a deeper, background independent formulation. This missing background independent formulation is not just a different language for the theory, it is hoped to be the statement of the principles and laws that define the theory, from which everything studied so far would be derived as an approximation."[132]

39.8.3 Sociological issues

Since the superstring revolutions of the 1980s and 1990s, string theory has become the dominant paradigm of high energy theoretical physics.[133] Some string theorists have expressed the view that there does not exist an equally successful alternative theory addressing the deep questions of fundamental physics. In an interview from 1987, Nobel laureate David Gross made the following controversial comments about the reasons for the popularity of string theory:

> The most important [reason] is that there are no other good ideas around. That's what gets most people into it. When people started to get interested in string theory they didn't know anything about it. In fact, the first reaction of most people is that the theory is extremely ugly and unpleasant, at least that was the case a few years ago when the understanding of string theory was much less developed. It was difficult for people to learn about it and to be turned on. So I think the real reason why people have got attracted by it is because there is no other game in town. All other approaches of constructing grand unified theories, which were more conservative to begin with, and only gradually became more and more radical, have failed, and this game hasn't failed yet.[134]

Several other high profile theorists and commentators have expressed similar views, suggesting that there are no viable alternatives to string theory.[135]

Many critics of string theory have commented on this state of affairs. In his book criticizing string theory, Peter Woit views the status of string theory research as unhealthy and detrimental to the future of fundamental physics. He argues that the extreme popularity of string theory among theoretical physicists is partly a consequence of the financial structure of academia and the fierce competition for scarce resources.[136] In his book *The Road to Reality*, mathematical physicist Roger Penrose expresses similar views, stating "The often frantic competitiveness that this ease of communication engenders leads to 'bandwagon' effects, where researchers fear to be left behind if they do not join in."[137] Penrose also claims that the technical difficulty of modern physics forces young scientists to rely on the preferences of established researchers, rather than forging new paths of their own.[138] Lee Smolin expresses a slightly different position in his critique, claiming that string theory grew out of a tradition of particle physics which discourages speculation about the foundations of physics, while his preferred approach, loop quantum gravity, encourages more radical thinking. According to Smolin,

> String theory is a powerful, well-motivated idea and deserves much of the work that has been devoted to it. If it has so far failed, the principal reason is that its intrinsic flaws are closely tied to its strengths—and, of course, the story is unfinished, since string theory may well turn out to be part of the truth. The real question is not why we have expended so much energy on string theory but why we haven't expended nearly enough on alternative approaches.[139]

Smolin goes on to offer a number of prescriptions for how scientists might encourage a greater diversity of approaches to quantum gravity research.[140]

39.9   References

39.9.1   Notes

[1] For example, physicists are still working to understand the phenomenon of quark confinement, the paradoxes of black holes, and the origin of dark energy.

[2] For example, in the context of the AdS/CFT correspondence, theorists often formulate and study theories of gravity in unphysical numbers of spacetime dimensions.

[3] "Top Cited Articles during 2010 in hep-th". Retrieved 25 July 2013.

[4] More precisely, one cannot apply the methods of perturbative quantum field theory.

[5] Two independent mathematical proofs of mirror symmetry were given by Givental 1996, 1998 and Lian, Liu, Yau 1997, 1999, 2000.

[6] More precisely, a nontrivial group is called *simple* if its only normal subgroups are the trivial group and the group itself. The Jordan–Hölder theorem exhibits finite simple groups as the building blocks for all finite groups.

39.9.2   Citations

[1] Becker, Becker, and Schwarz 2007, p. 1

[2] Zwiebach 2009, p. 6

[3] Becker, Becker, and Schwarz 2007, pp. 2–3

[4] Becker, Becker, and Schwarz 2007, pp. 9–12

[5] Becker, Becker, and Schwarz 2007, pp. 14–15

[6] Klebanov and Maldacena 2009

[7] Merali 2011

[8] Sachdev 2013

[9] Becker, Becker, and Schwarz 2007, pp. 3, 15–16

[10] Becker, Becker, and Schwarz 2007, p. 8

[11] Becker, Becker, and Schwarz 13–14

[12] Woit 2006

[13] Zee 2010

[14] Becker, Becker, and Schwarz 2007, p. 2

[15] Becker, Becker, and Schwarz 2007, p. 6

[16] Zwiebach 2009, p. 12

[17] Becker, Becker, and Schwarz 2007, p. 4

[18] Zwiebach 2009, p. 324

[19] Wald 1984, p. 4

[20] Zee 2010, Parts V and VI

[21] Zwiebach 2009, p. 9

[22] Zwiebach 2009, p. 8

[23] Yau and Nadis 2010, Ch. 6

[24] Greene 2000, p. 186

[25] Yau and Nadis 2010, p. ix

[26] Randall and Sundrum 1999

[27] Becker, Becker, and Schwarz 2007

[28] Zwiebach 2009, p. 376

[29] Moore 2005, p. 214

[30] Moore 2005, p. 215

[31] Aspinwall et al. 2009

[32] Kontsevich 1995

[33] Kapustin and Witten 2007

[34] Duff 1998

[35] Duff 1998, p. 64

[36] Nahm 1978

[37] Cremmer, Julia, and Scherk 1978

[38] Duff 1998, p. 65

[39] Sen 1994a

[40] Sen 1994b

[41] Hull and Townsend 1995

[42] Duff 1998, p. 67

[43] Bergshoeff, Sezgin, and Townsend 1987

[44] Duff et al. 1987

[45] Duff 1998, p. 66

[46] Witten 1995

[47] Duff 1998, pp. 67–68

[48] Becker, Becker, and Schwarz 2007, p. 296

[49] Hořava and Witten 1996

[50] Duff 1996, sec. 1

[51] Banks et al. 1997

[52] Connes 1994

[53] Connes, Douglas, and Schwarz 1998

[54] Nekrasov and Schwarz 1998

[55] Seiberg and Witten 1999

[56] de Haro et al. 2013, p. 2

[57] Yau and Nadis 2010, p. 187–188

[58] Bekenstein 1973

[59] Hawking 1975

[60] Wald 1984, p. 417

[61] Yau and Nadis 2010, p. 189

[62] Strominger and Vafa 1996

[63] Yau and Nadis 2010, pp. 190–192

[64] Maldacena, Strominger, and Witten 1997

[65] Ooguri, Strominger, and Vafa 2004

[66] Yau and Nadis 2010, pp. 192–193

[67] Yau and Nadis 2010, pp. 194–195

[68] Strominger 1998

[69] Guica et al. 2009

[70] Castro, Maloney, and Strominger 2010

[71] Maldacena 1998

[72] Gubser, Klebanov, and Polyakov 1998

[73] Witten 1998

[74] Klebanov and Maldacena 2009, p. 28

[75] Maldacena 2005, p. 60

[76] Maldacena 2005, p. 61

[77] Zwiebach 2009, p. 552

[78] Maldacena 2005, pp. 61–62

[79] Susskind 2008

[80] Zwiebach 2009, p. 554

[81] Maldacena 2005, p. 63

[82] Hawking 2005

[83] Zwiebach 2009, p. 559

[84] Kovtun, Son, and Starinets 2001

[85] Merali 2011, p. 303

[86] Luzum and Romatschke 2008

[87] Sachdev 2013, p. 51

[88] Candelas et al. 1985

[89] Yau and Nadis 2010, pp. 147–150

[90] Becker, Becker, and Schwarz 2007, pp. 530–531

[91] Becker, Becker, and Schwarz 2007, p. 531

[92] Becker, Becker, and Schwarz 2007, p. 538

[93] Becker, Becker, and Schwarz 2007, p. 533

[94] Becker, Becker, and Schwarz 2007, pp. 539–543

[95] Deligne et al. 1999, p. 1

[96] Hori et al. 2003, p. xvii

[97] Aspinwall et al. 2009, p. 13

[98] Hori et al. 2003

[99] Yau and Nadis 2010, p. 167

[100] Yau and Nadis 2010, p. 166

[101] Yau and Nadis 2010, p. 169

[102] Candelas et al. 1991

[103] Yau and Nadis 2010, p. 171

[104] Hori et al. 2003, p. xix

[105] Strominger, Yau, and Zaslow 1996

[106] Dummit and Foote 2004

[107] Dummit and Foote 2004, pp. 102–103

[108] Klarreich 2015

[109] Gannon 2006, p. 2

[110] Gannon 2006, p. 4

[111] Conway and Norton 1979

[112] Gannon 2006, p. 5

[113] Gannon 2006, p. 8

[114] Borcherds 1992

[115] Frenkel, Lepowsky, and Meurman 1988

[116] Gannon 2006, p. 11

[117] Eguchi, Ooguri, and Tachikawa 2010

[118] Cheng, Duncan, and Harvey 2013

[119] Duncan, Griffin, and Ono 2015

[120] Witten 2007

[121] Woit 2006, pp. 240–242

[122] Woit 2006, p. 242

[123] Weinberg 1987

[124] Woit 2006, p. 243

[125] Susskind 2005

[126] Woit 2006, pp. 242–243

[127] Woit 2006, p. 240

[128] Woit 2006, p. 249

[129] Smolin 2006, p. 81

[130] Smolin 2006, p. 184

[131] Polchinski 2007

[132] Lee Smolin, April 2007:"Archived copy". Archived from the original on November 5, 2015. Retrieved December 31, 2015. Response to review of The Trouble with Physics by Joe Polchinski

[133] Penrose 2004, p. 1017

[134] Woit 2006, pp. 224–225

[135] Woit 2006, Ch. 16

[136] Woit 2006, p. 239

[137] Penrose 2004, p. 1018

[138] Penrose 2004, pp. 1019–1020

[139] Smolin 2006, p. 349

[140] Smolin 2006, Ch. 20

39.9.3 Bibliography

- Aspinwall, Paul; Bridgeland, Tom; Craw, Alastair; Douglas, Michael; Gross, Mark; Kapustin, Anton; Moore, Gregory; Segal, Graeme; Szendrői, Balázs; Wilson, P.M.H., eds. (2009). *Dirichlet Branes and Mirror Symmetry*. American Mathematical Society. ISBN 978-0-8218-3848-8.

- Banks, Tom; Fischler, Willy; Schenker, Stephen; Susskind, Leonard (1997). "M theory as a matrix model: A conjecture". *Physical Review D* **55** (8): 5112–5128. arXiv:hep-th/9610043. Bibcode:1997PhRvD..55.5112B. doi:10.1103/physrevd.55.5112.

- Becker, Katrin; Becker, Melanie; Schwarz, John (2007). *String theory and M-theory: A modern introduction*. Cambridge University Press. ISBN 978-0-521-86069-7.

- Bekenstein, Jacob (1973). "Black holes and entropy". *Physical Review D* **7** (8): 2333–2346. Bibcode:1973PhRvD...7.2333B. doi:10.1103/PhysRevD.7.2333.

- Bergshoeff, Eric; Sezgin, Ergin; Townsend, Paul (1987). "Supermembranes and eleven-dimensional supergravity". *Physics Letters B* **189** (1): 75–78. Bibcode:1987PhLB..189...75B. doi:10.1016/0370-2693(87)91272-X.

- Borcherds, Richard (1992). "Monstrous moonshine and Lie superalgebras". *Inventiones Mathematicae* **109** (1): 405–444. Bibcode:1992InMat.109..405B. doi:10.1007/BF01232032.

- Candelas, Philip; de la Ossa, Xenia; Green, Paul; Parks, Linda (1991). "A pair of Calabi–Yau manifolds as an exactly soluble superconformal field theory". *Nuclear Physics B* **359** (1): 21–74. Bibcode:1991NuPhB.359...21C. doi:10.1016/0550-3213(91)90292-6.

- Candelas, Philip; Horowitz, Gary; Strominger, Andrew; Witten, Edward (1985). "Vacuum configurations for superstrings". *Nuclear Physics B* **258**: 46–74. Bibcode:1985NuPhB.258...46C. doi:10.1016/0550-3213(85)90602-9.

- Castro, Alejandra; Maloney, Alexander; Strominger, Andrew (2010). "Hidden conformal symmetry of the Kerr black hole". *Physical Review D* **82** (2). arXiv:1004.0996. Bibcode:2010PhRvD..82b4008C. doi:10.1103/PhysRevD.82.024008.

- Cheng, Miranda; Duncan, John; Harvey, Jeffrey (2013). "Umbral Moonshine". arXiv:1204.2779.

- Connes, Alain (1994). *Noncommutative Geometry*. Academic Press. ISBN 978-0-12-185860-5.

- Connes, Alain; Douglas, Michael; Schwarz, Albert (1998). "Noncommutative geometry and matrix theory". *Journal of High Energy Physics*. 19981 (2): 003. arXiv:hep-th/9711162. Bibcode:1998JHEP...02..003C. doi:10.1088/1126-6708/1998/02/003.

- Conway, John; Norton, Simon (1979). "Monstrous moonshine". *Bull. London Math. Soc.* **11** (3): 308–339. doi:10.1112/blms/11.3.308.

- Cremmer, Eugene; Julia, Bernard; Scherk, Joel (1978). "Supergravity theory in eleven dimensions". *Physics Letters B* **76** (4): 409–412. Bibcode:1978PhLB...76..409C. doi:10.1016/0370-2693(78)90894-8.

- de Haro, Sebastian; Dieks, Dennis; 't Hooft, Gerard; Verlinde, Erik (2013). "Forty Years of String Theory Reflecting on the Foundations". *Foundations of Physics* **43** (1): 1–7. Bibcode:2013FoPh...43....1D. doi:10.1007/s10701-012-9691-3.

- Deligne, Pierre; Etingof, Pavel; Freed, Daniel; Jeffery, Lisa; Kazhdan, David; Morgan, John; Morrison, David; Witten, Edward, eds. (1999). *Quantum Fields and Strings: A Course for Mathematicians* 1. American Mathematical Society. ISBN 978-0821820124.

- Duff, Michael (1996). "M-theory (the theory formerly known as strings)". *International Journal of Modern Physics A* **11** (32): 6523–41. arXiv:hep-th/9608117. Bibcode:1996IJMPA..11.5623D. doi:10.1142/S0217751X96002583.

- Duff, Michael (1998). "The theory formerly known as strings". *Scientific American* **278** (2): 64–9. doi:10.1038/scientificamerican0298-64.

- Duff, Michael; Howe, Paul; Inami, Takeo; Stelle, Kellogg (1987). "Superstrings in *D*=10 from supermembranes in *D*=11". *Nuclear Physics B* **191** (1): 70–74. Bibcode:1987PhLB..191...70D. doi:10.1016/0370-2693(87)91323-2.

- Dummit, David; Foote, Richard (2004). *Abstract Algebra*. Wiley. ISBN 978-0-471-43334-7.

- Duncan, John; Griffin, Michael; Ono, Ken (2015). "Proof of the Umbral Moonshine Conjecture". arXiv:1503.01472.

- Eguchi, Tohru; Ooguri, Hirosi; Tachikawa, Yuji (2011). "Notes on the K3 surface and the Mathieu group M_{24}". *Experimental Mathematics* **20** (1): 91–96. doi:10.1080/10586458.2011.544585.

- Frenkel, Igor; Lepowsky, James; Meurman, Arne (1988). *Vertex Operator Algebras and the Monster*. Pure and Applied Mathematics **134**. Academic Press. ISBN 0-12-267065-5.

- Gannon, Terry. *Moonshine Beyond the Monster: The Bridge Connecting Algebra, Modular Forms, and Physics*. Cambridge University Press.

- Givental, Alexander (1996). "Equivariant Gromov-Witten invariants". *International Mathematics Research Notices* **1996** (13): 613–663. doi:10.1155/S1073792896000414.

- Givental, Alexander (1998). "A mirror theorem for toric complete intersections". *Topological field theory, primitive forms and related topics*: 141–175. doi:10.1007/978-1-4612-0705-4_5. ISBN 978-1-4612-6874-1.

- Gubser, Steven; Klebanov, Igor; Polyakov, Alexander (1998). "Gauge theory correlators from non-critical string theory". *Physics Letters B* **428**: 105–114. arXiv:hep-th/9802109. Bibcode:1998PhLB..428..105G. doi:10.1016/S0370-2693(98)00377-3.

- Guica, Monica; Hartman, Thomas; Song, Wei; Strominger, Andrew (2009). "The Kerr/CFT Correspondence". *Physical Review D* **80** (12). arXiv:0809.4266. Bibcode:2009PhRvD..80l4008G. doi:10.1103/PhysRevD.80.124008.

- Hawking, Stephen (1975). "Particle creation by black holes". *Communications in Mathematical Physics* **43** (3): 199–220. Bibcode:1975CMaPh..43..199H. doi:10.1007/BF02345020.

- Hawking, Stephen (2005). "Information loss in black holes". *Physical Review D* **72** (8). arXiv:hep-th/0507171. Bibcode:2005PhRvD..72h4013H. doi:10.1103/PhysRevD.72.084013.

- Hořava, Petr; Witten, Edward (1996). "Heterotic and Type I string dynamics from eleven dimensions". *Nuclear Physics B* **460** (3): 506–524. arXiv:hep-th/9510209. Bibcode:1996NuPhB.460..506H. doi:10.1016/0550-3213(95)00621-4.

- Hori, Kentaro; Katz, Sheldon; Klemm, Albrecht; Pandharipande, Rahul; Thomas, Richard; Vafa, Cumrun; Vakil, Ravi; Zaslow, Eric, eds. (2003). *Mirror Symmetry* (PDF). American Mathematical Society. ISBN 0-8218-2955-6.

- Hull, Chris; Townsend, Paul (1995). "Unity of superstring dualities". *Nuclear Physics B* **4381** (1): 109–137. arXiv:hep-th/9410167. Bibcode:1995NuPhB.438..109H. doi:10.1016/0550-3213(94)00559-W.

- Kapustin, Anton; Witten, Edward (2007). "Electric-magnetic duality and the geometric Langlands program". *Communications in Number Theory and Physics* **1** (1): 1–236. arXiv:hep-th/0604151. Bibcode:2007CNTP....1....1K. doi:10.4310/cntp.2007.v1.n1.a1.

- Klarreich, Erica. "Mathematicians chase moonshine's shadow". *Quanta Magazine*. Retrieved March 2015.

- Klebanov, Igor; Maldacena, Juan (2009). "Solving Quantum Field Theories via Curved Spacetimes" (PDF). *Physics Today* **62**: 28–33. Bibcode:2009PhT....62a..28K. doi:10.1063/1.3074260. Archived from the original (PDF) on July 2, 2013. Retrieved May 2013.

- Kontsevich, Maxim (1995). "Homological algebra of mirror symmetry". *Proceedings of the International Congress of Mathematicians*: 120–139. arXiv:alg-geom/9411018. Bibcode:1994alg.geom.11018K.

- Kovtun, P. K.; Son, Dam T.; Starinets, A. O. (2001). "Viscosity in strongly interacting quantum field theories from black hole physics". *Physical Review Letters* **94** (11): 111601. arXiv:hep-th/0405231. Bibcode:2005PhRvL..94k1601K. doi:10.1103/PhysRevLett.94.111601. PMID 15903845.

- Lian, Bong; Liu, Kefeng; Yau, Shing-Tung (1997). "Mirror principle, I". *Asian Journal of Mathematics* **1**: 729–763. arXiv:alg-geom/9712011. Bibcode:1997alg.geom.12011L.

- Lian, Bong; Liu, Kefeng; Yau, Shing-Tung (1999a). "Mirror principle, II". *Asian Journal of Mathematics* **3**: 109–146. arXiv:math/9905006. Bibcode:1999math......5006L.

- Lian, Bong; Liu, Kefeng; Yau, Shing-Tung (1999b). "Mirror principle, III". *Asian Journal of Mathematics* **3**: 771–800. arXiv:math/9912038. Bibcode:1999math.....12038L.

- Lian, Bong; Liu, Kefeng; Yau, Shing-Tung (2000). "Mirror principle, IV". *Surveys in Differential Geometry* **7**: 475–496. arXiv:math/0007104. Bibcode:2000math......7104L. doi:10.4310/sdg.2002.v7.n1.a15.

- Luzum, Matthew; Romatschke, Paul (2008). "Conformal relativistic viscous hydrodynamics: Applications to RHIC results at $\sqrt{s_{NN}}$=200 GeV". *Physical Review C* **78** (3). arXiv:0804.4015. doi:10.1103/PhysRevC.78.034915.

- Maldacena, Juan (1998). "The Large N limit of superconformal field theories and supergravity". *Advances in Theoretical and Mathematical Physics* **2**: 231–252. arXiv:hep-th/9711200. Bibcode:1998AdTMP...2..231M. doi:10.1063/1.59653.

- Maldacena, Juan (2005). "The Illusion of Gravity" (PDF). *Scientific American* **293** (5): 56–63. Bibcode:2005SciAm.293e..56M. doi:10.1038/scientificamerican1105-56. PMID 16318027. Archived from the original (PDF) on November 1, 2014. Retrieved July 2013.

- Maldacena, Juan; Strominger, Andrew; Witten, Edward (1997). "Black hole entropy in M-theory". *Journal of High Energy Physics* **1997** (12). arXiv:hep-th/9711053. Bibcode:1997JHEP...12..002M. doi:10.1088/1126-6708/1997/12/002.

- Merali, Zeeya (2011). "Collaborative physics: string theory finds a bench mate". *Nature* **478** (7369): 302–304. Bibcode:2011Natur.478..302M. doi:10.1038/478302a. PMID 22012369.

- Moore, Gregory (2005). "What is ... a Brane?" (PDF). *Notices of the AMS* **52**: 214. Retrieved June 2013.

- Nahm, Walter (1978). "Supersymmetries and their representations". *Nuclear Physics B* **135** (1): 149–166. Bibcode:1978NuPhB.135..149N. doi:10.1016/0550-3213(78)90218-3.

- Nekrasov, Nikita; Schwarz, Albert (1998). "Instantons on noncommutative $\mathbf{R}^4$ and (2,0) superconformal six dimensional theory". *Communications in Mathematical Physics* **198** (3): 689–703. arXiv:hep-th/9802068. Bibcode:1998CMaPh.198..689N. doi:10.1007/s002200050490.

- Ooguri, Hirosi; Strominger, Andrew; Vafa, Cumrun (2004). "Black hole attractors and the topological string". *Physical Review D* **70** (10). arXiv:hep-th/0405146. Bibcode:2004PhRvD..70j6007O. doi:10.1103/physrevd.70.106007.

- Polchinski, Joseph (2007). "All Strung Out?". *American Scientist*. Retrieved April 2015.

- Penrose, Roger (2005). *The Road to Reality: A Complete Guide to the Laws of the Universe*. Knopf. ISBN 0-679-45443-8.

- Randall, Lisa; Sundrum, Raman (1999). "An alternative to compactification". *Physical Review Letters* **83** (23): 4690–4693. arXiv:hep-th/9906064. Bibcode:1999PhRvL..83.4690R. doi:10.1103/PhysRevLett.83.4690.

- Sachdev, Subir (2013). "Strange and stringy". *Scientific American* **308** (44): 44–51. Bibcode:2012SciAm.308a..44S. doi:10.1038/scientificamerican0113-44.

- Seiberg, Nathan; Witten, Edward (1999). "String Theory and Noncommutative Geometry". *Journal of High Energy Physics* **1999** (9): 032. arXiv:hep-th/9908142. Bibcode:1999JHEP...09..032S. doi:10.1088/1126-6708/1999/09/032.

- Sen, Ashoke (1994a). "Strong-weak coupling duality in four-dimensional string theory". *International Journal of Modern Physics A* **9** (21): 3707–3750. arXiv:hep-th/9402002. Bibcode:1994IJMPA...9.3707S. doi:10.1142/S0217751X94001497.

- Sen, Ashoke (1994b). "Dyon-monopole bound states, self-dual harmonic forms on the multi-monopole moduli space, and $SL(2,\mathbf{Z})$ invariance in string theory". *Physics Letters B* **329** (2): 217–221. arXiv:hep-th/9402032. Bibcode:1994PhLB..329..217S. doi:10.1016/0370-2693(94)90763-3.

- Smolin, Lee (2006). *The Trouble with Physics: The Rise of String Theory, the Fall of a Science, and What Comes Next*. New York: Houghton Mifflin Co. ISBN 0-618-55105-0.

- Strominger, Andrew (1998). "Black hole entropy from near-horizon microstates". *Journal of High Energy Physics* **1998** (2): 009. arXiv:hep-th/9712251. Bibcode:1998JHEP...02..009S. doi:10.1088/1126-6708/1998/02/009.

- Strominger, Andrew; Vafa, Cumrun (1996). "Microscopic origin of the Bekenstein–Hawking entropy". *Physics Letters B* **379** (1): 99–104. arXiv:hep-th/9601029. Bibcode:1996PhLB..379...99S. doi:10.1016/0370-2693(96)00345-0.

- Strominger, Andrew; Yau, Shing-Tung; Zaslow, Eric (1996). "Mirror symmetry is T-duality". *Nuclear Physics B* **479** (1): 243–259. arXiv:hep-th/9606040. Bibcode:1996NuPhB.479..243S. doi:10.1016/0550-3213(96)00434-8.

- Susskind, Leonard (2005). *The Cosmic Landscape: String Theory and the Illusion of Intelligent Design*. Back Bay Books. ISBN 978-0316013338.

- Susskind, Leonard (2008). *The Black Hole War: My Battle with Stephen Hawking to Make the World Safe for Quantum Mechanics*. Little, Brown and Company. ISBN 978-0-316-01641-4.

- Wald, Robert (1984). *General Relativity*. University of Chicago Press. ISBN 978-0-226-87033-5.

- Weinberg, Steven (1987). *Anthropic bound on the cosmological constant* **59**. Physical Review Letters. p. 2607.

- Witten, Edward (1995). "String theory dynamics in various dimensions". *Nuclear Physics B* **443** (1): 85–126. arXiv:hep-th/9503124. Bibcode:1995NuPhB.443...85W. doi:10.1016/0550-3213(95)00158-O.

- Witten, Edward (1998). "Anti-de Sitter space and holography". *Advances in Theoretical and Mathematical Physics* **2**: 253–291. arXiv:hep-th/9802150. Bibcode:1998AdTMP...2..253W.

- Witten, Edward (2007). "Three-dimensional gravity revisited". arXiv:0706.3359 [hep-th].

- Woit, Peter (2006). *Not Even Wrong: The Failure of String Theory and the Search for Unity in Physical Law*. Basic Books. p. 105. ISBN 0-465-09275-6.

- Yau, Shing-Tung; Nadis, Steve (2010). *The Shape of Inner Space: String Theory and the Geometry of the Universe's Hidden Dimensions*. Basic Books. ISBN 978-0-465-02023-2.

- Zee, Anthony (2010). *Quantum Field Theory in a Nutshell* (2nd ed.). Princeton University Press. ISBN 978-0-691-14034-6.

- Zwiebach, Barton (2009). *A First Course in String Theory*. Cambridge University Press. ISBN 978-0-521-88032-9.

39.10   Further reading

39.10.1   Popularizations

General

- Greene, Brian (2003). *The Elegant Universe: Superstrings, Hidden Dimensions, and the Quest for the Ultimate Theory*. New York: W.W. Norton & Company. ISBN 0-393-05858-1.

- Greene, Brian (2004). *The Fabric of the Cosmos: Space, Time, and the Texture of Reality*. New York: Alfred A. Knopf. ISBN 0-375-41288-3.

Critical

- Penrose, Roger (2005). *The Road to Reality: A Complete Guide to the Laws of the Universe*. Knopf. ISBN 0-679-45443-8.

- Smolin, Lee (2006). *The Trouble with Physics: The Rise of String Theory, the Fall of a Science, and What Comes Next*. New York: Houghton Mifflin Co. ISBN 0-618-55105-0.

- Woit, Peter (2006). *Not Even Wrong: The Failure of String Theory And the Search for Unity in Physical Law*. London: Jonathan Cape &: New York: Basic Books. ISBN 978-0-465-09275-8.

39.10.2   Textbooks

For physicists

- Becker, Katrin; Becker, Melanie; Schwarz, John (2007). *String Theory and M-theory: A Modern Introduction*. Cambridge University Press. ISBN 978-0-521-86069-7.

- Green, Michael; Schwarz, John; Witten, Edward (2012). *Superstring theory. Vol. 1: Introduction*. Cambridge University Press. ISBN 978-1107029118.

- Green, Michael; Schwarz, John; Witten, Edward (2012). *Superstring theory. Vol. 2: Loop amplitudes, anomalies and phenomenology*. Cambridge University Press. ISBN 978-1107029132.

- Polchinski, Joseph (1998). *String Theory Vol. 1: An Introduction to the Bosonic String*. Cambridge University Press. ISBN 0-521-63303-6.

- Polchinski, Joseph (1998). *String Theory Vol. 2: Superstring Theory and Beyond*. Cambridge University Press. ISBN 0-521-63304-4.

- Zwiebach, Barton (2009). *A First Course in String Theory*. Cambridge University Press. ISBN 978-0-521-88032-9.

For mathematicians

- Deligne, Pierre; Etingof, Pavel; Freed, Daniel; Jeffery, Lisa; Kazhdan, David; Morgan, John; Morrison, David; Witten, Edward, eds. (1999). *Quantum Fields and Strings: A Course for Mathematicians, Vol. 2*. American Mathematical Society. ISBN 978-0821819883.

39.11   External links

- *The Elegant Universe*—A three-hour miniseries with Brian Greene by *NOVA* (original PBS Broadcast Dates: October 28, 8–10 p.m. and November 4, 8–9 p.m., 2003). Various images, texts, videos and animations explaining string theory.

- Not Even Wrong—A blog critical of string theory

- The Official String Theory Web Site

- Why String Theory—An introduction to string theory.

Chapter 40

Compton wavelength

The **Compton wavelength** is a quantum mechanical property of a particle. It was introduced by Arthur Compton in his explanation of the scattering of photons by electrons (a process known as Compton scattering). The Compton wavelength of a particle is equivalent to the wavelength of a photon whose energy is the same as the rest-mass energy of the particle.

The standard Compton wavelength, λ, of a particle is given by

$$\lambda = \frac{h}{mc},$$

where h is the Planck constant, m is the particle's rest mass, and c is the speed of light. The significance of this formula is shown in the derivation of the Compton shift formula.

The CODATA 2010 value for the Compton wavelength of the electron is 2.4263102389(16)×10^{-12} m.[1] Other particles have different Compton wavelengths.

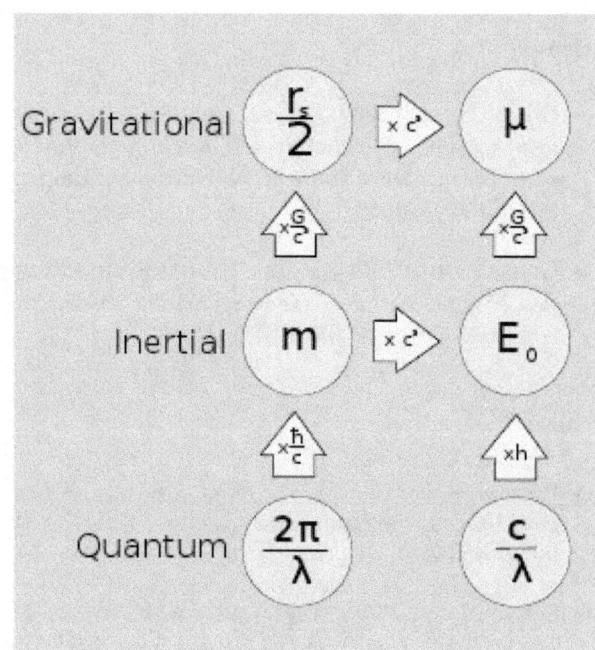

The relation between properties of mass and their associated physical constants. Every massive object is believed to exhibit all five properties. However, due to extremely large or extremely small constants, it is generally impossible to verify more than two or three properties for any object.
The Schwarzschild radius (rs) represents the ability of mass to cause curvature in space and time.
The standard gravitational parameter (μ) represents the ability of a massive body to exert Newtonian gravitational forces on other bodies.
Inertial mass (m) represents the Newtonian response of mass to forces.
Rest energy (E₀) represents the ability of mass to be converted into other forms of energy.
The Compton wavelength (λ) represents the quantum response of mass to local geometry.

40.1 Reduced Compton wavelength

When the Compton wavelength is divided by 2π, one obtains the "reduced" Compton wavelength λ (barred lambda), i.e. the Compton wavelength for 1 radian instead of 2π radians:

$$\lambda = \lambda/2\pi = \hbar/mc,$$

where $\hbar$ is the "reduced" Planck constant.

40.2 Role in equations for massive particles

The reduced Compton wavelength is a natural representation for mass on the quantum scale, and as such, it appears in many of the fundamental equations of quantum mechanics. The reduced Compton wavelength appears in the relativistic Klein–Gordon equation for a free particle:

$$\nabla^2\psi - \frac{1}{c^2}\frac{\partial^2}{\partial t^2}\psi = \left(\frac{mc}{\hbar}\right)^2\psi.$$

It appears in the Dirac equation (the following is an explicitly covariant form employing the Einstein summation convention):

$$-i\gamma^\mu\partial_\mu\psi + \left(\frac{mc}{\hbar}\right)\psi = 0.$$

The reduced Compton wavelength also appears in Schrödinger's equation, although its presence is obscured in traditional representations of the equation. The following is the traditional representation of Schrödinger's equation for an electron in a hydrogen-like atom:

$$i\hbar\frac{\partial}{\partial t}\psi = -\frac{\hbar^2}{2m}\nabla^2\psi - \frac{1}{4\pi\epsilon_0}\frac{Ze^2}{r}\psi.$$

Dividing through by $\hbar c$, and rewriting in terms of the fine structure constant, one obtains:

$$\frac{i}{c}\frac{\partial}{\partial t}\psi = -\frac{1}{2}\left(\frac{\hbar}{mc}\right)\nabla^2\psi - \frac{\alpha Z}{r}\psi.$$

40.3   Relationship between the reduced and non-reduced Compton wavelength

The reduced Compton wavelength is a natural representation for mass on the quantum scale. Equations that pertain to inertial mass like Klein-Gordon and Schrödinger's, use the reduced Compton wavelength. The non-reduced Compton wavelength is a natural representation for mass that has been converted into energy. Equations that pertain to the conversion of mass into energy, or to the wavelengths of photons interacting with mass, use the non-reduced Compton wavelength.

A particle of rest mass m has a rest energy of $E = mc^2$. The non-reduced Compton wavelength for this particle is the wavelength of a photon of the same energy. For photons of frequency f, energy is given by

$$E = hf = \frac{hc}{\lambda} = mc^2,$$

which yields the non-reduced or standard Compton wavelength formula if solved for λ.

40.4   Limitation on measurement

The Compton wavelength expresses a fundamental limitation on measuring the position of a particle, taking into account quantum mechanics and special relativity.[2]

This limitation depends on the rest mass m of the particle. To see how, note that we can measure the position of a particle by bouncing light off it – but measuring the position accurately requires light of short wavelength. Light with a short wavelength consists of photons of high energy. If the energy of these photons exceeds mc^2, when one hits the particle whose position is being measured the collision may yield enough energy to create a new particle of the same type. This renders moot the question of the original particle's location.

This argument also shows that the reduced Compton wavelength is the cutoff below which quantum field theory – which can describe particle creation and annihilation – becomes important. The above argument can be made a bit more precise as follows. Suppose we wish to measure the position of a particle to within an accuracy Δx. Then the uncertainty relation for position and momentum says that

$$\Delta x\,\Delta p \geq \frac{\hbar}{2},$$

so the uncertainty in the particle's momentum satisfies

$$\Delta p \geq \frac{\hbar}{2\Delta x}.$$

Using the relativistic relation between momentum and energy $E^2 = (pc)^2 + (mc^2)^2$, when Δp exceeds mc then the uncertainty in energy is greater than mc^2, which is enough energy to create another particle of the same type. But we must exclude this. It follows that there is a fundamental limitation on Δx:

$$\Delta x \geq \frac{1}{2}\left(\frac{\hbar}{mc}\right).$$

Thus the uncertainty in position must be greater than half of the reduced Compton wavelength $\hbar/mc$.

The Compton wavelength can be contrasted with the de Broglie wavelength, which depends on the momentum of a particle and determines the cutoff between particle and wave behavior in quantum mechanics.

40.5   Relationship to other constants

Typical atomic lengths, wave numbers, and areas in physics can be related to the reduced Compton wavelength for the electron ($\bar{\lambda}_e \equiv \frac{\lambda_e}{2\pi} \simeq 386$ fm) and the electromagnetic fine structure constant ($\alpha \simeq \frac{1}{137}$)

The Bohr radius is related to the Compton wavelength by:

$$a_0 = \frac{1}{\alpha}\left(\frac{\lambda_e}{2\pi}\right) \simeq 137 \times \bar{\lambda}_e \simeq 5.29 \times 10^4 \text{ fm}$$

The classical electron radius is about 3 times larger than the proton radius, and is written:

$$r_e = \alpha\left(\frac{\lambda_e}{2\pi}\right) \simeq \frac{\bar{\lambda}_e}{137} \simeq 2.82 \text{ fm}$$

The Rydberg constant is written:

$$R_\infty = \frac{\alpha^2}{2\lambda_e}$$

For fermions, the reduced Compton wavelength sets the cross-section of interactions. For example, the cross-section for Thomson scattering of a photon from an electron is equal to

$$\sigma_T = \frac{8\pi}{3}\alpha^2\bar{\lambda}_e^2 \simeq 66.5 \text{ fm}^2$$

which is roughly the same as the cross-sectional area of an iron-56 nucleus. For gauge bosons, the Compton wavelength sets the effective range of the Yukawa interaction: since the photon has no rest mass, electromagnetism has infinite range.

Typical lengths and areas in gravitational physics can be related to the Compton wavelength and the gravitational coupling constant (α_G which is the gravitational analog of the fine structure constant):

The Planck mass is special because the Compton wavelength and the Schwarzschild radius for this mass are equal. Their common value is called the Planck length (ℓ_P). This is a simple case of dimensional analysis: the Schwarzschild radius is proportional to the mass, whereas the Compton wavelength is proportional to the inverse of the mass. The Planck length is written:

$$\ell_P = \lambda_e \frac{\sqrt{\alpha_G}}{2\pi}$$

40.6 References

[1] CODATA 2010 value for Compton wavelength for the electron from NIST

[2] Garay, Luis J. "Quantum Gravity And Minimum Length." International Journal of Modern Physics A 10.02 (1995): 145-65. Arxiv.org. Web. 3 June 2014. <http://arxiv.org/pdf/gr-qc/9403008v2.pdf>.

40.7 External links

- Length Scales in Physics: the Compton Wavelength

- B.G. Sidharth, Planck scale to Compton scale, International Institute for Applicable Mathematics, Hyderabad (India) & Udine (Italy), Aug 2006.

Chapter 41

Holographic principle

The **holographic principle** is a property of string theories and a supposed property of quantum gravity that states that the description of a volume of space can be thought of as encoded on a boundary to the region—preferably a light-like boundary like a gravitational horizon. First proposed by Gerard 't Hooft, it was given a precise string-theory interpretation by Leonard Susskind[1] who combined his ideas with previous ones of 't Hooft and Charles Thorn.[1][2] As pointed out by Raphael Bousso,[3] Thorn observed in 1978 that string theory admits a lower-dimensional description in which gravity emerges from it in what would now be called a holographic way.

In a larger sense, the theory suggests that the entire universe can be seen as two-dimensional information on the cosmological horizon, the event horizon from which information may still be gathered and not lost due to the natural limitations of spacetime supporting a black hole, an observer and a given setting of these specific elements, such that the three dimensions we observe are an effective description only at macroscopic scales and at low energies. Cosmological holography has not been made mathematically precise, partly because the particle horizon has a non-zero area and grows with time.[4][5]

The holographic principle was inspired by black hole thermodynamics, which conjectures that the maximal entropy in any region scales with the radius *squared*, and not cubed as might be expected. In the case of a black hole, the insight was that the informational content of all the objects that have fallen into the hole might be entirely contained in surface fluctuations of the event horizon. The holographic principle resolves the black hole information paradox within the framework of string theory.[6] However, there exist classical solutions to the Einstein equations that allow values of the entropy larger than those allowed by an area law, hence in principle larger than those of a black hole. These are the so-called "Wheeler's bags of gold". The existence of such solutions conflicts with the holographic interpretation, and their effects in a quantum theory of gravity including the holographic principle are not yet fully understood.[7]

41.1 Black hole entropy

Main article: Black hole thermodynamics

An object with relatively high entropy is microscopically random, like a hot gas. A known configuration of classical fields has zero entropy: there is nothing random about electric and magnetic fields, or gravitational waves. Since black holes are exact solutions of Einstein's equations, they were thought not to have any entropy either.

But Jacob Bekenstein noted that this leads to a violation of the second law of thermodynamics. If one throws a hot gas with entropy into a black hole, once it crosses the event horizon, the entropy would disappear. The random properties of the gas would no longer be seen once the black hole had absorbed the gas and settled down. One way of salvaging the second law is if black holes are in fact random objects, with an enormous entropy whose increase is greater than the entropy carried by the gas.

Bekenstein assumed that black holes are maximum entropy objects—that they have more entropy than anything else in the same volume. In a sphere of radius R, the entropy in a relativistic gas increases as the energy increases. The only known limit is gravitational; when there is too much energy the gas collapses into a black hole. Bekenstein used this to put an upper bound on the entropy in a region of space, and the bound was proportional to the area of the region. He concluded that the black hole entropy is directly proportional to the area of the event horizon.[8]

Stephen Hawking had shown earlier that the total horizon area of a collection of black holes always increases with time. The horizon is a boundary defined by light-like geodesics; it is those light rays that are just barely unable to escape. If neighboring geodesics start moving toward each other they eventually collide, at which point their extension is inside the black hole. So the geodesics are always moving apart, and the number of geodesics which generate the boundary, the area of the horizon, always increases. Hawking's result was called the second law of black hole thermodynamics, by analogy with the law of entropy increase, but at first, he did not take the analogy too seriously.

Hawking knew that if the horizon area were an actual entropy, black holes would have to radiate. When heat is added to a thermal system, the change in entropy is the increase in mass-energy divided by temperature:

$$\mathrm{d}S = \frac{\mathrm{d}M}{T}.$$

If black holes have a finite entropy, they should also have a finite temperature. In particular, they would come to equilibrium with a thermal gas of photons. This means that black holes would not only absorb photons, but they would also have to emit them in the right amount to maintain detailed balance.

Time independent solutions to field equations do not emit radiation, because a time independent background conserves energy. Based on this principle, Hawking set out to show that black holes do not radiate. But, to his surprise, a careful analysis convinced him that they do, and in just the right way to come to equilibrium with a gas at a finite temperature. Hawking's calculation fixed the constant of proportionality at 1/4; the entropy of a black hole is one quarter its horizon area in Planck units.[9]

The entropy is proportional to the logarithm of the number of microstates, the ways a system can be configured microscopically while leaving the macroscopic description unchanged. Black hole entropy is deeply puzzling — it says that the logarithm of the number of states of a black hole is proportional to the area of the horizon, not the volume in the interior.[10]

Later, Raphael Bousso came up with a covariant version of the bound based upon null sheets.

41.2 Black hole information paradox

Main article: Black hole information paradox

Hawking's calculation suggested that the radiation which black holes emit is not related in any way to the matter that they absorb. The outgoing light rays start exactly at the edge of the black hole and spend a long time near the horizon, while the infalling matter only reaches the horizon much later. The infalling and outgoing mass/energy only interact when they cross. It is implausible that the outgoing state would be completely determined by some tiny residual scattering.

Hawking interpreted this to mean that when black holes absorb some photons in a pure state described by a wave function, they re-emit new photons in a thermal mixed state described by a density matrix. This would mean that quantum mechanics would have to be modified, because in quantum mechanics, states which are superpositions with probability amplitudes never become states which are probabilistic mixtures of different possibilities.[note 1]

Troubled by this paradox, Gerard 't Hooft analyzed the emission of Hawking radiation in more detail. He noted that when Hawking radiation escapes, there is a way in which incoming particles can modify the outgoing particles. Their gravitational field would deform the horizon of the black hole, and the deformed horizon could produce different outgoing particles than the undeformed horizon. When a particle falls into a black hole, it is boosted relative to an outside observer, and its gravitational field assumes a universal form. 't Hooft showed that this field makes a logarithmic tent-pole shaped bump on the horizon of a black hole, and like a shadow, the bump is an alternate description of the particle's location and mass. For a four-dimensional spherical uncharged black hole, the deformation of the horizon is similar to the type of deformation which describes the emission and absorption of particles on a string-theory world sheet. Since the deformations on the surface are the only imprint of the incoming particle, and since these deformations would have to completely determine the outgoing particles, 't Hooft believed that the correct description of the black hole would be by some form of string theory.

This idea was made more precise by Leonard Susskind, who had also been developing holography, largely independently. Susskind argued that the oscillation of the horizon of a black hole is a complete description[note 2] of both the infalling and outgoing matter, because the world-sheet theory of string theory was just such a holographic description. While short strings have zero entropy, he could identify long highly excited string states with ordinary black holes. This was a deep advance because it revealed that strings have a classical interpretation in terms of black holes.

This work showed that the black hole information paradox is resolved when quantum gravity is described in an unusual string-theoretic way assuming the string-theoretical description is complete, unambiguous and non-redundant.[12] The space-time in quantum gravity would emerge as an effective description of the theory of oscillations of a lower-dimensional black-hole horizon, and suggest that any black hole with appropriate properties, not just strings, would serve as a basis for a description of string theory.

In 1995, Susskind, along with collaborators Tom Banks, Willy Fischler, and Stephen Shenker, presented a formulation of the new M-theory using a holographic description in terms of charged point black holes, the D0 branes of type IIA string theory. The Matrix theory they proposed was first suggested as a description of two branes in 11-dimensional supergravity by Bernard de Wit, Jens Hoppe, and Hermann Nicolai. The later authors reinterpreted the same matrix models as a description of the dynamics of point black holes in particular limits. Holography allowed them to conclude that the dynamics of these black holes give a complete non-perturbative formulation of M-theory. In 1997, Juan Maldacena gave the first holographic descriptions of a higher-dimensional object, the 3+1-dimensional type IIB membrane, which resolved a long-standing problem of finding a string description which describes a gauge theory. These developments simultaneously explained how string theory is related to some forms of supersymmetric quantum field theories.

41.3   Limit on information density

Entropy, if considered as information (see information entropy), is measured in bits. The total quantity of bits is related to the total degrees of freedom of matter/energy.

For a given energy in a given volume, there is an upper limit to the density of information (the Bekenstein bound) about the whereabouts of all the particles which compose matter in that volume, suggesting that matter itself cannot be subdivided infinitely many times and there must be an ultimate level of fundamental particles. As the degrees of freedom of a particle are the product of all the degrees of freedom of its sub-particles, were a particle to have infinite subdivisions into lower-level particles, the degrees of freedom of the original particle would be infinite, violating the maximal limit of entropy density. The holographic principle thus implies that the subdivisions must stop at some level, and that the fundamental particle is a bit (1 or 0) of information.

The most rigorous realization of the holographic principle is the AdS/CFT correspondence by Juan Maldacena. However, J.D. Brown and Marc Henneaux had rigorously proved already in 1986, that the asymptotic symmetry of 2+1 dimensional gravity gives rise to a Virasoro algebra, whose corresponding quantum theory is a 2-dimensional conformal field theory.[13]

41.4   High-level summary

The physical universe is widely seen to be composed of "matter" and "energy". In his 2003 article published in Scientific American magazine, Jacob Bekenstein summarized a current trend started by John Archibald Wheeler, which suggests scientists may *regard the physical world as made of information, with energy and matter as incidentals.*

Bekenstein asks "Could we, as William Blake memorably penned, 'see a world in a grain of sand,' or is that idea no more than 'poetic license,'"[14] referring to the holographic principle.

41.4.1   Unexpected connection

Bekenstein's topical overview "A Tale of Two Entropies"[15] describes potentially profound implications of Wheeler's trend, in part by noting a previously unexpected connection between the world of information theory and classical physics. This connection was first described shortly after the seminal 1948 papers of American applied mathematician Claude E. Shannon introduced today's most widely used measure of information content, now known as Shannon entropy. As an objective measure of the quantity of information, Shannon entropy has been enormously useful, as the design of all modern communications and data storage devices, from cellular phones to modems to hard disk drives and DVDs, rely on Shannon entropy.

In thermodynamics (the branch of physics dealing with heat), entropy is popularly described as a measure of the "disorder" in a physical system of matter and energy. In 1877 Austrian physicist Ludwig Boltzmann described it more precisely in terms of the *number of distinct microscopic states* that the particles composing a macroscopic "chunk" of matter could be in while still *looking* like the same macroscopic "chunk". As an example, for the air in a room, its thermodynamic entropy would equal the logarithm of the count of all the ways that the individual gas molecules could be distributed in the room, and all the ways they could be moving.

41.4.2   Energy, matter, and information equivalence

Shannon's efforts to find a way to quantify the information contained in, for example, an e-mail message, led him unexpectedly to a formula with the same form as Boltzmann's. In an article in the August 2003 issue of Scientific American titled "Information in the Holographic Universe", Bekenstein summarizes that *"Thermodynamic entropy and Shannon entropy are conceptually equivalent: the number of arrangements that are counted by Boltzmann entropy reflects the amount of Shannon information one would need to implement any particular arrangement..."* of matter and energy. The only salient difference between the thermodynamic entropy of physics and Shannon's entropy of information is in the units of measure; the former is expressed in units of energy divided by temperature, the latter in *essentially dimensionless* "bits" of information.

The holographic principle states that the entropy of *ordinary mass* (not just black holes) is also proportional to surface area and not volume; that volume itself is illusory and the universe is really a hologram which is isomorphic to the information "inscribed" on the surface of its boundary.[10]

41.5 Experimental tests

The Fermilab physicist Craig Hogan claims that the holographic principle would imply quantum fluctuations in spatial position[16] that would lead to apparent background noise or "holographic noise" measurable at gravitational wave detectors, in particular GEO 600.[17] However these claims have not been widely accepted, or cited, among quantum gravity researchers and appear to be in direct conflict with string theory calculations.[18]

Analyses in 2011 of measurements of gamma ray burst GRB 041219A in 2004 by the INTEGRAL space observatory launched in 2002 by the European Space Agency shows that Craig Hogan's noise is absent down to a scale of 10^{-48} meters, as opposed to scale of 10^{-35} meters predicted by Hogan, and the scale of 10^{-16} meters found in measurements of the GEO 600 instrument.[19] Research continues at Fermilab under Hogan as of 2013.[20]

Jacob Bekenstein also claims to have found a way to test the holographic principle with a tabletop photon experiment.[21]

41.6 Tests of Maldacena's conjecture

Main article: Maldacena conjecture

Hyakutake et al. in 2013/4 published two papers[22] that bring computational evidence that Maldacena's conjecture is true. One paper computes the internal energy of a black hole, the position of its event horizon, its entropy and other properties based on the predictions of string theory and the effects of virtual particles. The other paper calculates the internal energy of the corresponding lower-dimensional cosmos with no gravity. The two simulations match. The papers are not an actual proof of Maldacena's conjecture for all cases but a demonstration that the conjecture works for a particular theoretical case and a verification of the AdS/CFT correspondence for a particular situation.[23]

41.7 See also

- Bekenstein bound
- Beyond black holes
- Bousso's holographic bound
- Brane cosmology
- Entropic gravity
- Implicate and explicate order according to David Bohm
- Margolus–Levitin theorem
- Physical cosmology
- Quantum foam
- Simulated reality

41.8 Notes

[1] except in the case of measurements, which the black hole should not be performing

[2] "Complete description" means all the *primary* qualities. For example, John Locke (and before him Robert Boyle) determined these to be *size, shape, motion, number,* and *solidity.* Such *secondary quality* information as *color, aroma, taste* and *sound,*[11] or internal quantum state is not information that is implied to be preserved in the surface fluctuations of the event horizon. (See however "path integral quantization")

41.9 References

General

- Bousso, Raphael (2002). "The holographic principle". *Reviews of Modern Physics* **74** (3): 825–874. arXiv:hep-th/0203101. Bibcode:2002RvMP...74..825B. doi:10.1103/RevModPhys.74.825.

- 't Hooft, Gerard (1993). "Dimensional Reduction in Quantum Gravity". arXiv:gr-qc/9310026.. 't Hooft's original paper.

Citations

[1] Susskind, Leonard (1995). "The World as a Hologram". *Journal of Mathematical Physics* **36** (11): 6377–6396. arXiv:hep-th/9409089. Bibcode:1995JMP....36.6377S. doi:10.1063/1.531249.

[2] Thorn, Charles B. (27–31 May 1991). *Reformulating string theory with the 1/N expansion*. International A.D. Sakharov Conference on Physics. Moscow. pp. 447–54. arXiv:hep-th/9405069. ISBN 978-1-56072-073-7.

[3] Bousso, Raphael (2002). "The Holographic Principle". *Reviews of Modern Physics* 74 (3): 825–874. arXiv:hep-th/0203101. Bibcode:2002RvMP...74..825B. doi:10.1103/RevModPhys.74.825.

[4] Lloyd, Seth (2002-05-24). "Computational Capacity of the Universe". *Physical Review Letters* 88 (23): 237901. arXiv:quant-ph/0110141. Bibcode:2002PhRvL..88w7901L. doi:10.1103/PhysRevLett.88.237901. PMID 12059399.

[5] Davies, Paul. "Multiverse Cosmological Models and the Anthropic Principle". *CTNS*. Retrieved 2008-03-14.

[6] Susskind, L. (2008). *The Black Hole War – My Battle with Stephen Hawking to Make the World Safe for Quantum Mechanics*. Little, Brown and Company.

[7] Marolf, Donald (April 2009). "Black Holes, AdS, and CFTs". *General Relativity and Gravitation* 41 (4): 903–17. arXiv:0810.4886. Bibcode:2009GReGr..41..903M. doi:10.1007/s10714-008-0749-7.

[8] Bekenstein, Jacob D. (January 1981). "Universal upper bound on the entropy-to-energy ratio for bounded systems". *Physical Review D* 23 (215): 287–298. Bibcode:1981PhRvD..23..287B. doi:10.1103/PhysRevD.23.287.

[9] Majumdar, Parthasarathi (1998). "Black Hole Entropy and Quantum Gravity" 73: 147. arXiv:gr-qc/9807045. Bibcode:1999InJPB..73..147M.

[10] Bekenstein, Jacob D. (August 2003). "Information in the Holographic Universe — Theoretical results about black holes suggest that the universe could be like a gigantic hologram". *Scientific American*. p. 59.

[11] Dennett, Daniel (1991). *Consciousness Explained*. New York: Back Bay Books. p. 371. ISBN 0-316-18066-1.

[12] Susskind, L. (February 2003). "The Anthropic landscape of string theory". arXiv:hep-th/0302219.

[13] Brown, J. D. & Henneaux, M. (1986). "Central charges in the canonical realization of asymptotic symmetries: an example from three-dimensional gravity". *Communications in Mathematical Physics* 104 (2): 207–226. Bibcode:1986CMaPh.104..207B. doi:10.1007/BF01211590..

[14] Information in the Holographic Universe

[15] http://webcache.googleusercontent.com/search?q=cache:E360V697cvgJ:ref-sciam.livejournal.com/1190.html&hl=en&gl=us&strip=1

[16] Hogan, Craig J. (2008). "Measurement of quantum fluctuations in geometry". *Physical Review D* 77 (10): 104031. arXiv:0712.3419. Bibcode:2008PhRvD..77j4031H. doi:10.1103/PhysRevD.77.104031..

[17] Chown, Marcus (15 January 2009). "Our world may be a giant hologram". *NewScientist*. Retrieved 2010-04-19.

[18] "Consequently, he ends up with inequalities of the type... Except that one may look at the actual equations of Matrix theory and see that none of these commutators is nonzero... The last displayed inequality above obviously can't be a consequence of quantum gravity because it doesn't depend on G at all! However, in the G→0 limit, one must reproduce non-gravitational physics in the flat Euclidean background spacetime. Hogan's rules don't have the right limit so they can't be right." – Lubos Motl, Hogan's holographic noise doesn't exist, Feb 7, 2012

[19] "Integral challenges physics beyond Einstein". European Space Agency. 30 June 2011. Retrieved 3 February 2013.

[20] "Frequently Asked Questions for the Holometer at Fermilab". 6 July 2013. Retrieved 14 February 2014.

[21] Cowen, Ron (22 November 2012). "Single photon could detect quantum-scale black holes". *Nature*. Retrieved 3 February 2013.

[22] Cowen, Ron (10 December 2013). "Simulations back up theory that Universe is a hologram". *Nature News*. doi:10.1038/nature.2013.14328. Hyakutake, Yoshifumi (March 2014). "Quantum near-horizon geometry of a black 0-brane". *Progress of Theoretical and Experimental Physics* 2014 (3): 033B04. arXiv:1311.7526. Bibcode:2014PTEP.2014c3B04H. doi:10.1093/ptep/ptu028. Hanada, Masanori; Hyakutake, Yoshifumi; Ishiki, Goro; Nishimura, Jun (23 May 2014). "Holographic description of a quantum black hole on a computer". *Science* 344 (6186): 882–5. arXiv:1311.5607. Bibcode:2014Sci...344..882H. doi:10.1126/science.1250122.

[23] Yirka, Bob (December 13, 2013). "New work gives credence to theory of universe as a hologram". Phys.org.

41.10 External links

- UC Berkeley's Raphael Bousso gives an introductory lecture on the holographic principle - Video.

- *Scientific American* article on holographic principle by Jacob Bekenstein

41.11 Text and image sources, contributors, and licenses

41.11.1 Text

- **Quintessence (physics)** *Source:* https://en.wikipedia.org/wiki/Quintessence_(physics)?oldid=719861774 *Contributors:* AxelBoldt, Bryan Derksen, Ted Longstaffe, Bth, Montrealais, Michael Hardy, Nixdorf, Muriel Gottrop~enwiki, Loren Rosen, Emperorbma, Alex S, Timwi, Reddi, Markhurd, DW40, Jeffq, Barbara Shack, Snowdog, Beland, Bbbl67, Thorwald, Rich Farmbrough, Pjacobi, Gianluigi, Dbachmann, Eric Forste, El C, I9Q79oL78KiL0QTFHgyc, Dirac1933, Richard Weil, GregorB, Joke137, Rjwilmsi, RE, Diza, Hairy Dude, NawlinWiki, Dbfirs, 2over0, Thnidu, Petri Krohn, KasugaHuang, SmackBot, Rentier, Cybercobra, LoveEncounterFlow, Ossipewsk, Kurtan~enwiki, Jerald Frazier (aka DJ Adrenaline), Penbat, Thijs!bot, N5iln, Peter Gulutzan, Jpod2, Tarotcards, VolkovBot, Don4of4, Lamro, EoGuy, Andwor, AbJ32, Agentxyz, Rreagan007, Addbot, Legobot, Yobot, Baxxterr, AnomieBOT, Nighthawk008, Citation bot, Finncarey, Lsj, Tom.Reding, RjwilmsiBot, Mmpcq, Klbrain, Hhhippo, Khestwol, ClueBot NG, P0lise, Mr Gearloose, MerlIwBot, Bibcode Bot, BG19bot, Mediran, Andyhowlett, Prokaryotes, Blanclar, Monkbot, Diloshwan, Tetra quark, The Rocky Road and Anonymous: 59

- **Fifth force** *Source:* https://en.wikipedia.org/wiki/Fifth_force?oldid=691274632 *Contributors:* The Epopt, Vicki Rosenzweig, XJaM, Robert Foley, Looxix~enwiki, Ideyal, Reddi, Lumos3, Robbot, Wikibot, Icairns, Lycurgus, Miss Madeline, GregorB, Joke137, Quale, Długosz, SmackBot, Ohnoitsjamie, TimBentley, Jgwacker, Khukri, APRCooper, AmeriCan, Cydebot, Thijs!bot, WinBot, .anacondabot, TimRI, Cuzkatzimhut, DumZiBoT, Addbot, Ekojekoj, Lightbot, Gameseeker, Yobot, AnomieBOT, Are you ready for IPv6?, Gsard, FrescoBot, Brichard37, ZéroBot, Cogiati, Quondum, QuantumSquirrel, DASHBotAV, Gary Dee, BG19bot, Khazar2, Tmfs10, Anderson, 1Minow, Cyberderp, BerFinelli, YooperBill and Anonymous: 28

- **Extra dimensions** *Source:* https://en.wikipedia.org/wiki/Extra_dimensions?oldid=705556627 *Contributors:* Michael Hardy, Lumidek, Jon Awbrey, A876, Mojo Hand, R'n'B, Mild Bill Hiccup, Addbot, Davdde, 🅰🅰, Dadonene89, ZéroBot, D.Lazard, Invadibot, SoledadKabocha, CAPTAIN RAJU and Anonymous: 2

- **Quintom scenario** *Source:* https://en.wikipedia.org/wiki/Quintom_scenario?oldid=715283601 *Contributors:* HangingCurve, SamuelRiv, Peter Gulutzan, Naturalnumber, Jovianeye, ZooFari, Dreamer08, AnomieBOT, Finncarey, I dream of horses, Yukterez, Andyhowlett, Tetra quark, Equinox and Anonymous: 2

- **Scale factor (cosmology)** *Source:* https://en.wikipedia.org/wiki/Scale_factor_(cosmology)?oldid=720359458 *Contributors:* XJaM, Patrick, Fropuff, Dbachmann, El C, I9Q79oL78KiL0QTFHgyc, Keenan Pepper, Sjakkalle, Rjwilmsi, Caco de vidro, SmackBot, Jbergquist, Lambiam, JHunterJ, Hypnosifl, Stanlekub, Stebbins, Michael C Price, Peter Gulutzan, Neilljones, Vreezkid, Brews ohare, Panos84, Addbot, Yobot, Ht686rg90, AnomieBOT, 🅰🅰, Tom.Reding, Helpful Pixie Bot, Bibcode Bot, Cyberbot II, Monkbot, Tetra quark and Anonymous: 25

- **Dark energy** *Source:* https://en.wikipedia.org/wiki/Dark_energy?oldid=719869333 *Contributors:* The Anome, Dachshund, Roadrunner, Schewek, Stevertigo, Thesteve, Nealmcb, Michael Hardy, Tim Starling, FrankH, Bobby D. Bryant, SebastianHelm, Ahoerstemeier, Glenn, Tristanb, Reddi, Wik, DW40, Dragons flight, Anupamsr, Pierre Boreal, BenRG, Jeffq, Donarreiskoffer, Robbot, Zandperl, Korath, Scott McNay, Vespristiano, Peak, Gandalf61, Rursus, Mlaine, UtherSRG, SC, Mattflaschen, Acm, Ancheta Wis, Giftlite, Graeme Bartlett, Awolf002, Jyril, Art Carlson, Herbee, Perl, Curps, Henry Flower, Gzornenplatz, Manuel Anastácio, Andycjp, BruceR, LucasVB, Antandrus, Beland, Karol Langner, Kevin B12, Bbbl67, Urvabara, JimJast, Discospinster, Rich Farmbrough, Pjacobi, Vsmith, D-Notice, Dbachmann, Bender235, Eric Forste, RJHall, JustinWick, Omnibus, El C, Lycurgus, Jomel, Kwamikagami, Frankenschulz, RoyBoy, Stesmo, Reuben, Russ3Z, I9Q79oL78KiL0QTFHgyc, Diego Moya, Keenan Pepper, Slugmaster, Axl, Benna, Wtmitchell, RainbowOfLight, Mikeo, Vuo, Freyr, DV8 2XL, Kazvorpal, Falcorian, Velho, Batintherain, Hottscubbard, OwenX, Mindmatrix, FeanorStar7, Velvetsmog, Uncle G, Netdragon, Jeff3000, GregorB, Isnow, SDC, 🅰🅰🅰🅰🅰🅰, Joke137, Abd, Christopher Thomas, Sneakums, Dysepsion, BD2412, Doc Savage, Malangthon, RadioActive~enwiki, Drbogdan, Loris Bennett, Rjwilmsi, Strait, TheRingess, Salleman, HappyCamper, Sohmc, Ems57fcva, DonJuan~enwiki, BitterMan, Tomer Ish Shalom, Srleffler, Smithbrenon, CJLL Wright, Chobot, DVdm, Wavelength, RobotE, SamuelR, Diliff, Bhny, Stephenb, CambridgeBayWeather, Merick, NawlinWiki, Msikma, FFLaguna, LiamE, SCZenz, FoolsWar, Bota47, Rwxrwxrwx, Daniel C. Enormousdude, 2over0, Helge Rosé, Pb30, Dr.alf, Joedixon, Rlove, Geoffrey.landis, Ilmari Karonen, Moonsleeper7, Kungfuadam, Bernd in Japan, GrinBot~enwiki, Treesmill, SmackBot, Ashill, Saravask, Bayardo, Tom Lougheed, InverseHypercube, KnowledgeOfSelf, Melchoir, J.Sarfatti, Nickst, Silverhand, Edgar181, Vixus, Gilliam, Skizzik, Jlsilva, Andy M. Wang, Tyciol, Sirex98, Oli Filth, DHN-bot~enwiki, Sbharris, Colonies Chris, Jdthood, Can't sleep, clown will eat me, ThePromenader, PoiZaN, Chlewbot, Joema, Cybercobra, Lpgeffen, Rpf, Kendrick7, Byelf2007, Rory096, Boradis, Richard L. Peterson, Xerxesx18, Writtenonsand, JorisvS, Mgiganteus1, Ckatz, Hypnosifl, Megane~enwiki, Ryulong, Quaeler, Dan Gluck, Spebudmak, Paul venter, Cxat, UncleDouggie, Courcelles, Tawkerbot2, JRSpriggs, Atomobot, Trevor.tombe, JForget, CRGreathouse, Lavateraguy, Nadyes, Mlsmith10, Arnavion, Logical2u, Rob Maguire, Cydebot, Stebbins, Gmusser, 879(CoDe), Rracecarr, Soetermans, Michael C Price, Chrislk02, Kozuch, Landroo, Thijs!bot, Headbomb, Marek69, Electron9, Second Quantization, Chris goulet, Davidhorman, Turelli, Dawnseeker2000, AntiVandalBot, Orionus, Gnixon, Fayenatic london, Tim Shuba, Empyrius, Archmagusrm, AstroPaul, Bagster, JAnDbot, Carl1011, Davewho2, MER-C, CosineKitty, Rkomatsu, Michael Wood-Vasey, Felix116, Acroterion, Bongwarrior, VoABot II, Tripbeetle, LordCémOnur, Seleucus, Kevinwiatrowski, Ours18, DerHexer, Nevit, Simplizissimus, NatureA16, Johann1870, Jimmilu, ARCG, Nikpapag, TechnoFaye, Christian424, Tgeairn, J.delanoy, Trusilver, Maurice Carbonaro, Natty4bumpo, Komowkwa, OttoMäkelä, Jlechem, Tsuite, SJP, Videokunst~enwiki, Malerin, Jorfer, Potatoswatter, Cmichael, DorganBot, Jcmargeson, Ja 62, JHussein, Jjabellar, Sheliak, Johnassassin, Caribbean H.Q., VolkovBot, ColdCase, JohnBlackburne, D A Patriarche, AlnoktaBOT, Fences and windows, Philip Trueman, Darren22, HowardFrampton, TXiKiBoT, Oshwah, Dwight666, Zanardm, Someguy1221, Oxfordwang, Jackfork, UnitedStatesian, Mazarin07, Venny85, Goaliemaster121, SwordSmurf, Lamro, RayNorris, Fourthark, Wanchung Hu, Obsidianmile, Radical Robert, Noncompliant one, Donauland~enwiki, PlanetStar, TrulyBlue, Murad.Shibli, Likebox, Flyer22 Reborn, Hotdiggity, Avidallred, Faradayplank, Poindexter Propellerhead, OKBot, Aquijex, Loren.wilton, Martarius, BillWilliam, ClueBot, Dead10ck, The Thing That Should Not Be, Rodhullandemu, SuperHamster, Andwor, Tms9, Jusdafax, Da rulz07, Barbarinaz, Kentgen1, Razorflame, Stevecrye, AC+79 3888, Pillar of Babel, TimothyRias, Gwark, Ost316, PL290, MikeSmith10, Parejkoj, Andreaprins, Dgirl1723, HexaChord, D.M. from Ukraine, Addbot, Gravitophoton, Uruk2008, DOI bot, Nernom, LaaknorBot, Adfellin, Glane23, Delaszk, ChenzwBot, Sophia8891, Combatman~enwiki, Craigsjones, Arbitrarily0, Gurusoft2, Cosmos72, Luckas-bot, Yobot, Cosoce, Systemizer, Aldebaran66, Fulcanelli, Amble, AnomieBOT, Iluziat, Materialscientist, Citation bot, Icosmology, ArthurBot, Xqbot, S h i v a (Visnu), Sionus, Driinoth, Wperdue, Tomwsulcer, BLP-outrageous move logs, ProtectionTaggingBot, Mathonius, Shadowjams, Finncarey, PrimeMatter, FrescoBot, Paine Ellsworth, Tobby72, Sławomir Biały, Zero Thrust, Kvgyarmati, Woodingdean, Alpha plus (a+), Citation bot 1, Redrose64, Pinethicket, I dream of horses, Jonesey95, Three887, Tom.Reding, Shahidur Rahman Sikder, Efficiency1101e, Casimir9999, Aknochel, IVAN3MAN, Meier99, BradTheBadWiki, Trappist the monk, TADEET, Jordgette, Heurisko, Michael9422, Adi4094, Earthandmoon, Wellsmax, RjwilmsiBot, Alph Bot, EmausBot, Mmpcq, Grrow, Quantanew, RA0808, Slightsmile, Italia2006, NicatronTg, H3llBot, Suslindisambiguator, Paulstarpaulstar, Frigotoni, Colin.campbell.27, Iiar, HCPotter, Tunborough, RockMagnetist, Herk1955, Deathglass, DASHBotAV, Fire Vortex, Mjbmrbot, Yceren Loq, ClueBot NG, Ccalen, Chester Markel, Matias Pocobi, Jj1236, Frietjes, Helvitica Bold, Curb Chain, Bibcode Bot, BG19bot, Gordonben, Cheeseray1, FiveColourMap, Hippokrateszholdacskai, Yizlpku, Snow Blizzard, Gerhardtschmerhardt, Migrainus, Mcspaans, Szczureq, Unclejoe0306, Guanghuilin, Akshay Lattimardi, Dexbot, CityOfUr, CuriousMind01, Wjs64, JustAMuggle, WorldWideJuan, Epicgenius, Yheyma, MiceEater, LindaYeah, DavidLeighEllis, Federicoturner, Babitaarora, Isateach, Onecreation, Prokaryotes, Christophe1946, BerdanII, Anrnusna, Stamptrader, Suelru, Monkbot, Mlsmith55, Haxxorz596, THemanRE$%S23, Jnojha007, Richard.drapeau, UrDreamViola, MF22, ChamithN, Larsyxa, EpicLX, Tibenas, Mediavalia, ScrapIronIV, 39Debangshu, Anunaki truth, Tetra quark, Isambard Kingdom, Anand2202, GeneralizationsAreBad, Jman135, Grammarian3.14159265359, KasparBot, ShankZeTank, Tgorewic, Sir Cumference, ShiningSword, Esadri21, Phseek, Srisri19962003, TychosElk, Buckbill10, Alopresti777, Themalina, SireWonton, Adithya2804, Khrpr, Kigaei, Gerald wish, Dr. Hung M. Choi, Soopdish and Anonymous: 533

- **Dark matter** *Source:* https://en.wikipedia.org/wiki/Dark_matter?oldid=720735175 *Contributors:* AxelBoldt, Chenyu, Derek Ross, CYD, BF, Bryan Derksen, The Anome, Tarquin, Taw, XJaM, Arvindn, William Avery, Roadrunner, Mintguy, Bth, Stevertigo, Edward, Nealmcb, Boud, FrankH, Cprompt, DopefishJustin, Bobby D. Bryant, Ixfd64, SebastianHelm, Alfio, CesarB, Looxix~enwiki, Mkweise, William M. Connolley, JWSchmidt, Glenn, Mxn, Charles Matthews, Timwi, Fuzheado, Rednblu, Haukurth, DW40, Dragons flight, Furrykef, Saltine, Dogface, Populus, Jusjih, Finlay McWalter, Bearcat, Robbot, Zandperl, Korath, Nurg, Naddy, Arkuat, Gandalf61, Pingveno, Rursus, Rtfisher, Wereon, Diberri, Adam78, Aasim75, Marc Venot, Ancheta Wis, Giftlite, Graeme Bartlett, Laudaka, Barbara Shack, Herbee, Fropuff, Xerxes314, Dratman, Curps, Joconnor, Jdavidb, Unconcerned, Eequor, Bobblewik, Andycjp, Alexf, Geni, Antandrus, HorsePunchKid, Melikamp, PDH, Rdsmith4, Anythingyouwant, Bosmon, Bbbl67, Icairns, Sam Hocevar, Cynical, Lumidek, Iantresman, Burschik, Joyous!, Adashiel, Urvabara, Discospinster, Rich Farmbrough, Oliver Lineham, Vsmith, Jpk, ArnoldReinhold, Murtasa, D-Notice, JPX7, KaiSeun, SpookyMulder, Bender235, Kjoonlee, Kaisershatner, Pk2000, PsychoDave, RJHall, Mr. Billion, El C, Bletch, PhilHibbs, Shanes, Frankenschulz, Art LaPella, RoyBoy, Themusicgod1, Bobo192, Smalljim, Shenme, Cmdrjameson, Reuben, Kmaguire, I9Q79oL78KiL0QTFHgyc, Zelda~enwiki, Mr. Brownstone, E is for Ian, Jumbuck, Storm Rider, Alansohn, Gary, Anthony Appleyard, Guy Harris, Eric Kvaalen, Arthena, Keenan Pepper, Kocio, Bart133, RPellessier, Benna, ClockworkSoul, Cal 1234, Count Iblis, Guthrie, H2g2bob, Bsadowski1, GabrielF, Pauli133, Leondz, DV8 2XL, Gene Nygaard, Feline1, Oleg Alexandrov, Brookie, Natalya, Flying fish, WilliamKF, Yeastbeast, Mindmatrix, RHaworth, Plek, BillC, JPFlip, Benbest, JFG, ^demon, WadeSimMiser, Gxojo, MONGO, Jwanders, Torqueing, 猫猫猫猫猫, Joke137, Wisq, Christopher Thomas, Palica, Mandarax, RedBLACKandBURN, Aarghdvaark, RichardWeiss, Ashmoo, Graham87, Malangthon, Mamling, Jclemens, Drbogdan, Loris Bennett, Rjwilmsi, Lars T., Strait, Patrick Gill, Tangotango, Tawker, Smithfarm, Stevenscollege, Mike Peel, HappyCamper, SeanMack, ScottJ, Krash, Dermeister, Rangek, Madcat87, FlaBot, Ian Pitchford, PlatypeanArchcow, A scientist, Margosbot~enwiki, Gark, Nivix, Gparker, Pathoschild, Gurch, Stevenfruitsmaak, Goudzovski, Tomer Ish Shalom, Smithbrenon, Chobot, Moocha, DVdm, Bgwhite, Gwernol, The Rambling Man, YurikBot, Wavelength, RobotE, Koveras, Hairy Dude, Huw Powell, Phmer, Hillman, RussBot, Michael Slone, Ohwilleke, Bhny, JabberWok, GLaDOS, DanMS, Zelmerszoetrop, Stephenb, Eleassar, Merick, Big Brother 1984, NawlinWiki, Alpertron, Długosz, Schlafly, FFLaguna, BlackAndy, Dbmag9, SCZenz, Haoie, Raven4x4x, Ospalh, Durval, Bota47, Supspirit, Pegship, Noosfractal, Charlie Wiederhold, WAS 4.250, Smoggyrob, Reyk, Tvaughan, Joedixon, Eric TF Bat, Emc2, Ilmari Karonen, Allens, Bernd in Japan, InsayneWrapper, Bclayabt, Attilios, MacsBug, SmackBot, Cubs Fan, Ashill, IddoGenuth, Tomer yaffe, Stellea, InverseHypercube, KnowledgeOfSelf, Clpo13, Nickst, RedSpruce, Nightbat, Doc Strange, Herbm, Edgar181, HalfShadow, Flux.books, Dheerajkakar, Yamaguchi先生, Richmeister, Gilliam, Folajimi, The Gnome, Oscarthecat, Skizzik, Kmarinas86, Chris the speller, SuperBuuBuu, Quinsareth, Persian Poet Gal, Sirex98, MalafayaBot, Silly rabbit, Sangrolu, Villarinho, DHN-bot~enwiki, Sbharris, Hongooi, Jdthood, CheerLeone, Gtkysor, Can't sleep, clown will eat me, Nick Levine, Tamfang, Kelvin Case, Vladislav, Vanished User 0001, Rrburke, Jgoulden, Auvii, Krich, Wen D House, Radagast83, Engwar, Nakon, VegaDark, John D. Croft, Alexander110, KimO, Adrigon, SpiderJon, Ultraexactzz, Zadignose, Tesseran, Byelf2007, L337p4wn, K7lim, SashatoBot, Mchavez, Swatjester, Leftydan6, Minaker, Attys, Brillow, John, Ashoat, Scientizzle, Acitrano, Linnell, JoshuaZ, James.S, JorisvS, Coredesat, Goodnightmush, ICBB, Plunge, JHunterJ, Hypnosifl, Silverthorn, Descubes, Freederick, Dr.K., Vanished user, Iridescent, Darkerprojects, Astrobayes, Newone, MOBle, Igoldste, CapitalR, AGK, Courcelles, Tawkerbot2, Dlohcierekim, Chetvorno, Hammer Raccoon, Owen214, Eastlaw, Peledre, Pukkie, Anakata, Banedon, Runningonbrains, DKOH, NickW557, Gregbard, MikeWren, Vttoth, Necessary Evil, Ryan, Viciouspiggy, Gogo Dodo, Anonymi, Xxanthippe, A Softer Answer, Odie5533, Tawkerbot4, DumbBOT, Robertinventor, Kozuch, Mtpaley, Philza85, Starship Trooper, UberScienceNerd, Crum375, Thijs!bot, Epbr123, Astroceltica, Passaggio, Barbarina, Mbell, Eugenespeed, N5iln, Mojo Hand, Carlif, Headbomb, Tonyle, Marek69, RickinBaltimore, Lars Lindberg Christensen, OtterSmith, SusanLesch, Dawnseeker2000, Mmortal03, Hmrox, Hires an editor, AntiVandalBot, Seaphoto, Orionus, Opelio, Shirt58, Rehnn83, Joehodge, AaronY, Jj137, TTN, Dylan Lake, Chill doubt, Spencer, Yellowdesk, Sniktaw, Lfstevens, CPitt76, Gökhan, Jcarter1, Res2216firestar, JAnDbot, Leuko, Husond, MER-C, CosineKitty, Plantsurfer, Mcorazao, Therealintellectual, Folkform, Balbers, 100110100, Autotheist, Wasell, Magioladitis, Bongwarrior, VoABot II, Timothy McVeigh, Charlesrkiss, AuburnPilot, Krkaiser, Mbarbier, Kaivosukeltaja, Foroa, Swpb, Stigmj, T a y l o s, Ekantik, Brusegadi, Bubba hotep, Fabrictramp, Catgut, Lilian.Kaufmann, Zhanghia, Acornwithwings, BatteryIncluded, Vssun, LtHija, Whisky5, DerHexer, Prisca6023, PeteSF, KenyaSong, Rickard Vogelberg, NatureA16, DancingPenguin, MartinBot, Schmloof, STBot, Pagw, Fs644, Nikpapag, Anaxial, CommonsDelinker, Jean-Pierre Petit~enwiki, PrestonH, WelshMatt, Chrishyman, Tgeairn, J.delanoy, Pharaoh of the Wizards, Trusilver, Adavidb, Kpvats, Kudpung, Rod57, Arion 3x3, PedEye1, McSly, Tarotcards, Davy p, HiLo48, NewEnglandYankee, Ohms law, Jorfer, Blckavnger, Potatoswatter, KylieTastic, Joshua Issac, Infiniteglitch, Remember the dot, Pitpif, Vanished user 39948282, Neekap, Natl1, Ldebain, BernardZ, SoCalSuperEagle, Squids and Chips, Borat fan, CardinalDan, Idioma-bot, Sheliak, Funandtrvl, Lights, Hammersoft, VolkovBot, Craigheinke, Itsfullofstars, ColdCase, Jeff G., JohnBlackburne, Mocirne, AlnoktaBOT, Scikid, Grammarmonger, Leojohns, Larry R. Holmgren, Philip Trueman, TXiKiBoT, Oshwah, Docanton, Authorized User, Theophilus reed, Drestros power, Strichek, MarekMahut, Monkey Bounce, Lradrama, Sintaku, Carillonatreides, Martin451, Broadbot, Wiae, Mazarin07, Inductiveload, Knightshield, Telecineguy, Spiral5800, Kurowoofwoof111, Greswik, RobertFritzius, SwordSmurf, Falcon8765, Hellothere17, Enviroboy, Littlehollah, Wanchung Hu, Illumini85, SonOfMog Worf, Jazzman123, PGWG, 19merlin69, FlyingLeopard2014, Neparis, Bkppage, S-n-ushakov, SieBot, Calliopejen1, Tresiden, Wibubba48, Tachyonics, Pallab1234, Paradoctor, KGyST, Bentogoa, Jimlester51, Battlepace, Oda Mari, Aaarnooo, Suomichris, Crowstar, PromX1, Lightmouse, Tombomp, Cyberplasm, Diego Grez-Cañete, Spartan-James, Thinghy, Mygerardromance, Hamiltondaniel, Superbeecat, Denisarona, JL-Bot, Escape Orbit, Starcluster, Troy 07, Atif.t2, ArepoEn, Ak47gforce, Ratemonth, Sfan00 IMG, ClueBot, Phoenix-wiki, GorillaWarfare, The Thing That Should Not Be, ArdClose, Rodhullandemu, Cptmurdok, Drmies, Frmorrison, Uncle Milty, Iuhkjhk87y678, Niceguyedc, MrBosnia, Bhaskarns, Andwor, Ktr101, Excirial, Dombom12, Cromescythe, Barbarinaz,

FOARP, Brews ohare, Jotterbot, Iohannes Animosus, R.Andrae, Kentgen1, Ordovico, Mastertek, Rgoogin, Thehelpfulone, 1ForTheMoney, Versus22, Palmer666palmer, PCHS-NJROTC, Burner0718, Pillar of Babel, SoxBot III, Erodium, Vanished user uih38riiw4hjlsd, 1ofhissheep, TimothyRias, Arianewiki1, XLinkBot, DCCougar, Oldnoah, Rror, Gwark, Feinoha, Ost316, Avoided, Webmaster369, Gthomson, Tugrul irmak, Noctibus, Ploversegg, ZooFari, Parejkoj, Tayste, Addbot, Xp54321, Grayfell, Experimental Hobo Infiltration Droid, Willking1979, Some jerk on the Internet, Uruk2008, 04aeverington, DOI bot, Tcncv, Nohomers48, CharlesChandler, Gmeyerowitz, Haasfelix, Download, Proxima Centauri, Ashirgo, RTG, Redheylin, Luke Maurer, Glane23, Darkmatter654, SamatBot, Nanzilla, Lzkelley, Clone 209, Tassedethe, Numbo3-bot, Peridon, Chinchinthehun, Evildeathmath, Tide rolls, Lightbot, OlEnglish, Qemist, Gail, North Polaris, Legobot, Artichoke-Boy, Luckas-bot, Yobot, WikiDan61, Cosoce, Dov Henis, Aldebaran66, KillYourLove, CzechFalcon, Amble, Mmxx, CinchBug, Perusnarpk, IW.HG, Einstein vs Dark energys, Eric-Wester, Tempodivalse, Synchronism, AnomieBOT, Letuño, Girl Scout cookie, IRP, JackieBot, RBM 72, AdjustShift, Nicolaas Vroom, Henrykandrup, Iluziat, Materialscientist, Dendlai, ImperatorExercitus, The High Fin Sperm Whale, Citation bot, Ternity0127, Maxis ftw, Frankenpuppy, Quebec99, LilHelpa, Aksel89, Xqbot, Stlwebs, Random astronomer, Sionus, Cureden, Jradis1337, Capricorn42, Wperdue, Julianhyde, Deleance, Raspw, Tomwsulcer, Magicxcian, Gap9551, AbigailAbernathy, Srich32977, NOrbeck, Artemis6234, Almabot, Abell 1367, Feldhaus, False vacuum, RibotBOT, Waleswatcher, Mikedr, Kongkokhaw, Rvnieuwe, Shadowjams, MeDrewNotYou, A. di M., Peter470, Sageman7, 蝈蝈, Luminique, Captain-n00dle, Imyfujita, FrescoBot, Andyradke0, Ag allstar, Paine Ellsworth, Originalwana, Styxpaint, Mark Renier, VS6507, PhysicsExplorer, Dbirkhofer, Steve Quinn, Nestlefolife, Adrian Akau, 1414rwbt, SF88, Citation bot 1, Redrose64, DUUJEEGWEEM, Tyler6298, Pinethicket, I dream of horses, Grammarspellchecker, Danlof, 10metreh, Jonesey95, Tom.Reding, Pmokeefe, A8UDI, For.a.limited.time.only, Elentirno, TedderBot, Aknochel, SkyMachine, IVAN3MAN, Kgrad, Nieuwenh, Trappist the monk, Puzl bustr, Fama Clamosa, Domeinthebumhole, Michael9422, UrukHaiLoR, Allen4names, JLincoln, Jeffrd10, Lovemybluetooth, Diannaa, Fastilysock, Innotata, DrCrisp, Whisky drinker, Onel5969, RjwilmsiBot, 5mgoblue5, Blakelewis122, Dorri, Mathewsyriac, Leandro.lelas, Mserard313, Mdznr, Sbugnon, Ultima821, EmausBot, Francophile124, Grrow, Super48paul, GoingBatty, RA0808, Gimmetoo, Solarra, Jmencisom, Slightsmile, Tommy2010, Winner 42, SusanaMultidark, Gocows2, Wikipelli, Serketan, Krifferjel, Zurich Astro, Hhhippo, Mz7, Mhatthei, Svolin, Micahqgecko, JSquish, Josve05a, Trojanmice, MithrandirAgain, Edwinkaren, Devilaza, Arbnos, Oraclan, Suslindisambiguator, Spork-Bot, AlbertusmagnusOP, Mcmatter, Tolly4bolly, L1A1 FAL, Ancient Anomaly, L Kensington, Maj den, Corabilek, Donner60, Aldnonymous, Ihardlythinkso, RockMagnetist, Terra Novus, TYelliot, DASHBotAV, Kroupap, D Phoesheezey, Mannix Chan, Travies10, Jxraynor, TheTimesAreAChanging, ClueBot NG, Rich Smith, Gareth Griffith-Jones, Afjvanraan, Lyla1205, Crystal7878, Catinthehat93, Bped1985, Infinifold, Wiggit002, Jj1236, PapaMike, MonEyshOt42069210, Muon, Esdacosta, Asukite, Masssly, Ph.d Carl edenburgh, Widr, Gavin.perch, Helpful Pixie Bot, Curb Chain, Calabe1992, Bibcode Bot, BG19bot, Dualus, Kishanparekh, Stevenwilkins, NacowY, Cheeseray1, Cyberguy5, Darkmatter adam, Yomomma8102, Hza a 9, Rarelight, Cyberpower678, Cosmologist77, MusikAnimal, தென்காசி சுப்பிரமணியன், Dahliamtl, Dodshe, Mark Arsten, Darkmatterotheruniverses, Samcstewart, Cadiomals, Trevayne08, Zedshort, Achowat, Rolandwilliamson, A2Die, Clint55555, Mgka79, NotEither, BattyBot, Millennium bug, Ronin712, Babymushrooms, Davidmexican, Drphilmarshall, Dilaton, Quin71901, U-95, Chris-Gualtieri, Npmay, Kvark92, Lukasz.astrus, Ducknish, JYBot, Davidlwinkler, Astrohap, Hunterf12, Dexbot, Caroline1981, Gravityking100, Junavia, Fredrikdn, CuriousMind01, Lugia2453, Wjs64, Andwor42, Frosty, Honneydewp243, Junjunone, DrHowzer73, JustAMuggle, Me, Myself, and I are Here, WadiElNatrun, Reatlas, Rfassbind, Acetotyce, I am One of Many, DirkXcal, Melonkelon, Ybidzian, Gig9876, M.ashrafinia, Trolololman12, Ilikedeletingstufffromhere, DavidLeighEllis, Onecreation, LahmacunKebab, The Herald, Zenibus, Jernahthern, Hipposaregrey, Frinthruit, Derekdoth, Stamptrader, Cyberalchemyst, Aaronknowsitall, FelixRosch, Darkmer, Doubleknockout, Monkbot, Yikkayaya, Wardinstrument, Leegrc, Vikas Rauniyar, Apipia, Upsalla, Jkvaternik, Lol kaptyn troll, HMSLavender, Mohammedshukoor, Callum92, Stefania.deluca, Ashweigh, Mrpiecjohnson, Oldstone James, Astezar, 39Debangshu, YoYoDude012, Anunaki truth, Pyrotle, Tetra quark, Carazmatic, God of matterrr, Silversparkcontributions, Isambard Kingdom, VexorAbVikipædia, Rizi0909, Absolutelypuremilk, Anand2202, Kbap2002, Kb2002, DN-boards1, Yohoona, Denniscabrams, KasparBot, I love trains sooo much, Sir Cumference, Ricardo A. Olea, Id6040, MHolland85, TychosElk, Boowiebear, Stephane Le Corre, Mustachman71, Abdelrahmam shawky, Huritisho, Jeffman257, Outedexits, Reg7d88, Maka Tree, Eslam nsr, Incendiary Iconoclasm, GSS-1987, Boobety boop boop boop, Atharv4321, Zhakhan9er, XABHISHEK23x, Not a creative person, The Voidwalker, Shane ducharme, Tyler kwejew, Igojamon, Iamcoolswag, Riki71144, Kdkskdfj, Soopdish, Mr ST 142, Portal da Ciência, WikiEditor79, Wiki62838351936, MoneyMonkey112 and Anonymous: 1304

- **Hot dark matter** *Source:* https://en.wikipedia.org/wiki/Hot_dark_matter?oldid=700297812 *Contributors:* Mav, Lexor, Cassini~enwiki, Looxix~enwiki, Theresa knott, AnthonyQBachler, Korath, Naddy, Eequor, DragonflySixtyseven, Sam Hocevar, Burschik, Kate, Tom, Mac Davis, RJFJR, Rjwilmsi, Strait, Marasama, Zaak, Rangek, Margosbot~enwiki, Chobot, DVdm, DanMS, Salsb, Kungfuadam, Nickst, MalafayaBot, Lenoxus, Alaibot, OEYoung, BatteryIncluded, Idioma-bot, BotKung, AlleborgoBot, MystBot, Addbot, LaaknorBot, Yobot, Romul~enwiki, AnomieBOT, Kikuyu3, LittleWink, Tom.Reding, Zurich Astro, Omyojj, Helpful Pixie Bot, Drunken Thinker, PuruMuthal and Anonymous: 14

- **Cold dark matter** *Source:* https://en.wikipedia.org/wiki/Cold_dark_matter?oldid=706630235 *Contributors:* Bryan Derksen, Roadrunner, Tim Starling, Llywrch, Looxix~enwiki, Schneelocke, Doradus, Rho~enwiki, Eequor, Christopherlin, Keith Edkins, Balcer, Karl Dickman, Rich Farmbrough, StephanKetz, I9Q79oL78KiL0QTFHgyc, Fwb22, Lysdexia, Uris, Rjwilmsi, Chobot, 2over0, Kungfuadam, KnightRider~enwiki, Nickst, Nightbat, MalafayaBot, Cmanser, Farseer, OhioFred, John, Robofish, Eridani, Joeyfox10, Alaibot, Mbell, Headbomb, CosineKitty, Magioladitis, Ryan WMD, Idioma-bot, 0-Jenny-0, Michael H 34, BotKung, Mazarin07, Paradoctor, Niceguyedc, Auntof6, Addbot, Lightbot, Sebas310, Yobot, Systemizer, AnomieBOT, GrouchoBot, Waleswatcher, IAP Astro, Erik9bot, Kikuyu3, Sae1962, Citation bot 4, Pinethicket, Jonesey95, Tom.Reding, Trappist the monk, Gwyneth99, RA0808, Zurich Astro, AvicAWB, SporkBot, AThinkingScientist, ClueBot NG, CocuBot, BBCDM, Helpful Pixie Bot, Bibcode Bot, BG19bot, Dualus, Winston Trechane, Willoakley, Prokaryotes, Sjzaslaw, Monkbot, Sofia Koutsouveli, Astronome de Meudon, Stefania.deluca, Jmesser5, TychosElk and Anonymous: 48

- **Accelerating expansion of the universe** *Source:* https://en.wikipedia.org/wiki/Accelerating_expansion_of_the_universe?oldid=720723619 *Contributors:* AxelBoldt, Bth, Dbundy, Tim Starling, EddEdmondson, Alfio, Looxix~enwiki, William M. Connolley, Julesd, Evercat, Hashar, Timwi, Dysprosia, Mw66, The Anomebot, DW40, Dragons flight, Phys, Peak, Lowellian, Rursus, JerryFriedman, Giftlite, Graeme Bartlett, DavidCary, Barbara Shack, Joe Kress, ConradPino, Bbbl67, Burschik, Eep², JimJast, Guanabot, Vsmith, LeeHunter, Dbachmann, Bender235, AdamSolomon, RJHall, Mr. Billion, Sietse Snel, Bobo192, Foobaz, I9Q79oL78KiL0QTFHgyc, Hackwrench, Eric Kvaalen, Vuo, Falcorian, Bobrayner, Zanaq, Richard Arthur Norton (1958-), Woohookitty, OCNative, Joke137, Wisq, Driftwoodzebulin, Drbogdan, Rjwilmsi, Seandop, Kolbasz, ThunderPeel2001, Marcperkel, Chase me ladies, I'm the Cavalry, Closedmouth, Dutch-Bostonian, SmackBot, Melchoir, Vald, Nickst, AnOddName, Dreadstar, Pulu, Wkerney, Ckatz, Hypnosifl, Xxxiv34, Geral Corasjo, CRGreathouse, Kjknohw, Meno25, Michael C Price, Doug Weller, DumbBOT, Peter Gulutzan, Dawnseeker2000, KrakatoaKatie, Seaphoto, Obeattie, Steelpillow, Arch dude, Bpmullins, Cardamon, Cgingold, Robin S, NatureA16, Hedwig in Washington, McSly, Mannhoodd, Tarotcards, VolkovBot, UnitedStatesian, Insanity Incarnate, Nihil novi, Puzhok, Lightmouse, Coldcreation, Gevgiorbran, Agge1000, Dr. Leif Rongved, Dmyersturnbull, Kentgen1, El bot de la dieta, DanielPharos, DumZiBoT, XLinkBot, Ladsgroup, Ost316, Addbot, Ridgepg, Simonm223, Iliketitz93, AkhtaBot, Marx01, Verbal, Legobot, Cosmos72, Luckas-bot, Yobot, Systemizer, Fraggle81, Amirobot, Amble, Synchronism, AnomieBOT, Jim1138, Citation bot, Xqbot, Gap9551, GrouchoBot, RibotBOT, Waleswatcher, Bellerophon, Michael93555, DivineAlpha, Citation bot 4, Jonesey95, Tom.Reding, Issuesixty souls-great, Trappist the monk, Lotje, RjwilmsiBot, John of Reading, Primefac, Tinss, Vanjka-ivanych, Solomonfromfinland, ZéroBot, Brandmeister, ChuispastonBot, MrChandmari, Mechachomp, ClueBot NG, Astrocog, Bibcode Bot, BG19bot, Yizlpku, Anukool.rajoriya, Minsbot, Pheng13, RiseUpAgain, Mediran, Kozmokonstans, Makecat-bot, Wjs64, Illuusio, Rfassbind, Yheyma, Blackbombchu, The Herald, Leegrc, Jsaur, Igby Kollektiv, Garfield Garfield, HannahFord428, Tetra quark, Isambard Kingdom, MusikBot, Sir Cumference, Youknowwhatimsayin, ICameHere-ToEditNotToFeel, Xx Cool Guy7202 xX and Anonymous: 93

- **Big Rip** *Source:* https://en.wikipedia.org/wiki/Big_Rip?oldid=718018077 *Contributors:* AxelBoldt, Vicki Rosenzweig, Bryan Derksen, Tarquin, Dcljr, Cyp, Glenn, Nikai, Jeandré du Toit, Schneelocke, Reddi, Maximus Rex, BenRG, Lumos3, Academic Challenger, Rursus, Tea2min, Smjg, Wolfkeeper, Herbee, Henry Flower, Matthäus Wander, Nickptar, Burschik, Discospinster, Guanabot, StephanKetz, Dbachmann, Nard the Bard, RJHall, Art LaPella, Rpresser, I9Q79oL78KiL0QTFHgyc, La goutte de pluie, HasharBot~enwiki, Free Bear, Arthena, Super-Magician, Versageek, Thryduulf, Scjessey, Aristotle Pagaltzis, Joke137, Banpei~enwiki, Drbogdan, Rjwilmsi, Pjetter, Koavf, Mike Peel, Maxim Razin, Crazy-computers, Wars, Angus Lepper, Hairy Dude, Bhny, Trovatore, Omega 13, Roman Soldier, ColinMcMillen, SmackBot, AaronM, C.Fred, Bluebot, Hibernian, Darthgriz98, Richard001, J 1982, Don't give an Ameriflag, Kirbytime, Danilot, Dan Gluck, Iridescent, Tawkerbot2, Henrickson, JForget, Kjknohw, Gregbard, Cydebot, Gogo Dodo, Peter Gulutzan, MarkGyver, AgentPeppermint, Noclevername, Northumbrian, EmRunTon-RespNin, WikiSlasher, Spartaz, LinkinPark, RadicalPi, Swpb, Nyttend, DGG, Sm8900, Monkeybait99, R'n'B, Mbweissman, J.delanoy, Foober, Framhein, Juliancolton, Lights, VolkovBot, RingtailedFox, Kinneytj, HowardFrampton, TXiKiBoT, Bentley4, W.a.tas, CorusTV, Wasted Sapience, Seresin, Bobo The Ninja, SieBot, Lalala98, Lalala95, Wing gundam, Flyer22 Reborn, Format43, Duae Quartunciae, Hamiltondaniel, Atilla98, Michael Petersen, Ratemonth, ClueBot, Estevoaei, AndyFielding, Staticshakedown, Addbot, Jncraton, OlEnglish, MattDredd, Luckas-bot, Yobot, Legobot II, Amble, KamikazeBot, LED65, Obersachsebot, Capricorn42, RibotBOT, CaZeRillo, LucienBOT, Io Herodotus, Citation bot 1, MusBeaumont, 70virginousc, Trappist the monk, NerdyScienceDude, EmausBot, WikitanvirBot, ZéroBot, Teapeat, Rememberway, ClueBot NG, Pennykohl, Bibcode Bot, Kvark92, Ikjyotsingh, Epicgenius, Madreterra, Amrellithy, Jwratner1, Kogge, Urmomisachode, Phleg1, Christolav, Wikipedian 2, Tonathan100, Tetra quark, Isambard Kingdom, Pulkitmidha, Sir Cumference and Anonymous: 126

- **Standard Model** *Source:* https://en.wikipedia.org/wiki/Standard_Model?oldid=719880398 *Contributors:* AxelBoldt, Derek Ross, CYD, Bryan Derksen, The Anome, Ed Poor, Andre Engels, Roadrunner, Anthere, David spector, Isis~enwiki, Youandme, Ram-Man, Stevertigo, Edward, Patrick, Boud, Michael Hardy, SebastianHelm, Looxix~enwiki, Julesd, Glenn, AugPi, Mxn, Raven in Orbit, Reddi, Phr, Tpbradbury, Populus, Haoherb428, Phys, Floydian, Bevo, Pierre Boreal, AnonMoos, BenRG, Jeffq, Dmytro, Drxenocide, Robbot, Nurg, Securiger, Texture, Roscoe x, Fuelbottle, Mattflaschen, Tea2min, Alan Liefting, Ancheta Wis, Giftlite, Dbenbenn, Harp, Herbee, Monedula, LeYaYa, Xerxes314, Dratman, Alison, JeffBobFrank, Dmmaus, Pharotic, Brockert, Bodhitha, Andycjp, Sonjaaa, HorsePunchKid, APH, Icairns, AmarChandra, Gscshoyru, Kate, Arivero, FT2, Rama, Vsmith, David Schaich, Xezbeth, D-Notice, Dfan, Bender235, Pt, El C, Laurascudder, Shanes, Drhex, Fogger~enwiki, Brim, Rbj, Jeodesic, Jumbuck, Alansohn, Gary, ChristopherWillis, Guy Harris, Axl, Sligocki, Kocio, Stillnotelf, Alinor, Wtmitchell, Egg, TenOfAllTrades, H2g2bob, Killing Vector, Linas, Mindmatrix, Benbest, Dodiad, Mpatel, Faethon, TPickup, Faethon34, Palica, Dysepsion, Faethon36, Qwertyca, Drbogdan, Rjwilmsi, Zbxgscqf, Macumba, Strangethingintheland, Dstudent, R.e.b., Bubba73, Drrngrvy, Agasicles, FlaBot, Naraht, Agasides, DannyWilde, Dave1g, Itinerant1, Gparker, Jrtayloriv, Goudzovski, Chobot, Bgwhite, FrankTobia, YurikBot, Bambaiah, Ohwilleke, VoxMoose, Bhny, JabberWok, Bovineone, Krbabu, SCZenz, JulesH, Davemck, Lomn, E2mb0t~enwiki, Jrf, Dv82matt, Tetracube, Hirak 99, Arthur Rubin, Netrapt, JLaTondre, Caco de vidro, RG2, GrinBot~enwiki, That Guy, From That Show!, Hal peridol, SmackBot, YellowMonkey, Tom Lougheed, Melchoir, Bazza 7, KocjoBot~enwiki, Jagged 85, Thunderboltz, Setanta747 (locked), Skizzik, Dauto, Chris the speller, Bluebot, TimBentley, Sirex98, Silly rabbit, Complexica, Metacomet, DHN-bot~enwiki, MovGP0, QFT, Kittybrewster, Addshore, Jmnbatista, Cybercobra, Jgwacker, BullRangifer, Soarhead77, Daniel.Cardenas, Yevgeny Kats, Byelf2007, TriTertButoxy, Craig Bolon, Ajnosek, Vgy7ujm, Ekjon Lok, Bjankuloski06, Tarcieri, Waggers, JarahE, Michaelbusch, Lottamiata, Newone, Twas Now, IanOfNorwich, Srain, Patrickwooldridge, J Milburn, Mosaffa, Gatortpk, Vessels42, Geremia, Van helsing, Harrigan, Phatom87, Cydebot, David edwards, Verdy p, Michael C Price, Xantharius, Crum375, JamesAM, Thijs!bot, Epbr123, Headbomb, Phy1729, Stannered, Tariqhada, Seaphoto, Orionus, Voyaging, Gnixon, Jbaranao, Jrw@pobox.com, Len Raymond, Narssarssuaq, Bakken, CattleGirl, Davidoaf, Vanished user ty12kl89jq10, Lvwarren, Taborgate, Leyo, HEL, J.delanoy, Hans Dunkelberg, Stephanwehner, Wbellido, Aoosten, Jacksonwalters, The Transliterator, Dada-Neem, Student7, Joshmt, WJBscribe, Jozwolf, Hexane2000, BernardZ, Awren, Sheliak, Physicist brazuca, Schucker, Goop Goop, Fences and windows, Dextrose, Mcewan, Swamy g, TXiKiBoT, Sharikkamur, Thrawn562, Voorlandt, Escalona, Setreset, PDFbot, Pleroma, UnitedStatesian, Piyush Sriva, Kacser, Billinghurst, Francis Flinch, Moose-32, Ptrslv72, David Barnard, SieBot, ShiftFn, Robdunst, Jim E. Black, SheepNotGoats, Gerakibot, Nozzer42, Mr swordfish, Wing gundam, Bamkin, Likebox, MaynardClark, Arthur Smart, HungarianBarbarian, Commutator, KathrynLybarger, Iomesus, C0nanPayne, Crazz bug 5, ClueBot, Superwj5, Wwheaton, Garyzx, SuperHamster, Elsweyn, Maldmac, DragonBot, Djr32, Diagramma Della Verita, Nymf, Eeekster, Brews ohare, NuclearWarfare, PhySusie, Ordovico, Mastertek, DumZiBoT, BodhisattvaBot, Guarracino, Mitch Ames, Truthnlove, Stephen Poppitt, Tayste, Addbot, Deepmath, Eric Drexler, DWHalliday, Mjamja, Leszek Jańczuk, NjardarBot, Mwoldin, Bassbonerocks, Barak Sh, AgadaUrbanit, Lightbot, Smeagol 17, Abjiklam, Ve744, Luckas-bot, Yobot, Orion11M87, AnomieBOT, JackieBot, Icalanise, Citation bot, ArthurBot, Northryde, LilHelpa, Xqbot, Sionus, Professor J Lawrence, Tomwsulcer, Edsegal, GrouchoBot, Trongphu, QMarion II, Ernsts, A. di M., Bytbox, FrescoBot, Paine Ellsworth, Aliotra, Steve Quinn, Citation bot 1, Rameshngbot, MJ94, RedBot, MastiBot, Aknochel, Sijothankam, Puzl bustr, Beta Orionis, Physics therapist, Bj norge, Innotata, Jesse V., RjwilmsiBot, Mathewsyriac, Afteread, EmausBot, Bookalign, WikitanvirBot, Wilhelm-physiker, Bdijkstra, DerNeedle, Kenmint, Dbraize, Tanner Swett, HeptishHotik, ﺩﺍﺮﺒﻧ ﻰﻨﺸﻣﻮﻫ, Suslindisambiguator, Quondum, Webbeh, UniversumExNihilo, Vanished user fijw983kjaslkekfhj45, Maschen, Rock-Magnetist, Stormymountain, Ζeτα ξ, Whoop whoop pull up, Isocliff, ClueBot NG, Smtchahal, Snotbot, Tonypak, O.Koslowski, CharleyQuinton, Dsperlich, Theopolisme, ZakMarksbury, Helpful Pixie Bot, Bibcode Bot, BG19bot, Tirebiter78, AvocatoBot, Lukys~enwiki, Stapletongrey, Ownedroad9, Chip123456, ChrisGualtieri, Khazar2, Billyfesh399, Rhlozier, JYBot, Dexbot, Doom636, Rongended, Cerabot~enwiki, CuriousMind01, Cjean42, Jayanta mallick, Joeinwiki, Kowtje, JPaestpreornJeolhlna, Eyesnore, Euan Richard, Nigstomper, Particle physicist, Prokaryotes, Jernahthern, Ginsuloft, Dimension10, JNrgbKLM, Krabaey, 1codesterS, FelixRosch, Monkbot, Delbert7, Trackteur, BradNorton1979, Lathamboyle, Tetra quark, Isambard Kingdom, I enjoy sandwiches, KasparBot, Buckbill10, Huritisho, S3rr8s and Anonymous: 368

- **Fundamental interaction** *Source:* https://en.wikipedia.org/wiki/Fundamental_interaction?oldid=713122554 *Contributors:* AxelBoldt, Zun-

Garos, Pigalle, Washingtonlerias, Ubuthustra, D.H. Nick Number, Klausness, Sam42, DarthNemesis, Northumbrian, Escarbot, WikiSlasher, AntiVandalBot, Seaphoto, Maxibons, Tim Shuba, Braindrain0000, Tempest115, Jrw@pobox.com, Narssarssuaq, Husond, MER-C, Andrewdolby, RogierBrussee, Bongwarrior, VoABot II, Bakken, Appraiser, Faizhaider, Cuardin, Stijn Vermeeren, Trebor1, Catgut, Cardamon, NMarkRoberts, IkonicDeath, MetsBot, Mwasim1, JaGa, GuelphGryphon98, NatureA16, FisherQueen, Flowanda, MartinBot, TechnoFaye, Wikeepeedier, Player 03, Tgeairn, HEL, J.delanoy, Bobvinson, Maurice Carbonaro, Foober, 3halfinchfloppy, Lantonov, NewEnglandYankee, LeighvsOptimvsMaximvs, KylieTastic, Ja 62, Vinsfan368, Izno, Idioma-bot, Makewater, 28bytes, VolkovBot, XCelam, JohnBlackburne, AlnoktaBOT, Philip Trueman, Oshwah, Zidonuke, Red Act, Anonymous Dissident, Yilloslime, Fizzackerly, PaulTanenbaum, PhilyG, Wingedsubmariner, Hotmoklet, Eubulides, Zhongsan, SmileToday, Falcon8765, Cubed mass, LachlanSosa, StevenJohnston, Hunter826242, PSSnyder, Hobojaks, YohanN7, SieBot, ShiftFn, Paradoctor, Dawn Bard, Vanished user 82345ijgeke4tg, Flyer22 Reborn, Csblack, Henry Delforn (old), Nuttycoconut, MrWikiMiki, Hjelmerus, Dposte46, Jeanlovecomputers, Mátyás, OKBot, Fedosin, Coldcreation, Fuddle, Mike2vil, Anchor Link Bot, MarkMLl, VanishedUser sdu9aya9fs787sads, ImageRemovalBot, Martarius, De728631, ClueBot, The Thing That Should Not Be, EoGuy, Exploto, Razimantv, IshanAlmazi, Shinpah1, JFlav, Noca2plus, Eeekster, Tam 66 7, DPCU, Cenarium, Mozart21, Mentor364, Themantyke, McXX, Tin Whistle Man, Galor612, MalWilley, NERIC-Security, Rror, Avoided, Richard-of-Earth, Whizmd, Addbot, Derivator, Gravitophoton, DOI bot, Gul e, Startstop123, Gustavo José Meano Brito, Vishnava, CanadianLinuxUser, CarsracBot, RTG, Monypooh12, DFS454, Glane23, Tod.davidson, Mcsploogerson, AnnaFrance, Favonian, Doniago, West.andrew.g, Numbo3-bot, Lightbot, Zorrobot, Legobot, Yobot, Ht686rg90, THEN WHO WAS PHONE?, Allowgolf~enwiki, Synchronism, AnomieBOT, Jim1138, Materialscientist, Citation bot, Maxis ftw, ChristianH, Expooz, Xqbot, Tjcheckley, Δζ, Anna Frodesiak, Shindamaru, False vacuum, Omnipaedista, Frankie0607, RibotBOT, Gsard, MerlLinkBot, Bearnfæder, 7575474087ALBERT, CES1596, FrescoBot, H.W. Clihor, Paine Ellsworth, Dogposter, Tj2691, Majopius, Mouselarry, Haeinous, Vbrcat, Citation bot 1, Deadtotruth, Pinethicket, Hypernovic, Lesath, 10metreh, Tom.Reding, Smuckola, Rushbugled13, A8UDI, RedBot, Sjb13, Elvis633, December21st2012Freak, Fredkinfollower, IVAN3MAN, Double sharp, Euriditi, TobeBot, 0x30114, LogAntiLog, Jonkerz, Dinamik-bot, Capt. James T. Kirk, Aribashka, Brianann MacAmhlaidh, Seahorseruler, Bj norge, The Utahraptor, Scipioafricanus, Reagster, Thslackliner19, Theslackliner19, EmausBot, WikitanvirBot, AlexUT, Pekka.virta, Stewiefool, RA0808, Slightsmile, Wikipelli, Scalable Vector Raccoon, Thecheesykid, Hhhippo, JSquish, Fæ, StringTheory11, Int21hexster, Quondum, Ellie Rickett, Donner60, Nelium12, Maheshbahadur, Ihardlythinkso, AndyTheGrump, RockMagnetist, Winston7, Rmashhadi, ClueBot NG, MelbourneStar, Irisrune, Lanthanum-138, Doorsrocklikerocks, Frietjes, Helpful Pixie Bot, HMSSolent, Gob Lofa, Bibcode Bot, Transscientific, Bobc3, BG19bot, Hz.tiang, Fw0116, Davidiad, Piguy101, Giarcea, Naveedyaykhan, Cadiomals, Kaaalbert, MrBill3, Wiki2103, Penguinstorm300, RiseUpAgain, The1337gamer, Nirmal kumar 9, SteveBM, Stigmatella aurantiaca, Cyberbot II, Zachhansonhart, U5ard, Nkzf, Khazar2, Ekren, Larskk101, Vanished user 23i4hjwrjfiij4t, Kevinfrank17, Twhitguy14, CuriousMind01, Mike.leivers, Kingcircle, DmVdx, Telfordbuck, Cadillac000, Asad shayan, Epicgenius, Bantennyson, Eachandall, AnthonyJ Lock, B14709, Ericgermate, Wedgeline, RogrMexico, Jwratner1, NottNott, Frinthruit, Iliketrains1234567890, Skyshad4w, AnonymousAuthority, TuxLibNit, Da.pro1, Monkbot, Yikkayaya, BethNaught, Zacwill, 97dc, Neeraj Bhakta, KH-1, Vedic Earthian, AHusain3141, Aaronfranke, 39Debangshu, Tetra quark, Isambard Kingdom, Epictacotree, Anand2202, Kbap2002, Djniew, Hriton, KasparBot, Edgar-lausanne, Alout, Baking Soda, Tramrattan, Chemistry1111, Sundaram108, Boomer Vial, Spaceviewer, Tolibonboni06, Virtumanity and Anonymous: 576

- **Neutrino** *Source:* https://en.wikipedia.org/wiki/Neutrino?oldid=720259911 *Contributors:* AxelBoldt, Chenyu, Bryan Derksen, Zundark, The Anome, Tarquin, Andre Engels, Xaonon, XJaM, William Avery, Roadrunner, DrBob, Heron, Cwitty, MimirZero, Spiff~enwiki, Edward, Patrick, Ken Arromdee, EddEdmondson, Ezra Wax, Gdarin, Meekohi, Bcrowell, Cyde, Arpingstone, Alfio, Looxix~enwiki, Strebe, JWSchmidt, Julesd, Glenn, Nikai, Andres, Evercat, Rob Hooft, TheSeez, Crissov, Wikiborg, Reddi, Lfh, Cos111, Tpbradbury, Fibonacci, Warofdreams, Twang, Donarreiskoffer, Drxenocide, Robbot, Findel, Zandperl, Nurg, Masao, Merovingian, Bobunf, Rursus, Meelar, Matty j, Intangir, Wikibot, Wereon, Duien, Jimduck, Bbx, David Gerard, Giftlite, Graeme Bartlett, DocWatson42, Laudaka, Mikez, Harp, Lethe, HangingCurve, Xerxes314, Anville, Dratman, Curps, Jorge Stolfi, Eequor, Mdob, Espetkov, ChicXulub, LiDaobing, Elroch, Icairns, Doug Danner, Nickptar, Fg2, Lrenh, Deglr6328, Hmmm~enwiki, Mattman723, Helohe, Rich Farmbrough, Hydrox, Cacycle, Pjacobi, Vsmith, Dbachmann, Mani1, Pavel Vozenilek, Bender235, Ralfoide, Sunborn, Neko-chan, Kharhaz, RJHall, Charm, Haxwell, RoyBoy, Smalljim, Cje~enwiki, Viriditas, Unquietwiki, Foobaz, I9Q79oL78KiL0QTFHgyc, La goutte de pluie, Thewayforward, Thuktun, Fleurot~enwiki, Quaoar, Alansohn, Anthony Appleyard, ChristopherWillis, Calton, Axl, Mac Davis, Hdeasy, RJFJR, Dirac1933, TenOfAllTrades, Vuo, Cmprince, Pauli133, Gene Nygaard, Lyuokdea, Flying fish, Richard Arthur Norton (1958-), Woohookitty, Swamp Ig, Insaneinside, Benhocking, Nakos2208~enwiki, GregorB, SDC, Joke137, Fxer, Palica, RedBLACKandBURN, Ashmoo, Graham87, Qwertyus, Raymond Hill, Drbogdan, Rjwilmsi, Coemgenus, Strait, John187, Staecker, Jmcc150, Salix alba, Mike Peel, Vegaswikian, Oblivious, Ligulem, R.e.b., Jehochman, The wub, FlaBot, Ian Pitchford, DannyWilde, Itinerant1, RexNL, Gurch, Kolbasz, Fresheneesz, Goudzovski, Sperxios, Scythe33, Smithbrenon, Chobot, Nagytibi, DVdm, YurikBot, Bambaiah, Vuvar1, Phmer, RussBot, Ohwilleke, Arado, Witan, Xihr, Bhny, Chris Capoccia, JabberWok, Gaius Cornelius, Salsb, Grafen, Dlugosz, Gillis, SCZenz, Ravedave, Abb3w, CecilWard, Santaduck, Bota47, Maunus, Dna-webmaster, Ms2ger, Rhynchosaur, Wsiegmund, DrWorm, Alias Flood, Ilmari Karonen, Nimbex, Phr en, Otto ter Haar, Fragman, That Guy, From That Show!, AndrewWTaylor, Palapa, Morgan wascko, SmackBot, Trainbrain27, Reedy, Tom Lougheed, Melchoir, The Monster, Arbe, Mscuthbert, BiT, Ohnoitsjamie, Dauto, Kmarinas86, Marc Kupper, GregRM, Decowski, Deathanatos, Yurigerhard, DHN-bot~enwiki, Sbharris, Colonies Chris, Hengsheng120, Nap~enwiki, Sergio.ballestrero, Милан Јелисавчић, Cophus, Mayrel, Wen D House, Engwar, Jdlambert, Webmaster Pete, TheMaster42, Pwjb, Akriasas, DenisRS, Rjn~enwiki, Kukini, Yevgeny Kats, Ged UK, Jjpcondor, Gobonobo, ThorAvaTahr, JorisvS, Makyen, Libera~enwiki, Aeluwas, Dicklyon, Mets501, MTSbot~enwiki, Galactor213, Fredil Yupigo, Masoninman, Newone, Richard75, Chalnoth, Mssgill, Valoem, Abeneal, Rszasz, CmdrObot, Calmargulis, Olaf Davis, Vyznev Xnebara, MrFizyx, Thubsch, KanysLupys, Myasuda, Alton, Astralusenet, Icek~enwiki, Szdori~enwiki, Hyperdeath, Gogo Dodo, HPaul, RC Master, Q43, Michael C Price, Quibik, Christian75, DumbBOT, Joe Chick, PamD, Thijs!bot, Martin Hogbin, Naucer, Headbomb, Dtgriscom, WVhybrid, Esemono, James086, Second Quantization, Davidhorman, Weasel5i2, Jonny-mt, Hcobb, D.H. Greg L, Mentifisto, Luna Santin, Widefox, Guy Macon, Seaphoto, Alphachimpbot, Astavats, Parande, DagosNavy, JAnDbot, Deflective, MER-C, CosineKitty, Savant13, Magioladitis, WolfmanSF, VoABot II, Nyq, Websterwebfoot, Bakken, DMcanada, Christoph Scholz~enwiki, Seleucus, Dirac66, Adrian J. Hunter, LorenzoB, NJR ZA, Khalid Mahmood, Squidonius, Pavel Jelínek, Gwern, Denis tarasov, Glrx, R'n'B, Fatka, Maurice Carbonaro, MrBell, Aqwis, Salih, Nalumc, Plasticup, Warut, Nwbeeson, Rosenknospe, Juliancolton, Mike Clough, Bonadea, Lseixas, Sheliak, Cuzkatzimhut, Jharris1993, VolkovBot, Camrn86, AlnoktaBOT, DrJohnPCostella, TXiKiBoT, The Original Wildbear, Nxavar, MinotAuruS, Cgr1123, Awl, Michael H 34, HannesHultgren~enwiki, SuperLonghorn, BotKung, SwordSmurf, Norbu19, Richwil, Tomaxer, The assassin 47, Morangm, AlleborgoBot, Thunderbird2, Angelastic, SieBot, Fredelige, PlanetStar, Zelab, Laoris, Ergateesuk, Stratman07, Jerryobject, RadicalOne, Aaarnooo, ScAvenger lv, John fromer, Thehotelambush, ShadowPhox, Ergo4sum, AWeishaupt, Jahilia, Extensive~enwiki, Sagredo, Martarius, ClueBot, Justin W Smith, Aldek, Plastikspork, Apparentslug, Der Golem, Frmorrison, Mild Bill Hic-

Michael C Price, Thijs!bot, Headbomb, Marek69, Hcobb, Len Raymond, SHCarter, David Eppstein, Avirab, LedgendGamer, Maurice Carbonaro, Gill110951, Nwbeeson, Tweisbach, Ross Fraser, TXiKiBoT, Red Act, Maxw41, Wing gundam, Martarius, Agge1000, Leobold1, HarrivBOT, DumZiBoT, Stephen Poppitt, Addbot, Ehrenkater, Ettrig, Traitor, Yobot, Bility, AnomieBOT, Götz, Xqbot, NOrbeck, Paine Ellsworth, Machine Elf 1735, Orenburg1, EmausBot, Dcirovic, Wiggles007, Snotbot, Bibcode Bot, BG19bot, Contact '97, CarrieVS, Doquynh2k7, Monkbot, Nerissamorrow, Jrsousa2, Daimonie, FedezCompa and Anonymous: 55

- **Brans–Dicke theory** *Source:* https://en.wikipedia.org/wiki/Brans%E2%80%93Dicke_theory?oldid=708458291 *Contributors:* The Anome, XJaM, Maury Markowitz, Michael Hardy, Charles Matthews, Greenrd, Xanzzibar, Jason Quinn, FT2, Pjacobi, MuDavid, Dmr2, Pt, Reinyday, John Vandenberg, Foobaz, Count Iblis, Falcorian, Mpatel, Pdn~enwiki, Joke137, Torquil~enwiki, BD2412, Rjwilmsi, Zbxgscqf, MarSch, Ligulem, Ems57fcva, BradBeattie, Chobot, Bgwhite, Hillman, Gaius Cornelius, ErkDemon, RG2, SmackBot, Clpo13, Silly rabbit, Colonies Chris, Syrcro, Ligulembot, Titus III, Garthbarber, Freederick, DogsBreakfast, CmdrObot, Gregbard, Michael C Price, Headbomb, Apon, JaGa, Hair Commodore, Lseixas, Jackfork, Neparis, WurmWoode, General Epitaph, MelonBot, Addbot, DOI bot, Yobot, Anypodetos, AnomieBOT, Citation bot, Armbrust, Robert sides, Steve Quinn, BenzolBot, Maschen, Zueignung, Fluctuating metric, Itai33, Bibcode Bot, Gergely.Szekely, BG19bot, Elenceq, Monkbot, Jerodlycett, Almonaster, I am not really dead and Anonymous: 24

- **General relativity** *Source:* https://en.wikipedia.org/wiki/General_relativity?oldid=720537337 *Contributors:* AxelBoldt, Mav, Bryan Derksen, The Anome, AstroNomer, Ap, RK, Andre Engels, XJaM, Chrislintott, JeLuF, Christian List, William Avery, Roadrunner, Ktsquare, B4hand, Stevertigo, Frecklefoot, Patrick, Boud, Michael Hardy, Menchi, Ixfd64, Bcrowell, Nimrod~enwiki, TakuyaMurata, Mcarling, Minesweeper, Alfio, Looxix~enwiki, ArnoLagrange, Ellywa, Ahoerstemeier, Stevenj, William M. Connolley, Snoyes, Angela, Mark Foskey, Julesd, Salsa Shark, AugPi, Andres, Evercat, Hectorthebat, Hick ninja, A.Tigges~enwiki, Gingekerr, Jitse Niesen, Gutza, Rednblu, Doradus, Wik, Dragons flight, Tero~enwiki, Phys, Shizhao, Elwoz, Jerzy, BenRG, Banno, Northgrove, Phil Boswell, Robbot, Craig Stuntz, Sdedeo, Bvc2000, Goethean, Altenmann, Romanm, Lowellian, Mayooranathan, Gandalf61, Blainster, Diderot, DHN, Hadal, Alba, Johnstone, Fuelbottle, Isopropyl, Xanzzibar, Carnildo, Tea2min, Enochlau, Ancheta Wis, Tosha, Giftlite, JamesMLane, Graeme Bartlett, Mikez, BenFrantzDale, Lethe, Tom harrison, Fropuff, Everyking, Physman, Curps, Michael Devore, Jason Quinn, Alvestrand, SWAdair, Glengarry, Bobblewik, Edcolins, DefLog~enwiki, Pgan002, Knutux, GeneralPatton, HorsePunchKid, Robert Brockway, Kaldari, MadIce, Karol Langner, Rjpetti, Rdsmith4, JimWae, Anythingyouwant, Martin Wisse, Thincat, Euphoria, Icairns, Zfr, AmarChandra, Zondor, Econrad, JimJast, Discospinster, Rich Farmbrough, Guanabot, Pak21, ThomasK, Masudr, Pjacobi, Vsmith, Cdyson37, Jowr, Paul August, SpookyMulder, Dmr2, Bender235, Dcabrilo, Ground, Ben Standeven, Nabla, Livajo, El C, Worldtraveller, Shanes, Etimbo, Causa sui, Bobo192, Robotje, Smalljim, Rbj, JW1805, ParticleMan, I9Q79oL78KiL0QTFHgyc, Mr2001, Matt McIrvin, PWilkinson, Haham hanuka, Schnolle, Varuna, Jumbuck, Jérôme, Alansohn, Hackwrench, Cctoide, Crebbin, Wikidea, SlimVirgin, Benefros, Alexwg, Wtmitchell, Orionix, CloudNine, Bsadowski1, DV8 2XL, LordLoki, HenryLi, Oleg Alexandrov, Kelly Martin, Linas, FeanorStar7, Sabejias, Moneky, Kzollman, Cleonis, Mpatel, Jok2000, Schzmo, Pdn~enwiki, GregorB, Plrk, Wayward, Joke137, Christopher Thomas, Mandarax, Colodia, Canderson7, Rjwilmsi, WCFrancis, MarSch, Eyu100, JoshuacUK, JHMM13, Mike Peel, SanitysEdge, R.e.b., Ems57fcva, Bubba73, Gringo300, Ian Pitchford, RobertG, Mishuletz, Arnero, Mathbot, Nihiltres, Vsion, Perfect Tommy~enwiki, Itinerant1, Alfred Centauri, Gparker, Slant, Carrionluggage, Srleffler, Chobot, DVdm, Bgwhite, Dresdnhope, Manscher, PointedEars, Roboto de Ajvol, YurikBot, Wavelength, Bcarm1185, Splintercellguy, Hillman, EDG, MattWright, RussBot, Loom91, AVM, KSmrq, DanMS, SpuriousQ, Shawn81, Eleassar, Shanel, Syth, Madcoverboy, Tailpig, Schlafly, Dputig07, Beanyk, Tony1, Dna-webmaster, Enormousdude, 2over0, KGasso, Petri Krohn, GraemeL, Rlove, Sambc, LeonardoRob0t, Geoffrey.landis, HereToHelp, Willtron, Caballero1967, Meegs, Bsod2, Finell, Luk, Sardanaphalus, SmackBot, Kurochka, Hydrogen Iodide, Pavlovič, Gnangarra, Unyoyega, Nickst, Delldot, Motorneuron, Cessator, Harald88, Edgar181, Shai-kun, Sectryan, Gilliam, Skizzik, Dauto, Saros136, Silly rabbit, Complexica, Colonies Chris, Zven, Abyssal, RProgrammer, Hve, RedHillian, BentSm, Phaedriel, Khoikhoi, Cybercobra, Downwards, Coolbho3000, Nakon, Peterwhy, SkyWriter, DMacks, Nairebis, Henning Makholm, UncleFester, Bidabadi~enwiki, Byelf2007, SashatoBot, Lambiam, Lapaz, Cronholm144, Gizzakk, CPMcE, JorisvS, Goodnightmush, Ckatz, Frokor, Garthbarber, SirFozzie, SandyGeorgia, Midnightblueowl, RichardF, Novangelis, Peter Horn, MTSbot~enwiki, Kvng, JarahE, Licorne, Quaeler, Fan-1967, Editor.singapore, MFago, JoeBot, ShyK, MOBle, RekishiEJ, CapitalR, MD:astronomer, Courcelles, Tawkerbot2, JRSpriggs, Kurtan~enwiki, Harold f, JForget, Sakurambo, Thermochap, Avanu, NickW557, MarsRover, Harrigan, Ian Beynon, Cydebot, Jasperdoomen, WillowW, Fl, MC10, Mato, Pascal.Tesson, Michael C Price, Christian75, DumbBOT, Biblbroks, Omicronpersei8, Crum375, N. Macchiavelli, Epbr123, Fisherjs, Markus Pössel, Martin Hogbin, MrXow, Oliver202, Headbomb, Pjvpjv, Tom Barlow, Davidhorman, D.H, AntiVandalBot, Abu-Fool Danyal ibn Amir al-Makhiri, Tkirkman, Gnixon, VectorPosse, TimVickers, Scepia, Dawz, Billevans~enwiki, Tim Shuba, Rico402, Archmagusrm, Jaredroberts, JAnDbot, Vorpal blade, Hut 8.5, YK Times, Acroterion, Pervect, Magioladitis, Connormah, RogierBrussee, WolfmanSF, JamesBWatson, Swpb, Ling.Nut, Soulbot, Pixel ;-), KConWiki, WhatamIdoing, BatteryIncluded, Eldumpo, Allstarecho, User A1, Mollwollfumble, Chris G, Archen~enwiki, Thompson.matthew, STBot, Mermaid from the Baltic Sea, Shentino, Mschel, CommonsDelinker, Pbroks13, J.delanoy, DrKay, R. Baley, Numbo3, Leafsfan85, Aveh8, Lantonov, M C Y 1008, Mathlabster, Zedmelon, Aboutmovies, C quest000, Tcisco, Marrilpet, Nwbeeson, Aatomic1, Potatoswatter, Kolja21, Lseixas, Rémih, Caracalocelot, DemonicInfluence, Sheliak, Deor, Part Deux, JohnBlackburne, Philip Trueman, TXiKiBoT, Oshwah, Coder Dan, GimmeBot, Gombo, Hqb, Rei-bot, IPSOS, Qxz, T doffing, Molinogi, Fizzackerly, JhsBot, Leafyplant, Geometry guy, Ilyushka88, Thebigbendizzle, SwordSmurf, Andy Dingley, Gabrielsleitao, Lamro, Antixt, Vector Potential, James-Chin, Arcfrk, Ccheese4, StevenJohnston, Katzmik, YohanN7, Dnarby, SieBot, Tiddly Tom, Work permit, Yintan, RadicalOne, Wizzard2k, SteakNShake, Arbor to SJ, Babareddeer, JSpung, Phil Bridger, Wmpearl, Oxymoron83, Henry Delforn (old), Csloomis, Thehotelambush, Lightmouse, BrightRoundCircle, OpTioNiGhT, The-G-Unit-Boss, Emgg, AWeishaupt, Divinestuff, Coldcreation, Adam Cuerden, Duae Quartunciae, Heptarchy of teh Anglo-Saxons, baby, Randomblue, TFCforever, Danthewhale, Martarius, Sfan00 IMG, ClueBot, The Thing That Should Not Be, Rjd0060, Metaprimer, Wwheaton, Der Golem, JTBX, TheAmigo42, CounterVandalismBot, Viran, Blanchardb, Rotational, Agge1000, Itzguru, Tanketz, CohesionBot, Eeekster, Stealth500, Brews ohare, NuclearWarfare, PhySusie, SockPuppetForTomruen, SchreiberBike, Another Believer, RubenGarciaHernandez, AC+79 3888, MasterOfHisOwnDomain, He6kd, TimothyRias, Lazyrussian, PseudoOne, Skarebo, NellieBly, JinJian, Truthnlove, Everydayidiot, Tayste, Balungifrancis, Addbot, Mortense, Some jerk on the Internet, Fizzycyst, DOI bot, Mistyocean3, Metagraph, Stariki, Fluffernutter, Schmoolik, MrOllie, Download, EconoPhysicist, Delaszk, Favonian, LinkFA-Bot, Tuition, Tassedethe, Nnedass, Tide rolls, Lightbot, Knutls, Luckas-bot, Ptbotgourou, Legobot II, Julia W, Anypodetos, Trickyboarder93, Superamoeba, AnomieBOT, Kristen Eriksen, Giordano.ferdinandi, Jim1138, Jo3sampl, Materialscientist, Wandering Courier, The High Fin Sperm Whale, Citation bot, Xqbot, Stlwebs, Sionus, Amareto2, Unigfjkl, Nickkid5, Stsang, Coretheapple, GrouchoBot, Collin21594, RibotBOT, Rucko123, GhalyBot, Acannas, LucienBOT, Paine Ellsworth, Lagelspeil, Steve Quinn, Knowandgive, Pokyrek, Citation bot 1, Citation bot 4, Electrozity8, Pinethicket, LittleWink, Jonesey95, A412, Tom.Reding, Yougeeaw, Barras, Jauhienij, Meier99, Citator, Comet Tuttle, Hughston, Defender of torch, Duoduoduo, Aribashka, Iibbmm, Diannaa, Earth-

Masterpiece2000, Canis Lupus, EverettYou, AnonyScientist, Albambot, Addbot, Gravitophoton, Protonk, Lightbot, Matěj Grabovský, Yobot, Turul2, Jo3sampl, Citation bot, Omnipaedista, RibotBOT, Paine Ellsworth, Quiden711, Tom.Reding, RockSolidCosmo, Crabhiggins, Bj norge, David.c.stone, Arbnos, Quondum, TonyMath, Helpful Pixie Bot, Bibcode Bot, BG19bot, Zerothat, Ownedroad9, Metsfreak2121, MSUGRA, Mogism, Lianatajo, MuonRay, Orderofmagnitudeapproximation, Frinthruit, Monkbot, ManitouLance, Claudio Orzalesi and Anonymous: 87

- **String theory** *Source:* https://en.wikipedia.org/wiki/String_theory?oldid=720334602 *Contributors:* AxelBoldt, Sodium, Mav, Bryan Derksen, Zundark, The Anome, Tarquin, Taw, Eean, Malcolm Farmer, Hephaestos, Olivier, Drseudo, Stevertigo, Spiff~enwiki, Edward, PhilipMW, Michael Hardy, Bewildebeast, Dante Alighieri, Gabbe, Graue, Tgeorgescu, Mcarling, CesarB, Looxix~enwiki, Ahoerstemeier, Theresa knott, Suisui, Angela, Den fjättrade ankan~enwiki, Jdforrester, Julesd, Salsa Shark, Schneelocke, Charles Matthews, Timwi, Bemoeial, Jitse Niesen, 4lex, Greenrd, ErikStewart, Furrykef, Saltine, Phys, Omegatron, Bevo, Topbanana, Trent, Nufy8, Robbot, Craig Stuntz, Fredrik, Chris 73, R3m0t, COGDEN, Mirv, Wjhonson, Sverdrup, Academic Challenger, DHN, Hadal, Khlo, ElBenevolente, HaeB, Xanzzibar, Tea2min, Giftlite, DocWatson42, Christopher Parham, Awolf002, Mporter, Amorim Parga, Mikez, Harp, Kim Bruning, Tom harrison, Ferkelparade, Leflyman, Fropuff, No Guru, Anville, Moyogo, Curps, Pashute, Nomad~enwiki, Mboverload, Solipsist, SWAdair, DemonThing, Wmahan, Btphelps, MSTCrow, Decoy, Chowbok, Gadfium, Steuard, Pgan002, Quadell, Carandol~enwiki, Antandrus, Beland, JoJan, Khaosworks, Tothebarricades.tk, Thincat, Tomruen, Shidobu, Icairns, Lumidek, NoPetrol, Avihu, Fanghong~enwiki, Trevor MacInnis, Lacrimosus, Zro, Mike Rosoft, D6, Urvabara, Felix Wan, Jkl, Discospinster, ElTyrant, Rich Farmbrough, Rhobite, Pjacobi, Alien life form, Vapour, Silence, Kzzl, LindsayH, Mani1, Pavel Vozenilek, Paul August, Bender235, Kjoonlee, Mashford, Kelvinc, Perlman10s, Panu~enwiki, Brian0918, Dpotter, Livajo, El C, Laurascudder, Shanes, Zegoma beach, RoyBoy, Causa sui, Bobo192, Directorstratton, Janna Isabot, Smalljim, John Vandenberg, Flxmghvgvk, I9Q79oL78KiL0QTFHgyc, Physicistjedi, Bongoo, 4v4l0n42, Merope, Geschichte, Linuxlad, Phils, Merenta, Alansohn, Gary, JYolkowski, Enirac Sum, Ryanmcdaniel, Arthena, Borisblue, Rd232, Plumbago, Axl, R Calvete, Lightdarkness, Kocio, Bart133, Wtmitchell, Isaac, Tycho, Cal 1234, Fadereu, CloudNine, Sciurinæ, Computerjoe, Kusma, DV8 2XL, Pwqn, Gene Nygaard, Ringbang, Ceyockey, Falcorian, Bobrayner, Joriki, Mel Etitis, Linas, BillC, Jacobolus, HFarmer, Before My Ken, Netdragon, MONGO, GeorgeOrr, Mpatel, Bbatsell, GregorB, 慕尼黑啤酒, Joke137, Christopher Thomas, Dysepsion, GSlicer, Jan.bannister, Graham87, Magister Mathematicae, Hillbrand, BD2412, Elvey, Galwhaa, Raymond Hill, JIP, RxS, Athelwulf, Edison, Sjakkalle, Rjwilmsi, Xgamer4, Jake Wartenberg, Arabani, MarSch, TheRingess, Jmcc150, Aero66, Crazynas, Juan Marquez, R.e.b., Bubba73, DoubleBlue, Zelos, AlisonW, Asafavi, Lionelbrits, Conorific, Zunz, Mathbot, Crazycomputers, RexNL, Gurch, Algri, TeaDrinker, Zifnabxar, XAXISx, Erik4, Phoenix2~enwiki, Antimatter15, Ggb667, Chobot, Visor, DVdm, Mhking, VolatileChemical, Bgwhite, Algebraist, Ben Tibbetts, YurikBot, Ugha, Wavelength, Borgx, NuclearFusion~enwiki, Angus Lepper, Hairy Dude, Jimp, Hillman, Cyferx, Wolfmankurd, Pip2andahalf, RussBot, Moronoman, Crazytales, Pippo2001, Bhny, Pigman, SpuriousQ, Branman515, Stephenb, Gaius Cornelius, Eleassar, Rsrikanth05, Bovineone, Cheesus, Shanel, NawlinWiki, Tong~enwiki, Mike18xx, SCZenz, Cleared as filed, Bdiah, Pym98, SColombo, Haemo, FF2010, Closedmouth, Reyk, Brina700, Chris Brennan, Vicarious, Brianlucas, Geoffrey.landis, Hitchhiker89, Spliffy, Pred, ArielGold, Roy Fultun, Ilmari Karonen, Katieh5584, Pentasyllabic, Lunch, DVD R W, WikiFew, That Guy, From That Show!, Street Scholar, AndrewWTaylor, QSquared, Sardanaphalus, Vanka5, MacsBug, Hvitlys, SmackBot, Kurochka, Zazaban, Tom Lougheed, Prodego, KnowledgeOfSelf, Hydrogen Iodide, Melchoir, Vald, Skrewtape, Atomota, Canthusus, GaeusOctavius, Cool3, Andyvn22, Gilliam, Skizzik, RobertM525, Dauto, Bluebot, SSJ 5, Keegan, Aidan Croft, Thumperward, Oli Filth, Silly rabbit, Timneu22, SchfiftyThree, Moshe Constantine Hassan Al-Silverburg, Complexica, Rediahs, RayAYang, Aero77, Adamstevenson, Ikiroid, Epastore, Baronnet, Ned Scott, Sbharris, Colonies Chris, Konstable, Sct72, Scwlong, Can't sleep, clown will eat me, Timothy Clemans, Onorem, Neilanderson, EvelinaB, TKD, KerathFreeman, Addshore, UU, The tooth, Pepsidrinka, Somebody2292, --=The Doctor=--, Fuhghettaboutit, Cybercobra, Irish Souffle, Nakon, Jdlambert, James McNally, MichaelBillington, Lostart, Insineratehymn, Drphilharmonic, SpiderJon, DMacks, Ihatetoregister, Where, Michael IFA, Yevgeny Kats, Vasiliy Faronov, Byelf2007, Angela26, Visium, Rory096, Zymurgy, Harryboyles, Mdl53711, T-dot, Titus III, Ergative rlt, MagnaMopus, UberCryxic, Vgy7ujm, Lazylaces, Linnell, Mgiganteus1, Nonsuch, IronGargoyle, Ckatz, DoItAgain, AstroGod, Kirbytime, Jimbo Mahoney, FredrickS, Invisifan, Ryulong, Ryanjunk, MathStuf, Mike Doughney, Norm mit, Hindol, Dan Gluck, Huntscorpio, Iridescent, K, Sunoco, You? Me? Us?, CzarB, Rabinzkaman, JoeBot, Lottamiata, Tony Fox, Vrkaul, Torrazzo, Gil Gamesh, Areldyb, Courcelles, Tawkerbot2, Gebrah, Shamvil, Fdssdf, DKqwerty, Lbr123, Harold f, Heqs, Devourer09, Duduong, Sarvagnya, Dewayne76, JForget, Cg-realms, InvisibleK, CRGreathouse, CmdrObot, Earthlyreason, Van helsing, Olaf Davis, CBM, Rawling, Jibal, Witten Is God, Nunquam Dormio, Giko, KnightLago, Thubsch, Leujohn, SlashDot, TheTito, Karenjc, Myasuda, Emarv, Cydebot, Gmusser, Gogo Dodo, Jkokavec, Kahananite, Quajafrie, Michael C Price, Doug Weller, DumbBOT, Narayanese, AlphaNumeric, SRoughsedge, Vanished User jdksfajlasd, Woland37, Zalgo, Daniel Olsen, UberScienceNerd, Bkazaz, DJBullfish, Thijs!bot, Epbr123, Rwmnau, Babemachine, Pimpin101, Mbell, O, Faigl.ladislav, Ucanlookitup, Andyjsmith, Headbomb, Tcturner2002, Marek69, Brahmajnani, Arthurcprado~enwiki, Y.t., D3gtrd, Babemonkey, Dark dude, Duncan McB, EdJohnston, MichaelMaggs, Ancientanubis, Natalie Erin, Hempfel, Jomoal99, Mmortal03, Mentifisto, Geekdom04, AntiVandalBot, Luna Santin, Seaphoto, Ed270791, Opelio, Doc Tropics, David136a, NithinBekal, Dotdotdotdash, Helicoptor, Poshzombie, MontanNito, Dylan Lake, Maximilian77, Shlomi Hillel, Db63376, SamIAmNot, Knotwork, Res2216firestar, Superior IQ Genius, MER-C, Andonic, Sitethief, 100110100, TallulahBelle, Nestamachine, Savant13, Daynightrader, Goldenglove, Charibdis, Acroterion, Ophion, Aigisthos, Editmyhandman, Aruben537, Magioladitis, WolfmanSF, Bongwarrior, VoABot II, Yandman, JamesBWatson, باسم, Qutt, Jespinos, Kevinmon, Aka042, Froid, DAGwyn, Catgut, Panser Born, Ensign beedrill, Perspectival, JJ Harrison, Dirac66, Justanother, Aziz1005, Cpl Syx, ChazBeckett, Teardrop onthefire, WLU, Stephen shenker, Robin S, SkepticVK, Joshua Davis, Mkroh, B9 hummingbird hovering, S3000, Hdt83, MartinBot, FlieGerFaUstMe262, Ytomem, Shimwell, Arjun01, KrishSundaresan, Anaxial, Jay Litman, Alexcalamaro, Andrej.westermann, Smokizzy, LedgendGamer, Cyrus Andiron, Peteryoung144, Tgeairn, Artaxiad, HEL, AlphaEta, J.delanoy, AstroHurricane001, Maurice Carbonaro, Yonidebot, Morris729, M C Y 1008, 69gangsta420, It Is Me Here, Shawn in Montreal, Janus Shadowsong, Bailo26, Fredsie, Madagaskar07, Duchesserin, AntiSpamBot, CHIAGEHYANG, Chiswick Chap, Watsup1313, Belovedfreak, HaloInverse, NewEnglandYankee, Scott1329m, Thesis4Eva, Policron, Jrcla2, KylieTastic, WJBscribe, Rnricklefs, Jamesofur, Eyelidlessness, Jonnyk aus, Kvdveer, JavierMC, Izno, Xiahou, CardinalDan, Sheliak, HamatoKameko, Malik Shabazz, Concertmusic, JohnBlackburne, JustinHagstrom, Fences and windows, Wooba doob, Philip Trueman, DoorsAjar, HowardFrampton, TXiKiBoT, Zidonuke, Red Act, Kriak, Calwiki, Technopat, Hqb, Andrius.v, Anonymous Dissident, Crohnie, AlysTarr, Qxz, Vanished user ikijeirw34iuaeolaseriffic, Impunv, Seraphim, Martin451, Don4of4, ABigGreenHippo, Huperphuff, LeaveSleaves, Kaenneth, StringyGuy, Maxim, Erth64net, Meters, Lamro, Rickstauduhar, Enviroboy, Turgan, Anna512, PhysPhD, Northfox, NPguy, Matthew Sanders, Luke Walkerson, Newbyguesses, MissMJ, SieBot, Escher26, J.A.Ireland, BA (IHPST), 4wajzkd02, Robdunst, Dreamafter, Pallab1234, Dbelange, MTHarden, Lemonflash, Kylemew, Yintan, GlassCobra, Discrete, Bentogoa, Likebox, Flyer22 Reborn, Exert, ProGeek314, Arbor to SJ, Babawhitemoose, Caidh, Dhatfield, Audree, Oxymoron83, Pretty Green, Weaselstomp, Manway, Alex.muller, Taco Manipulator, Tschach, Manheat84, Anchor Link Bot, Mikebernstein, ImperialismGo, Nergaal, Ionfield, Ayleuss, Sh4wz0r, Naturespace, ImageRemovalBot, Martarius, Phyte, ClueBot, The Thing That Should Not Be, String4d, Illusion96, Polyamorph, Mpd1989, Alexdeburca18, Wiggl3sLimited, Excirial,

Kzl.zawlin, Isocliff, ClueBot NG, Raidr, Helpful Pixie Bot, Bibcode Bot, Gordonben, Bmusician, Snow Rise, CitationCleanerBot, Shirtbrigade, Caypartisbot, BattyBot, Jimw338, Damrvrhunter, Epicgenius, Christophe1946, Polytope24, JaconaFrere, Almaionescu, Monkbot, Stringer63, Hardkhora, Atreus57, AnimikhRoy967, Toes1111111231111231111123, Nøkkenbuer, Christos Theopoulos, Npmats, The 1editr, Universal-Dude0.27 and Anonymous: 128

41.11.2 Images

- **File:1919_eclipse_negative.jpg** *Source:* https://upload.wikimedia.org/wikipedia/commons/d/da/1919_eclipse_negative.jpg *License:* Public domain *Contributors:* F. W. Dyson, A. S. Eddington, and C. Davidson, "A Determination of the Deflection of Light by the Sun's Gravitational Field, from Observations Made at the Total Eclipse of May 29, 1919" *Philosophical Transactions of the Royal Society of London. Series A. Containing Papers of a Mathematical or Physical Character* (1920): 291-333, on 332. *Original artist:* F. W. Dyson, A. S. Eddington, and C. Davidson

- **File:1e0657_scale.jpg** *Source:* https://upload.wikimedia.org/wikipedia/commons/a/a8/1e0657_scale.jpg *License:* Public domain *Contributors:* Chandra X-Ray Observatory: 1E 0657-56 *Original artist:* NASA/CXC/M. Weiss

- **File:ALSEP_AS15-85-11468.jpg** *Source:* https://upload.wikimedia.org/wikipedia/commons/2/21/ALSEP_AS15-85-11468.jpg *License:* Public domain *Contributors:* ? *Original artist:* ?

- **File:AdS3.svg** *Source:* https://upload.wikimedia.org/wikipedia/commons/4/47/AdS3.svg *License:* CC BY-SA 3.0 *Contributors:* This file was derived from: AdS3 (new).png
Original artist:

- derivative work: Alex Dunkel (Maky)

- **File:Albert_Einstein_portrait.jpg** *Source:* https://upload.wikimedia.org/wikipedia/en/f/f7/Albert_Einstein_portrait.jpg *License:* PD-US *Contributors:*

  http://images.google.com/hosted/life/628e99cf2e26233d.html *Original artist:*
  E. O. Hoppe. (1878-1972) Published on LIFE

- **File:Ambox_current_red.svg** *Source:* https://upload.wikimedia.org/wikipedia/commons/9/98/Ambox_current_red.svg *License:* CC0 *Contributors:* self-made, inspired by Gnome globe current event.svg, using Information icon3.svg and Earth clip art.svg *Original artist:* Vipersnake151, penubag, Tkgd2007 (clock)

- **File:Ambox_important.svg** *Source:* https://upload.wikimedia.org/wikipedia/commons/b/b4/Ambox_important.svg *License:* Public domain *Contributors:* Own work, based off of Image:Ambox scales.svg *Original artist:* Dsmurat (talk · contribs)

- **File:Apollo_11_Lunar_Laser_Ranging_Experiment.jpg** *Source:* https://upload.wikimedia.org/wikipedia/commons/0/08/Apollo_11_Lunar_Laser_Ranging_Experiment.jpg *License:* Public domain *Contributors:* NASA Apollo Archive http://www.hq.nasa.gov/office/pao/History/alsj/a11/AS11-40-5952.jpg *Original artist:* NASA

- **File:Apollo_15_feather_and_hammer_drop.ogg** *Source:* https://upload.wikimedia.org/wikipedia/commons/3/3c/Apollo_15_feather_and_hammer_drop.ogg *License:* Public domain *Contributors:* Taken from Spacecraftfilms.com DVD "Apollo 15: The Great Explorations Begin" *Original artist:* NASA

- **File:Artist's_impression_of_the_expected_dark_matter_distribution_around_the_Milky_Way.ogv** *Source:* https://upload.wikimedia.org/wikipedia/commons/0/03/Artist%E2%80%99s_impression_of_the_expected_dark_matter_distribution_around_the_Milky_Way.ogv *License:* CC BY 4.0 *Contributors:* ESO *Original artist:* ESO/L. Calçada

- **File:Artist's_impression_of_the_pulsar_PSR_J0348+0432_and_its_white_dwarf_companion.jpg** *Source:* https://upload.wikimedia.org/wikipedia/commons/2/26/Artist%E2%80%99s_impression_of_the_pulsar_PSR_J0348%2B0432_and_its_white_dwarf_companion.jpg *License:* CC BY 4.0 *Contributors:* http://www.eso.org/public/images/eso1319c/ *Original artist:* ESO/L. Calçada

- **File:Asymmetric_Ashes_(artist'[}s_impression).jpg** *Source:* https://upload.wikimedia.org/wikipedia/commons/d/db/Asymmetric_Ashes_%28artist%27s_impression%29.jpg *License:* CC BY 4.0 *Contributors:* http://www.eso.org/public/images/eso0644a/ *Original artist:* ESO

- **File:BBH_gravitational_lensing_of_gw150914.webm** *Source:* https://upload.wikimedia.org/wikipedia/commons/a/a4/BBH_gravitational_lensing_of_gw150914.webm *License:* CC BY-SA 4.0 *Contributors:* https://www.ligo.caltech.edu/video/ligo20160211v3 (video link); see also http://www.black-holes.org/gw150914 *Original artist:* Simulating eXtreme Spacetimes Lensing

- **File:BH_LMC.png** *Source:* https://upload.wikimedia.org/wikipedia/commons/5/5e/BH_LMC.png *License:* CC BY-SA 2.5 *Contributors:* Own work *Original artist:* User:Alain r

- **File:Baryon-decuplet-small.svg** *Source:* https://upload.wikimedia.org/wikipedia/commons/7/78/Baryon-decuplet-small.svg *License:* Public domain *Contributors:* Own work *Original artist:* Trassiorf

- **File:Baryon-octet-small.svg** *Source:* https://upload.wikimedia.org/wikipedia/commons/b/b5/Baryon-octet-small.svg *License:* Public domain *Contributors:* Own work *Original artist:* Trassiorf

- **File:Baryon_decuplet_w_mass.png** *Source:* https://upload.wikimedia.org/wikipedia/en/c/c1/Baryon_decuplet_w_mass.png *License:* PD *Contributors:*
  self-made
  Original artist:
  Venny85 (talk)

- **File:Edit-clear.svg** *Source:* https://upload.wikimedia.org/wikipedia/en/f/f2/Edit-clear.svg *License:* Public domain *Contributors:* The *Tango! Desktop Project. Original artist:*

  The people from the Tango! project. And according to the meta-data in the file, specifically: "Andreas Nilsson, and Jakub Steiner (although minimally)."

- **File:Edward_Witten.jpg** *Source:* https://upload.wikimedia.org/wikipedia/commons/9/97/Edward_Witten.jpg *License:* Public domain *Contributors:* Own work *Original artist:* Ojan

- **File:Eg1.png** *Source:* https://upload.wikimedia.org/wikipedia/en/b/be/Eg1.png *License:* PD *Contributors:*

  self-made

  Original artist:

  Venny85 (talk)

- **File:Eg2.png** *Source:* https://upload.wikimedia.org/wikipedia/en/3/3e/Eg2.png *License:* PD *Contributors:*

  self-made

  Original artist:

  Venny85 (talk)

- **File:Eg3.png** *Source:* https://upload.wikimedia.org/wikipedia/en/1/15/Eg3.png *License:* PD *Contributors:*

  self-made

  Original artist:

  Venny85 (talk)

- **File:Eg4.png** *Source:* https://upload.wikimedia.org/wikipedia/en/6/65/Eg4.png *License:* PD *Contributors:*

  self-made

  Original artist:

  Venny85 (talk)

- **File:Einstein_cross.jpg** *Source:* https://upload.wikimedia.org/wikipedia/commons/c/c8/Einstein_cross.jpg *License:* Public domain *Contributors:* http://hubblesite.org/newscenter/archive/releases/1990/20/image/a/ *Original artist:* NASA, ESA, and STScI

- **File:Electron_self_energy.svg** *Source:* https://upload.wikimedia.org/wikipedia/commons/e/e4/Electron_self_energy.svg *License:* Public domain *Contributors:* Transferred from en.wikipedia to Commons by Sreejithk2000 using CommonsHelper. *Original artist:* DnetSvg at English Wikipedia

- **File:Elementary_particle_interactions_in_the_Standard_Model.png** *Source:* https://upload.wikimedia.org/wikipedia/commons/a/a7/Elementary_particle_interactions_in_the_Standard_Model.png *License:* CC0 *Contributors:* Own work *Original artist:* Eric Drexler

- **File:Elevator_gravity.svg** *Source:* https://upload.wikimedia.org/wikipedia/commons/1/11/Elevator_gravity.svg *License:* CC BY-SA 3.0 *Contributors:*

- Elevator_gravity2.png *Original artist:*

- derivative work: Pbroks13 (talk)

- **File:Ergosphere.svg** *Source:* https://upload.wikimedia.org/wikipedia/commons/0/0c/Ergosphere.svg *License:* CC-BY-SA-3.0 *Contributors:* own work based on the graphic uploaded by IMeowbot *Original artist:* MesserWoland

- **File:Fermi_Observations_of_Dwarf_Galaxies_Provide_New_Insights_on_Dark_Matter.ogv** *Source:* https://upload.wikimedia.org/wikipedia/commons/a/a9/Fermi_Observations_of_Dwarf_Galaxies_Provide_New_Insights_on_Dark_Matter.ogv *License:* Public domain *Contributors:* Goddard Multimedia *Original artist:* NASA/Goddard Space Flight Center

- **File:FirstNeutrinoEventAnnotated.jpg** *Source:* https://upload.wikimedia.org/wikipedia/commons/5/57/FirstNeutrinoEventAnnotated.jpg *License:* Public domain *Contributors:* Image courtesy of Argonne National Laboratory *Original artist:* Argonne National Laboratory

- **File:Folder_Hexagonal_Icon.svg** *Source:* https://upload.wikimedia.org/wikipedia/en/4/48/Folder_Hexagonal_Icon.svg *License:* Cc-by-sa-3.0 *Contributors:* ? *Original artist:* ?

- **File:FullMoon2010.jpg** *Source:* https://upload.wikimedia.org/wikipedia/commons/e/e1/FullMoon2010.jpg *License:* CC BY-SA 3.0 *Contributors:* Own work *Original artist:* Gregory H. Revera

- **File:GPB_circling_earth.jpg** *Source:* https://upload.wikimedia.org/wikipedia/commons/d/d1/GPB_circling_earth.jpg *License:* Public domain *Contributors:* http://www.nasa.gov/mission_pages/gpb/gpb_012.html *Original artist:* NASA

- **File:GabrieleVeneziano.jpg** *Source:* https://upload.wikimedia.org/wikipedia/commons/9/95/GabrieleVeneziano.jpg *License:* CC BY-SA 2.5 *Contributors:* Taken by Betsythedevine *Original artist:* The original uploader was Betsythedevine at English Wikipedia

- **File:GalacticRotation2.svg** *Source:* https://upload.wikimedia.org/wikipedia/commons/b/b9/GalacticRotation2.svg *License:* CC-BY-SA-3.0 *Contributors:* Own work in Inkscape 0.42 *Original artist:* PhilHibbs

- **File:Gnome-searchtool.svg** *Source:* https://upload.wikimedia.org/wikipedia/commons/1/1e/Gnome-searchtool.svg *License:* LGPL *Contributors:* http://ftp.gnome.org/pub/GNOME/sources/gnome-themes-extras/0.9/gnome-themes-extras-0.9.0.tar.gz *Original artist:* David Vignoni

- **File:Gravitational-wave_detector_sensitivities_and_astrophysical_gravitational-wave_sources.png** *Source:* https://upload.wikimedia.org/wikipedia/commons/a/af/Gravitational-wave_detector_sensitivities_and_astrophysical_gravitational-wave_sources.png *License:* CC BY-SA 1.0 *Contributors:* http://rhcole.com/apps/GWplotter/ *Original artist:* Christopher Moore, Robert Cole and Christopher Berry

- **File:Gravitational_red-shifting.png** *Source:* https://upload.wikimedia.org/wikipedia/commons/5/5c/Gravitational_red-shifting.png *License:* CC-BY-SA-3.0 *Contributors:* ? *Original artist:* ?

- **File:Merge-arrows.svg** *Source:* https://upload.wikimedia.org/wikipedia/commons/5/52/Merge-arrows.svg *License:* Public domain *Contributors:* ? *Original artist:* ?

- **File:Milgrom_Mordechai.jpg** *Source:* https://upload.wikimedia.org/wikipedia/commons/b/bb/Milgrom_Mordechai.jpg *License:* Public domain *Contributors:* Weizmann Institute of Science *Original artist:* Weizmann Institute of Science

- **File:Momentum_exchange.svg** *Source:* https://upload.wikimedia.org/wikipedia/commons/1/1b/Momentum_exchange.svg *License:* CC0 *Contributors:* Own work *Original artist:* Krishnavedala

- **File:MontreGousset001.jpg** *Source:* https://upload.wikimedia.org/wikipedia/commons/4/45/MontreGousset001.jpg *License:* CC-BY-SA-3.0 *Contributors:* Self-published work by ZA *Original artist:* Isabelle Grosjean ZA

- **File:Moon-Mdf-2005.jpg** *Source:* https://upload.wikimedia.org/wikipedia/commons/4/4d/Moon-Mdf-2005.jpg *License:* CC-BY-SA-3.0 *Contributors:* No machine-readable source provided. Own work assumed (based on copyright claims). *Original artist:* No machine-readable author provided. WikedKentaur assumed (based on copyright claims).

- **File:NewtonsLawOfUniversalGravitation.svg** *Source:* https://upload.wikimedia.org/wikipedia/commons/0/0e/ NewtonsLawOfUniversalGravitation.svg *License:* CC BY 3.0 *Contributors:* Self-made by User:Dna-Dennis *Original artist:* User:Dna-Dennis

- **File:Noneto_mesónico_de_spin_0.png** *Source:* https://upload.wikimedia.org/wikipedia/commons/c/cd/Noneto_mes%C3%B4nico_de_ spin_0.png *License:* Public domain *Contributors:* ? *Original artist:* ?

- **File:Nuvola_apps_edu_mathematics_blue-p.svg** *Source:* https://upload.wikimedia.org/wikipedia/commons/3/3e/Nuvola_apps_edu_ mathematics_blue-p.svg *License:* GPL *Contributors:* Derivative work from Image:Nuvola apps edu mathematics.png and Image:Nuvola apps edu mathematics-p.svg *Original artist:* David Vignoni (original icon); Flamurai (SVG convertion); bayo (color)

- **File:Nuvola_apps_kalzium.svg** *Source:* https://upload.wikimedia.org/wikipedia/commons/8/8b/Nuvola_apps_kalzium.svg *License:* LGPL *Contributors:* Own work *Original artist:* David Vignoni, SVG version by Bobarino

- **File:Office-book.svg** *Source:* https://upload.wikimedia.org/wikipedia/commons/a/a8/Office-book.svg *License:* Public domain *Contributors:* This and myself. *Original artist:* Chris Down/Tango project

- **File:Open_and_closed_strings.svg** *Source:* https://upload.wikimedia.org/wikipedia/commons/5/56/Open_and_closed_strings.svg *License:* Public domain *Contributors:* Own work *Original artist:* Xoneca

- **File:Particle_overview.svg** *Source:* https://upload.wikimedia.org/wikipedia/commons/7/7f/Particle_overview.svg *License:* CC BY-SA 3.0 *Contributors:* Own work *Original artist:* Headbomb

- **File:Penrose.svg** *Source:* https://upload.wikimedia.org/wikipedia/commons/a/a8/Penrose.svg *License:* Public domain *Contributors:* Transferred from en.wikipedia to Commons by Andrei Stroe using CommonsHelper. *Original artist:* Cronholm144 at English Wikipedia

- **File:Portal-puzzle.svg** *Source:* https://upload.wikimedia.org/wikipedia/en/f/fd/Portal-puzzle.svg *License:* Public domain *Contributors:* ? *Original artist:* ?

- **File:PositronDiscovery.jpg** *Source:* https://upload.wikimedia.org/wikipedia/commons/6/69/PositronDiscovery.jpg *License:* Public domain *Contributors:* Anderson, Carl D. (1933). "The Positive Electron". *Physical Review* 43 (6): 491–494. DOI:10.1103/PhysRev.43.491. *Original artist:* Carl D. Anderson (1905–1991)

- **File:Proton_proton_cycle.svg** *Source:* https://upload.wikimedia.org/wikipedia/commons/a/ac/Proton_proton_cycle.svg *License:* CC BY-SA 2.5 *Contributors:* file:Proton proton cycle.png *Original artist:* Dorottya Szam

- **File:Psr1913+16-weisberg_en.png** *Source:* https://upload.wikimedia.org/wikipedia/commons/7/79/Psr1913%2B16-weisberg_en.png *License:* Public domain *Contributors:* M. Haynes et Lorimer (2001) (redrawn by Dantor as Image:Psr1913+16-weisberg.png, English labels added by mapos) *Original artist:* ?

- **File:Question_book-new.svg** *Source:* https://upload.wikimedia.org/wikipedia/en/9/99/Question_book-new.svg *License:* Cc-by-sa-3.0 *Contributors:*
Created from scratch in Adobe Illustrator. Based on Image:Question book.png created by User:Equazcion *Original artist:*
Tkgd2007

- **File:Question_dropshade.png** *Source:* https://upload.wikimedia.org/wikipedia/commons/d/dd/Question_dropshade.png *License:* Public domain *Contributors:* Image created by JRM *Original artist:* JRM

- **File:Recording_data_at_each_of_the_telescopes_in_a_VLBI_array.gif** *Source:* https://upload.wikimedia.org/wikipedia/en/3/3a/ Recording_data_at_each_of_the_telescopes_in_a_VLBI_array.gif *License:* PD *Contributors:* ? *Original artist:* ?

- **File:Relativistic_precession.svg** *Source:* https://upload.wikimedia.org/wikipedia/commons/2/28/Relativistic_precession.svg *License:* CC-BY-SA-3.0 *Contributors:* Own work, self-made using gnuplot with manual alterations *Original artist:* KSmrq

- **File:SN1994D.jpg** *Source:* https://upload.wikimedia.org/wikipedia/commons/a/a2/SN1994D.jpg *License:* CC BY 3.0 *Contributors:* http://www.spacetelescope.org/images/opo9919i/ *Original artist:* NASA/ESA, The Hubble Key Project Team and The High-Z Supernova Search Team

- **File:Scalar_Field.ogv** *Source:* https://upload.wikimedia.org/wikipedia/commons/0/01/Scalar_Field.ogv *License:* CC BY-SA 3.0 *Contributors:* Own work *Original artist:* Quintus

- **File:Scalar_field.png** *Source:* https://upload.wikimedia.org/wikipedia/commons/a/a8/Scalar_field.png *License:* Public domain *Contributors:* Own work *Original artist:* Lucas V. Barbosa

- **File:Science.jpg** *Source:* https://upload.wikimedia.org/wikipedia/commons/5/54/Science.jpg *License:* Public domain *Contributors:* ? *Original artist:* ?

- **File:Wikinews-logo.svg** *Source:* https://upload.wikimedia.org/wikipedia/commons/2/24/Wikinews-logo.svg *License:* CC BY-SA 3.0 *Contributors:* This is a cropped version of Image:Wikinews-logo-en.png. *Original artist:* Vectorized by Simon 01:05, 2 August 2006 (UTC) Updated by Time3000 17 April 2007 to use official Wikinews colours and appear correctly on dark backgrounds. Originally uploaded by Simon.

- **File:Wikiquote-logo.svg** *Source:* https://upload.wikimedia.org/wikipedia/commons/f/fa/Wikiquote-logo.svg *License:* Public domain *Contributors:* ? *Original artist:* ?

- **File:Wikisource-logo.svg** *Source:* https://upload.wikimedia.org/wikipedia/commons/4/4c/Wikisource-logo.svg *License:* CC BY-SA 3.0 *Contributors:* Rei-artur *Original artist:* Nicholas Moreau

- **File:Wikiversity-logo-Snorky.svg** *Source:* https://upload.wikimedia.org/wikipedia/commons/1/1b/Wikiversity-logo-en.svg *License:* CC BY-SA 3.0 *Contributors:* Own work *Original artist:* Snorky

- **File:Wikiversity-logo.svg** *Source:* https://upload.wikimedia.org/wikipedia/commons/9/91/Wikiversity-logo.svg *License:* CC BY-SA 3.0 *Contributors:* Snorky (optimized and cleaned up by verdy_p) *Original artist:* Snorky (optimized and cleaned up by verdy_p)

- **File:Wiktionary-logo-en.svg** *Source:* https://upload.wikimedia.org/wikipedia/commons/f/f8/Wiktionary-logo-en.svg *License:* Public domain *Contributors:* Vector version of Image:Wiktionary-logo-en.png. *Original artist:* Vectorized by Fvasconcellos (talk · contribs), based on original logo tossed together by Brion Vibber

- **File:World_lines_and_world_sheet.svg** *Source:* https://upload.wikimedia.org/wikipedia/commons/2/25/World_lines_and_world_sheet.svg *License:* Public domain *Contributors:* Point&string.png *Original artist:* Kurochka, svg version by Actam

- **File:Wz-z.jpg** *Source:* https://upload.wikimedia.org/wikipedia/commons/f/f0/Wz-z.jpg *License:* CC BY-SA 4.0 *Contributors:* Own work *Original artist:* Esadri21

- **File:Yukawa_coulomb_compare.svg** *Source:* https://upload.wikimedia.org/wikipedia/commons/1/19/Yukawa_coulomb_compare.svg *License:* CC BY-SA 3.0 *Contributors:* Own work *Original artist:* Phancy Physicist

- **File:Yukawa_m_compare.svg** *Source:* https://upload.wikimedia.org/wikipedia/commons/c/ce/Yukawa_m_compare.svg *License:* CC BY-SA 3.0 *Contributors:* Own work *Original artist:* Phancy Physicist

41.11.3   Content license